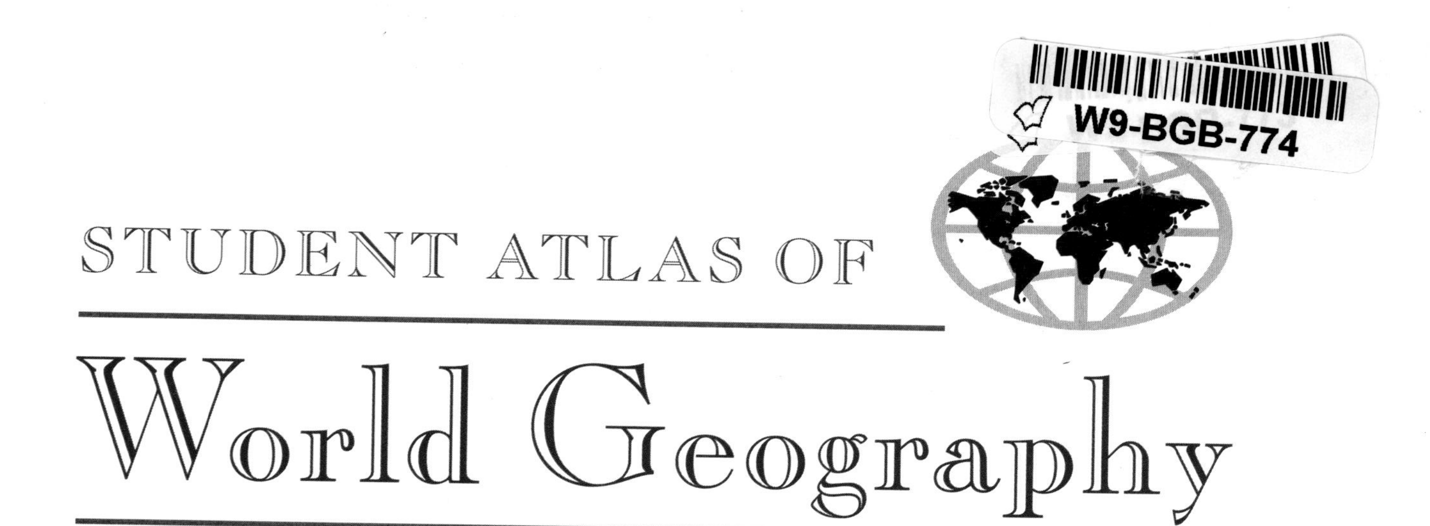

STUDENT ATLAS OF World Geography

Fifth Edition

John L. Allen
University of Wyoming

2460 Kerper Blvd., Dubuque, IA 52001

Visit us on the Internet
http://www.mhcls.com

Book Team

Managing Editor *Larry Loeppke*
Senior Developmental Editor *Susan Brusch*
Production Manager *Beth Kundert*
Project Manager *Jane Mohr*
Design Specialist *Tara McDermott*
Editorial Assistant *Nancy Meissner*
Senior Marketing Manager *Julie Keck*
Marketing Communications Specialist *Mary Klein*
Marketing Coordinator *Alice Link*
Senior Marketing Assistant *Tracie Kammerude*
Pemissions Coordinator *Lori Church*
Cartography *Carto-Graphics, Hudson, WI*
Compositor *Carlisle Publishing Services*
Cover *ImageState/Alamy*

We would like to thank Digital Wisdom Incorporated for allowing us to use their Mountain High Maps cartography software. This software was used to create maps 104, 106–109, 111, 113, 114, 116, 118–120, 122, 124, 125, 127, 129–132, 134, and 136–140.

MHID 0-07-352757-2 ISBN 978-0-07-352757-4 ISSN: 1531-0221

Printed in the United States of America

123456789QPDQPD987

A Note to the Student

The study of geography has become an increasingly important part of the curriculum in secondary schools and institutions of higher education over the last decade. This trend, a most welcome one from the standpoint of geographers, has begun to address the massive problem of "geographic illiteracy" that has characterized the United States, almost alone among the world's developed nations. When a number of international comparative studies on world geography were undertaken, beginning in the 1970s, it became apparent that most American students fell far short of their counterparts in Europe, Russia, Canada, Australia, and Japan in their abilities to recognize geographic location, to identify countries or regions on maps, or to explain the significance of such key geographic phenomena such as population distribution, economic or urban location, or the availability of natural resources. Indeed, many American students could not even locate the United States on world maps, let alone countries like France, or Indonesia, or Nigeria. This atlas, and the texts it is intended to accompany, is a small part of the process of attempting to increase the geographic literacy of American students. As the true meaning of "the global community" becomes more apparent, such an increase in geographic awareness is not only important but necessary. If the United States has learned any lesson from the tragic events at the World Trade Center and the Pentagon on September 11, 2001, these lessons would surely include the considerations that we are not isolated from events that transpire in other parts of the world; our boundaries do not make us secure; and we ignore the conditions of political, economic, cultural, and physical geography outside those boundaries at our great peril.

The maps in the *Student Atlas of World Geography* are designed to introduce you to the patterns or "spatial distribution" of the wide variety of human and physical features of the earth's surface and to help you understand the relationships between these patterns. We call such relationships "spatial correlation" and whenever you compare the patterns made by two or more phenomena that exist at or near the earth's surface–the distribution of human population and the types of climate, for example–you are engaging in spatial correlation. Like the maps, the data sets in the atlas are intended to enable you to make comparisons between the distributions of different geographic features (for example, population growth and literacy rates) and to understand the character of the geographic variation in a single geographic feature. In many instances, the data in the tables of your atlas are the same data that have been used to produce the maps. At the very outset of your study of this atlas, you should be aware of some limitations of the data tables. In some instances, there may be data missing from a table. In such cases, the cause may represent the failure of a country to report information to a central international body (like the United Nations or the World Bank), or it may mean that the shifting of political boundaries and changed responsibility for reporting data have caused some countries (for example, those countries that made up the former Soviet Union or the former Yugoslavia) to delay their reports. It is always our aim to use the most up-to-date data that is possible. Subsequent editions of this atlas will have increased data on countries like Slovenia, Ukraine, or Uzbekistan when it becomes available. In the meantime, as events continue to restructure our world, it's an exciting time to be a student of world geography!

You will find your study of this atlas more productive if you study the maps and tables on the following pages in the context of the five distinct themes that have been developed as part of the increasing awareness of the importance of geographic education:

1. *Location: Where Is It?* This theme offers a starting point from which you discover the precise location of places in both absolute terms (the latitude and longitude of a place) and in relative terms (the location of a place in relation to the location of other places). When you think of location, you should automatically think of both forms. Knowing something about absolute location will help you to understand a variety of features of physical geography, since such key elements are so closely related to their position on the earth. But it is equally important to think of location in relative terms. The location of places in relation to other places is often more important as a determinant of social, economic, and cultural characteristics than the factors of physical geography.
2. *Place: What Is It Like?* This theme investigates the political, economic, cultural, environmental, and other characteristics that give a place its identity. You should seek to understand the similarities and differences of places by exploring their basic characteristics. Why are some places with similar environmental characteristics so very different in economic, cultural, social, and political ways? Why are other places with such different environmental characteristics so seemingly alike in terms of their institutions, their economies, and their cultures?
3. *Human/Environment Interactions: How Is the Landscape Shaped?* This theme illustrates the ways in which people respond to and modify their environments. Certainly the environment is an important factor in influencing human activities and behavior. But the characteristics of the environment do not exert a controlling influence over human activities; they only provide a set of alternatives from which different cultures, in different times, make their choices. Observe the relationship between the basic elements of physical geography such as climate and terrain and the host of ways in which humans have used the land surfaces of the world.

4. *Movement: How Do People Stay in Touch?* This theme examines the transportation and communications systems that link people and places. Movement or "spatial interaction" is the chief mechanism for the spread of ideas and innovations from one place to another. It is spatial interaction that validates the old cliché, "the world is getting smaller." We find McDonald's restaurants in Tokyo and Honda automobiles in New York City because of spatial interaction. Advanced transportation and communications systems have transformed the world into which your parents were born. And the world your children will be born into will be very different from your world. None of this would happen without the force of movement or spatial interaction.
5. *Regions: Worlds Within a World.* This theme helps to organize knowledge about the land and its people. The world consists of a mosaic of "regions" or areas that are somehow different and distinctive from other areas. The region of Anglo-America (the United States and Canada) is, for example, different enough from the region of Western Europe that geographers clearly identify them as two unique and separate areas. Yet despite their differences, Anglo-Americans and Europeans share a number of similarities: common cultural backgrounds, comparable economic patterns, shared religious traditions, and even some shared physical environmental characteristics. Conversely, although the regions of Anglo-America and Eastern Asia are also easily distinguished as distinctive units of the earth's surface, they have a greater number of shared physical environmental characteristics. But those who live in Anglo-America and Eastern Asia have fewer similarities and more differences between them than is the case with Anglo-America and Western Europe: different cultural traditions, different institutions, different linguistic and religious patterns. An understanding of both the differences and similarities between regions like Anglo-America and Europe on the one hand, or Anglo-America and Eastern Asia on the other, will help you to understand the world around you. At the very least, an understanding of regional similarities and differences will help you to interpret what you read on the front page of your daily newspaper or view on the evening news report on your television set.

Not all of these themes will be immediately apparent on each of the maps and tables in this atlas. But if you study the contents of *Student Atlas of World Geography*, along with the reading of your text and think about the five themes, maps and tables and text will complement one another and improve your understanding of global geography.

John L. Allen

About the Author

John L. Allen is professor and chair of geography at the University of Wyoming and emeritus professor of geography at the University of Connecticut, where he taught from 1967 to 2000. He is a native of Wyoming. He received his bachelor's degree in 1963 and his M.A. in 1964 from the University of Wyoming, and in 1969 his Ph.D. from Clark University. His special areas of interest are perceptions of the environment and the impact of human societies on environmental systems. Dr. Allen is the author and editor of many books and articles as well as several other student atlases, including the best-selling *Student Atlas of World Politics*.

Acknowledgments

Nozar Alaolmolki
Hiram College

Barbara Batterson-Rossi
Palomar College

A. Steele Becker
University of Nebraska at Kearney

Koop Berry
Walsh University

Daniel A. Bunye
South Plains College

Winifred F. Caponigri
Holy Cross College

Femi Ferreira
Hutchinson Community College

Eric J. Fournier
Samford University

William J. Frazier
Columbus State College

Hari P. Garbharran
Middle Tennessee State University

Baher Gosheh
Edinboro University of Pennsylvania

Donald Hagan
Northwest Missouri State University

Robert Janiskee
University of South Carolina

David C. Johnson
University of Louisiana

Effie Jones
Crichton College

Cub Kahn
Marythurst University

Artimus Keiffer
Franklin College

Leonard E. Lancette
Mercer University

Donald W. Lovejoy
Palm Beach Atlantic College

Mark Maschhoff
Harris-Stowe State College

Richard Matthews
University of South Carolina

Madolia Mills
University of Colorado–Colorado Springs

Robert Mulcahy
Providence College

Otto H. Muller
Alfred University

J. Henry Owusu
University of Northern Iowa

Steven Parkansky
Morehead State University

William Preston
California Polytechnic State University, San Luis Obispo

Neil Reid
The University of Toledo

A. L. Rydant
Keene State College

Deborah Berman Santana
Mills College

Steven Slakey
University of La Verne

Rolf Sternberg
Montclair State University

Richard Ulack
University of Kentucky

David Woo
California State University, Haywood

Donald J. Zeigler
Old Dominion University

Table of Contents

A Note to the Student iii
Introduction: How to Read an Atlas xi

Part I Global Physical Patterns 1

Map 1 World Political Divisions 2
Map 2 World Physical Features 4
Map 3a Average Annual Precipitation 6
Map 3b Seasonal Average Precipitation, November Through April 7
Map 3c Seasonal Average Precipitation, May Through October 7
Map 3d Variation in Average Annual Precipitation 8
Map 4a Temperature Regions and Ocean Currents 9
Map 4b Average January Temperature 10
Map 4c Average July Temperature 10
Map 5a Atmospheric Pressure and Predominant Surface Winds, January 11
Map 5b Atmospheric Pressure and Predominant Surface Winds, July 12
Map 6 Climate Regions 13
Map 7 Vegetation Types 14
Map 8 Soil Orders 15
Map 9 Ecological Regions 16
Map 10 Plate Tectonics 18
Map 11 Topography 19
Map 12a Resources: Mineral Fuels 20
Map 12b Resources: Critical Metals 21
Map 13 Natural Hazards 22

Part II Global Human Patterns 23

Map 14 Past Population Distributions and Densities 24
Map 15 Population Density 26
Map 16 Land use, AD 1500 27
Map 17 Economic Activities 28
Map 18 Urbanization 29
Map 19 Transportation Patterns 30
Map 20 Religions 31
Map 21 Languages 32
Map 22 Linguistic Diversity 33
Map 23 External Migrations in Modern Times 34

Part III Global Demographic Patterns 35

Map 24 Population Growth Rates 36
Map 25 Total Fertility Rates 37
Map 26 Infant Mortality Rate 39
Map 27 Child Mortality Rate 40
Map 28 Child Malnutrition 41

Map 29 Primary School Enrollment 42
Map 30 Average Life Expectancy at Birth 43
Map 31 Population by Age Group 44
Map 32 International Migrant Populations 2005 45
Map 33 World Daily Per Capita Supply of Calories (Kilocalories) 46
Map 34 Proportion of Daily Food Intake from Animal Products 47
Map 35 Illiteracy Rates 48
Map 36 Female/Male Inequality in Education and Employment 49
Map 37 Global Scourges: Major Infectious Diseases 50
Map 38 Adult HIV Prevalence 2004 51
Map 39 The Index of Human Development 52
Map 40 Demographic Stress: The Youth Bulge 53
Map 41 Demographic Stress: Rapid Urban Growth 54
Map 42 Demographic Stress: Competition for Cropland 55
Map 43 Demographic Stress: Competition for Fresh Water 56
Map 44 Demographic Stress: Death in the Prime of Life 57
Map 45 Demographic Stress: Interactions of Demographic Stress 58

Part IV Global Economic Patterns 59

Map 46 Rich and Poor Countries: Gross National Income 60
Map 47 Gross National Income Per Capita 61
Map 48 Economic Growth: GDP Change Per Capita 62
Map 49 Total Labor Force 63
Map 50 Employment by Economic Activity 64
Map 51 Economic Output per Sector 65
Map 52 Agricultural Production Per Capita 66
Map 53 Exports of Primary Products 67
Map 54 Dependence on Trade 68
Map 55 The Indebtedness of States 69
Map 56 Global Flow of Investment Capital 70
Map 57 Central Government Expenditures Per Capita 71
Map 58 Purchasing Power Parity 72
Map 59 Private Per Capita Consumption 2003 73
Map 60 Energy Production Per Capita 74
Map 61 Energy Consumption Per Capita 75
Map 62 Energy Dependency 76
Map 63 Flows of Oil 77

Part V Global Patterns of Environmental Disturbance 79

Map 64 Global Air Pollution: Sources and Wind Currents 80
Map 65 The Acid Deposition Problem: Air, Water, Soil 81
Map 66 Major Polluters and Common Pollutants 82
Map 67 Global Carbon Dioxide Emissions 83
Map 68 Potential Global Temperature Change 84
Map 69 Water Resources: Availability of Renewable Water Per Capita 85

Map 70 Water Resources: Annual Withdrawal Per Capita 86
Map 71 Pollution of the Oceans 87
Map 72 Food Supply from Marine and Freshwater Systems 88C
Map 73 Cropland Per Capita: Changes 1996–2003 89
Map 74 World Pastureland 2005 90
Map 75 Fertilizer Use 2005 91
Map 76 Annual Deforestation Rates 1997–2003 92
Map 77 World Timber Production 2000 93
Map 78 The Loss of Biodiversity: Globally Threatened Animal Species 94
Map 79 The Loss of Biodiversity: Globally Threatened Plant Species 95
Map 80 Hotspots of Biodiversity 96
Map 81 The Risks of Desertification 97
Map 82 Soil Degradation 98
Map 83 The Degree of Human Disturbance 99

Part VI Global Political Patterns 101

Map 84 Political Systems 102
Map 85 Sovereign States: Duration of Independence 103
Map 86 The Emergence of the State 104
Map 87 Organized States and Chiefdoms, A.D. 1500 106
Map 88 European Colonialism 1500–2000 108
Map 89 Post–Cold War International Alliances 109
Map 90 International Conflicts in the Post–World War II World 110
Map 91 Political Realms: Regional Changes, 1945–2003 112
Map 92 The Geopolitical World at the Beginning of the Twenty-First Century 113
Map 93 Flashpoints 2006 114
Map 94 Nations with Nuclear Weapons 126
Map 95 Military Power 127
Map 96 International Terrorist Incidents, 1998–2003 128
Map 97 Distribution of Minority Populations 129
Map 98 The Political Geography of a Global Religion: The Islamic World 130
Map 99 World Refugee Population 131
Map 100 Women's Rights 132
Map 101 Political and Civil Liberties 2004 133
Map 102 A Decade of Risk of Civil Conflict 2005–2015 134

Part VII World Regions 137

Map 103 North America: Political Division 139
Map 104 North America: Physical Features 140
Map 105 North America: Thematic
Map 105a Environment and Economy: The Use of Land 141
Map 105b Population Density 142
Map 105c Population Distribution: Clustering 142
Map 105d Deforestation in the Americas 143
Map 106 Canada 144

Map 107 United States 145
Map 108 Middle America 146
Map 109 The Caribbean 147
Map 110 South America: Political Divisions 148
Map 111 South America: Physical Features 149
Map 112a Environment and Economy 150
Map 112b Population Density 151
Map 112c Environmental Problems 151
Map 112d Protection of Natural Areas 152
Map 113 Northern South America 153
Map 114 Southern South America 154
Map 115 Europe: Political Divisions 155
Map 116 Europe: Physical Features 156
Map 117 Europe: Thematic
Map 117a Environment and Economy: The Use of Land 157
Map 117b Population Density 158
Map 117c European Political Boundaries, 1914–1948 159
Map 117d Political Changes 1989–2004 161
Map 118 Western Europe 162
Map 119 Eastern Europe 163
Map 120 Northern Europe 164
Map 121 Africa: Political Divisions 165
Map 122 Africa: Physical Features 166
Map 123 Africa: Thematic
Map 123a Environment and Economy 167
Map 123b Population Density 167
Map 123c Colonial Patterns 168
Map 123d African Cropland and Dryland Degradation 169
Map 124 Northern Africa 170
Map 125 Southern Africa 171
Map 126 Asia: Political Divisions 172
Map 127 Asia: Physical Features 173
Map 128a Environment and Economy 174
Map 128b Population Density 175
Map 128c Industrialization: Manufacturing and Resources 175
Map 128d South, South-East, and East Asian Deforestation 176
Map 129 Northern Asia 177
Map 130 Southwestern Asia 178
Map 130a The Middle East: Territorial Change, 1918–2004 179
Map 130b South-Central Eurasia: An Ethnolinguistic Crazy Quilt 180
Map 131 East Asia 181
Map 132 Southeast Asia 182
Map 133 Australasia and Oceania: Political Divisions 183
Map 134 Australasia and Oceania: Physical Features 184
Map 135a Environment and Economy 185
Map 135b Population Density 186
Map 135c Climatic Patterns 186

Map 136 The Pacific Rim 187
Map 137 The Atlantic Ocean 188
Map 138 The Indian Ocean 189
Map 139 The Arctic 190
Map 140 Antarctica 190

Part VIII Tables 189

Table A World Countries: Area, Population, Population Density, 2003 190
Table B World Countries: Form of Government, Capital City, Major Languages 195
Table C World Countries: Basic Economic Indicators 200
Table D World Countries: Population Growth, 1950–2025 204
Table E World Countries: Basic Demographic Data, 1975–2001 208
Table F World Countries: Mortality, Health, and Nutrition 213
Table G World Countries: Education and Literacy, 1990–2002 218
Table H World Countries: Agricultural Operations, 1996–2002 223
Table I World Countries: Land Use and Deforestation, 1980–2000 227
Table J World Countries: Energy Production and Use, 1980–2000 231
Table K World Countries: Energy Efficiency and Emissions, 1980–1999 236
Table L World Countries: Power and Transportation 240
Table M World Countries: Communications, Information, and Science and Technology 245
Table N World Countries: Water Resources 250
Table O World Countries: Globally Threatened Plant and Animal Species 255

Part IX Geographic Index 261

Sources 295

Introduction: How to Read an Atlas

An atlas is a book containing maps which are "models" of the real world. By the term "model" we mean exactly what you think of when you think of a model: a representation of reality that is generalized, usually considerably smaller than the original, and with certain features emphasized, depending on the purpose of the model. A model of a car does not contain all of the parts of the original but it may contain enough parts that it is recognizable as a car and can be used to study principles of automotive design or maintenance. A car model designed for racing, on the other hand, may contain fewer parts but would have the mobility of a real automobile. Car models come in a wide variety of types containing almost anything you can think of relative to automobiles that doesn't require the presence of a full-size car. Since geographers deal with the real world, virtually all of the printed or published studies of that world require models. Unlike a mechanic in an automotive shop, we can't roll our study subject into the shop, take it apart, put it back together. We must use models. In other words, we must generalize our subject, and the way we do that is by using maps. Some maps are designed to show specific geographic phenomena, such as the climates of the world or the relative rates of population growth for the world's countries. We call these maps "thematic maps" and Parts I through VI of this atlas contain maps of this type. Other maps are designed to show the geographic location of towns and cities and rivers and lakes and mountain ranges and so on. These are called "reference maps" and they make up many of the maps in Part VII. All of these maps, whether thematic or reference, are models of the real world that selectively emphasize the features that we want to show on the map.

In order to read maps effectively—in other words, in order to understand the models of the world presented in the following pages—it is important for you to know certain things about maps: how they are made using what are called *projections;* how the level of mathematical proportion of the map or what geographers call *scale* affects what you see; and how geographers use *generalization* techniques such as simplification and symbols where it would be impossible to draw a small version of the real world feature. In this brief introduction, then, we'll explain to you three of the most important elements of map interpretation: projection, scale, and generalization.

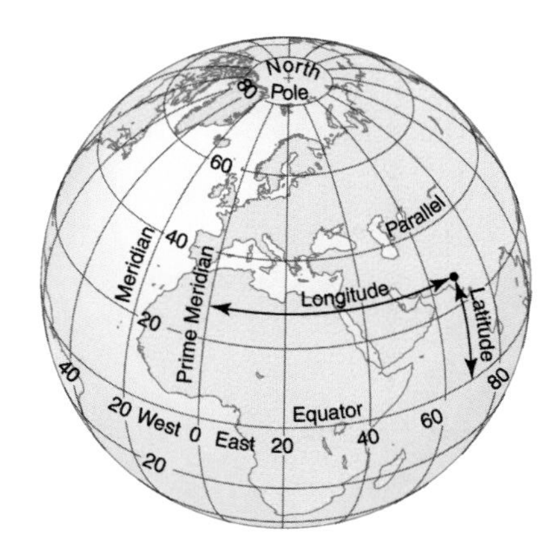

The Coordinate System

Map Projections

Perhaps the most basic problem in *cartography*, or the art and science of map-making, is the fact that the subject of maps—the earth's surface—is what is called by mathematicians "a non-developable surface." Since the world is a sphere (or nearly so—it's actually slightly flattened at the poles and bulges a tiny bit at the equator), it is impossible to flatten out the world or any part of its curved surface without producing some kind of distortion. This "near sphere" is represented by a geographic grid or coordinate system of lines of latitude or *parallels* that run east and west and are used to measure distance north and south on the globe, and lines of longitude or *meridians* that run north and south and are used to measure distance east and west. All the lines of longitude are half circles of equal length and they all converge at the poles. These meridians are numbered from 0 degrees (Prime or Greenwich Meridian) east and west to 180 degrees. The meridian of 0 degrees and the meridian of 180 degrees are halves of the same "great circle" or line representing a plane that bisects the globe into two equal hemispheres. All lines of longitude are halves of great circles. All the lines of latitude are complete circles that are parallel to one another and are spaced equidistant on the meridians. The circumference of these circles lessens as you move north or south from the equator. Parallels of latitude are numbered from 0 degrees at the equator north and south to 90 degrees at the North and South poles. The only line of latitude that is a great circle is the equator, which equally divides the world into a northern and southern hemisphere. In the real world, all these grid lines of latitude and longitude intersect at right angles. The problem for cartographers is to convert this spherical or curved grid into a geometrical shape that is "developable"; that is, it can be flattened (such as a cylinder or cone) or is already flat (a plane). The reason the results of the conversion process are called "projections" is that we imagine a world globe (or some part of it) that is made up of wires running north-south and east-west to represent the grid lines of latitude and longitude and other wires or even solid curved plates to represent the coastlines of continents or the continents themselves. We then imagine a light source at some location inside or outside the wire globe that can "project" or cast shadows of the wires representing grid lines onto a developable surface. Sometimes the basic geometric principles of projection may be modified by other mathematical principles to yield projections that are not truly geometric but have certain desirable features. We call these types of projections "arbitrary." The three most basic types of projections are named according to the type of developable surface: cylindrical, conic, or azimuthal (plane). Each type has certain characteristic features: they may be *equal area* projections in which the size of each area on the map is a direct proportional representation of that same area in the real world but shapes are distorted; they may be *conformal* projections in which area may be distorted but shapes are shown correctly; or they may be *compromise* projections in which both shape and area are distorted but the overall picture presented is fairly close to reality. It is important to remember that all maps distort the geographic grid and continental outlines in characteristic ways. The only representation of the world that does not distort either shape or area is a globe. You can see why we must use projections—can you imagine an atlas that you would have to carry back and forth across campus that would be made up entirely of globes?

Cylindrical Projections

Cylindrical projections are drawn as if the geographic grid were projected onto a cylinder. Cylindrical projections have the advantage of having all lines of latitude as true parallels or straight lines. This makes these projections quite useful for showing geographic relationships in which latitude or distance north-south is important (many physical features, such as climate, are influenced by latitude). Unfortunately, most cylindrical-type projections distort area significantly. One of the most famous is the Mercator projection shown above. This projection makes areas disproportionately large as you move toward the pole, making

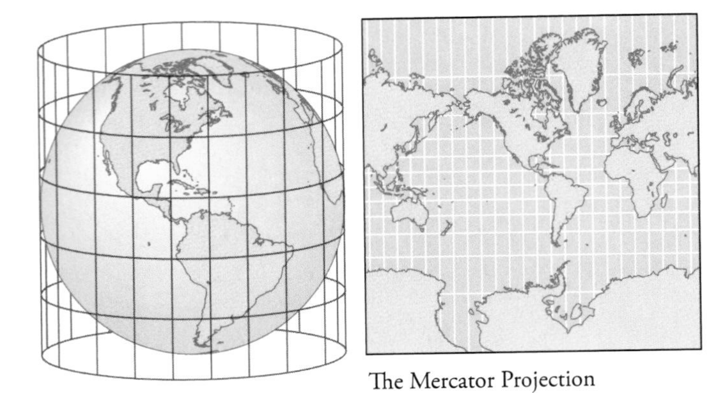

The Mercator Projection

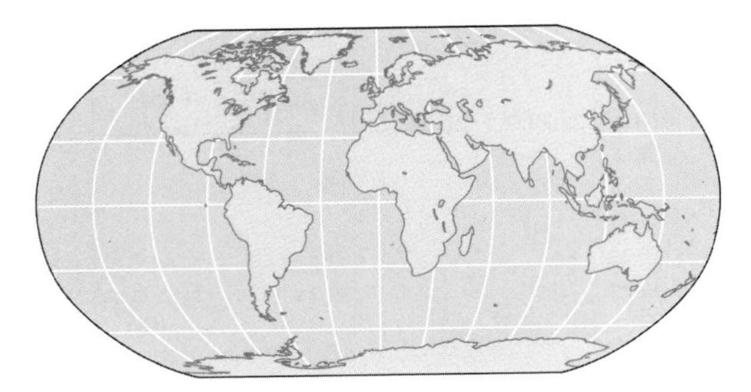

The Robinson Projection

Greenland, which is actually about one-seventh the size of South America, appear to be as large as the southern continent. But the Mercator projection has the quality of conformality: landmasses on the map are true in shape and thus all coastlines on the map intersect lines of latitude and longitude at the proper angles. This makes the Mercator projection, named after its inventor, a sixteenth-century Dutch cartographer, ideal for its original purpose as a tool for navigation—but not a good projection for attempting to show some geographical feature in which areal relationship is important. Unfortunately, the Mercator projection has often been used for wall maps for schoolrooms and the consequence is that generations of American school children have been "tricked" into thinking that Greenland is actually larger than South America. Much better cylindrical-type projections are those like the Robinson projection used in this atlas that is neither equal area nor conformal but a compromise that portrays the real world much as it actually looks, enough so that we can use it for areal comparisons.

Conic Projections

Conic projections are those that are imagined as being projected onto a cone that is tangent to the globe along a standard parallel, or a series of cones tangent along several parallels or even intersecting the globe.

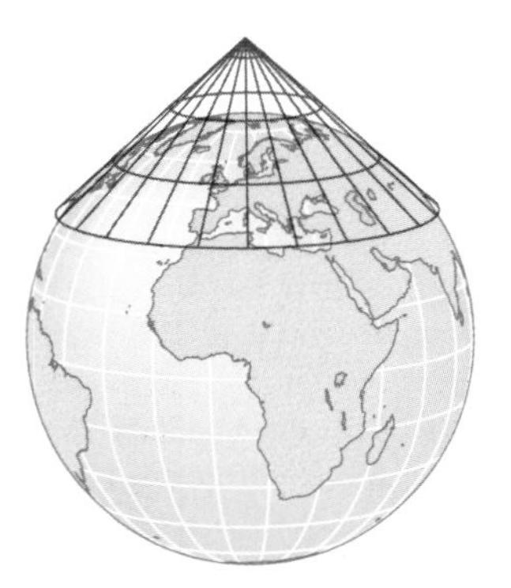

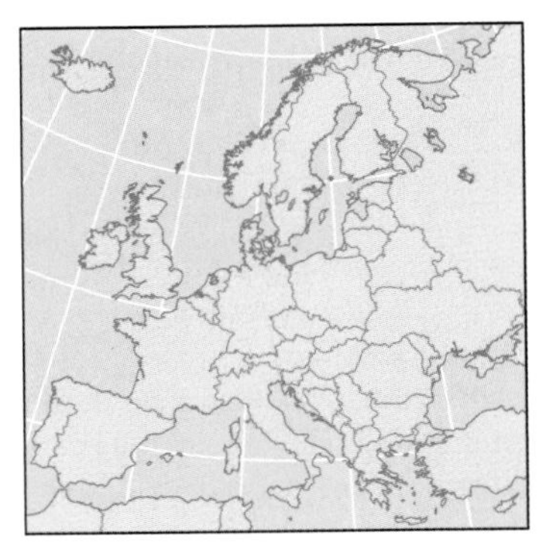

Conic Projection of Europe

Conic projections usually show latitude as curved lines and longitude as straight lines. They are good projections for areas with north-south extent, like the map of Europe to the right, and may be either conformal, equal area, or compromise, depending on how they are constructed. Many of the regional maps in the last map section of this atlas are conic projections.

Azimuthal Projections

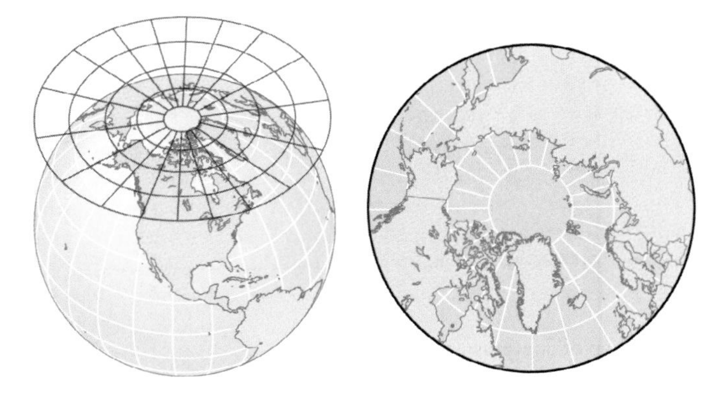

Azimuthal Projection of the North Polar Region

Azimuthal projections are those that are imagined as being projected onto a plane or flat surface. They are named for one of their essential properties. An "azimuth" is a line of compass bearing and azimuthal projections have the property of yielding true compass directions from the center of the map. This makes azimuthal maps useful for navigation purposes, particularly air navigation. But, because they distort area and shape so greatly, they are seldom used for maps designed to show geographic relationships. When they are used as illustrative rather than navigation maps, it is often in the "polar case" projection shown below where the plane has been made tangent to the globe at the North Pole.

Map Scale

Since maps are models of the real world, it follows that they are not the same size as the real world or any portion of it. Every map, then, is subject to generalization, which is another way of saying that maps are drawn to certain scales. The term *scale* refers to the mathematical quality of *proportional representation*, and is expressed as a ratio between an area of the real world or the distance between places on the real world and the same area or distance on the map. We show map scale on maps in three different ways. Sometimes we simply use the proportion and write what is called a *natural scale* or representative fraction": for example, we might show on a map the mathematical proportion of 1:62,500. A map at this scale is one that is one sixty-two thousand five-hundredth the size of the same area in the real world. Other times we convert the proportion to a written description that approximates the relationship between distance on the map and distance in the real world. Since there are nearly 62,500 inches in a mile, we would refer to a map having a natural scale of 1:62,500 as having an "inch-mile" scale of "1 inch represents 1 mile." If we draw a line one inch long on this map, that line represents a distance of approximately one mile in the real world. Finally, we usually use a graphic or linear scale: a bar or line, often graduated into miles or kilometers, that shows graphically the proportional representation. A graphic scale for our 1:62,500 map might be about five inches long, divided into five equal units clearly labeled as "1 mile," "2 miles," and so on. Our examples below show all three kinds of scales.

The most important thing to keep in mind about scale, and the reason why knowing map scale is important to being able to read a map correctly, is the relationship between proportional representation and generalization. A map that fills a page but shows the whole world is much

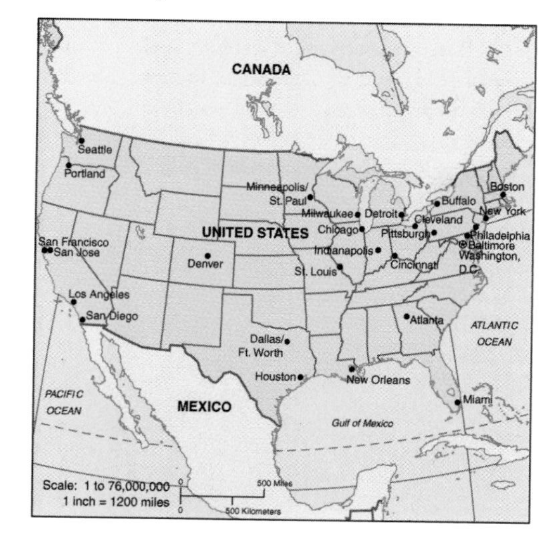

Map 1 Small Scale Map of the United States

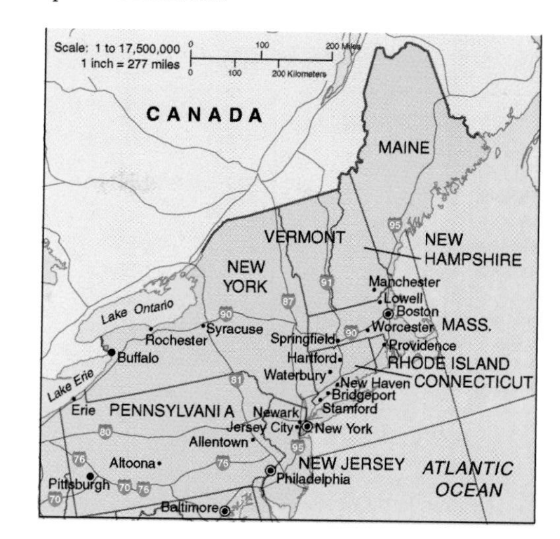

Map 2 Map of the Northeast

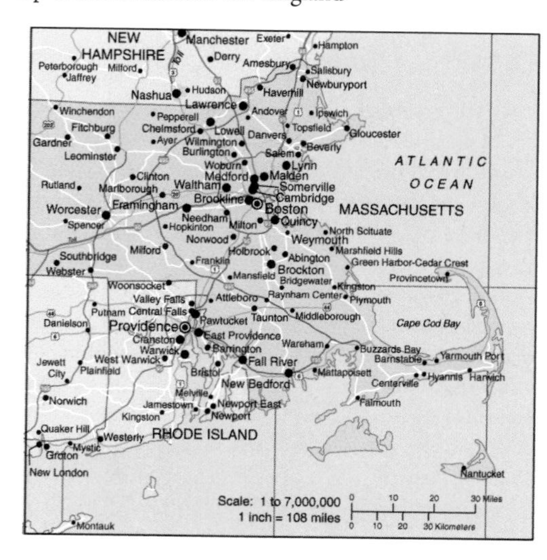

Map 3 Map of Southeastern New England

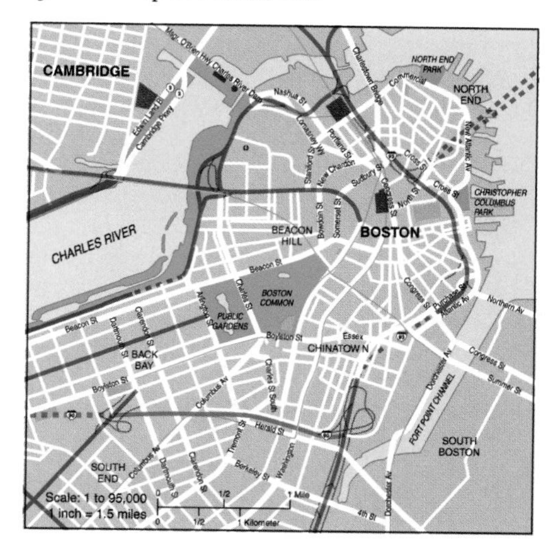

Map 4 Large Scale Map of Boston, MA

more highly generalized than a map that fills a page but shows a single city. On the world map, the city may appear as a dot. On the city map, streets and other features may be clearly seen. We call the first map, the world map, a *small-scale* map because the proportional representation is a small number. A page-size map showing the whole world may be drawn at a scale of 1:150,000,000. That is a very small number indeed—hence the term *small-scale* map even though the area shown is large. Conversely, the second map, a city map, may be drawn at a scale of 1:250,000. That is still a very small number but it is a great deal larger than 1:150,000,000! And so we'd refer to the city map as a *large scale* map, even though it shows only a small area. On our world map, geographical features are generalized greatly and many features can't even be shown at all. On the city map, much less generalization occurs—we can show specific features that we couldn't on the world map—but generalization still takes place. The general rule is that the smaller the map scale, the greater the degree of generalization; the larger the map scale, the less the degree of generalization. The only map that would not generalize would be a map at a scale of 1:1 and that map wouldn't be very handy to use. Examine the relationship between scale and generalization in the four maps on the previous page.

Generalization on Maps

A review of the four maps on the previous page should give you some indication of how cartographers generalize on maps. One thing that you should have noticed is that the first map, that of the United States, is much simpler than the other three and that the level of *simplification* decreases with each map. When a cartographer simplifies map data, information that is not important for the purposes of the map is just left off. For example, on the first map the objective may have been to show cities over 1 million in population. To do that clearly and effectively, it is not necessary to show and label rivers and lakes. The map has been simplified by leaving those items out. The final map, on the other hand, is more complex and shows and labels geographic features that are important to the character of the city of Boston; therefore, the Charles River is clearly indicated on the map.

Another type of generalization is *classification*. Map 1 on the previous page shows cities over 1 million in population. Map 2 shows cities of several different sizes and a different symbol is used for each size classification or category. Many of the thematic maps used in this atlas rely on classification to show data. A thematic map showing population growth rates (see Map 24) will use different colors to show growth rates in different classification levels or what are sometimes called *class intervals*. Thus, there will be one color applied to all countries with population growth rates between 1.0 percent and 1.4 percent, another color applied to all countries with population growth rates between 1.5 percent and 2.1 percent, and so on. Classification is necessary because it is impossible to find enough symbols or colors to represent precise values. Classification may also be used for qualitative data, such as the national or regional origin of migrating populations. Cartographers show both quantitative and qualitative classification levels or class intervals in important sections of maps called *legends*. These legends, as in the samples that follow, make it possible for the reader of the map to interpret the patterns shown.

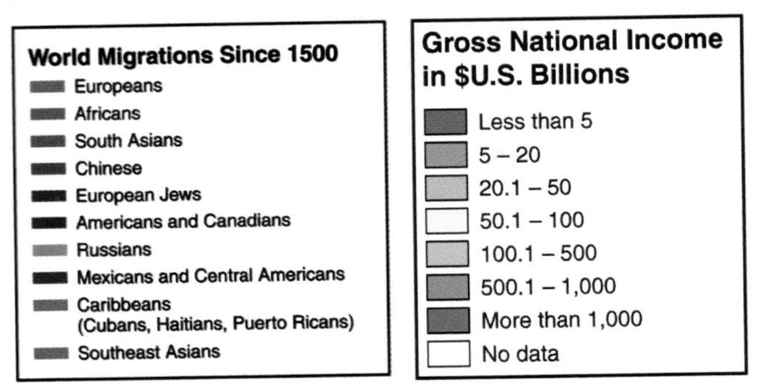

Map Legends from Maps 23 and 46

A third technique of generalization is *symbolization* and we've already noted several different kinds of symbols: those used to represent cities on the preceding maps, or the colors used to indicate population growth levels on Map 24. One general category of map symbols is quantitative in nature and this category can further be divided into a number of different types. For example, the symbols showing city size on Maps 1 and 2 on the preceding page can be categorized as *ordinal* in that they show relative differences in quantities (the size of cities). A cartographer might also use lines of different widths to express the quantities of movement of people or goods between two or more points as on Map 23.

The color symbols used to show rates of population growth can be categorized as *interval* in that they express certain levels of a mathematical quantity (the percentage of population growth). Interval symbols are often used to show physical geographic characteristics such as inches of precipitation, degrees of temperature, or elevation above sea level. The following sample, for example, shows precipitation (from Map 3a).

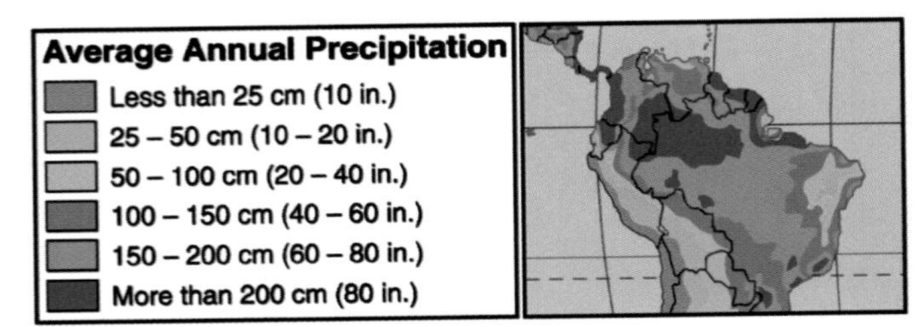

Interval Symbols

Still another type of mathematical symbolization is the *ratio* in which sets of mathematical quantities are compared: the number of persons per square mile (population density) or the growth in gross national product per capita (per person). The map above shows GDP change per capita (from Map 48).

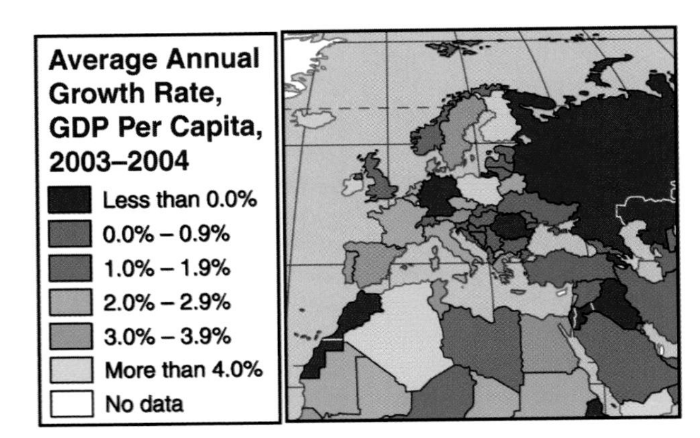

Ratio Symbols

Finally, there are a vast number of cartographic symbols that are not mathematical but show differences in the kind of information being portrayed. These symbols are called *nominal* and they range from the simplest differences such as land and water to more complex differences such as those between different types of vegetation. Shapes or patterns or colors or iconographic drawings may all be used as nominal symbols on maps. The sample map at the bottom of page xiii (Map 8) uses color to show the distribution of soil types.

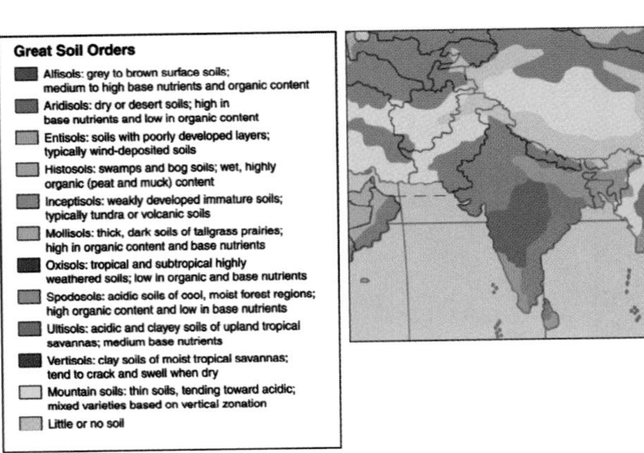

Nominal Symbols

The final technique of generalization is what cartographers refer to as *interpolation*. Here, the maker of a map may actually show more information on the map than is actually supplied by the original data. In understanding the process of interpolation is it necessary for you to visualize the quantitative data shown on maps as being three dimensional: *x* values provide geographic location along a north-south axis of the map; *y* values provide geographic location along the east-west axis of the map; and *z* values are those values of whatever data (for example, temperature) are being shown on the map at specific points. We all can imagine a real three-dimensional surface in which the *x* and y values are directions and the *z* values are the heights of mountains and the depths of valleys. On a topographic map showing a real three-dimensional surface, contour lines are used to connect points of equal elevation above sea level. These contour lines are not measured directly; they are estimated by interpolation on the basis of the elevation points that are provided.

Interpolation

It is harder to imagine the statistical surface of a temperature map in which the *x* and *y* values are directions and the *z* values represent degrees of temperature at precise points. But that is just what cartographers do. And to obtain the values between two or more specific points where *z* values exist, they interpolate based on a class interval they have decided is appropriate and use *isolines* (which are statistical equivalents of a contour line) to show increases or decreases in value. The following diagram shows an example of an interpolation process. Occasionally interpolation is referred to as *induction*. By whatever name, it is one of the most difficult parts of the cartographic process.

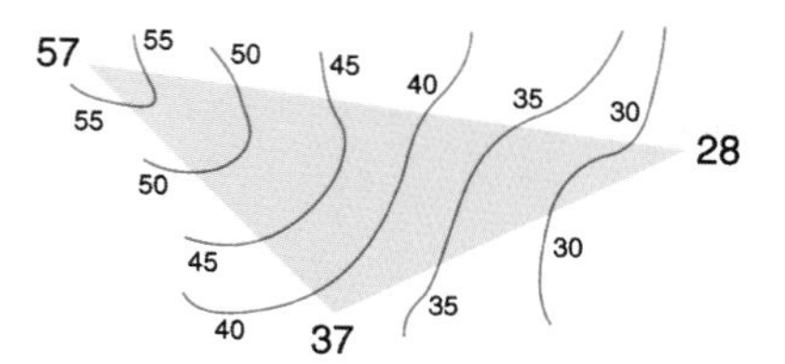

And you thought all you had to do to read an atlas was look at the maps! You've now learned that it is a bit more involved than that. As you read and study this atlas, keep in mind the principles of projection and scale and generalization (including simplification, classification, symbolization, and interpolation) and you'll do just fine. Good luck and enjoy your study of the world of maps as well as maps of the world!

Part I

Global Physical Patterns

Map 1 World Political Divisions
Map 2 World Physical Features
Map 3a Average Annual Precipitation
Map 3b Seasonal Average Precipitation, November through April
Map 3c Seasonal Average Precipitation, May through October
Map 3d Variation in Average Annual Precipitation
Map 4a Temperature Regions and Ocean Currents
Map 4b Average January Temperature
Map 4c Average July Temperature
Map 5a Atmospheric Pressure and Predominant Surface Winds, January
Map 5b Atmospheric Pressure and Predominant Surface Winds, July
Map 6 Climate Regions
Map 7 Vegetation Types
Map 8 Soil Orders
Map 9 Ecological Regions
Map 10 Plate Tectonics
Map 11 Topography
Map 12a Resources: Mineral Fuels
Map 12b Resources: Critical Metals
Map 13 Natural Hazards

The international system includes the political units called "states" or countries as the most important component. The boundaries of countries are the primary source of political division in the world and for most people nationalism is the strongest source of political identity. State boundaries are an important indicator of cultural, linguistic, economic, and other geographic divisions as well, and the states themselves normally serve as the base level

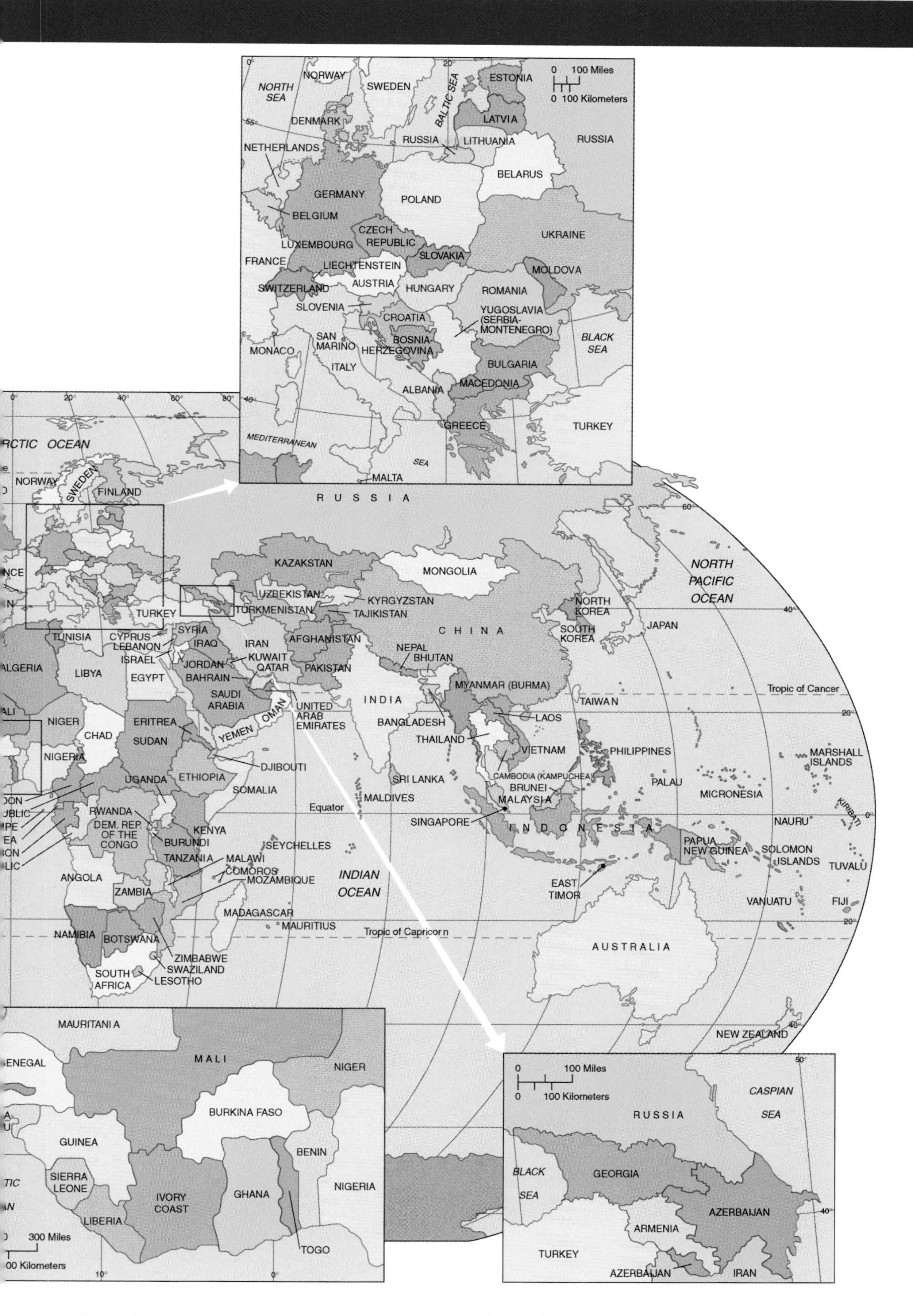

for which most global statistics are available. The subfield of geography known as "political geography" has as its primary concern the geographic or spatial character of this international system and its components.

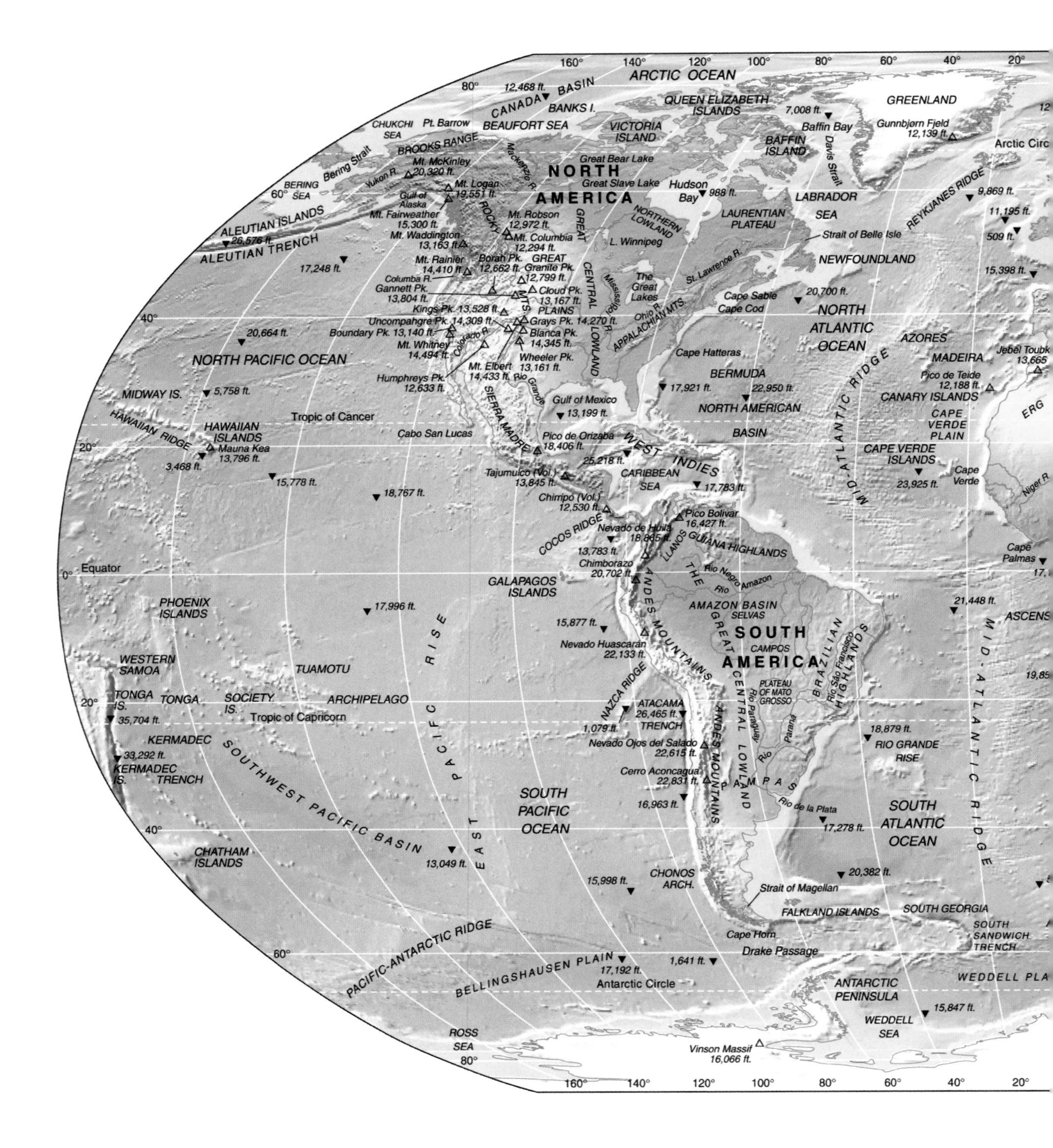
ARCTIC OCEAN
CANADA BASIN
12,468 ft.
BANKS I.
QUEEN ELIZABETH ISLANDS
GREENLAND
CHUKCHI SEA
Pt. Barrow
BEAUFORT SEA
VICTORIA ISLAND
7,008 ft.
Baffin Bay
Gunnbjørn Fjeld 12,139 ft.
BROOKS RANGE
BAFFIN ISLAND
Davis Strait
Arctic Circ
Bering Strait
Yukon R.
Mt. McKinley 20,320 ft.
Mackenzie R.
Great Bear Lake
NORTH AMERICA
BERING SEA
Mt. Logan 19,551 ft.
Great Slave Lake
Hudson Bay
988 ft.
LABRADOR SEA
REYKJANES RIDGE
9,869 ft.
Gulf of Alaska
Mt. Fairweather 15,300 ft.
Mt. Robson 12,972 ft.
GREAT
NORTHERN LOWLAND
LAURENTIAN PLATEAU
11,195 ft.
ALEUTIAN ISLANDS
26,576 ft.
Mt. Waddington 13,163 ft.
ROCKY
Mt. Columbia 12,294 ft.
L. Winnipeg
Strait of Belle Isle
509 ft.
ALEUTIAN TRENCH
17,248 ft.
Mt. Rainier 14,410 ft.
Borah Pk. 12,662 ft.
GREAT
Granite Pk. 12,799 ft.
CENTRAL LOWLAND
The Great Lakes
St. Lawrence R.
NEWFOUNDLAND
Columbia R.
Gannett Pk. 13,804 ft.
Cloud Pk. 13,167 ft.
Mississippi R.
Cape Sable
Cape Cod
20,700 ft.
15,398 ft.
Kings Pk. 13,528 ft.
MTS.
PLAINS
Ohio R.
APPALACHIAN MTS.
NORTH ATLANTIC OCEAN
Uncompahgre Pk. 14,309 ft.
Grays Pk. 14,270 ft.
20,664 ft.
Boundary Pk. 13,140 ft.
Colorado R.
Blanca Pk. 14,345 ft.
AZORES
Mt. Whitney 14,494 ft.
Wheeler Pk. 13,161 ft.
Cape Hatteras
MADEIRA
Jebel Toubk 13,665
NORTH PACIFIC OCEAN
Mt. Elbert 14,433 ft.
Rio Grande
BERMUDA
Pico de Teide 12,188 ft.
Humphreys Pk. 12,633 ft.
SIERRA MADRE
17,921 ft.
22,950 ft.
CANARY ISLANDS
MIDWAY IS.
5,758 ft.
Gulf of Mexico
13,199 ft.
NORTH AMERICAN BASIN
CAPE VERDE PLAIN
ERG
HAWAIIAN RIDGE
Tropic of Cancer
HAWAIIAN ISLANDS
Cabo San Lucas
Pico de Orizaba 18,406 ft.
WEST INDIES
MID-ATLANTIC RIDGE
Mauna Kea 13,796 ft.
25,218 ft.
CAPE VERDE ISLANDS
3,468 ft.
Tajumulco (Vol.) 13,845 ft.
CARIBBEAN SEA
Cape Verde
15,778 ft.
17,783 ft.
23,925 ft.
Niger R.
18,767 ft.
Chirripo (Vol.) 12,530 ft.
Pico Bolivar 16,427 ft.
COCOS RIDGE
Nevado de Huila 18,865 ft.
LLANOS
GUIANA HIGHLANDS
Cape Palmas
13,783 ft.
Chimborazo 20,702 ft.
Equator
Rio Negro
Amazon
GALAPAGOS ISLANDS
THE ANDES MOUNTAINS
AMAZON BASIN
SELVAS
21,448 ft.
PHOENIX ISLANDS
17,996 ft.
ASCENS
15,877 ft.
SOUTH AMERICA
CAMPOS
BRAZILIAN HIGHLANDS
Rio São Francisco
RISE
Nevado Huascarán 22,133 ft.
WESTERN SAMOA
TUAMOTU
GREAT CENTRAL LOWLAND
PLATEAU OF MATO GROSSO
NAZCA RIDGE
Rio Paraguay
TONGA IS.
TONGA
SOCIETY IS.
ARCHIPELAGO
ATACAMA 26,465 ft. TRENCH
Tropic of Capricorn
35,704 ft.
PACIFIC
1,079 ft.
Paraná
18,879 ft.
KERMADEC
Nevado Ojos del Salado 22,615 ft.
RIO GRANDE RISE
33,292 ft.
KERMADEC IS.
TRENCH
Cerro Aconcagua 22,831 ft.
ANDES MOUNTAINS
PAMPAS
SOUTHWEST PACIFIC BASIN
SOUTH PACIFIC OCEAN
16,963 ft.
Rio de la Plata
SOUTH ATLANTIC OCEAN
EAST
17,278 ft.
CHATHAM ISLANDS
13,049 ft.
CHONOS ARCH.
15,998 ft.
20,382 ft.
Strait of Magellan
FALKLAND ISLANDS
SOUTH GEORGIA
SOUTH SANDWICH TRENCH
PACIFIC-ANTARCTIC RIDGE
Cape Horn
Drake Passage
BELLINGSHAUSEN PLAIN
17,192 ft.
1,641 ft.
Antarctic Circle
ANTARCTIC PENINSULA
WEDDELL PLA
15,847 ft.
WEDDELL SEA
ROSS SEA
Vinson Massif 16,066 ft.
160°
140°
120°
100°
80°
60°
40°
20°

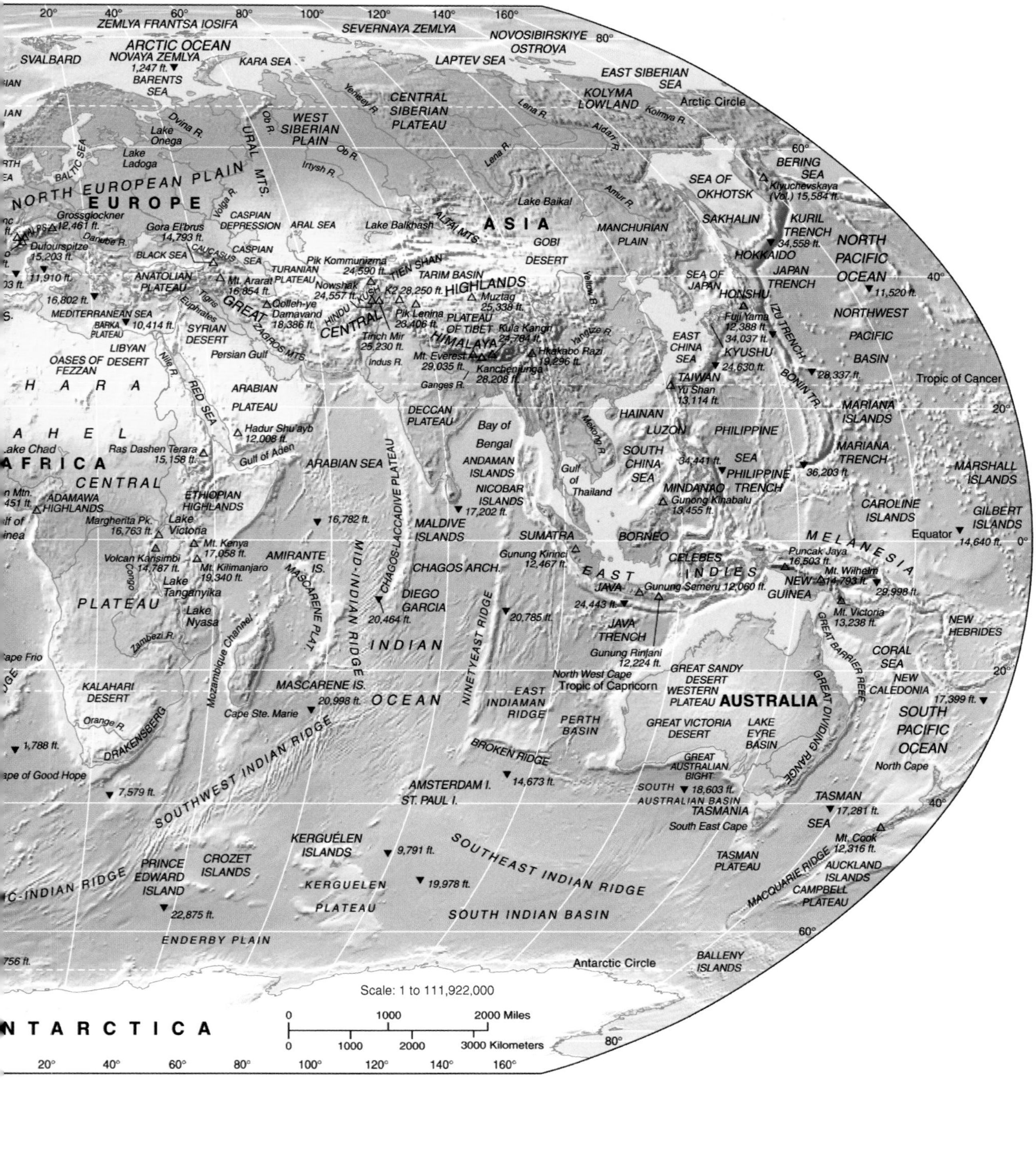

ZEMLYA FRANTSA IOSIFA
SEVERNAYA ZEMLYA
NOVOSIBIRSKIYE OSTROVA
ARCTIC OCEAN
SVALBARD
NOVAYA ZEMLYA
1,247 ft.
BARENTS SEA
KARA SEA
LAPTEV SEA
EAST SIBERIAN SEA
CENTRAL SIBERIAN PLATEAU
KOLYMA LOWLAND
Arctic Circle
WEST SIBERIAN PLAIN
URAL MTS.
Lake Onega
Lake Ladoga
BALTIC SEA
NORTH EUROPEAN PLAIN
EUROPE
BERING SEA
SEA OF OKHOTSK
Klyuchevskaya (Vol.) 15,584 ft.
Lake Baikal
ASIA
CASPIAN DEPRESSION
ARAL SEA
Lake Balkhash
ALTAI MTS.
MANCHURIAN PLAIN
SAKHALIN
KURIL TRENCH
34,558 ft.
Grossglockner 12,461 ft.
Gora El'brus 14,793 ft.
GOBI DESERT
HOKKAIDO
NORTH PACIFIC OCEAN
Dufourspitze 15,203 ft.
BLACK SEA
CAUCASUS
CASPIAN SEA
Pik Kommunizma 24,590 ft.
TIEN SHAN
TARIM BASIN
SEA OF JAPAN
JAPAN TRENCH
11,910 ft.
ANATOLIAN PLATEAU
Mt. Ararat 16,854 ft.
TURANIAN PLATEAU
Nowshak 24,557 ft.
K2 28,250 ft.
HIGHLANDS
Muztag 25,338 ft.
HONSHU
11,520 ft.
16,802 ft.
MEDITERRANEAN SEA
Qolleh-ye Damavand 18,386 ft.
HINDU KUSH
CENTRAL
Pik Lenina 23,406 ft.
PLATEAU OF TIBET
Kula Kangri 24,784 ft.
Fuji Yama 12,388 ft.
34,037 ft.
IZU TRENCH
NORTHWEST PACIFIC BASIN
BARKA PLATEAU
10,414 ft.
SYRIAN DESERT
GREAT ZAGROS MTS.
Tirich Mir 25,230 ft.
HIMALAYA
Hkakabo Razi 19,296 ft.
EAST CHINA SEA
KYUSHU
LIBYAN DESERT
Persian Gulf
Indus R.
Mt. Everest 29,035 ft.
Kanchenjunga 28,208 ft.
24,630 ft.
OASES OF FEZZAN
Ganges R.
TAIWAN
Yu Shan 13,114 ft.
28,337 ft.
BONIN TR.
Tropic of Cancer
ARABIAN PLATEAU
RED SEA
Nile R.
DECCAN PLATEAU
Bay of Bengal
HAINAN
MARIANA ISLANDS
LUZON
PHILIPPINE SEA
Lake Chad
Hadur Shu'ayb 12,008 ft.
SOUTH CHINA SEA
MARIANA TRENCH
Ras Dashen Terara 15,158 ft.
Gulf of Aden
ARABIAN SEA
ANDAMAN ISLANDS
34,441 ft.
PHILIPPINE TRENCH
36,203 ft.
MARSHALL ISLANDS
AFRICA
CENTRAL
ADAMAWA HIGHLANDS
ETHIOPIAN HIGHLANDS
CHAGOS-LACCADIVE PLATEAU
NICOBAR ISLANDS
Gulf of Thailand
MINDANAO
Gunong Kinabalu 13,455 ft.
CAROLINE ISLANDS
GILBERT ISLANDS
Margherita Pk. 16,763 ft.
Lake Victoria
16,782 ft.
17,202 ft.
MALDIVE ISLANDS
SUMATRA
BORNEO
MELANESIA
Equator
14,640 ft.
Mt. Kenya 17,058 ft.
AMIRANTE IS.
Gunung Kirinci 12,467 ft.
CELEBES
Puncak Jaya 16,503 ft.
Volcan Karisimbi 14,787 ft.
Congo R.
Mt. Kilimanjaro 19,340 ft.
MASCARENE PLAT.
CHAGOS ARCH.
EAST INDIES
Mt. Wilhelm 14,793 ft.
Lake Tanganyika
MID-INDIAN RIDGE
DIEGO GARCIA
JAVA
Gunung Semeru 12,060 ft.
NEW GUINEA
29,998 ft.
PLATEAU
Lake Nyasa
20,464 ft.
NINETYEAST RIDGE
20,785 ft.
24,443 ft.
Mt. Victoria 13,238 ft.
NEW HEBRIDES
Zambezi R.
Mozambique Channel
INDIAN
JAVA TRENCH
Gunung Rinjani 12,224 ft.
GREAT BARRIER REEF
CORAL SEA
Cape Frio
North West Cape
GREAT SANDY DESERT
KALAHARI DESERT
MASCARENE IS.
EAST INDIAMAN RIDGE
Tropic of Capricorn
WESTERN PLATEAU
AUSTRALIA
GREAT DIVIDING RANGE
NEW CALEDONIA
17,399 ft.
Orange R.
Cape Ste. Marie
20,998 ft.
OCEAN
PERTH BASIN
GREAT VICTORIA DESERT
LAKE EYRE BASIN
SOUTH PACIFIC OCEAN
1,788 ft.
DRAKENSBERG
SOUTHWEST INDIAN RIDGE
BROKEN RIDGE
GREAT AUSTRALIAN BIGHT
North Cape
Cape of Good Hope
7,579 ft.
AMSTERDAM I.
ST. PAUL I.
14,673 ft.
SOUTH AUSTRALIAN BASIN
18,603 ft.
TASMAN SEA
17,281 ft.
TASMANIA
South East Cape
KERGUÉLEN ISLANDS
9,791 ft.
Mt. Cook 12,316 ft.
PRINCE EDWARD ISLAND
CROZET ISLANDS
SOUTHEAST INDIAN RIDGE
TASMAN PLATEAU
MACQUARIE RIDGE
AUCKLAND ISLANDS
CAMPBELL PLATEAU
ATLANTIC-INDIAN RIDGE
KERGUELEN PLATEAU
19,978 ft.
22,875 ft.
SOUTH INDIAN BASIN
ENDERBY PLAIN
Antarctic Circle
BALLENY ISLANDS
ANTARCTICA
Scale: 1 to 111,922,000
0 1000 2000 Miles
0 1000 2000 3000 Kilometers

Map 3a Average Annual Precipitation

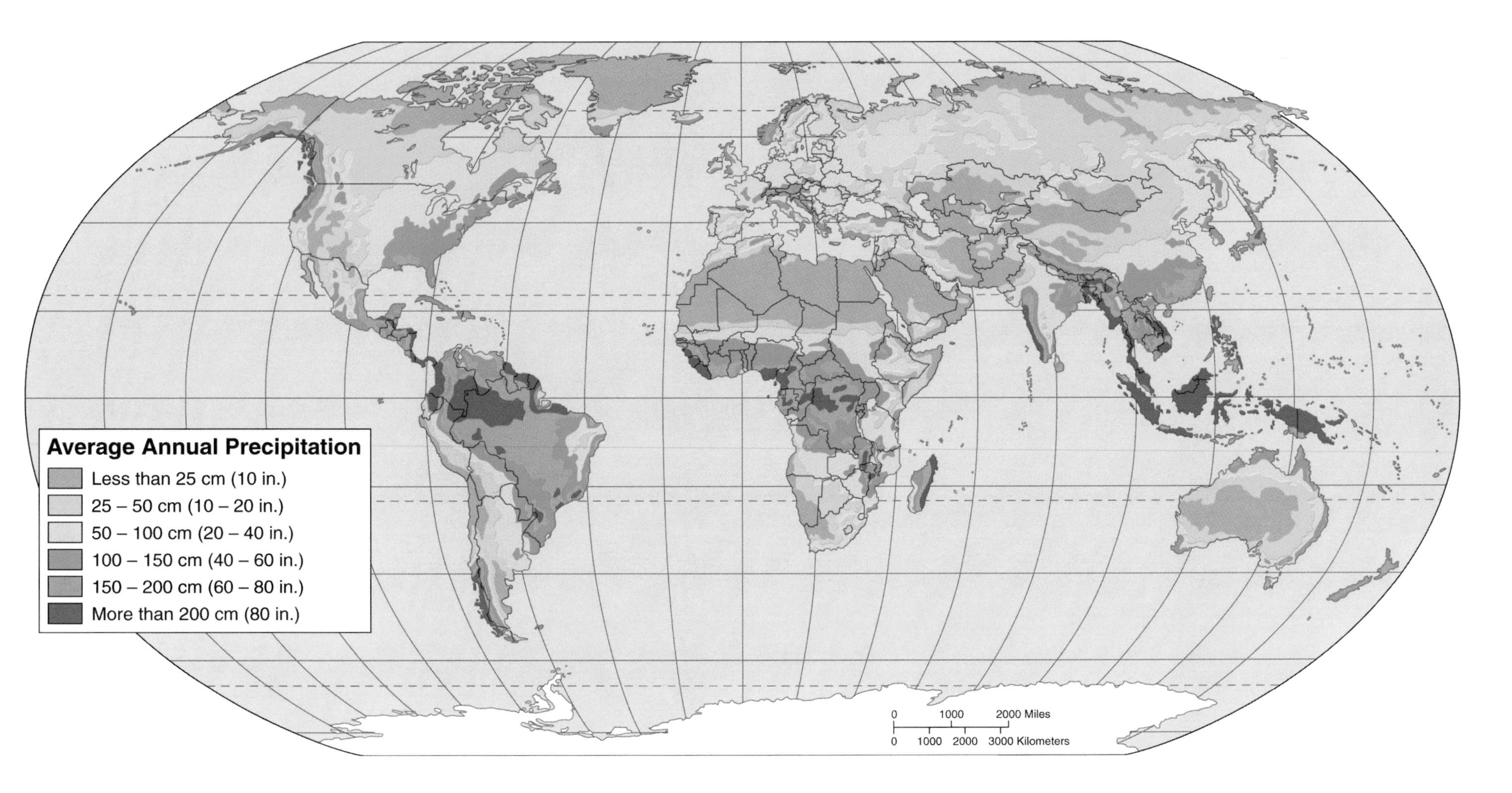

The two most important physical geographic variables are precipitation and temperature, the essential elements of weather and climate. Precipitation is a conditioner of both soil type and vegetation. More than any other single environmental element, it influences where people do or do not live. Water is the most precious resource available to humans, and water availability is largely a function of precipitation. Water availability is also a function of several precipitation variables that do not appear on this map: the seasonal distribution of precipitation (is precipitation or drought concentrated in a particular season?), the ratio between precipitation and temperature (how much of the water that comes to the earth in the form of precipitation is lost through mechanisms such as evaporation and transpiration that are a function of temperature?), and the annual variability of precipitation (how much do annual precipitation totals for a place or region tend to vary from the "normal" or average precipitation?). In order to obtain a complete understanding of precipitation, these variables should be examined along with the more general data presented on this map. The study of precipitation and other climatic elements is the concern of the branch of physical geography called "climatology."

Map 3b Seasonal Average Precipitation, November through April

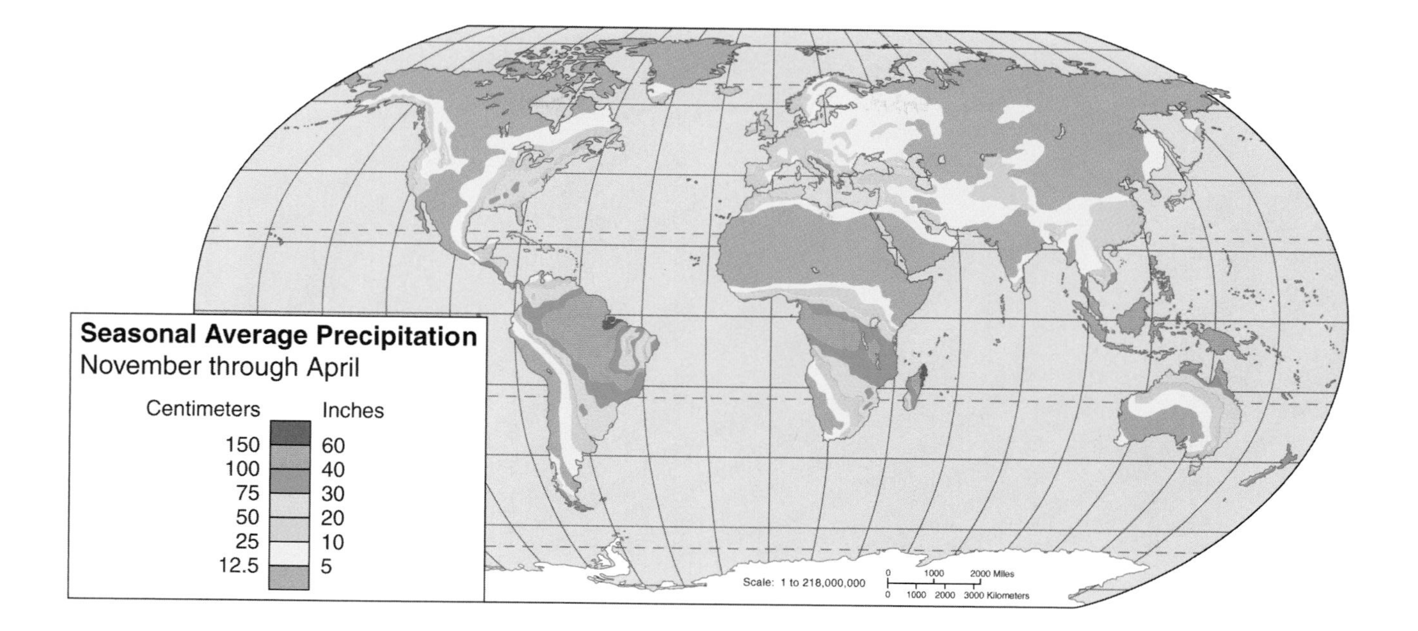

Map 3c Seasonal Average Precipitation, May through October

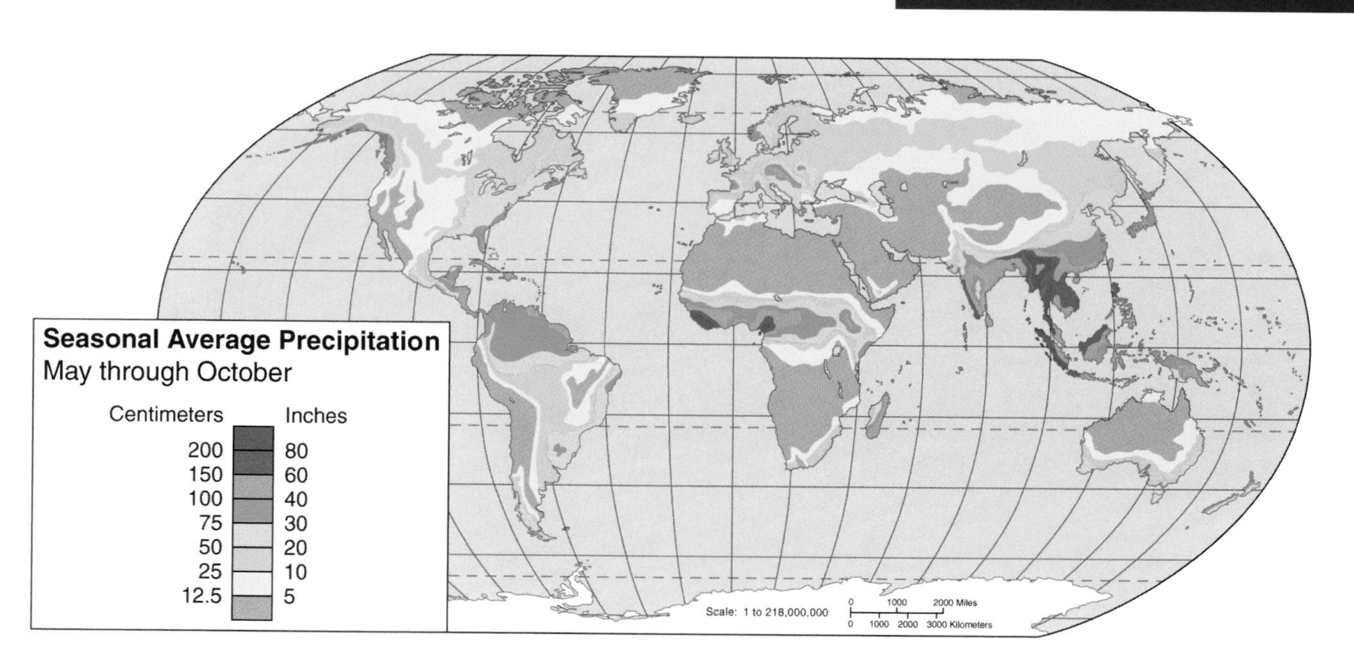

Seasonal average precipitation is nearly as important as annual precipitation totals in determining the habitability of an area. Critical factors are such things as whether precipitation coincides with the growing season and thus facilitates agriculture or during the winter when it is less effective in aiding plant growth, and whether precipitation occurs during summer with its higher water loss through evaporation and transpiration or during the winter when more of it can go into storage. Several of the world's great climate zones have pronounced seasonal precipitation rhythms. The tropical and subtropical savanna grasslands have a long winter dry season and abundant precipitation in the summer. The Mediterranean climate is the only major climate with a marked dry season during the summer, making agriculture possible only through irrigation or other adjustments to cope with drought during the period of plant growth. And the great monsoon climates of South and Southeast Asia have their winter dry season and summer rain that have conditioned the development of Asian agriculture and the rhythms of Asian life.

Map 3d Variation in Average Annual Precipitation

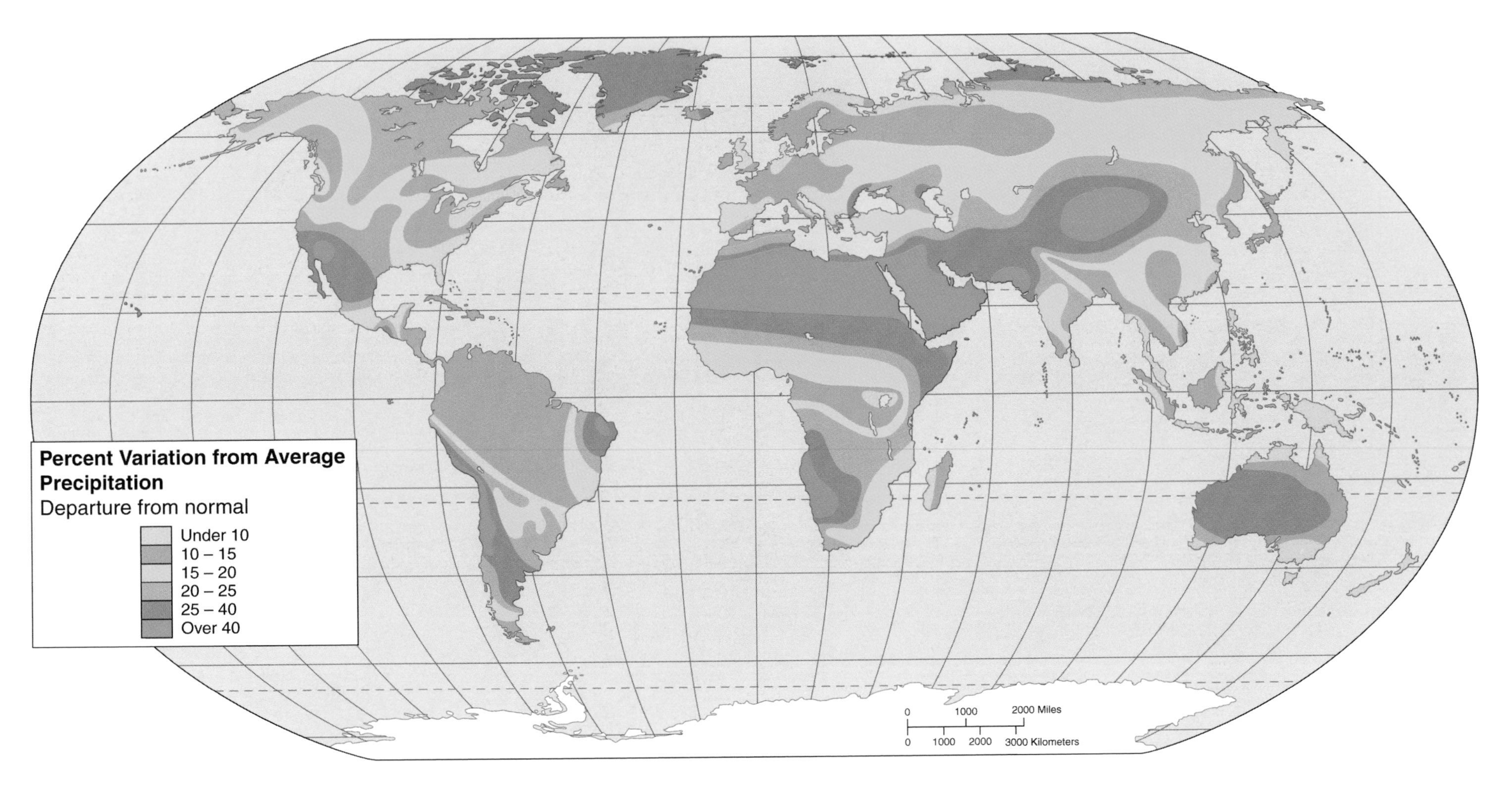

While annual precipitation totals and seasonal distribution of precipitation are important variables, the variability of precipitation from one year to the next may be even more critical. You will note from the map that there is a general spatial correlation between the world's drylands and the amount of annual variation in precipitation. Generally, the drier the climate, the more likely it is that there will be considerable differences in rainfall and/or snowfall from one year to the next. We might determine that the average precipitation of the mid-Sahara is 2 inches per year. What this really means is that a particular location in the Sahara during one year might receive .5", during the next year 3.5", and during a third year 2". If you add these together and divide by the number of years, the "average" precipitation is 2" per year. The significance of this is that much of the world's crucial agricultural output of cereals (grains) comes from dryland climates (the Great Plains of the United States, the Pampas of Argentina, the steppes of Ukraine and Russia, for example), and variations in annual rainfall totals can have significant impacts on levels of grain production and, therefore, important consequences for both economic and political processes.

Map 4a Temperature Regions and Ocean Currents

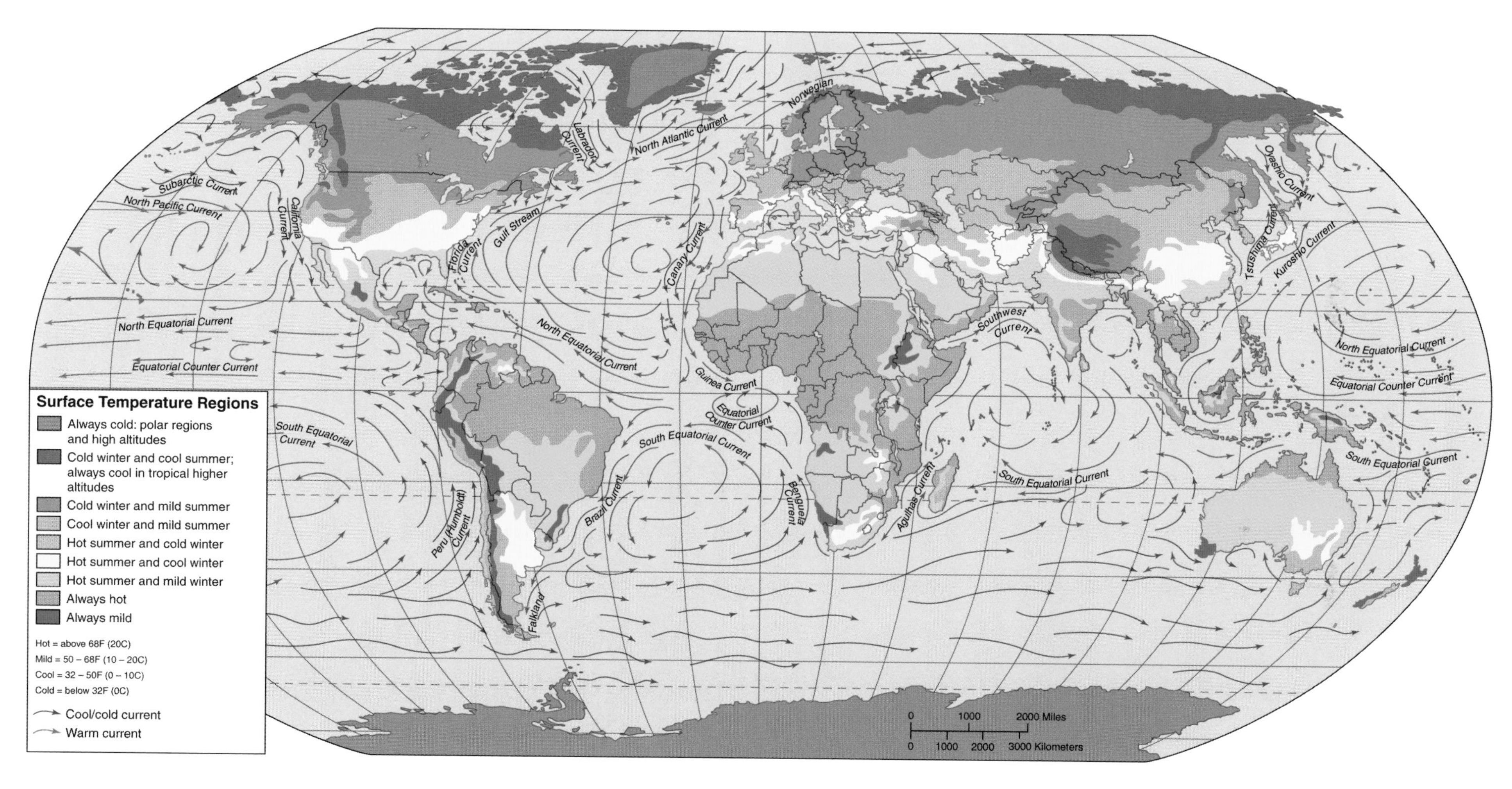

Along with precipitation, temperature is one of the two most important environmental variables, defining the climate conditions so essential for the distribution of such human activities as agriculture and the distribution of the human population. The seasonal rhythm of temperature, including such measures as the average annual temperature range (difference between the average temperature of the warmest month and that of the coldest month), is an additional variable not shown on the map but, like the seasonality of precipitation, should be a part of any comprehensive study of climate. The ocean currents illustrated exert a significant influence over the climate of adjacent regions and are the most important mechanism for redistributing surplus heat from the equatorial region into middle and high latitudes. Physical geographers known as "climatologists" study the phenomenon of temperature and related climatic characteristics.

Map 4b Average January Temperature

Map 4c Average July Temperature

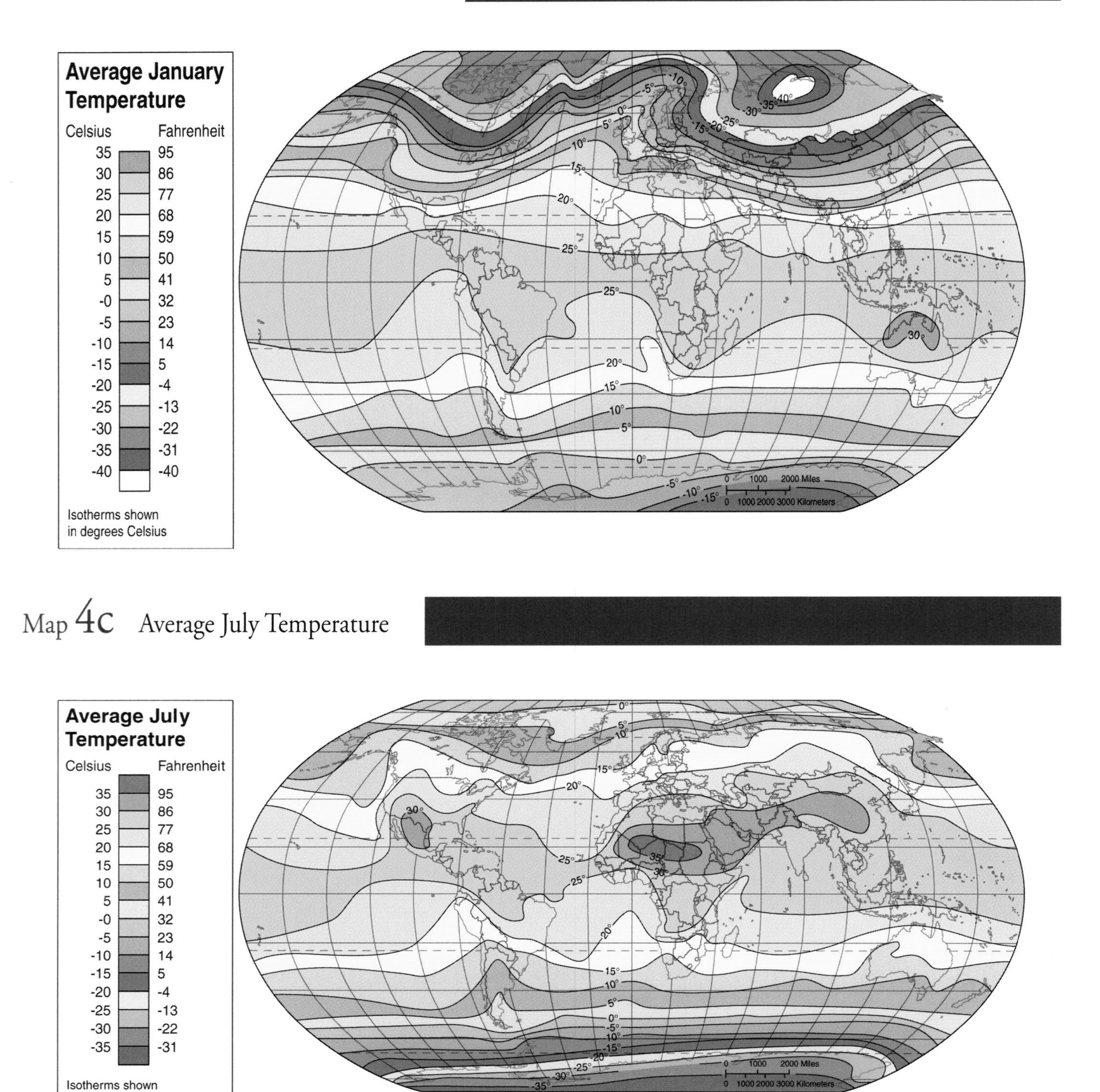

Where moisture availability tends to mark the seasons in the tropics and subtropics, in the mid-latitudes, seasons are marked by temperature. Temperature is determined by latitudinal transition, by altitude or elevation above sea level, and by location of a place relative to the world's landmasses and oceans. The most important of these controls is latitude, and temperatures generally become lower with increasing latitude. Proximity to water, however, tends to moderate temperature extremes, and "maritime" climates influenced by the oceans will be warmer in the winter and cooler in the summer than continental climates in the same general latitude. Maritime climates will also show smaller temperature ranges, the difference between January and July temperatures, while climates of the continental interiors, far from the moderating influences of the oceans, will tend to have greater temperature ranges. In the Northern Hemisphere, where there are both large landmasses and oceans, the range is great. But in the Southern Hemisphere, dominated by water and, hence, by the more moderate maritime air masses, the temperature range is comparatively small. Significant temperature departures from the "normal" produced by latitude may also be the result of elevation. With exceptions, lower temperatures produced by topography are difficult to see on maps of this scale.

Map 5a Atmospheric Pressure and Predominant Surface Winds, January

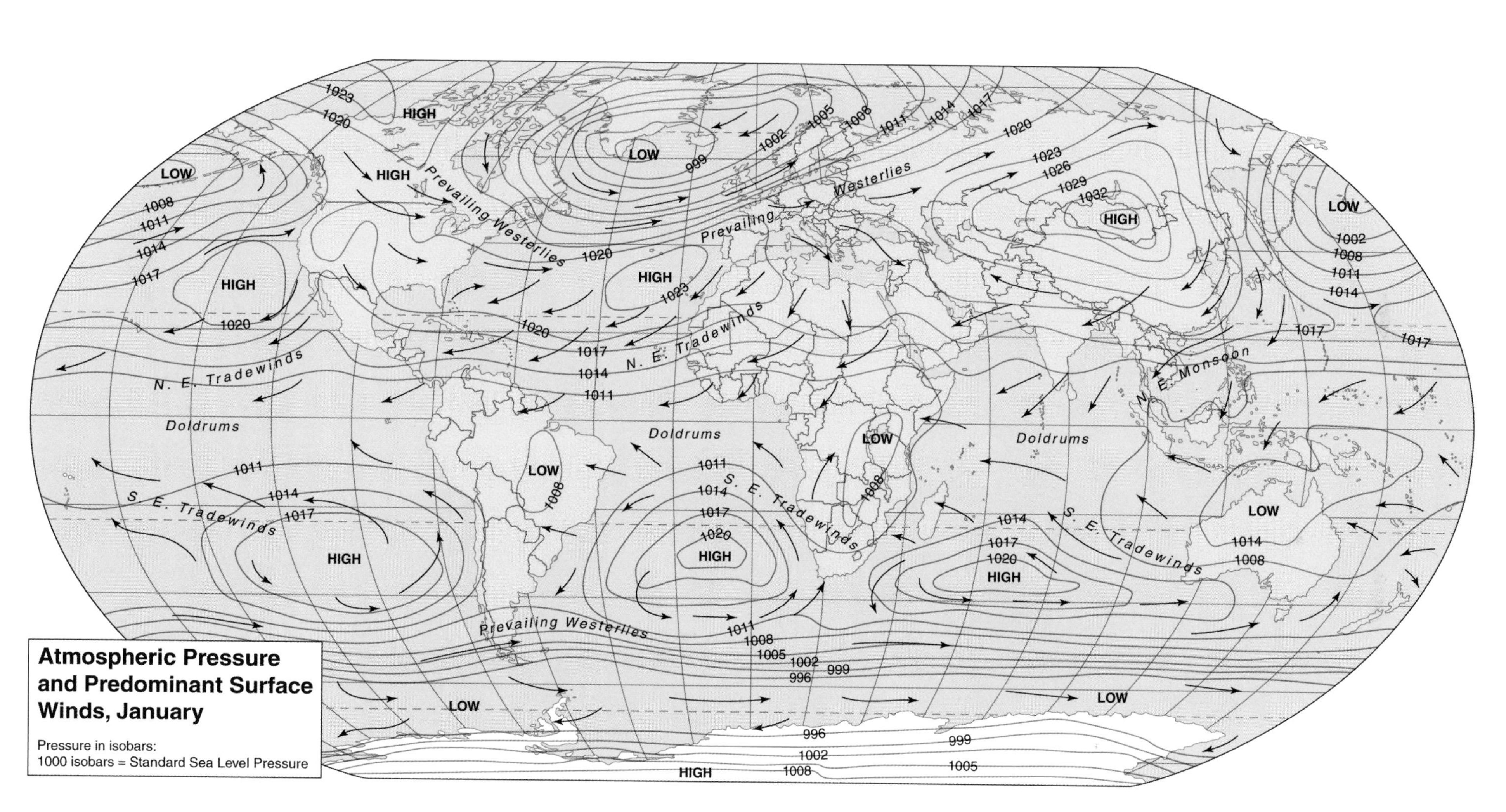

Map 5b Atmospheric Pressure and Predominant Surface Winds, July

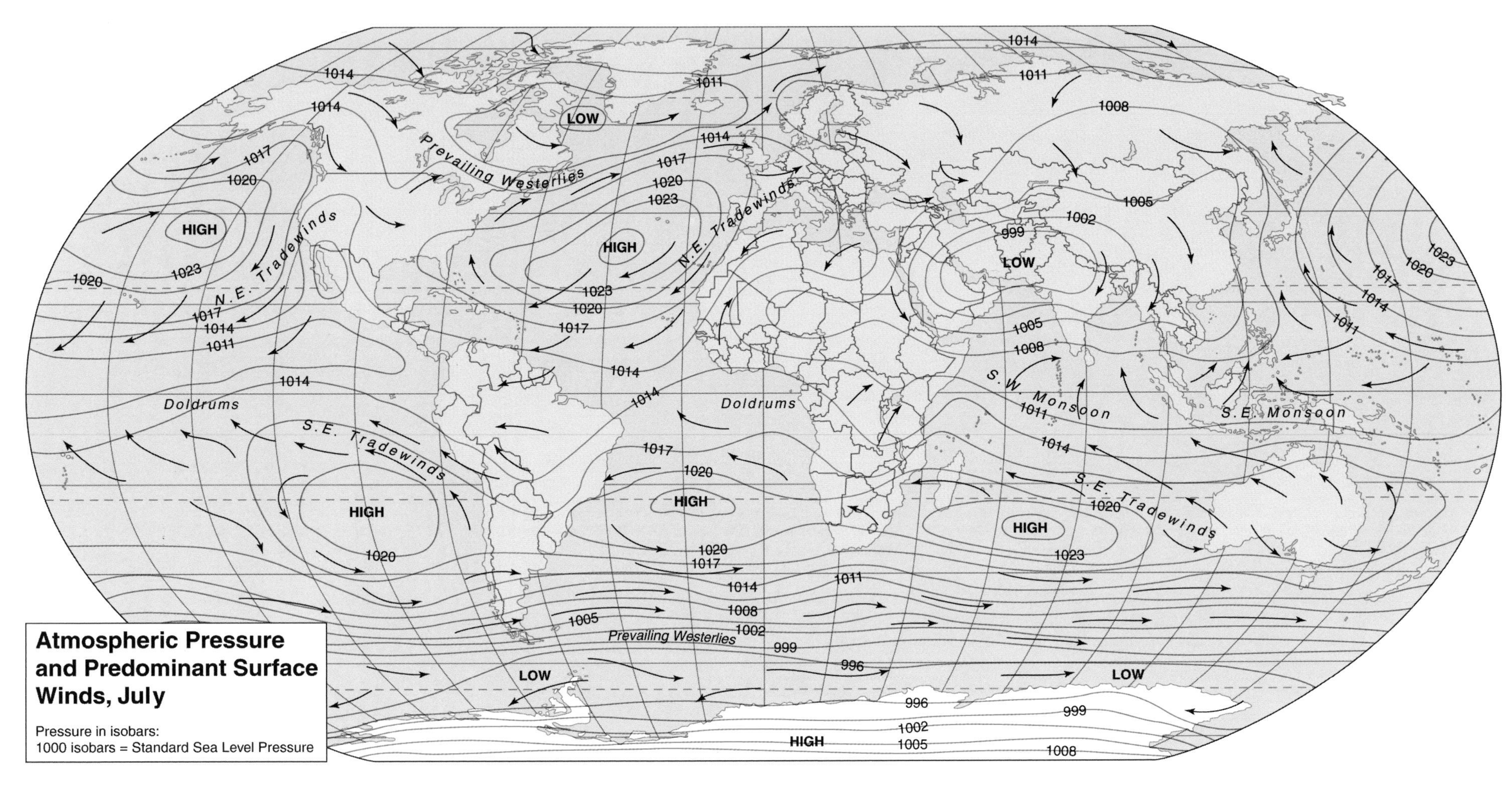

Atmospheric pressure, or the density of air, is a function largely of air temperature: the colder the air, the denser and heavier it is, hence the higher its pressure; the warmer the air, the lighter and less stable it is, hence the lower its pressure. Global pressure systems are the alternating low and high pressure systems that, from the equator north and south, include: the equatorial low (sometimes called the intertropical convergence) centered on the equator for much of the year; the subtropical highs with their centers near the 30th degrees of north and south latitude; the subpolar lows or polar fronts centered near the 60th parallel of north and south latitude; and the polar highs near the north and south poles. Air flows from high pressure to low pressure regions, and this air flow constitutes the earth's major surface winds such as the tropical tradewinds and the prevailing westerlies. This flow of air is one of the chief mechanisms by which surplus heat energy from the equatorial region is redistributed to higher latitudes. It is also the primary conditioner of the world's major precipitation belts, with rainfall and snowfall associated primarily with lower atmospheric pressure conditions.

Map 6 Climate Regions

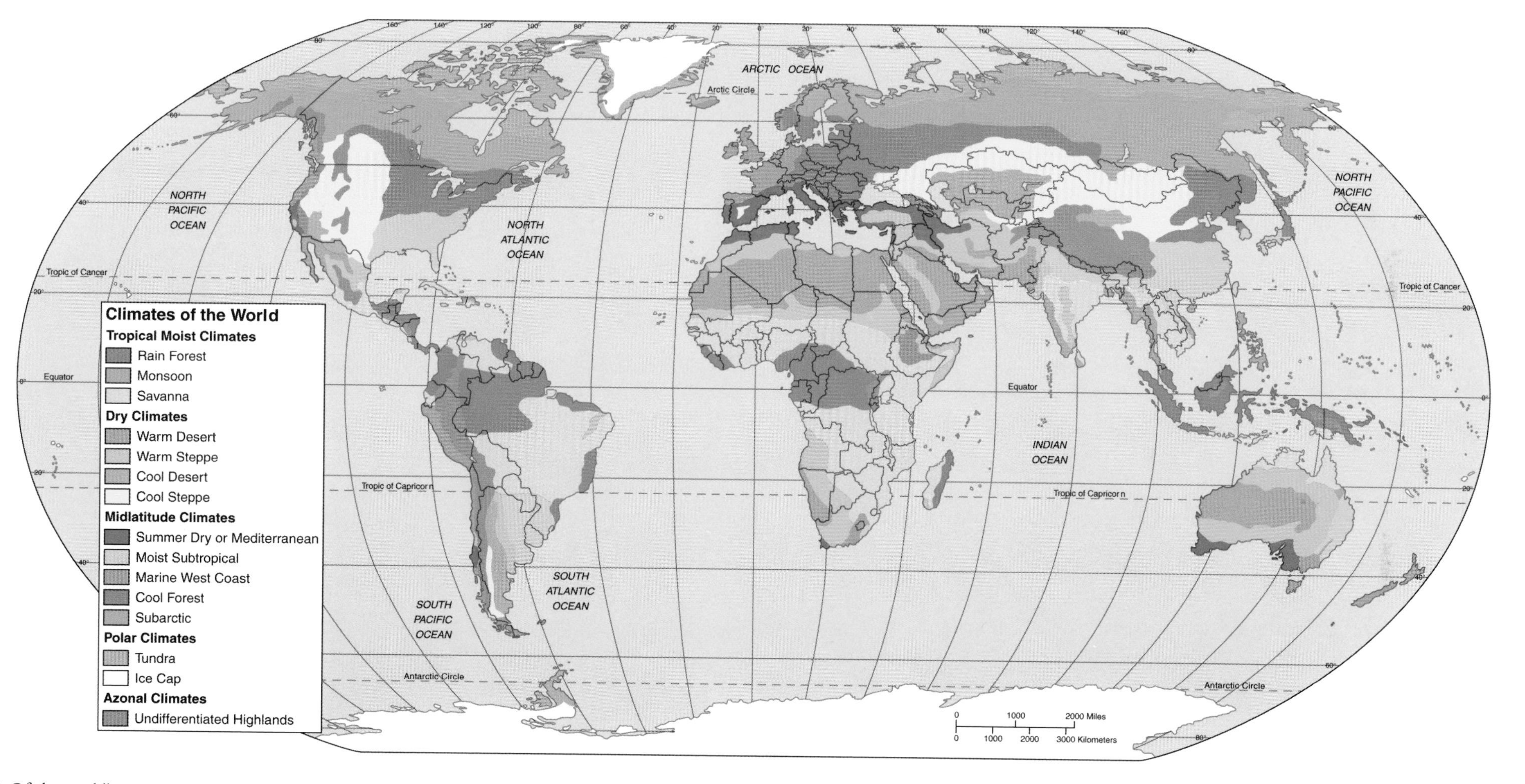

Of the world's many patterns of physical geography, climate or the long-term average of weather conditions such as temperature and precipitation is the most important. It is climate that conditions the distribution of natural vegetation and the types of soils that will exist in an area. Climate also influences the availability of our most crucial resource: water. From an economic standpoint, the world's most important activity is agriculture; no other element of physical geography is more important for agriculture than climate. Ultimately, it is agricultural production that determines where the bulk of human beings live, and therefore, climate is a basic determinant of the distribution of human populations as well. The study of climates or "climatology" is one of the most important branches of physical geography.

The climate classification system shown on this map is based on that developed by Wladimir Köppen. To establish his climate regions, Köppen used the climatic parameters of *precipitation, temperature,* and *evapotranspiration* as they impacted certain kinds of major vegetative associations. Hence the names for many of the climate regions are also the names of vegetative regions.

Map 7 Vegetation Types

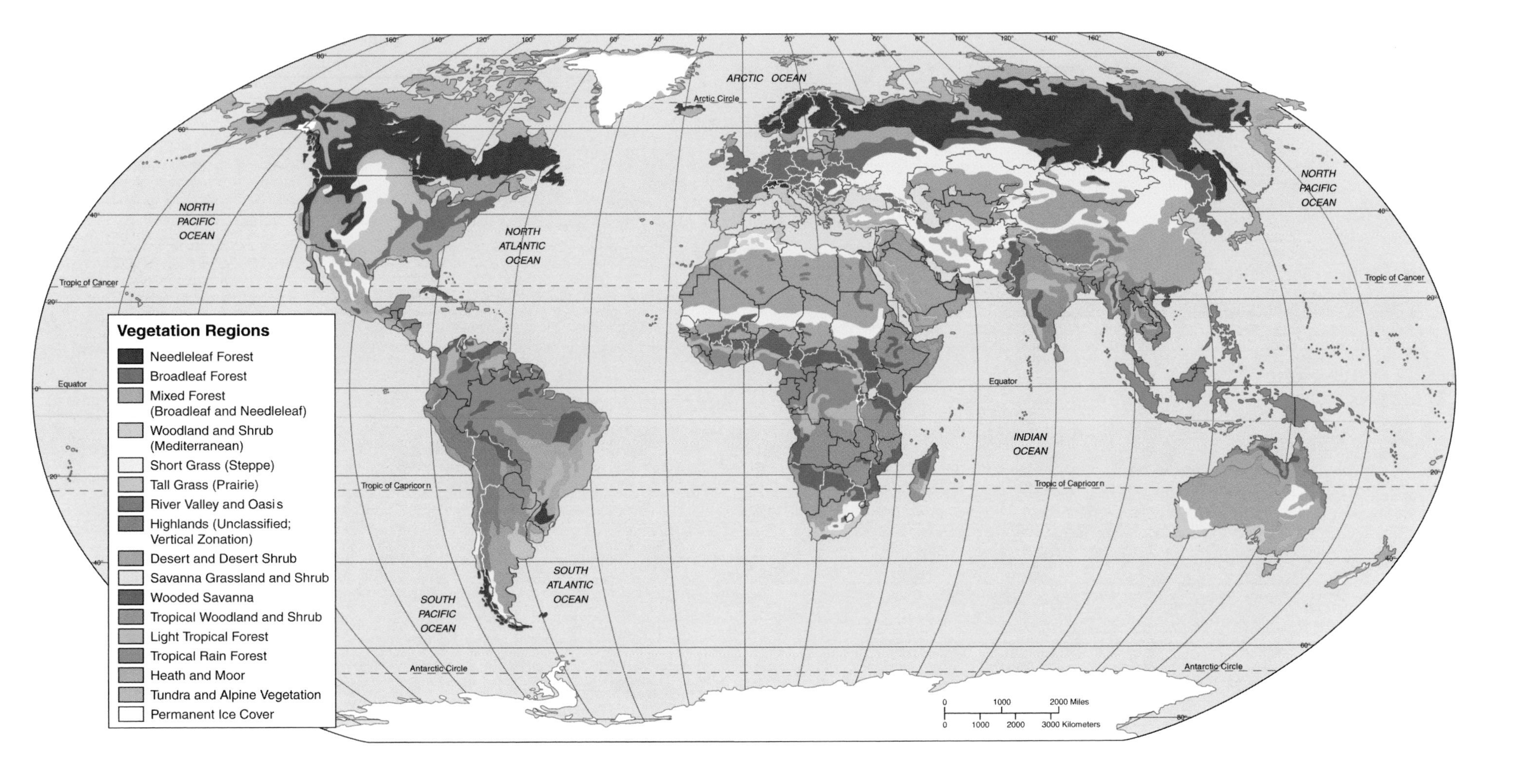

Vegetation is the most visible consequence of the distribution of temperature and precipitation. The global pattern of vegetative types or "habitat classes" and the global pattern of climate are closely related and make up one of the great global spatial correlations. But not all vegetation types are the consequence of temperature and precipitation or other climatic variables. Many types of vegetation in many areas of the world are the consequence of human activities, particularly the grazing of domesticated livestock, burning, and forest clearance. This map shows the pattern of natural or "potential" vegetation, or vegetation as it might be expected to exist without significant human influences, rather than the actual vegetation that results from a combination of environmental and human factors. Physical geographers who are interested in the distribution and geographic patterns of vegetation are "biogeographers."

Map 8 Soil Orders

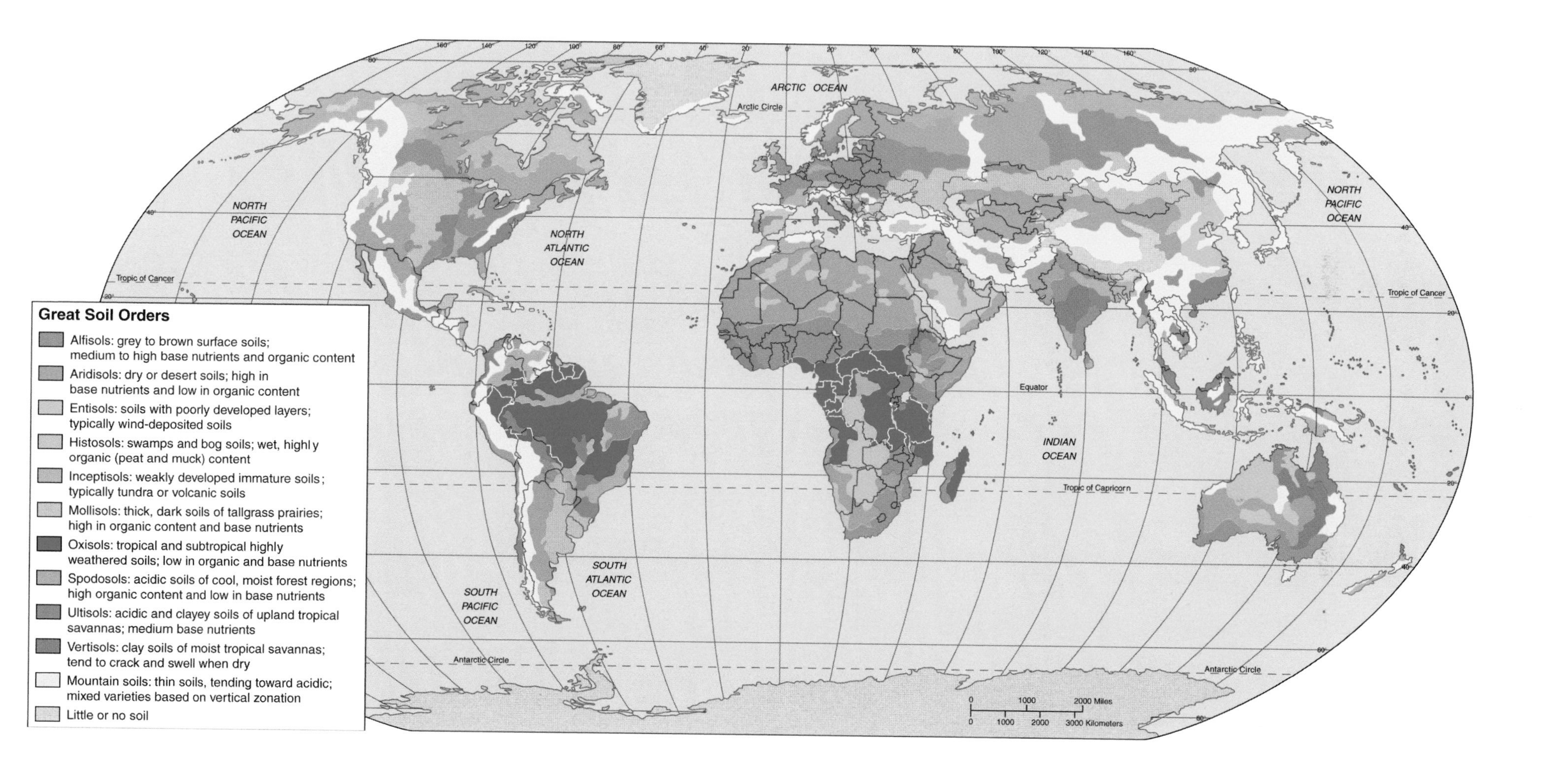

The characteristics of soil are one of the three primary physical geographic factors, along with climate and vegetation, that determine the habitability of regions for humans. In particular, soils influence the kinds of agricultural uses to which land is put. Since soils support the plants that are the primary producers of all food in the terrestrial food chain, their characteristics are crucial to the health and stability of ecosystems. Two types of soil are shown on this map: zonal soils, the characteristics of which are based on climatic patterns; and azonal soils, such as alluvial (water-deposited) or aeolian {wind-deposited) soils, the characteristics of which are derived from forces other than climate. However, many of the azonal soils, particularly those dependent upon drainage conditions, appear over areas too small to be readily shown on a map of this scale. Thus, almost none of the world's swamp or bog soils appear on this map. People who study the geographic characteristics of soils are most often "soil scientists," a discipline closely related to that branch of physical geography called "geomorphology."

Map 9 Ecological Regions

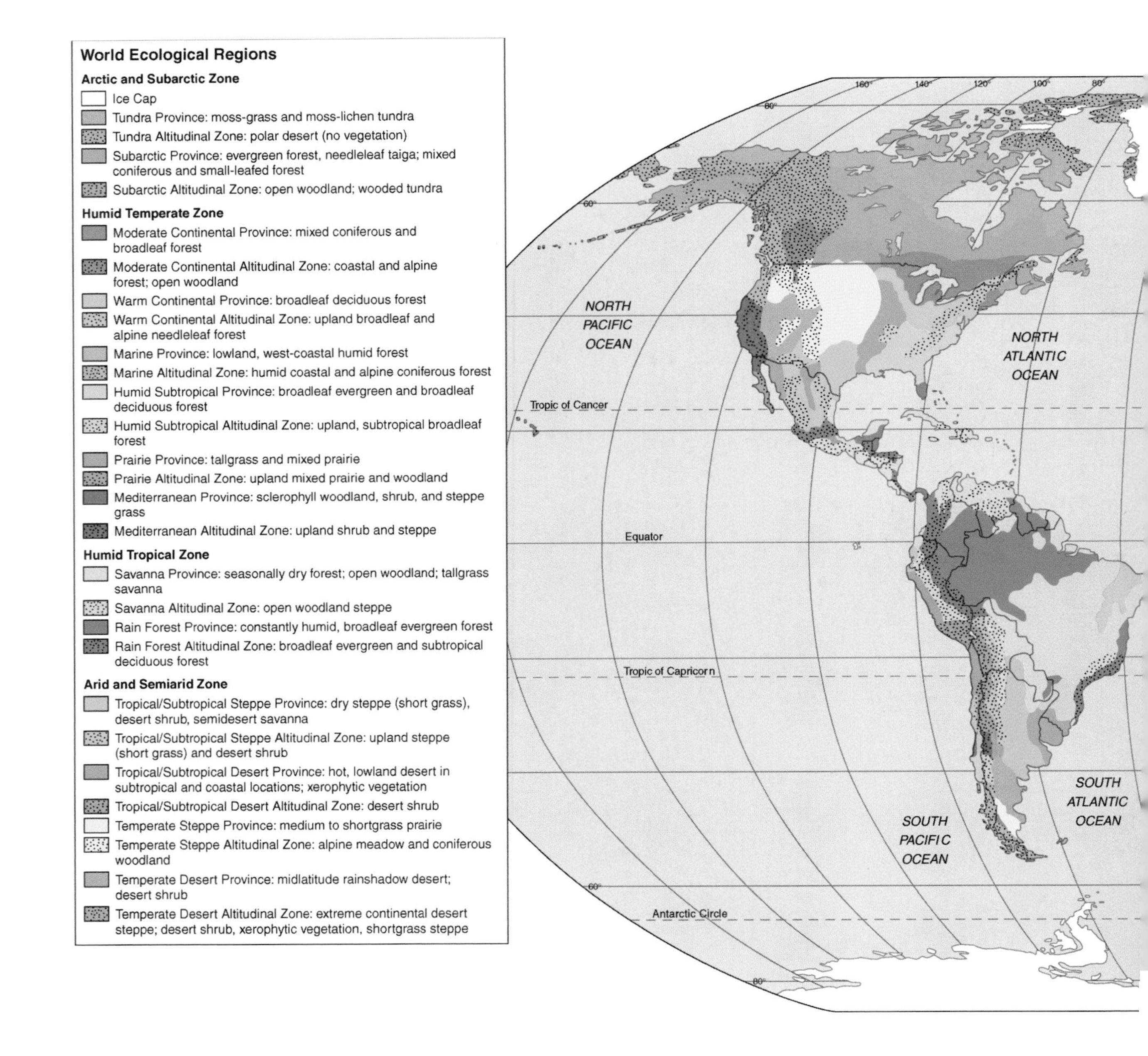

Ecological regions are distinctive areas within which unique sets of organisms and environments are found. We call the study of the relationships between organisms and their environmental surroundings "ecology." Within each of the ecological regions portrayed on the map, a particular combination of vegetation, wildlife, soil, water, climate, and terrain defines that region's habitability, or ability to support life, including human life. Like climate and landforms, ecological relationships are crucial to the existence of agriculture, the most basic of our economic activities, and important for many other kinds of economic activity as well. Biogeographers are especially concerned with the concept of ecological regions since such regions so clearly depend upon the geographic distribution of plants and animals in their environmental settings.

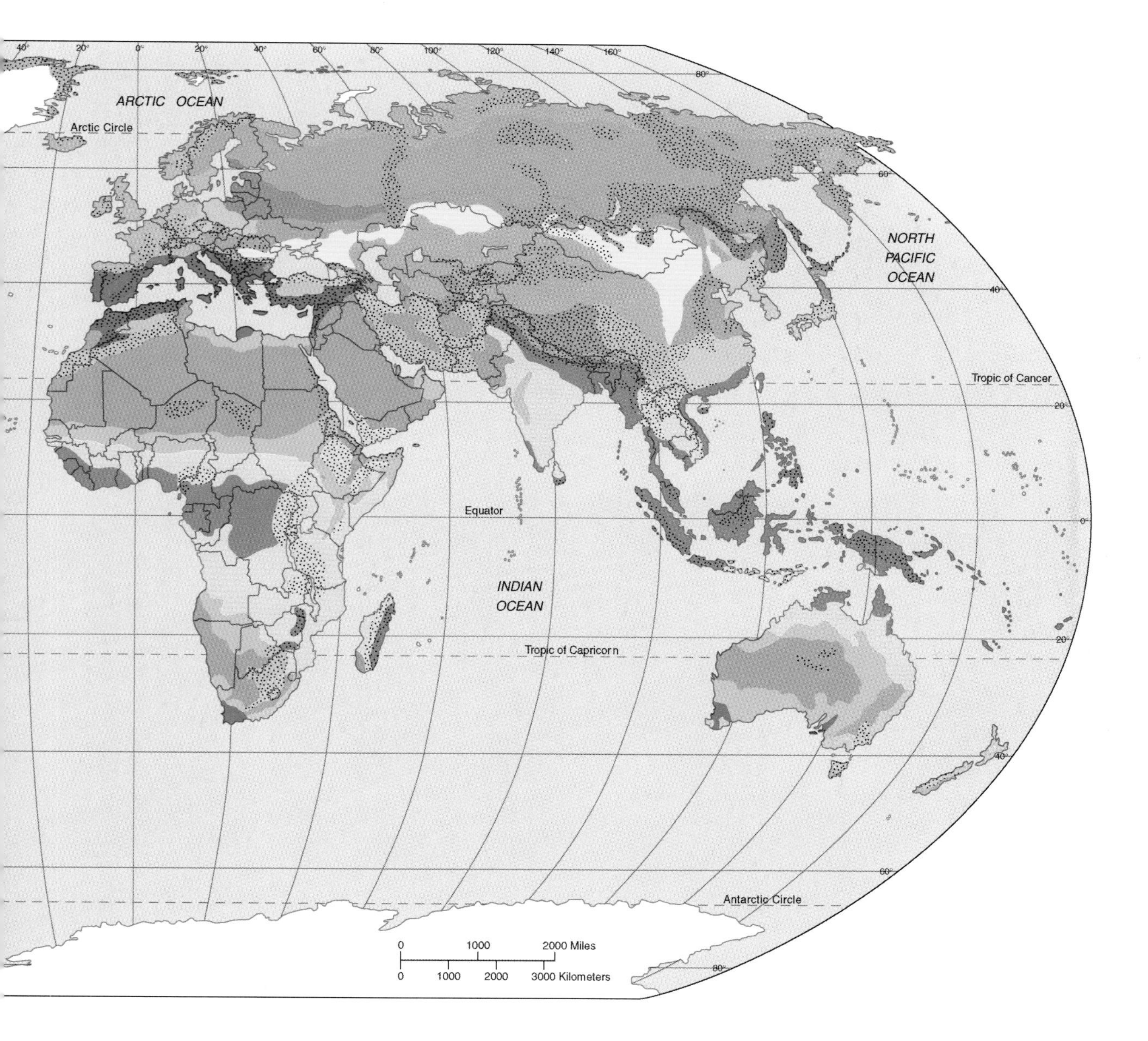

ARCTIC OCEAN
Arctic Circle
NORTH
PACIFIC
OCEAN
Tropic of Cancer
Equator
INDIAN
OCEAN
Tropic of Capricorn
Antarctic Circle
0
1000
2000 Miles
0
1000
2000
3000 Kilometers

Map 10 Plate Tectonics

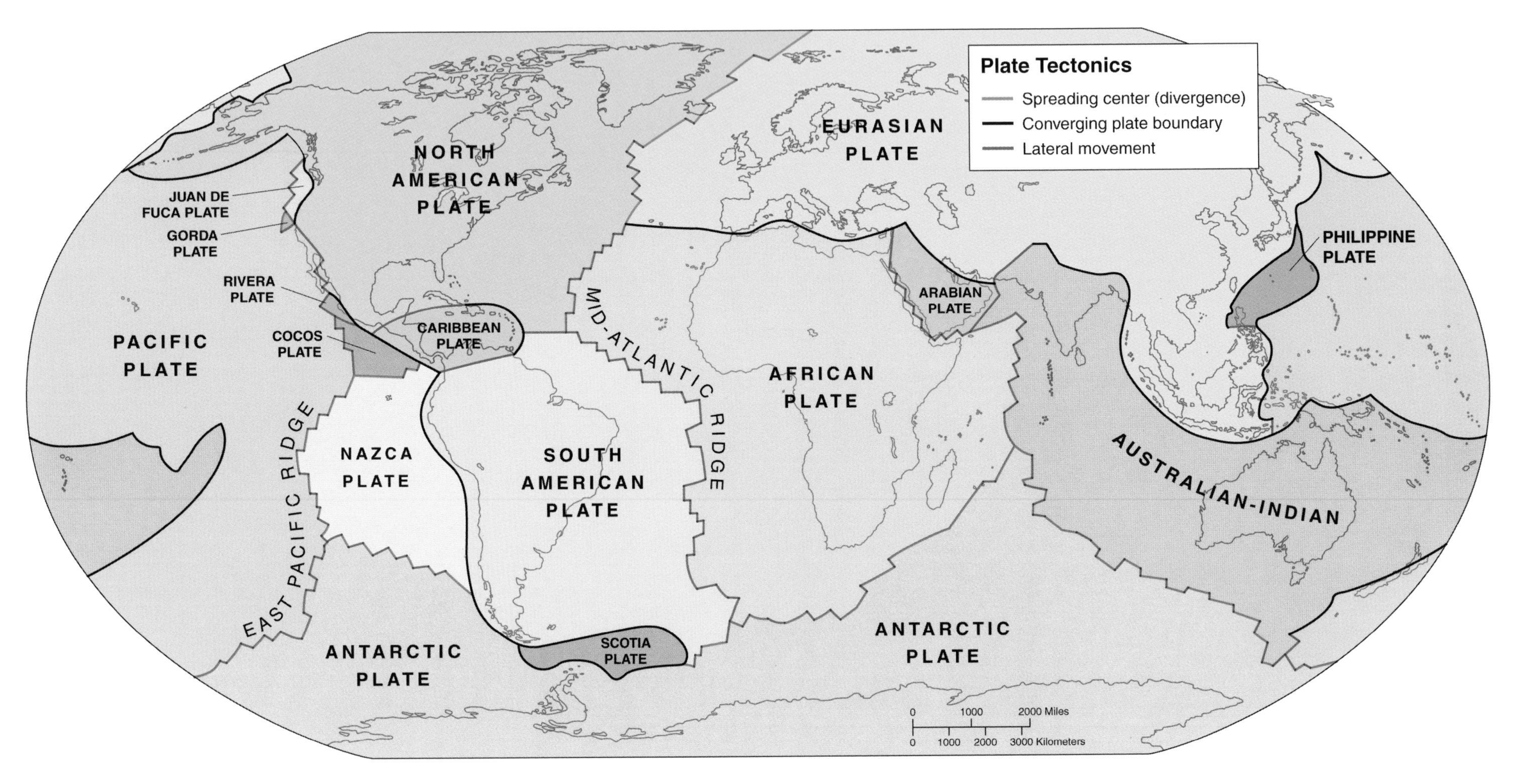

An understanding of the forces that shape the primary features of the earth's surface—the continents and ocean basins—requires a view of the earth's crust as fragments or "lithospheric plates" that shift position relative to one another. There are three dominant types of plate movement: *convergence,* in which plates move together, compressing former ocean floor or continental rocks together to produce mountain ranges, or producing mountain ranges through volcanic activity if one plate slides beneath another; *divergence,* in which the plates move away from one another, producing rifts in the earth's crust through which molten material wells up to produce new sea floors and mid-oceanic ridges; and *lateral shift,* in which plates move horizontally relative to one another, causing significant earthquake activity. All the major forms of these types of shifts are extremely slow and take place over long periods of geologic time. The movement of crustal plates, or what is known as "plate tectonics," is responsible for the present shape and location of the continents but is also the driving force behind some much shorter-term earth phenomena like earthquakes and volcanoes. A comparison of the map of plates with maps of hazards and terrain will reveal some interesting relationships.

Map 11 Topography

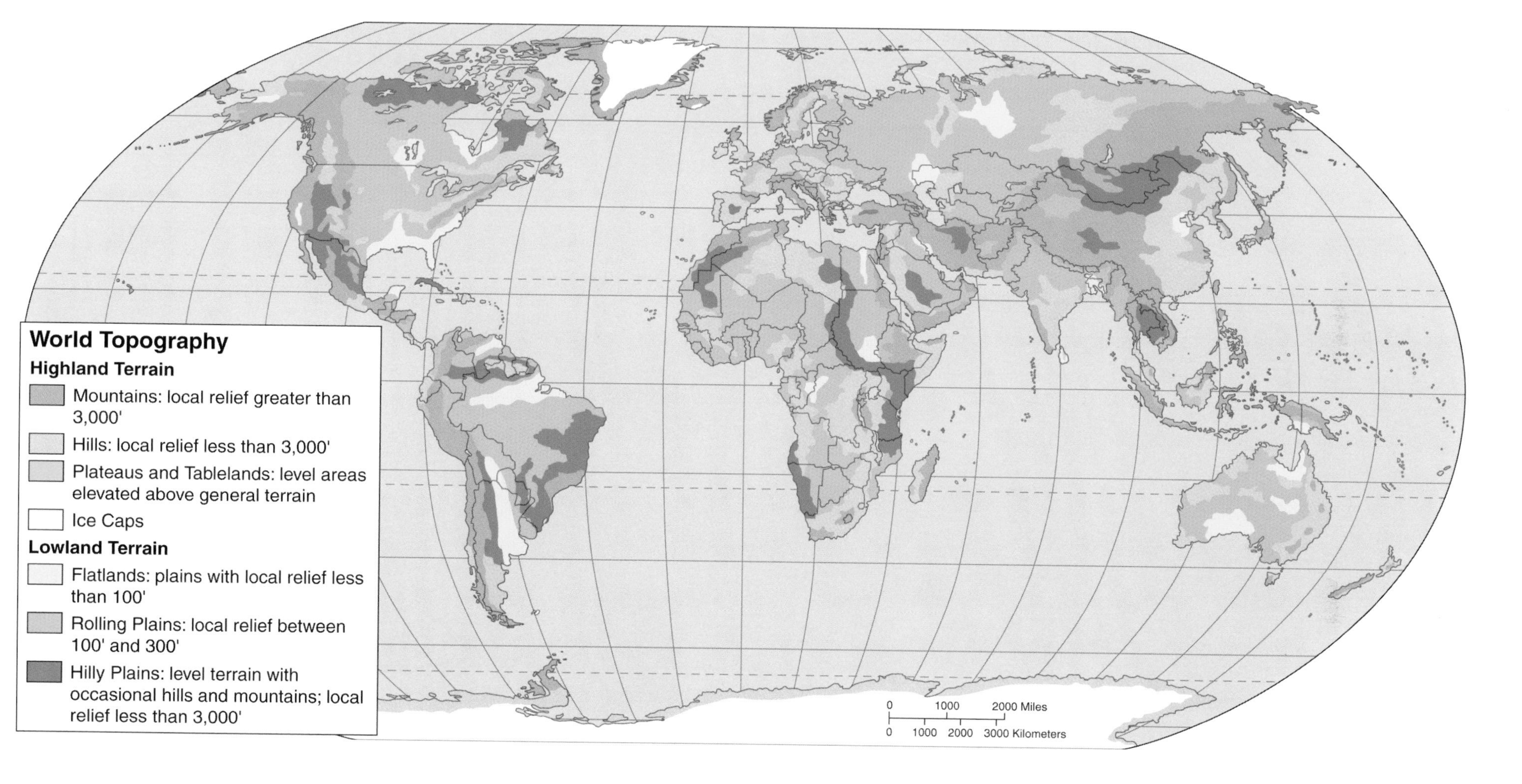

Topography or terrain, also called "landforms," is second only to climate as a conditioner of human activity, particularly agriculture but also the location of cities and industry. A comparison of this map of mountains, valleys, plains, plateaus, and other features of the earth's surface with a map of land use (Map 19) shows that most of the world's productive agricultural zones are located in lowland and relatively level regions. Where large regions of agricultural productivity are found, we also tend to find urban concentrations and, with cities, we find industry. There is also a good spatial correlation between the map of topography and the map showing the distribution and density of the human population (Map 14). Normally the world's major landforms are the result of extremely gradual primary geologic activity such as the long-term movement of crustal plates. This activity occurs over hundreds of millions of years. Also important is the more rapid (but still slow by human standards) geomorphological or erosional activity of water, wind, glacial ice, and waves, tides, and currents. Some landforms may be produced by abrupt or "cataclysmic" events such as a major volcanic eruption or a meteor strike, but such events are relatively rare and their effects are usually too minor to show up on a map of this scale. The study of the processes that shape topography is known as "geomorphology" and is an important branch of physical geography.

Map 12a Resources: Mineral Fuels

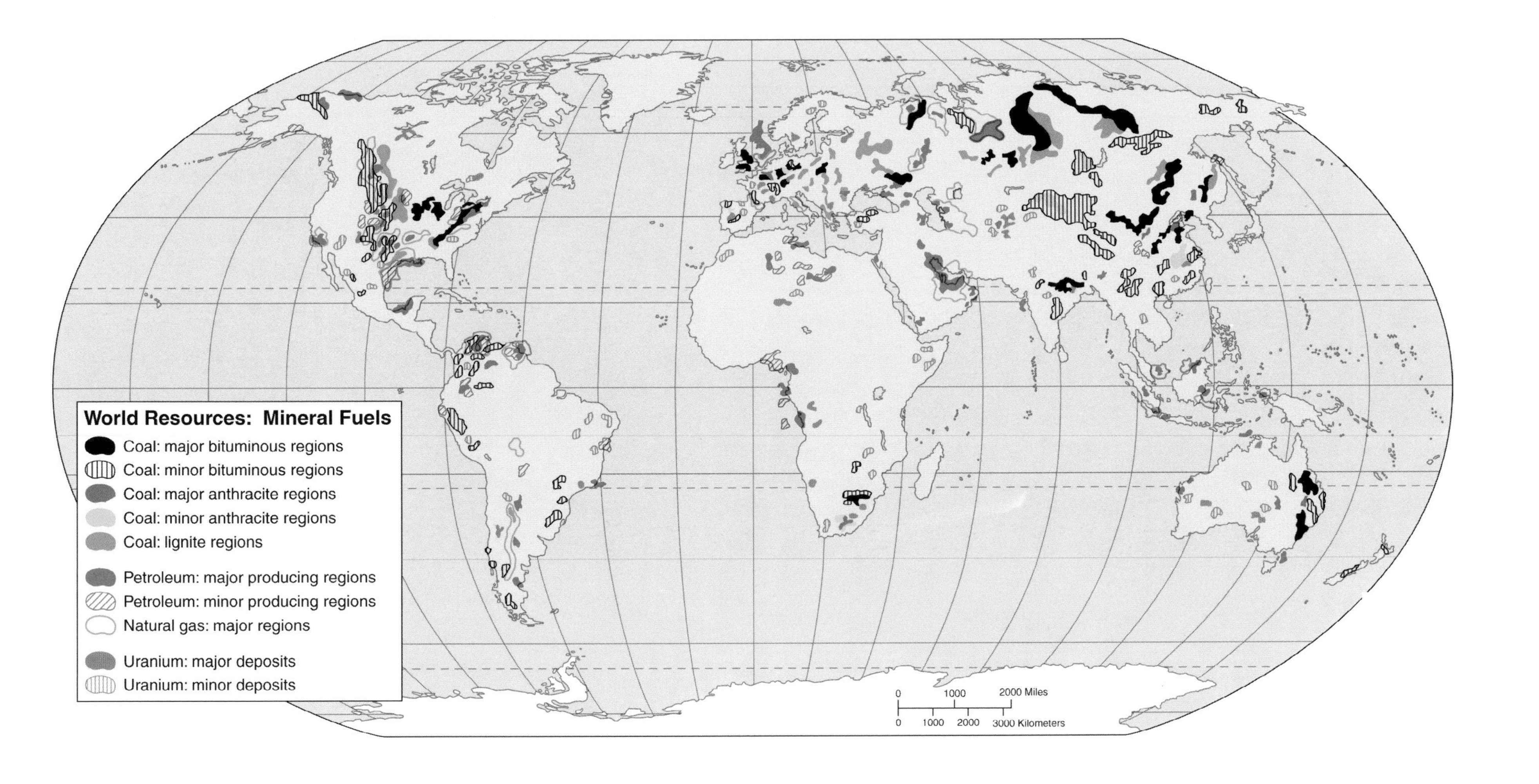

The extraction and transportation of mineral fuels rank with agriculture and forestry as "primary" human activities that impact the environment on a global scale. Nearly all of the most highly publicized environmental disasters of recent decades—the Prince William Sound oil spill or the Chernobyl nuclear accident, for example—have involved mineral fuels that were being stored, transported, or used. And the continuing extraction of mineral fuels like oil, natural gas, coal, and uranium produces high levels of atmospheric, soil, and water pollution. The location of mineral fuels tells us a great deal about where environmental degradation is likely to be occurring or to occur in the future. One need only look at the levels of atmospheric pollution and vegetative disruption in central and eastern Europe to recognize the damaging consequences of heavy reliance on coal as a domestic and industrial fuel. The location of mineral fuels also tells us something about existing or potential levels of economic development with those countries possessing abundant reserves of mineral fuels having more of a chance to maintain or attain higher levels of prosperity.

Map 12b Resources: Critical Metals

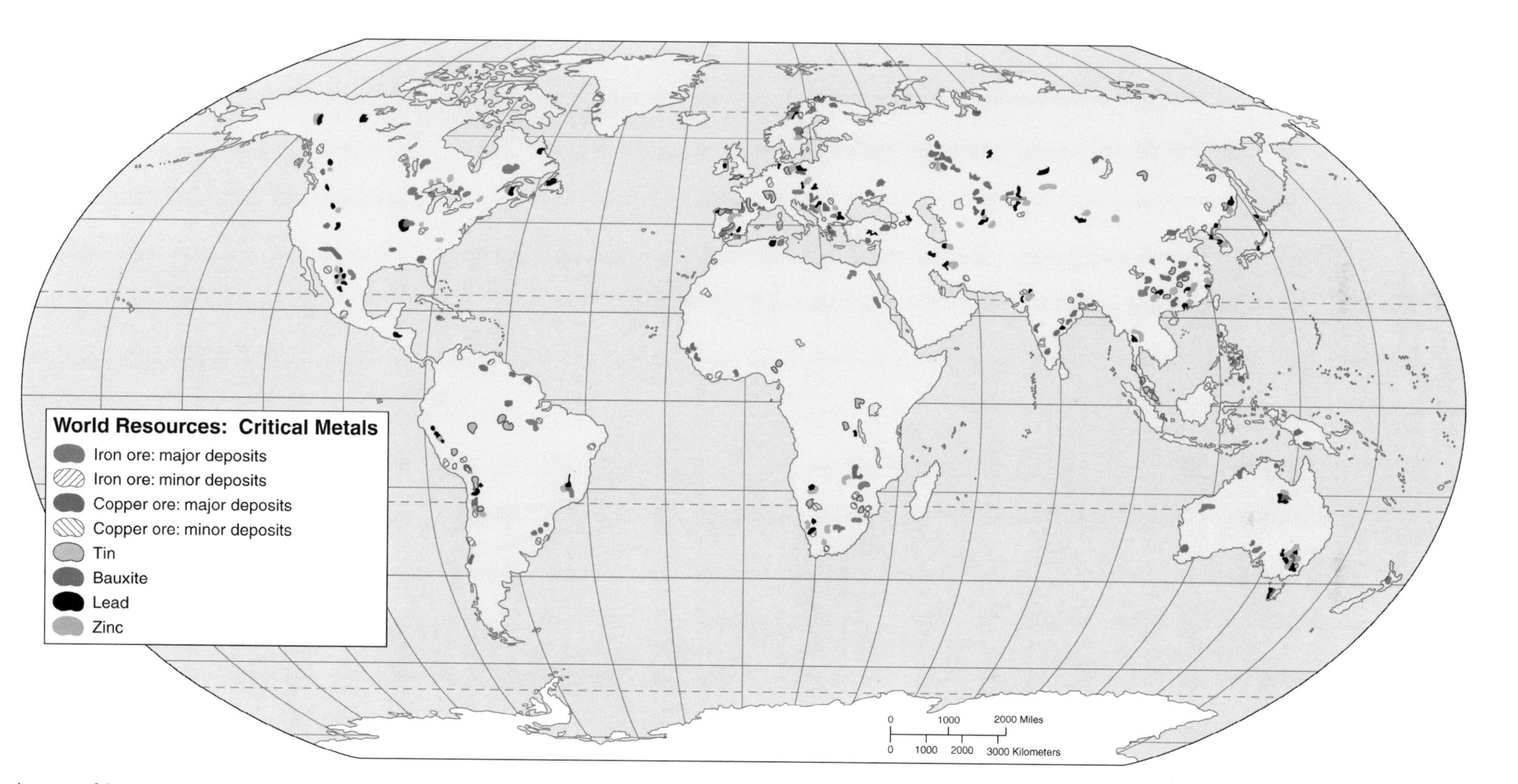

The location of deposits of critical metals such as iron, copper, tin, and others is an important determinant of the location of mining activities, like mineral fuel extraction, a "primary" economic activity. Also like mineral fuel extraction, mining for critical metallic ores makes significant environmental impact, particularly on vegetation, soils, and water resources. Some of the world's most dramatic examples of human modification of environments are located in areas of metallic ore extraction: the open pit copper mining areas of Arizona and Utah, for example. Environmental impact aside, those countries with significant critical metal deposits tend to stand a better chance of reaching higher levels of economic development, as long as they can extract and market the ores themselves rather than having the extraction process controlled by outside concerns. The average Bolivian, for example, does not benefit greatly from the fact that his/her country is an important producer of tin and other metals. Bolivia is a "colonial dependency" country and the wealth generated by metallic ore production there tends to flow out of the country to Europe and North America. On the other hand, another South American country, Brazil, is paying for much of its own current economic development by utilizing its reserves of iron and other metals and more of the wealth from the extraction of those resources stays within the country.

Map 13 Natural Hazards

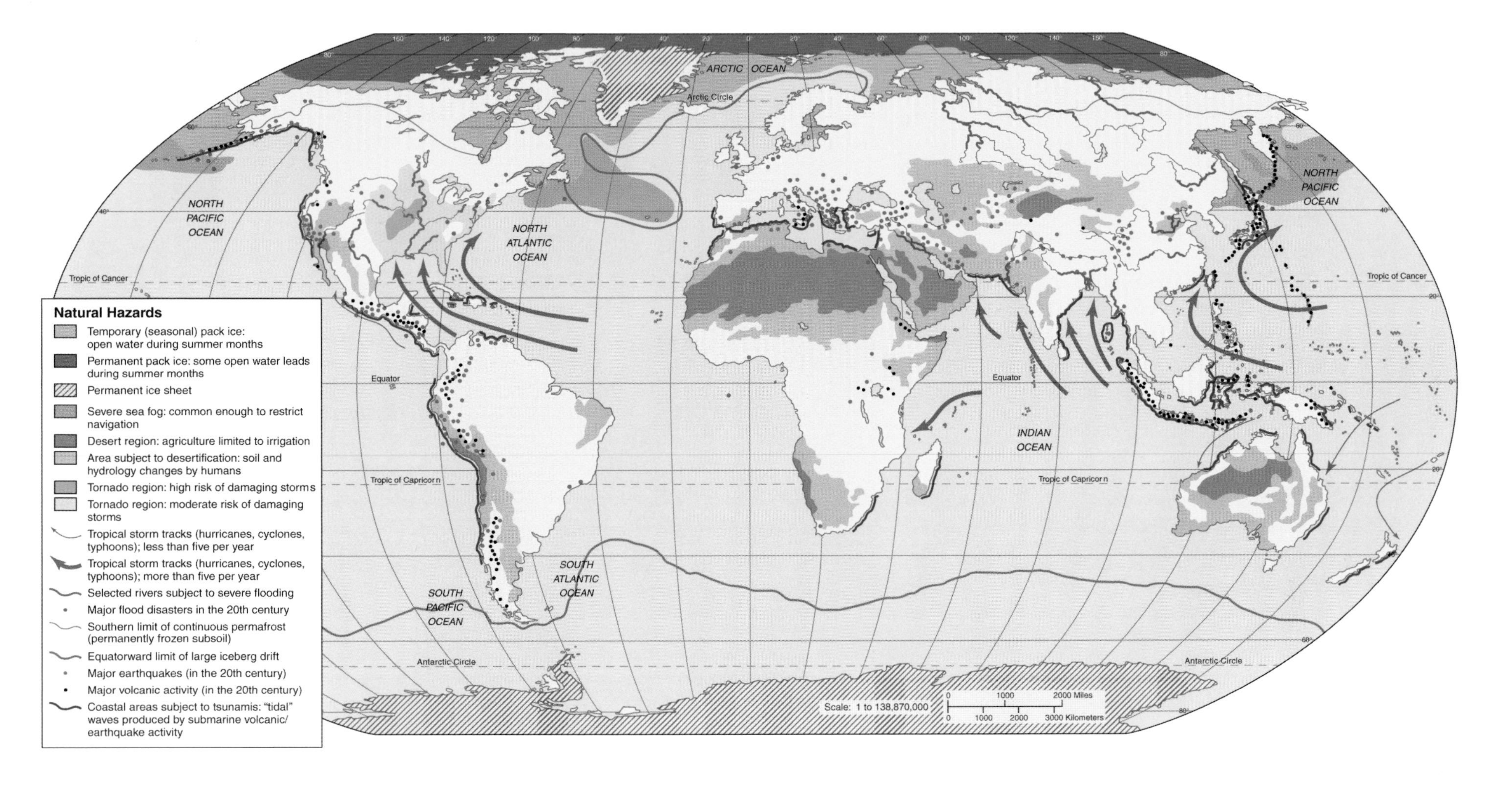

Unlike other elements of physical geography, most natural hazards are unpredictable. However, there are certain regions where the probability of the occurrence of a particular natural hazard is high. This map shows regions affected by major natural hazards at rates that are higher than the global norm. The presence of persistent natural hazards may influence the types of modifications that people make in the environment and certainly influence the styles of housing and other elements of cultural geography. Natural hazards may also undermine the utility of an area for economic purposes and some scholars suggest that regions of environmental instability may be regions of political instability as well. The study of natural hazards has become an important activity for "resource geographers" whose areas of interest overlap both human and physical fields of geography.

Part II

Global Human Patterns

Map 14 Past Population Distributions and Densities
Map 15 Population Density
Map 16 Land Use, A.D. 1500
Map 17 Economic Activities
Map 18 Urbanization
Map 19 Transportation Patterns
Map 20 Religions
Map 21 Languages
Map 22 Linguistic Diversity
Map 23 External Migrations in Modern Times

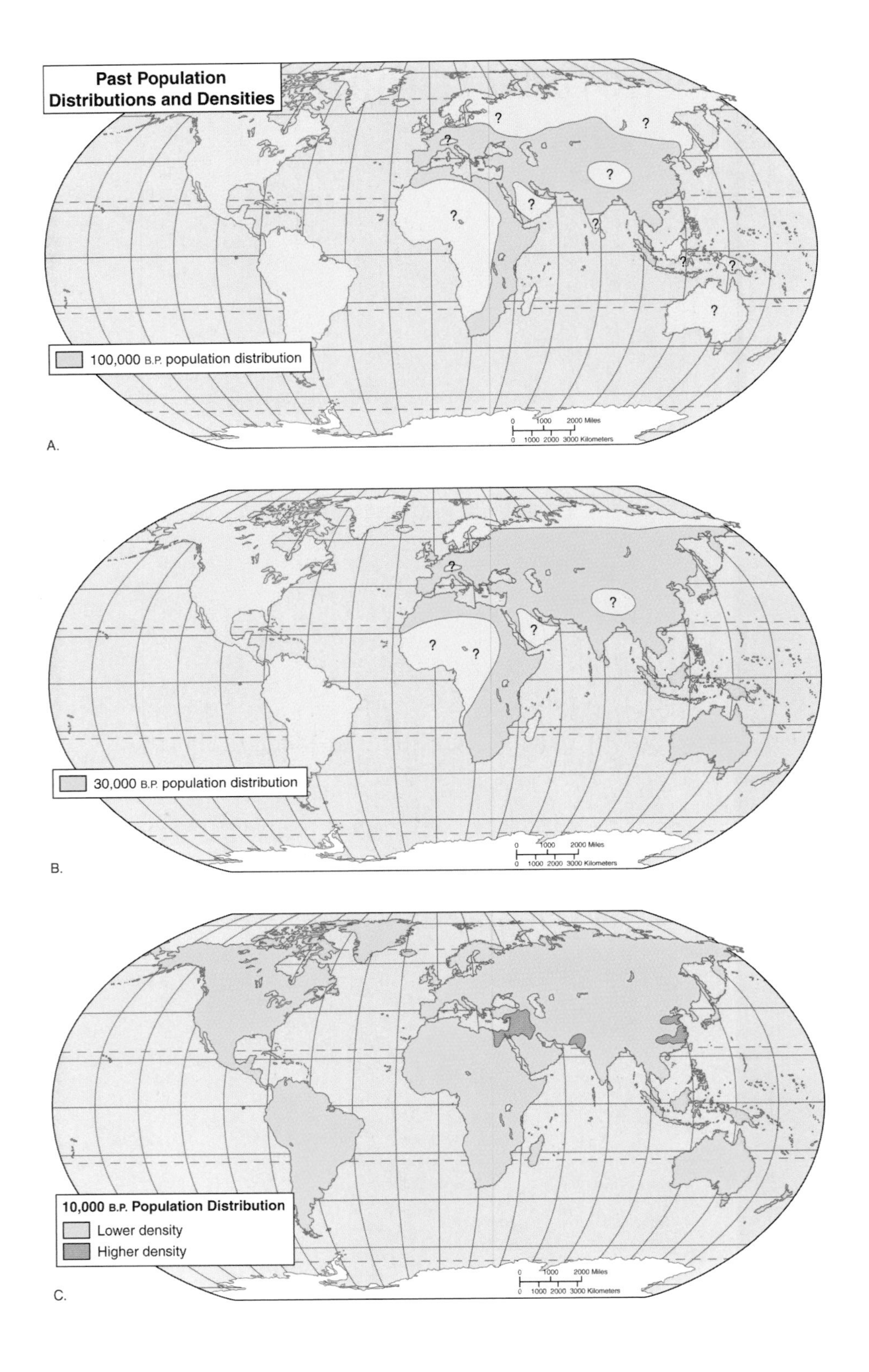

The map of the world at 100,000 B.P. (Before Present) shows the distributions of hominids who at that time had spread from their probable origin in Africa into parts of the Old World. At 30,000 B.P. few places in the Old World remained uninhabited. All the people then were hunters and gatherers who subsisted on wild foods. The environment probably was not at its carrying capacity for human populations until about 15,000 B.P.

By 10,000 B.P. hunting and gathering people had spread throughout the world. Plant and animal domestication (farming and pastoralism) had begun in some parts of the Old World perhaps as a response to behavioral changes necessitated by the population, which now exceeded the

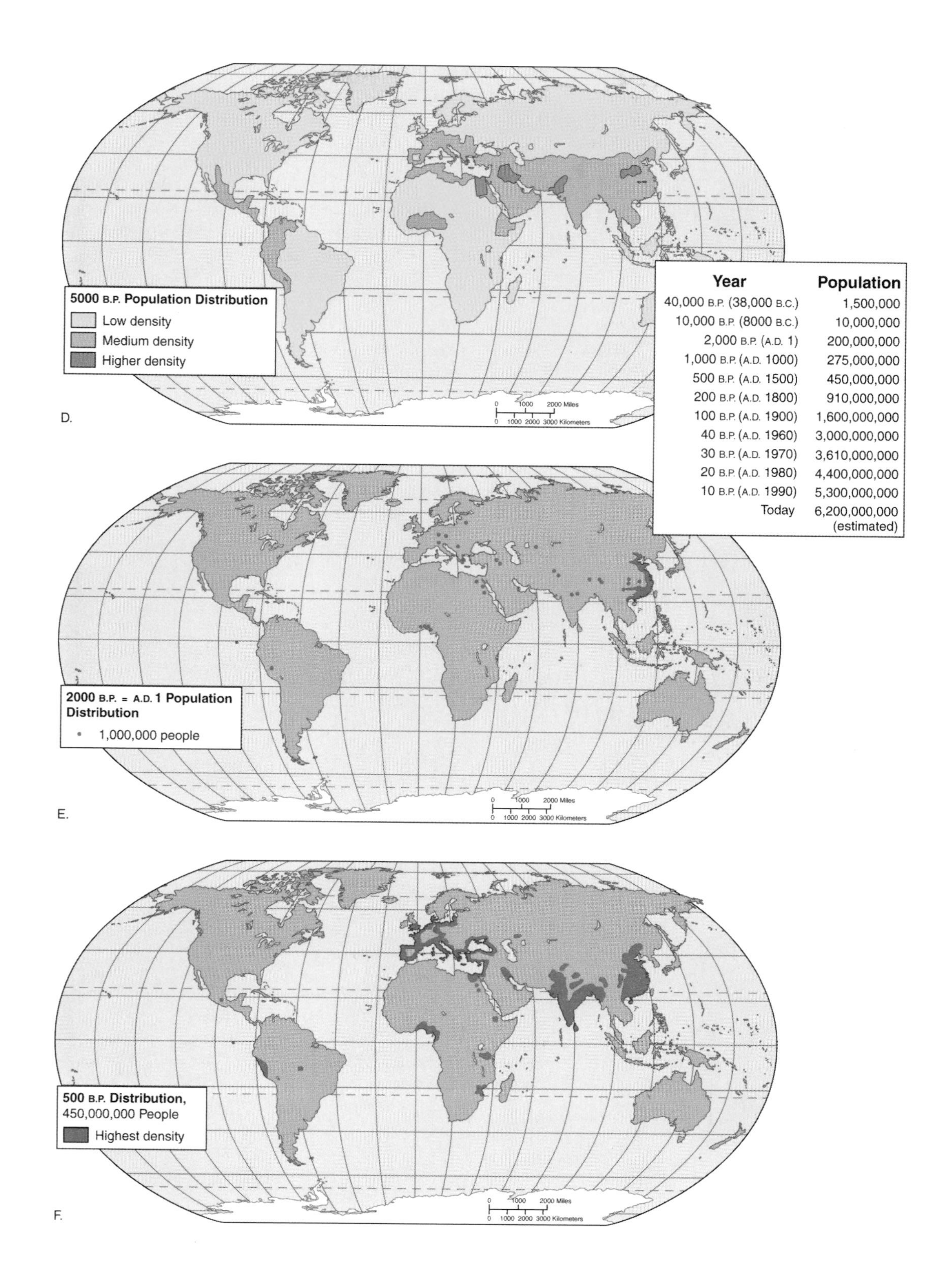

Year	Population
40,000 B.P. (38,000 B.C.)	1,500,000
10,000 B.P. (8000 B.C.)	10,000,000
2,000 B.P. (A.D. 1)	200,000,000
1,000 B.P. (A.D. 1000)	275,000,000
500 B.P. (A.D. 1500)	450,000,000
200 B.P. (A.D. 1800)	910,000,000
100 B.P. (A.D. 1900)	1,600,000,000
40 B.P. (A.D. 1960)	3,000,000,000
30 B.P. (A.D. 1970)	3,610,000,000
20 B.P. (A.D. 1980)	4,400,000,000
10 B.P. (A.D. 1990)	5,300,000,000
Today	6,200,000,000 (estimated)

environmental carrying capacity. Farming supports higher population densities than hunting and gathering. Urban civilization with cities dependent on their hinterlands had developed by 5000 B.P. The maps of A.D. 1 and A.D. 1500 approximate actual population density on a scale of one dot to every million people. A chart provides total world population figures for different time periods. Compare these maps to the contemporary population density map on the following page.

Map 15 Population Density

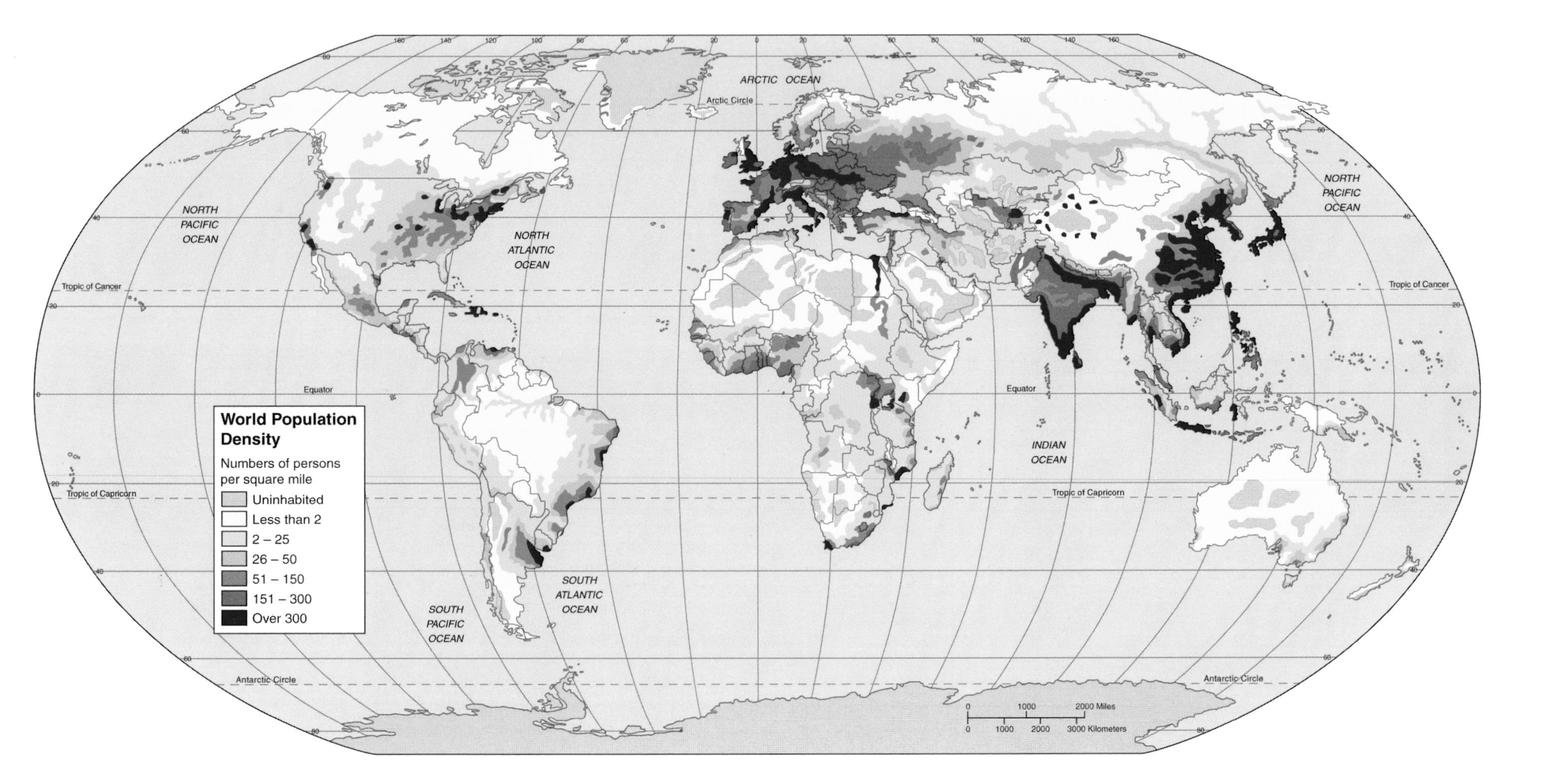

No feature of human activity is more reflective of geographic relationships than where people live. In the areas of densest populations, a mixture of natural and human factors has combined to allow maximum food production, maximum urbanization, and maximum centralization of economic activities. Three great concentrations of human population appear on the map—East Asia, South Asia, and Europe—with a fourth, lesser concentration in eastern North America. While population growth is relatively slow in three of these population clusters, in the fourth—South Asia—growth is still rapid and South Asia is expected to become even more densely populated in the early years of the twenty-first century, while density of the other regions is expected to remain about as it now appears. In Europe and North America, the relatively stable population growth rates are the result of economic development that has caused population growth to level off within the last century. In East Asia, the growth rates have also begun to decline. In the case of Japan, Taiwan, the Koreas, and other more highly developed nations of the Pacific Rim, the reduced growth is the result of economic development. In China, at least until recently, lowered population growth rates have resulted from strict family planning. The areas of future high density of population, in addition to those already existing, are likely to be in Middle and South America and in Central Africa, where population growth rates are well above the world average.

Map 16 Land Use, A.D. 1500

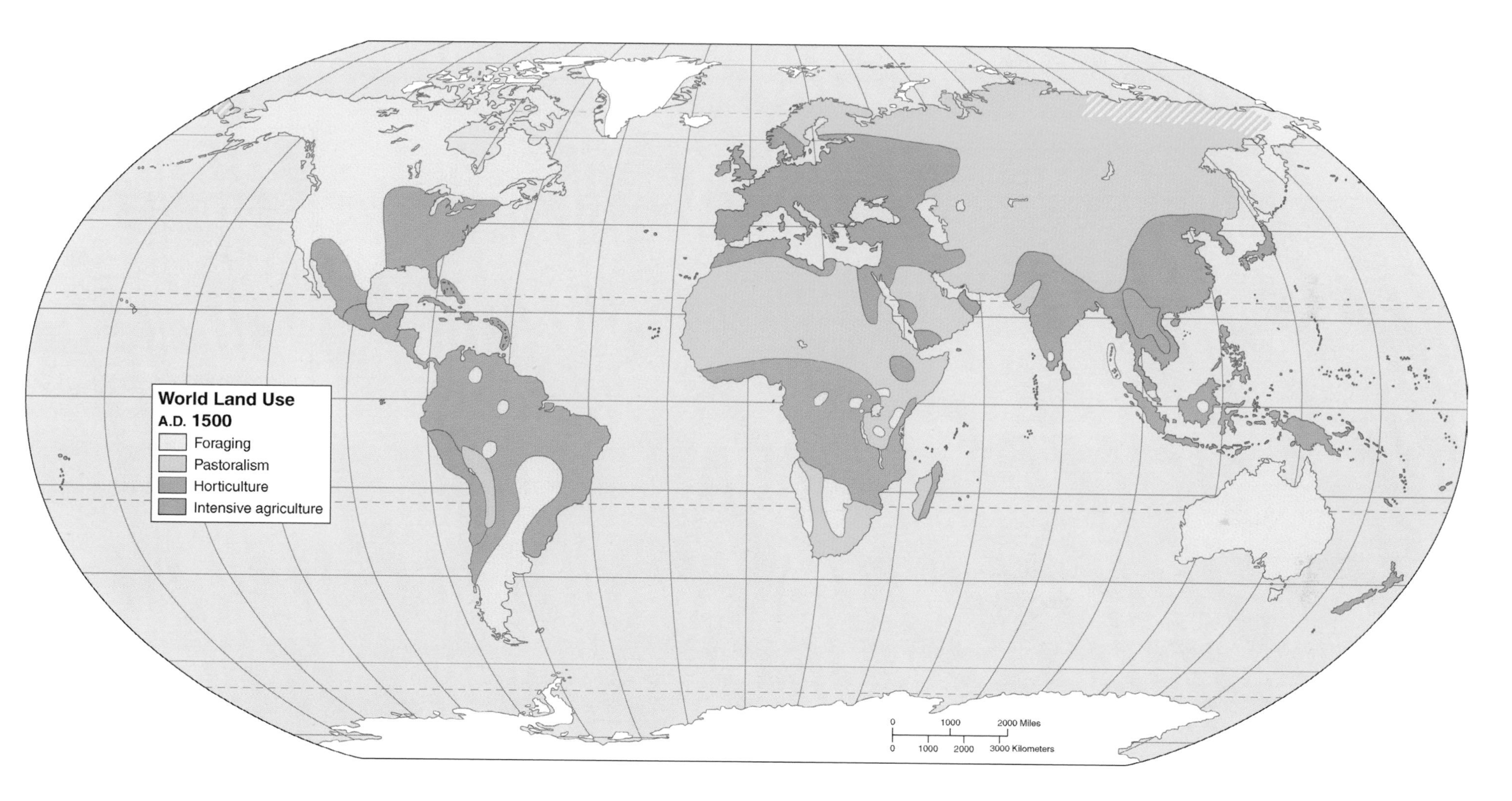

Europeans began to explore the world in the late 1400s. They encountered many independent people with self-sustaining economies at that time. Foraging people practiced hunting and gathering, utilizing the wild forms of plants and animals in their environments. Horticultural people practiced a simple form of agriculture using hoes or digging sticks as their basic tools. They sometimes cleared their land by burning and then planted crops. Pastoralists herded animals as their basic subsistence pattern. Complex state-level societies, such as the Mongols, had pastoralism as their base. Intensive agriculturalists based their societies on complicated irrigation systems and/or the plow and draft animals. Wheat and rice were two kinds of crops that supported large populations. In many of the areas of intensive agriculture—particularly in MesoAmerica, Europe, Southwest Asia, South Asia, and East Asia—complex patterns of market economies had begun to develop well before the 15th century and the beginnings of European expansion.

Map 17 Economic Activities

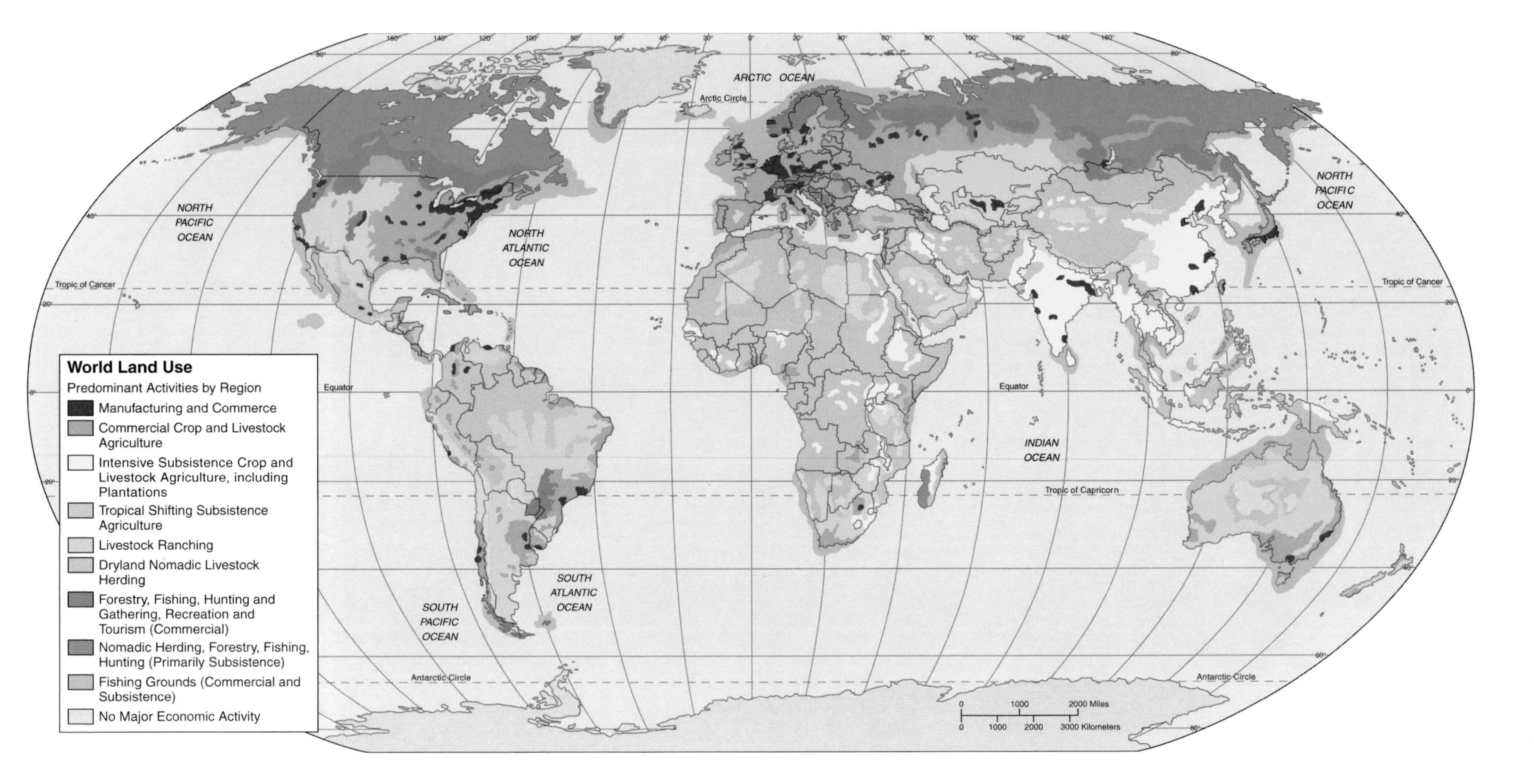

Land uses can be categorized as lying somewhere on a scale between extensive uses, in which human activities are dispersed over relatively large areas, and intensive uses, in which human activities are concentrated in relatively small areas. Many of the most important land use patterns of the world (such as urbanization, industry, mining, or transportation) are intensive and therefore relatively small in area and not easily seen on maps of this scale. Hence even in the areas identified as "Manufacturing and Commerce" on the map there are many land uses that are not strictly industrial or commercial in nature, and, in fact, more extensive land uses (farming, residential, open space) may actually cover more ground than the intensive industrial or commercial activities. On the other hand, the more extensive land uses like agriculture and forestry, tend to dominate the areas in which they are found. Thus, primary economic activities such as agriculture and forestry tend to dominate the world map of land use because of their extensive character. Much of this map is, therefore, a map that shows the global variations in agricultural patterns. Note, among other things, the differences between land use patterns in the more developed countries of the temperate zones and the less developed countries of the tropics.

Map 18 Urbanization

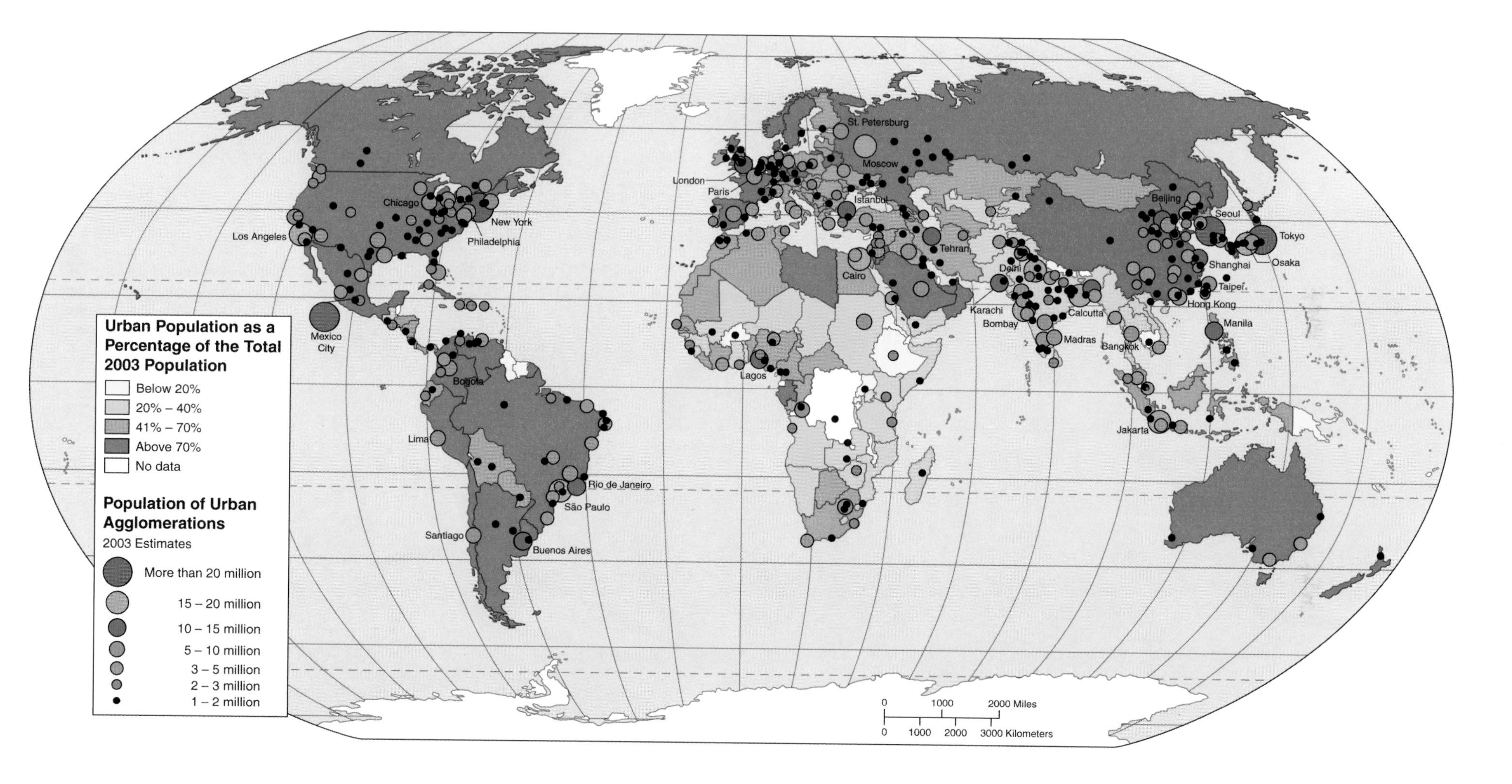

The degree to which a region's population is concentrated in urban areas is a major indicator of a number of things: the potential for environmental impact, the level of economic development, and the problems associated with human concentrations. Urban dwellers are rapidly becoming the norm among the world's people and rates of urbanization are increasing worldwide, with the greatest increases in urbanization taking place in developing regions. Whether in developed or developing countries, those who live in cities exert an influence on the environment, politics, economics, and social systems that goes far beyond the confines of the city itself. Acting as the focal points for the flow of goods and ideas, cities draw resources and people not just from their immediate hinterland but from the entire world. This process creates far-reaching impacts as resources are extracted, converted through industrial processes, and transported over great distances to metropolitan regions, and as ideas spread or *diffuse* along with the movements of people to cities and the flow of communication from them. The significance of urbanization can be most clearly seen, perhaps, in North America where, in spite of vast areas of relatively unpopulated land, well over 90 percent of the population lives in urban areas.

Map 19 Transportation Patterns

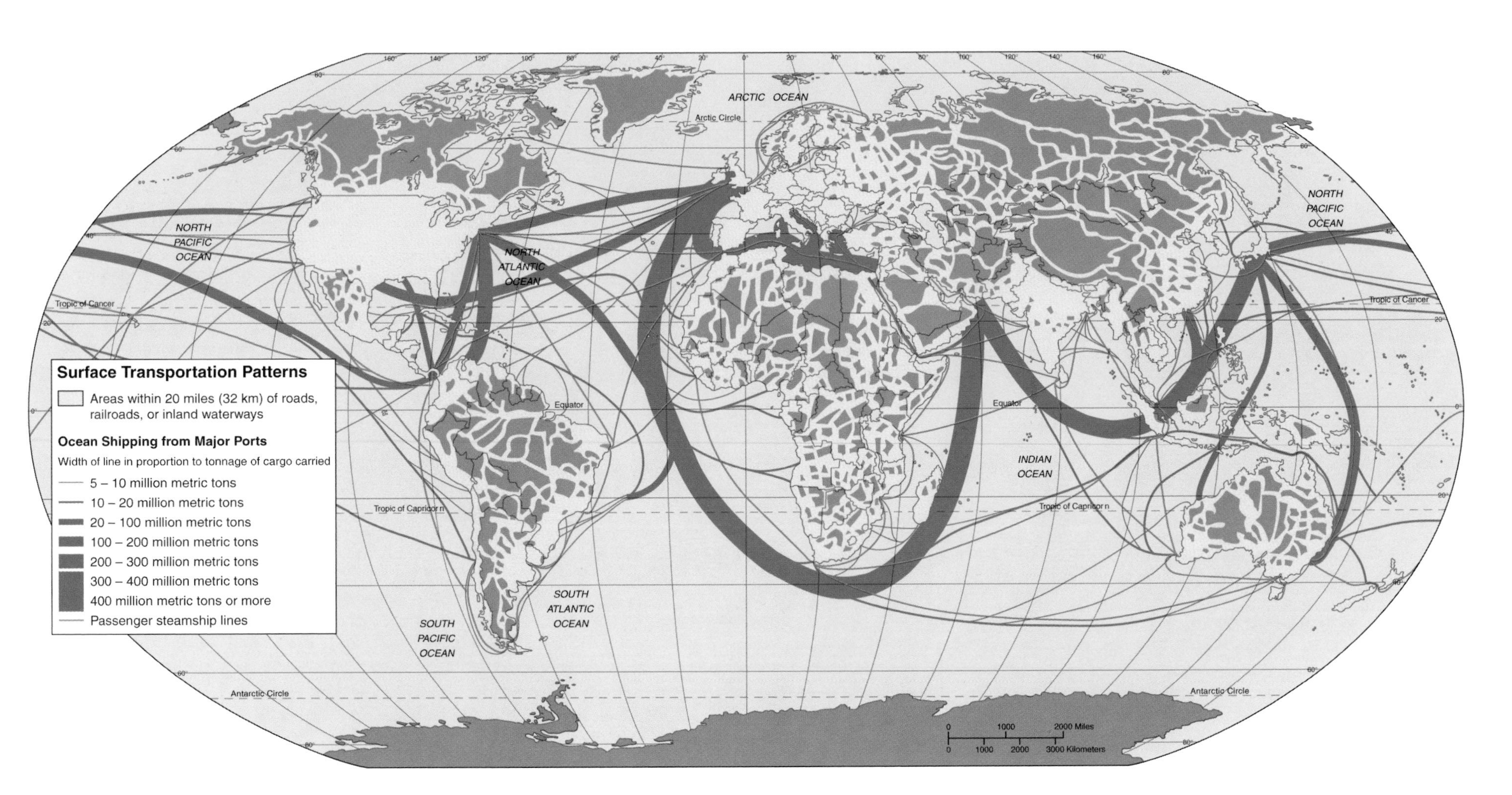

As a form of land use, transportation is second only to agriculture in its coverage of the earth's surface and is one of the clearest examples in the human world of a *network,* a linked system of lines allowing flows from one place to another. The global transportation network and its related communication web is responsible for most of the *spatial interaction,* or movement of goods, people, and ideas between places. As the chief mechanism of spatial interaction, transportation is linked firmly with the concept of a shrinking world and the development of a global community and economy.

Because transportation systems require significant modification of the earth's surface, transportation is also responsible for massive alterations in the quantity and quality of water, for major soil degradations and erosion, and (indirectly) for the air pollution that emanates from vehicles utilizing the transportation system. In addition, as improved transportation technology draws together places on the earth that were formerly remote, it allows people to impact environments a great distance away from where they live.

Map 20 Religions

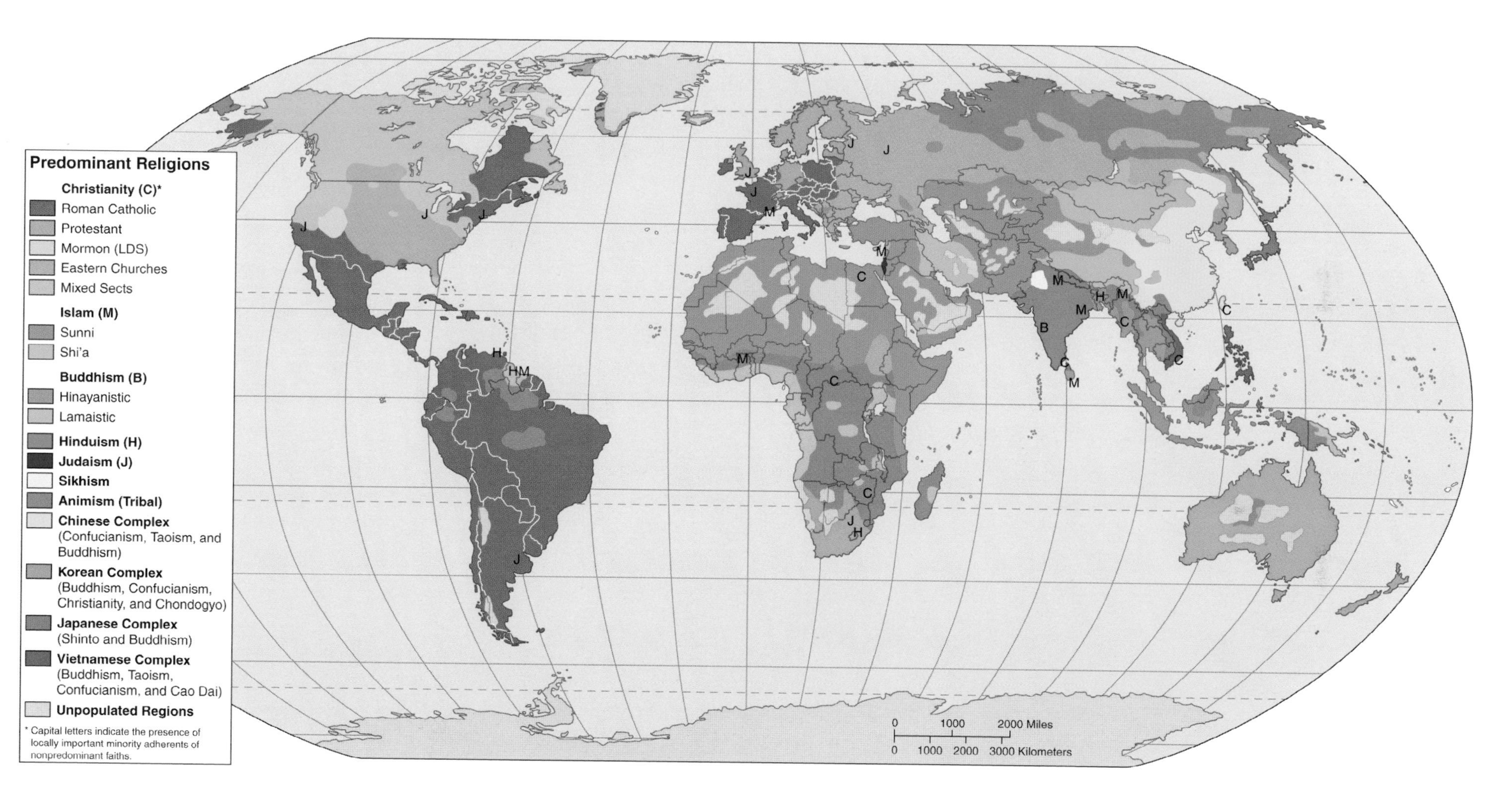

Religious adherence is one of the fundamental defining characteristics of human culture, the style of life adopted by a people and passed from one generation to the next. Because of the importance of religion for culture, a depiction of the spatial distribution of religions is as close as we can come to a map of cultural patterns. More than just a set of behavioral patterns having to do with worship and ceremony, religion is a vital conditioner of the ways that people deal with one another, with their institutions, and with the environments they occupy. In many areas of the world, the ways in which people make a living, the patterns of occupation that they create on the land, and the impacts that they make on ecosystems are the direct consequences of their adherence to a religious faith. An examination of the map in the context of international and intranational conflict will also show that tension between countries and the internal stability of states is also a function of the spatial distribution of religion.

Map 21 Languages

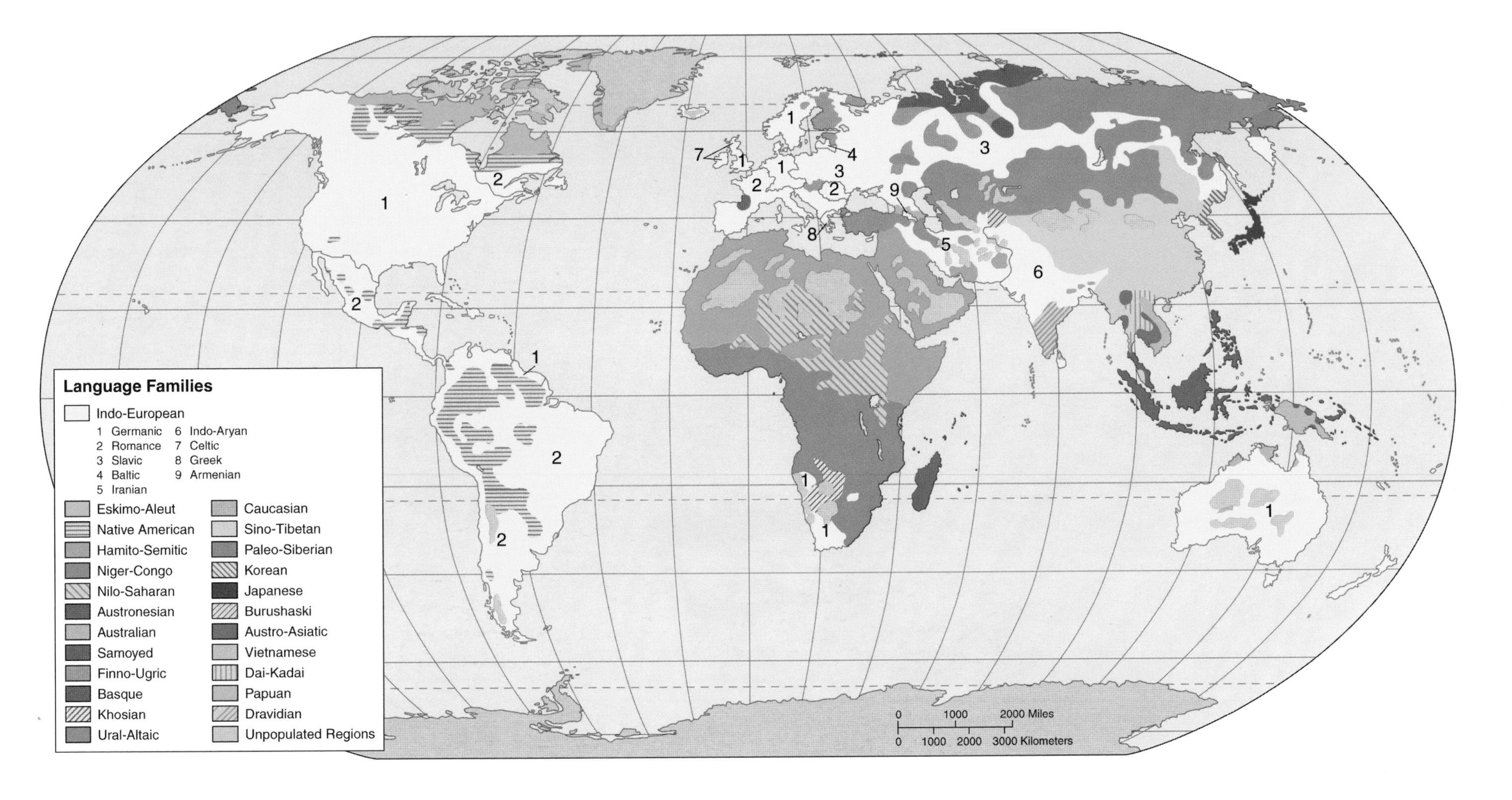

Language, like religion, is an important identifying characteristic of culture. Indeed, it is perhaps the most durable of all those identifying characteristics or *cultural traits:* language, religion, institutions, material technologies, and ways of making a living. After centuries of exposure to other languages or even conquest by speakers of other languages, the speakers of a specific tongue will often retain their own linguistic identity. Language helps us to locate areas of potential conflict, particularly in regions where two or more languages overlap, Many, if not most, of the world's conflict zones are also areas of linguistic diversity. Knowing the distribution of languages helps us to understand some of the reasons behind important current events: for example, linguistic identity differences played an important part in the disintegration of the Soviet Union in the early 1990s; and in areas emerging from recent colonial rule, such as Africa, the participants in conflicts over territory and power are often defined in terms of linguistic groups. Language distributions also help us to comprehend the nature of the human past by providing clues that enable us to chart the course of human migrations, as shown in the distribution of Indo-European, Austronesian, or Hamito-Semitic languages. Finally, because languages have a great deal to do with the way people perceive and understand the world around them, linguistic patterns help to explain the global variations in the ways that people interact.

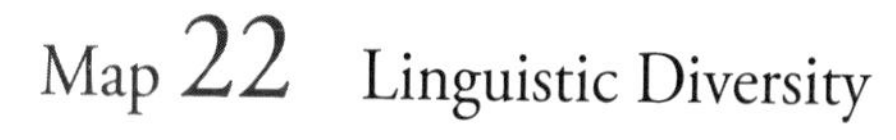

Map 22 Linguistic Diversity

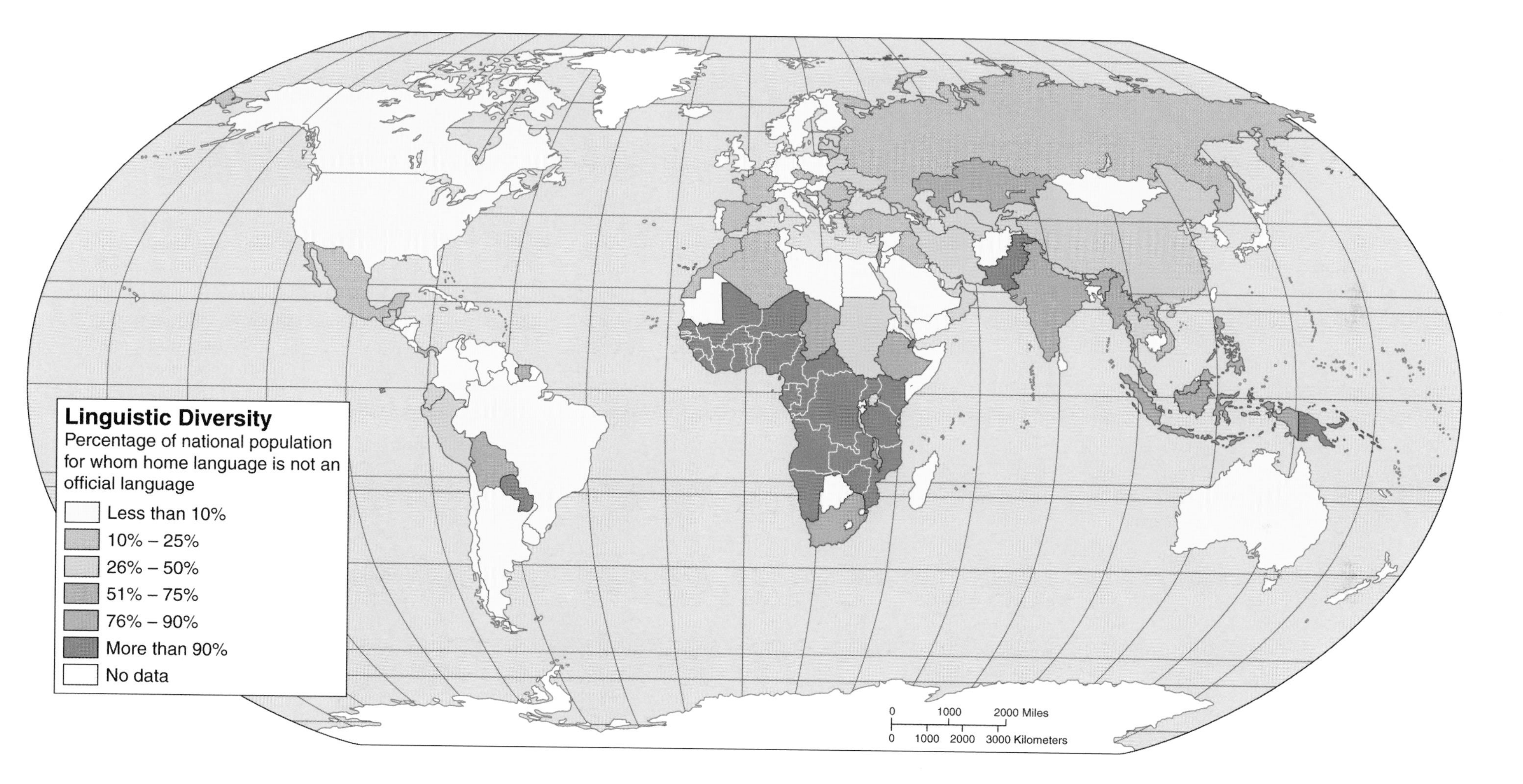

Of the world's approximately 6,000 languages, fewer than 100 are official languages, those designated by a country as the language of government, commerce, education, and information. This means that for much of the world's population, the language that is spoken in the home is different from the official language of the country of residence. The world's former colonial areas in Middle and South America, Africa, and South and Southeast Asia stand out on the map as regions in which there is significant disparity between home languages and official languages. To complicate matters further, for most of the world's population, the primary international languages of trade and tourism (French and English) are neither home nor official languages. China is a special case as the official language is the written form of Chinese while several spoken Chinese dialects such as Mandarin and Cantonese, most of them mutually unintelligible, are recognized as official languages. The formal language of government and business is Mandarin."

Map 23 External Migrations in Modern Times

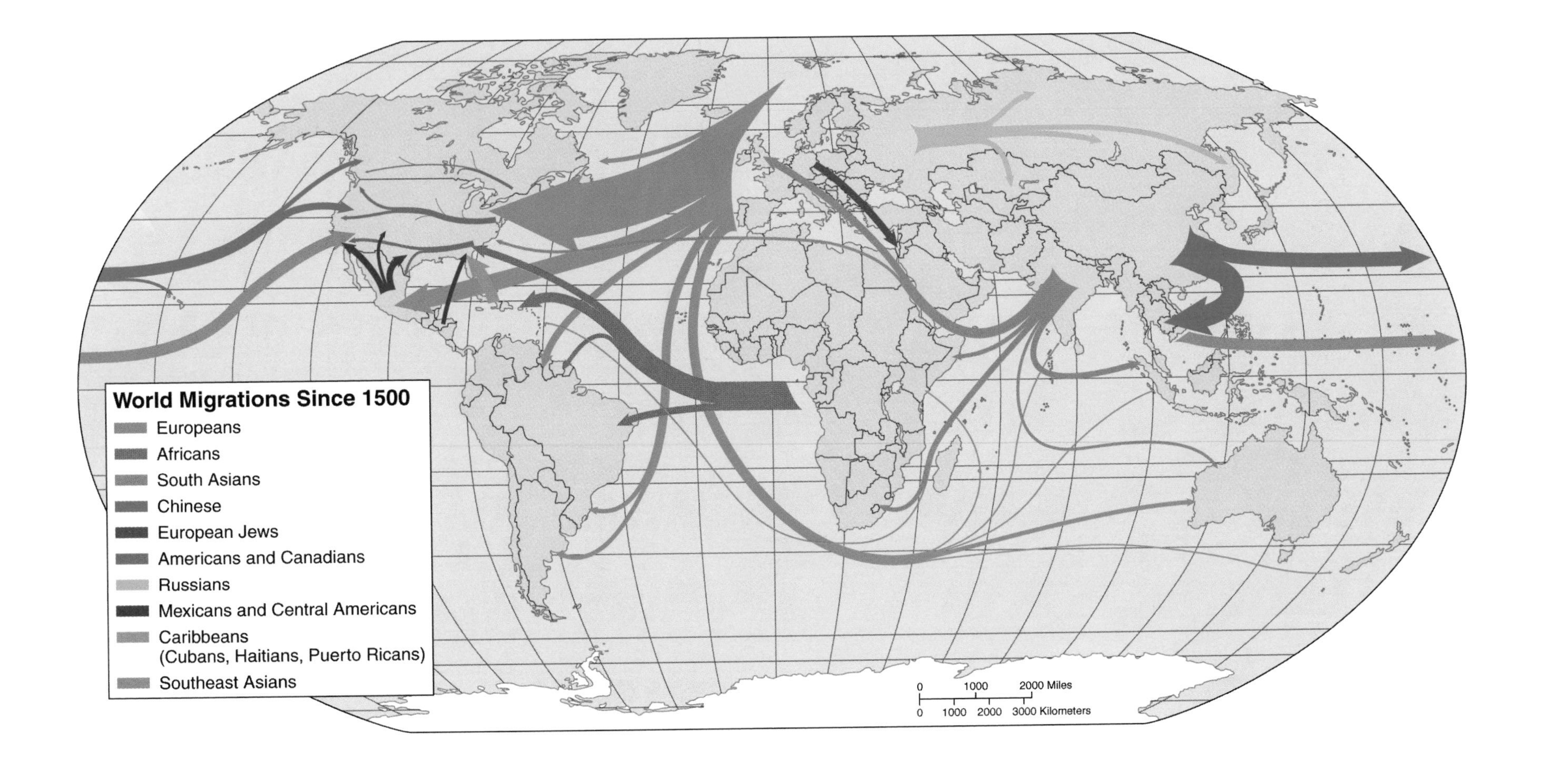

Migration has had a significant effect on world geography, contributing to cultural change and development, to the diffusion of ideas and innovations, and to the complex mixture of people and cultures found in the world today. *Internal migration* occurs within the boundaries of a country; *external migration* is movement from one country or region to another. Over the last 50 years, the most important migrations in the world have been internal, largely the rural-to-urban migration that has been responsible for the recent rise of global urbanization. Prior to the mid-twentieth century, three types of external migrations were most important: *voluntary,* most often in search of better economic conditions and opportunities; *involuntary* or *forced,* involving people who have been driven from their homelands by war, political unrest, or environmental disasters, or who have been transported as slaves or prisoners; and *imposed,* not entirely forced but which conditions make highly advisable. Human migrations in recorded history have been responsible for major changes in the patterns of languages, religions, ethnic composition, and economies. Particularly during the last 500 years, migrations of both the voluntary and involuntary or forced type have literally reshaped the human face of the earth.

Part III

Global Demographic Patterns

Map 24 Population Growth Rates
Map 25 Total Fertility Rates
Map 26 Infant Mortality Rate
Map 27 Child Mortality Rate
Map 28 Child Malnutrition
Map 29 Primary School Enrollment
Map 30 Average Life Expectancy at Birth
Map 31 Population by Age Group
Map 32 International Migrant Populations 2005
Map 33 World Daily per Capita Supply of Calories (Kilocalories)
Map 34 Proportion of Daily Food Intake from Animal Products
Map 35 Illiteracy Rates
Map 36 Female/Male Inequality in Education and Employment
Map 37 Global Scourges: Major Infectious Diseases
Map 38 Adult HIV Prevalence 2004
Map 39 The Index of Human Development
Map 40 Demographic Stress: The Youth Bulge
Map 41 Demographic Stress: Rapid Urban Growth
Map 42 Demographic Stress: Competition for Cropland
Map 43 Demographic Stress: Competition for Fresh Water
Map 44 Demographic Stress: Death in the Prime of Life
Map 45 Demographic Stress: Interactions of Demographic Stress

Map 24 Population Growth Rates

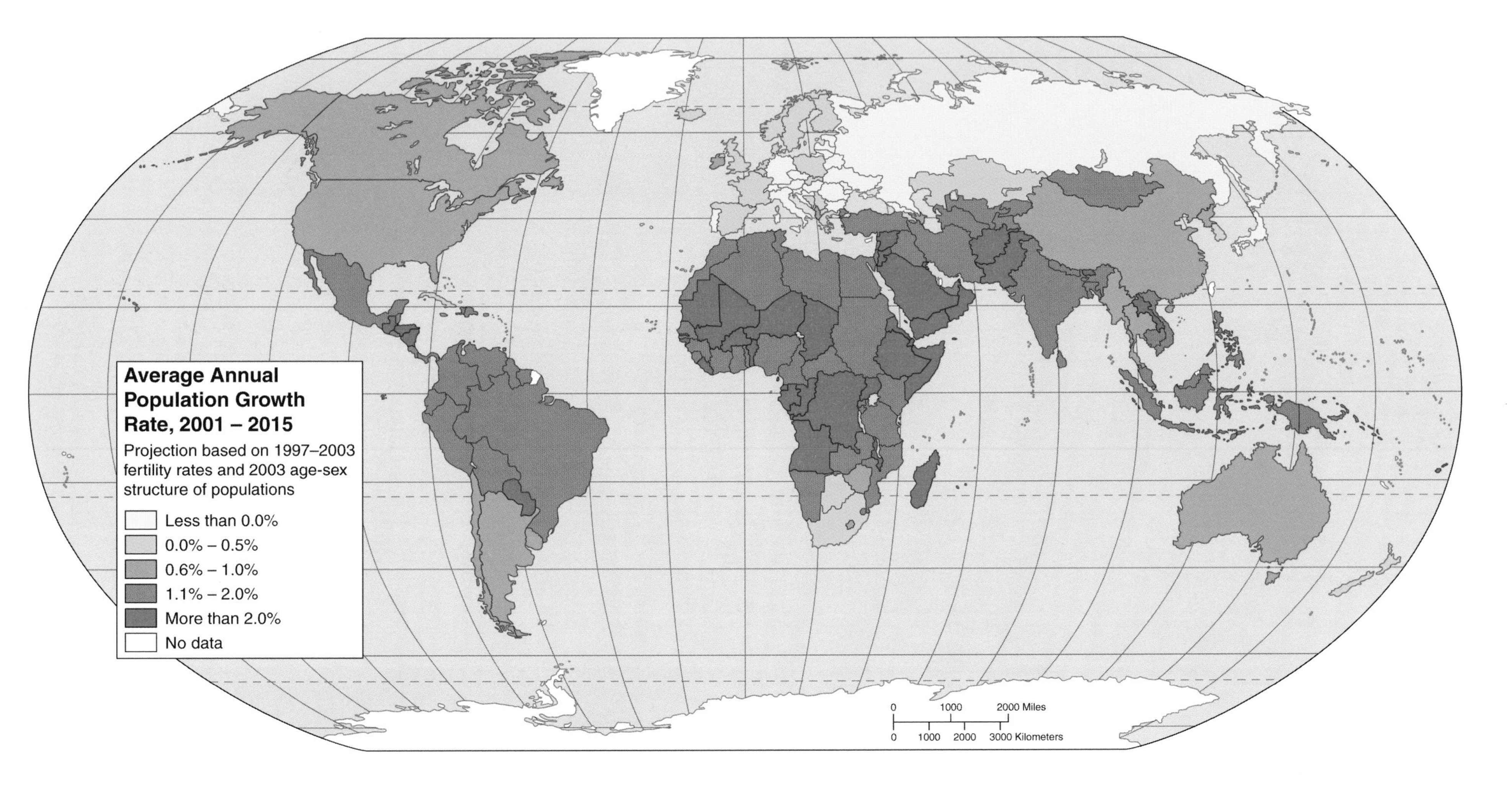

Of all the statistical measurements of human population, that of the rate of population growth is the most important. The growth rate of a population is a combination of natural change (births and deaths), in-migration, and out-migration; it is obtained by adding the number of births to the number of immigrants during a year and subtracting from that total the sum of deaths and emigrants for the same year. For a specific country, this figure will determine many things about the country's future ability to feed, house, educate, and provide medical services to its citizens. Some of the countries with the largest populations (such as India) also have high growth rates. Since these countries tend to be in developing regions, the combination of high population and high growth rates poses special problems for political stability and continuing economic development; the combination also carries heightened risks for environmental degradation. Many people believe that the rapidly expanding world population is a potential crisis that may cause environmental and human disaster by the middle of the twenty-first century.

Map 25a Total Fertility Rates

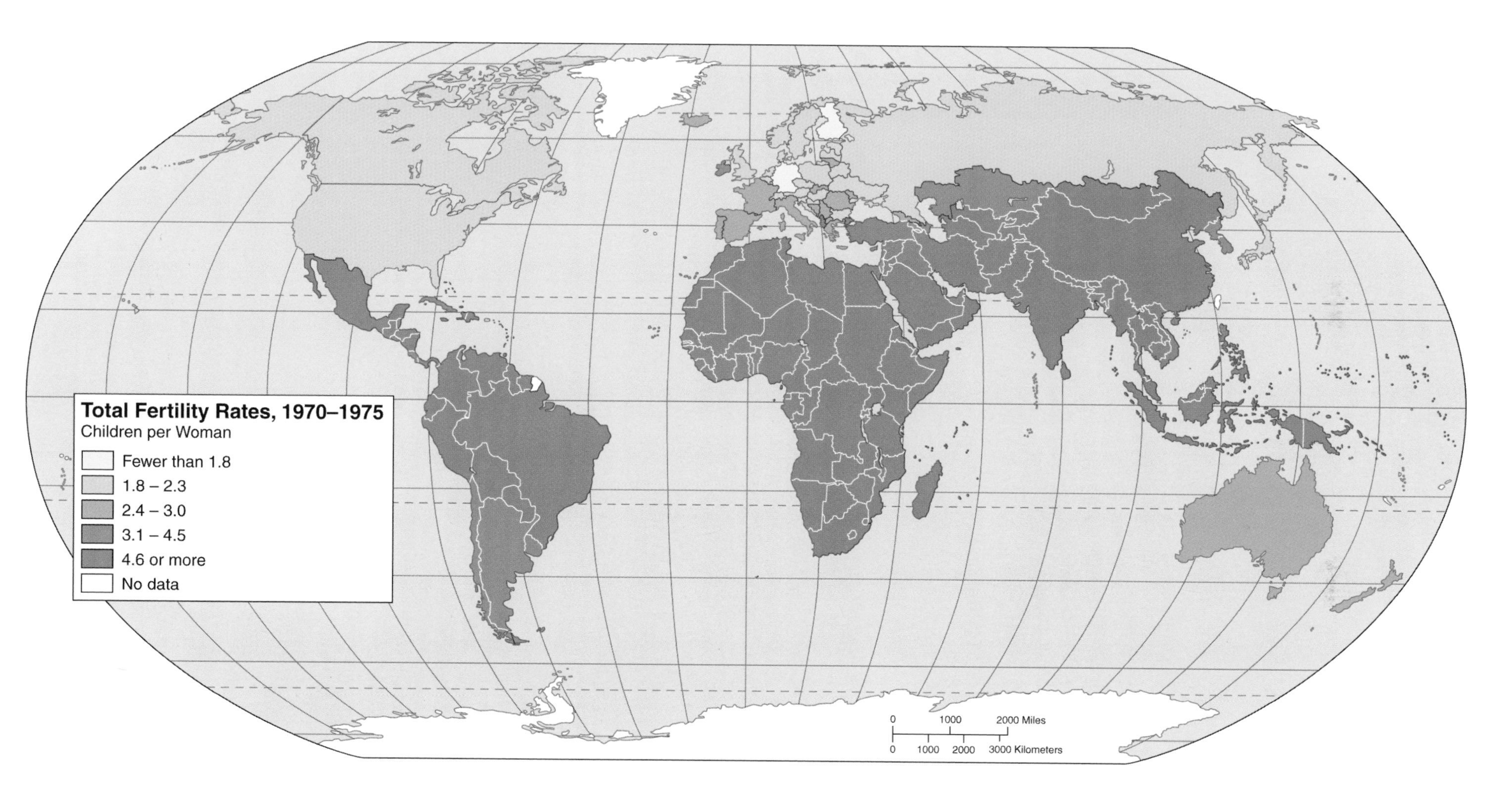

Fertility rates measure the number of children that might be expected to be given birth by a single woman during her child-bearing years. It is often a better predictor of potential population growth than such measures as birth rate. More important, it is a major indicator of the "demographic transition" or shift from a population characterized by large families and short lives to one characterized by small families and longer lives. For the last century, the first type of population profile (large families and short lives) has been evident in the lesser developed parts of the world while the developed nations had populations with the smaller family and longer life span profile. What these comparative maps show us is that over the last quarter century, the same kind of demographic transition begun in Europe and America in the nineteenth century is now beginning in Middle and South America, in Africa, and in Asia. There are still areas of concern—African fertility rates are, for example, still far too high. But particularly in South America, Southeast Asia, and East Asia, fertility rates have dropped dramatically, indicating the potential for much slower population growth in the next generation.

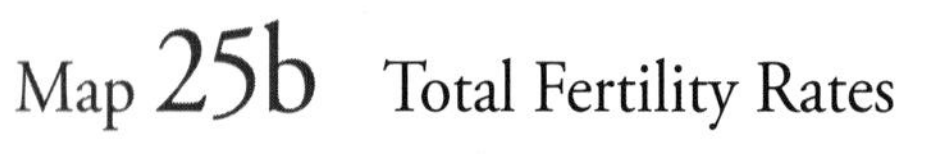

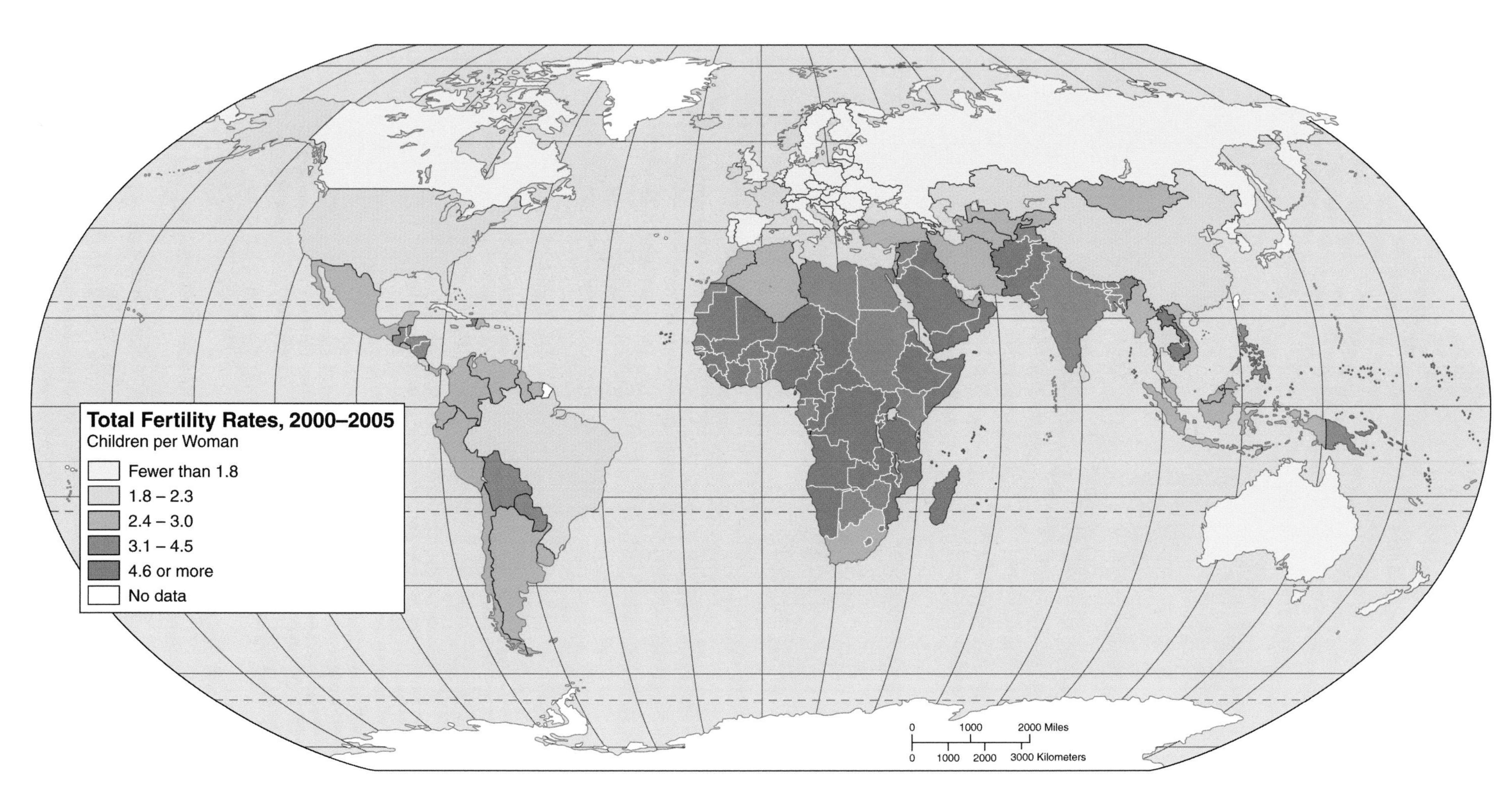
Total Fertility Rates, 2000–2005
Children per Woman
Fewer than 1.8
1.8 – 2.3
2.4 – 3.0
3.1 – 4.5
4.6 or more
No data
0 1000 2000 Miles
0 1000 2000 3000 Kilometers

Map 26 Infant Mortality Rate

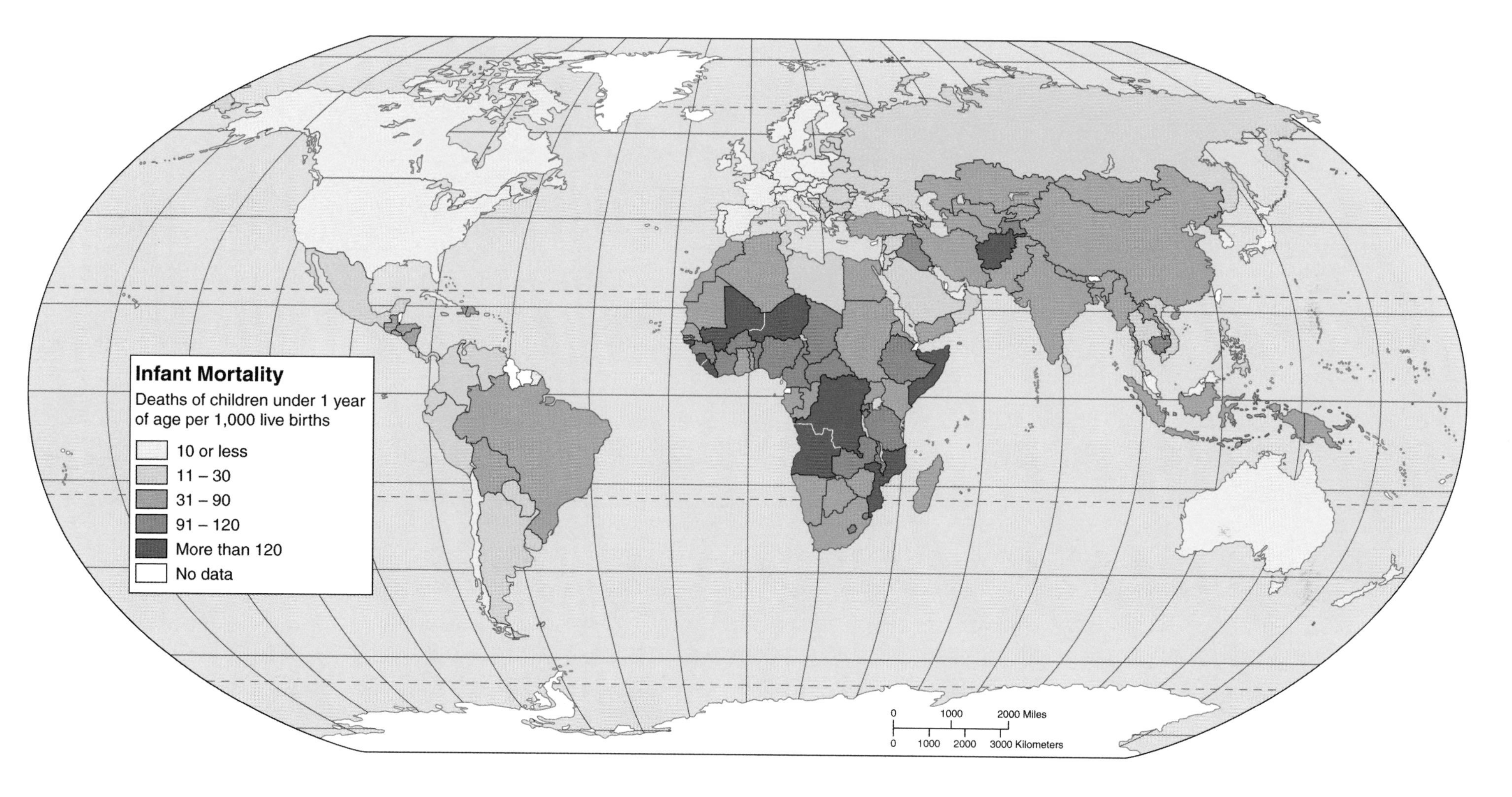

Infant mortality rates are calculated by dividing the number of children born in a given year who die before their first birthday by the total number of children born that year and then multiplying by 1,000; this shows how many infants have died for every 1,000 births. Infant mortality rates are prime indicators of economic development. In highly developed economies, with advanced medical technologies, sufficient diets, and adequate public sanitation, infant mortality rates tend to be quite low. By contrast, in less developed countries, with the disadvantages of poor diet, limited access to medical technology, and the other problems of poverty, infant mortality rates tend to be high.

Although worldwide infant mortality has decreased significantly during the last 2 decades, many regions of the world still experience infant mortality above the 10 percent level (100 deaths per 1,000 live births). Such infant mortality rates not only represent human tragedy at its most basic level, but also are powerful inhibiting factors for the future of human development. Comparing infant mortality rates in the midlatitudes and the tropics shows that children in most African countries are more than 10 times as likely to die within a year of birth as children in European countries.

Map 27 Child Mortality Rate

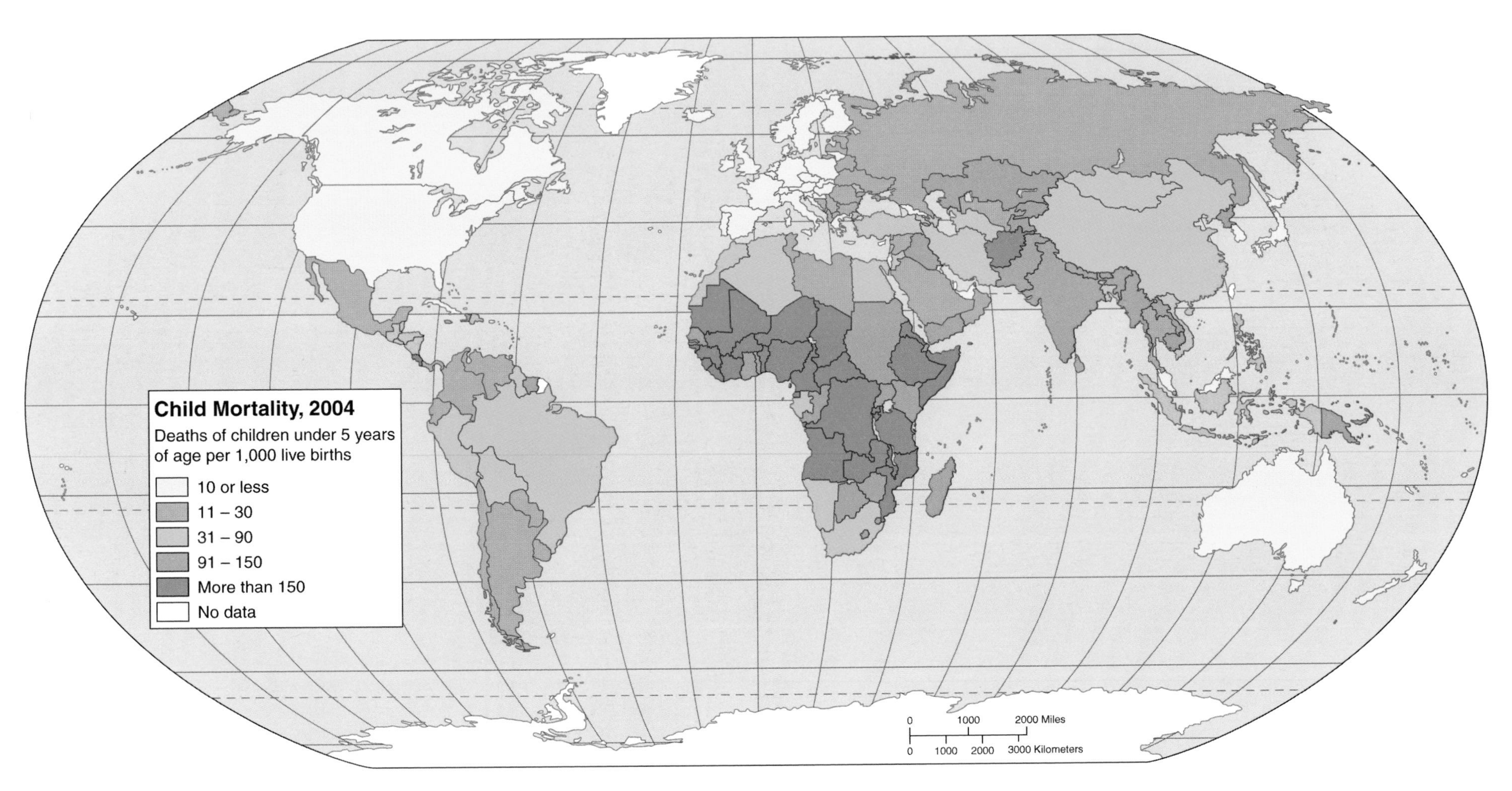

Child mortality rates are calculated by determining the probability that a child born in a specified year will die before reaching the age 5, using current age-specific mortality rates for a population. The major sources of mortality rates are vital registration systems and estimates made from surveys and/or census reports. Along with infant mortality and average life-expectancy rates, child mortality rates, according to the World Bank, "are probably the best general indicators of a community's current health status and are often cited as overall measures of a population's welfare or quality of life." Where infant mortality often reflects health care conditions, child mortality is usually a reflection of the inadequacy of nutrition, leading to early deaths from nutritionally related diseases. In some less developed countries in Africa and Asia, child mortality is also an indicator of the widespread presence of infectious diseases such as malaria, tuberculosis, and HIV/AIDS.

Map 28 Child Malnutrition

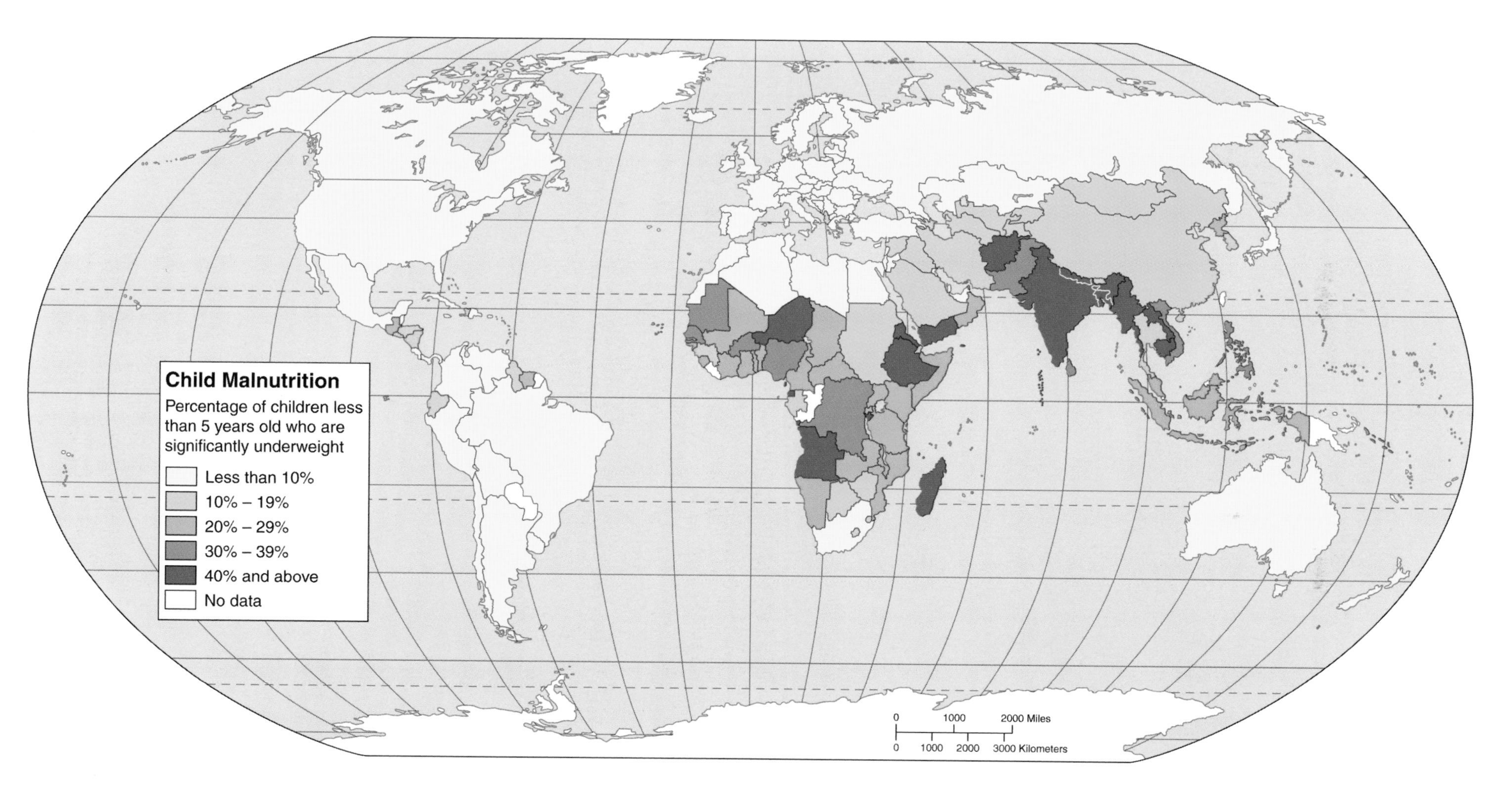

The weight of poverty is not evenly spread among the members of a population, falling disproportionately upon the weakest and most disadvantaged members of society. In most societies, these individuals are children, particularly female children. Children simply do not compete as successfully as adults for their (meager) share of the daily food supply. Where food shortages prevail, children tend to have the quality of their future lives severely compromised by poor nutrition, which, in a downward spiral, robs them of the energy necessary to compete more effectively for food. Children who are inadequately fed are less likely to do well in school, are more prone to debilitating disease, and will more often become a drain on scarce societal resources than well-fed children. Recently, health care officials in the more developed world have become concerned over the trend to "overnutrition," leading to obesity and related health problems in the world's economically developed countries. Nevertheless, child malnourishment remains one of the primary distinguishing factors between the "haves" and "have-nots."

Map 29 Primary School Enrollment

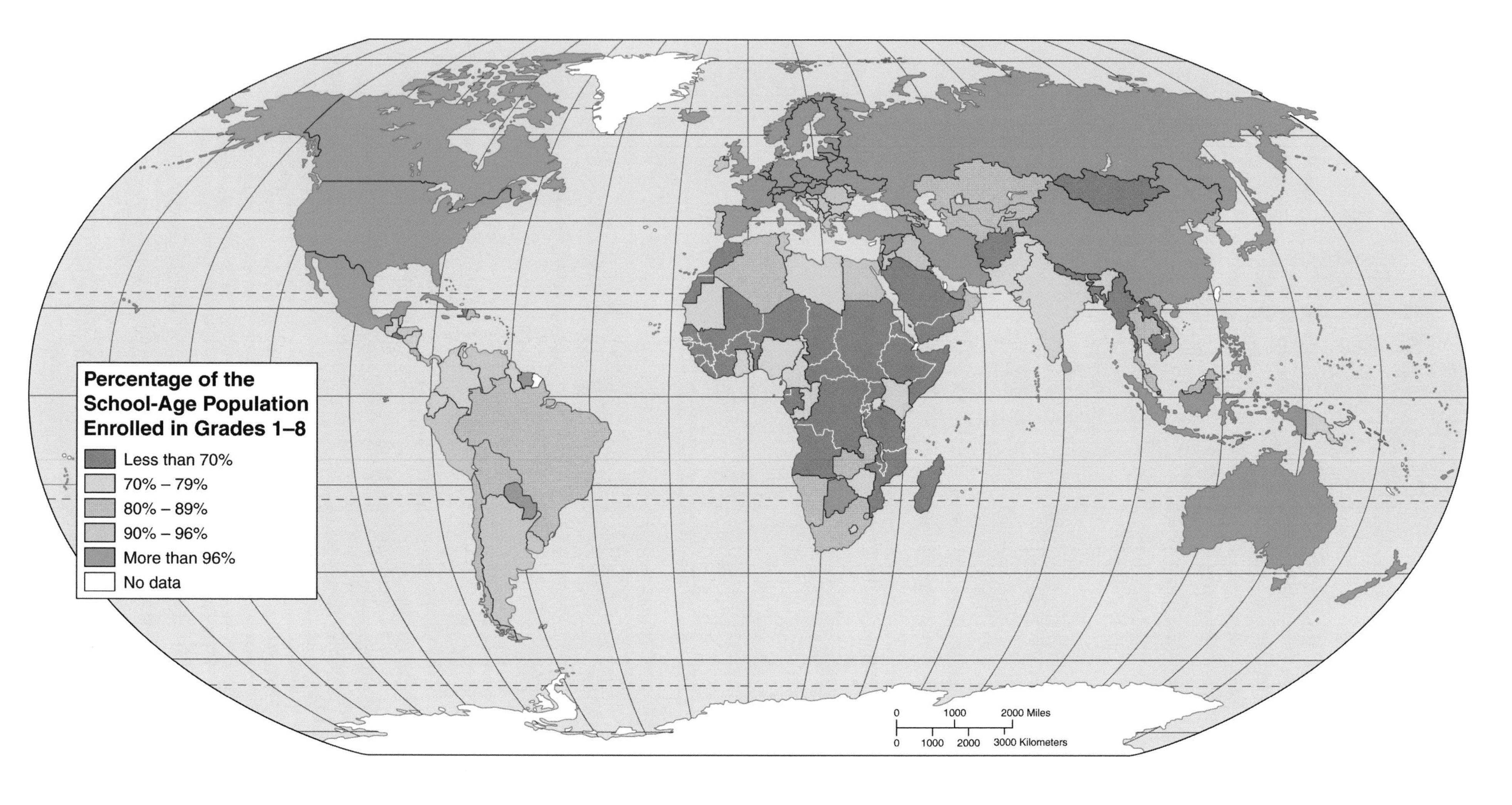

Like many of the other measures illustrated in this atlas, primary school enrollment is a clear reflection of the division of the world into "have" and "have-not" countries. It is also a measure that has changed more rapidly over the last decade than demographic and other indicators of development, as countries of even very modest means have made concerted attempts to attain relatively high percentages of primary school enrollment. That they have been able to do so is good evidence of the fact that reasonably respectable levels of human development are feasible at even modest income levels. High primary school enrollment is also a reflection of the worldwide opinion that a major element in economic development is a well-educated, literate population. The links between human progress, as typified by higher levels of education, and economic growth are not automatic, however, and those countries without programs for maintaining the headway gained by improved education may be on the road to failure in terms of economic development.

Map 30 Average Life Expectancy at Birth

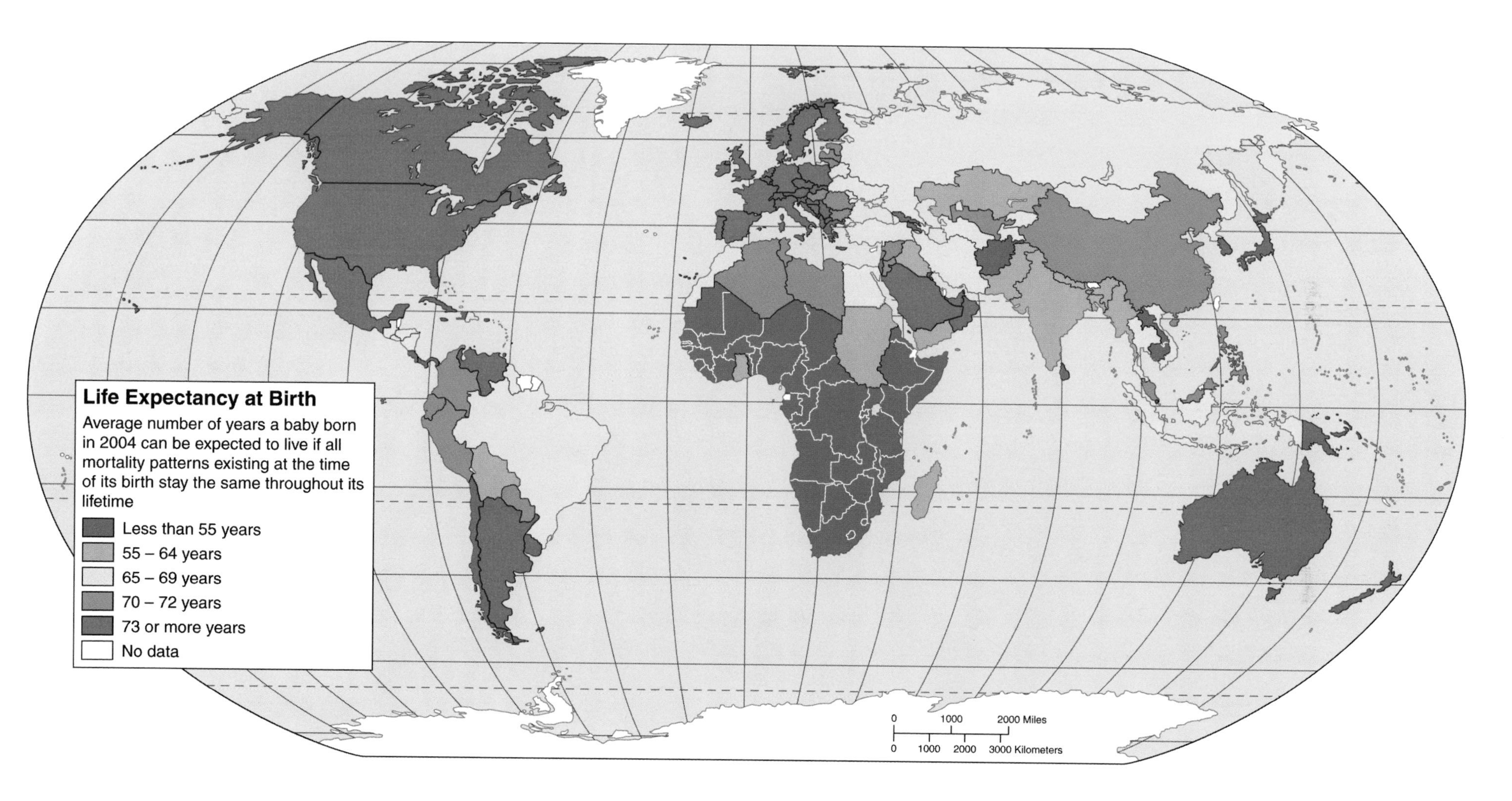

Average life expectancy at birth is a measure of the average longevity of the population of a country. Like all average measures, it is distorted by extremes. For example, a country with a high mortality rate among children will have a low average life expectancy. Thus, an average life expectancy of 45 years does not mean that everyone can be expected to die at the age of 45. More normally, what the figure means is that a substantial number of children die between birth and 5 years of age, thus reducing the average life expectancy for the entire population. In spite of the dangers inherent in misinterpreting the data, average life expectancy (along with infant mortality and several other measures) is a valid way of judging the relative health of a population. It reflects the nature of the health care system, public sanitation and disease control, nutrition, and a number of other key human need indicators. As such, it is a measure of well-being that is significant in indicating economic development and predicting political stability.

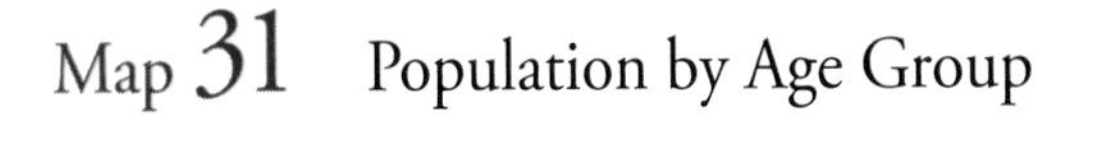

Map 31 Population by Age Group

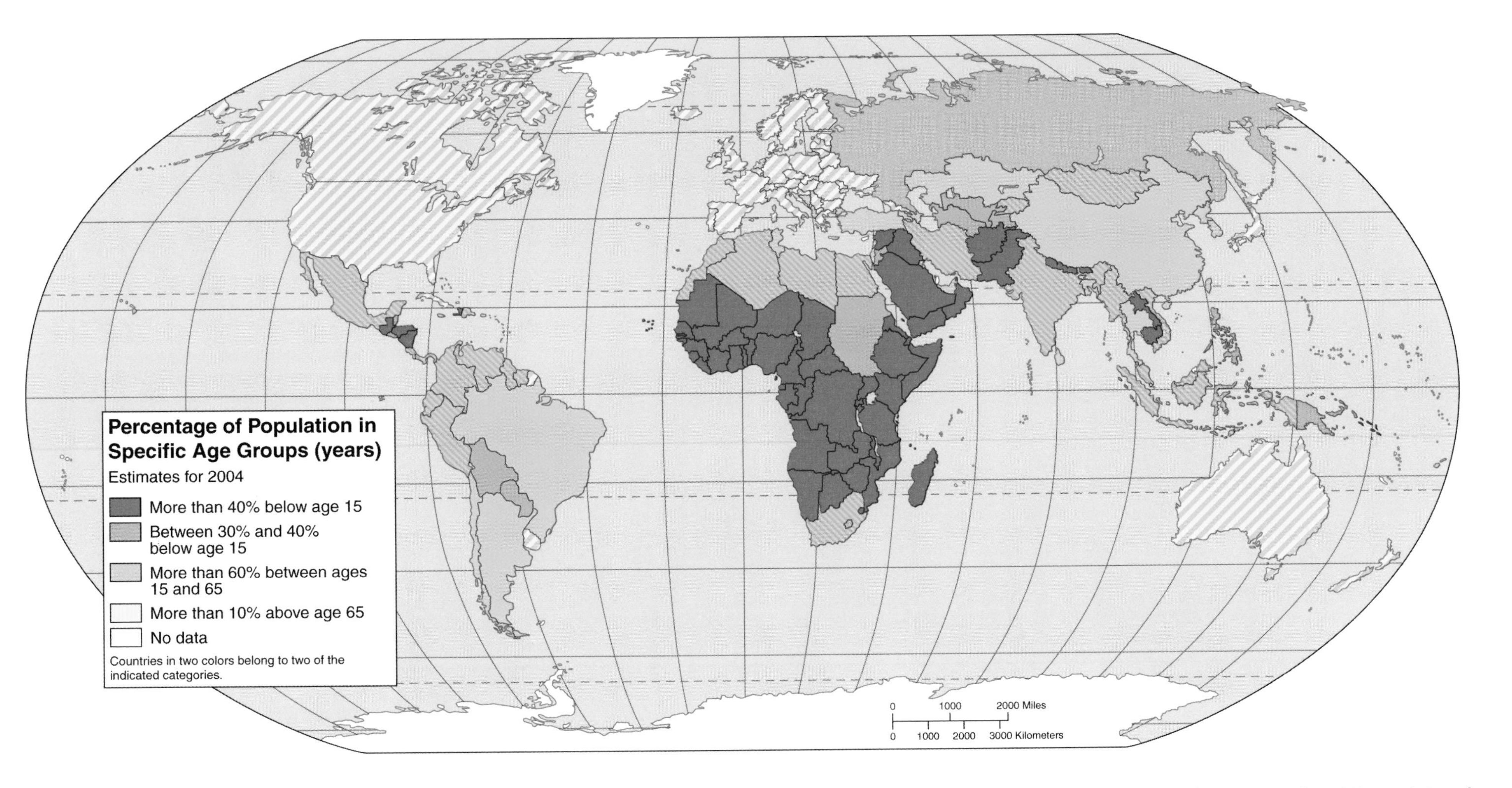

Of all the measurements that illustrate the dynamics of a population, age distribution may be the most significant, particularly when viewed in combination with average growth rates. The particular relevance of age distribution is that it tells us what to expect from a population in terms of growth over the next generation. If, for example, approximately 40–50 percent of a population is below the age of 15, that suggests that in the next generation about one-quarter of the total population will be women of childbearing age. When age distribution is combined with fertility rates (the average number of children born per woman in a population), an especially valid measurement of future growth potential may be derived. A simple example: Nigeria, with a 2002 population of 130 million, has 43.6 percent of its population below the age of 15 and a fertility rate of 5.5; the United States, with a 2002 population of 280 million, has 21 percent of its population below the age of 15 and a fertility rate of 2.07. During the period in which those women presently under the age of 15 are in their childbearing years, Nigeria can be expected to add a total of approximately 155 million persons to its total population. Over the same period, the United States can be expected to add only 61 million.

Map 32 International Migrant Populations 2005

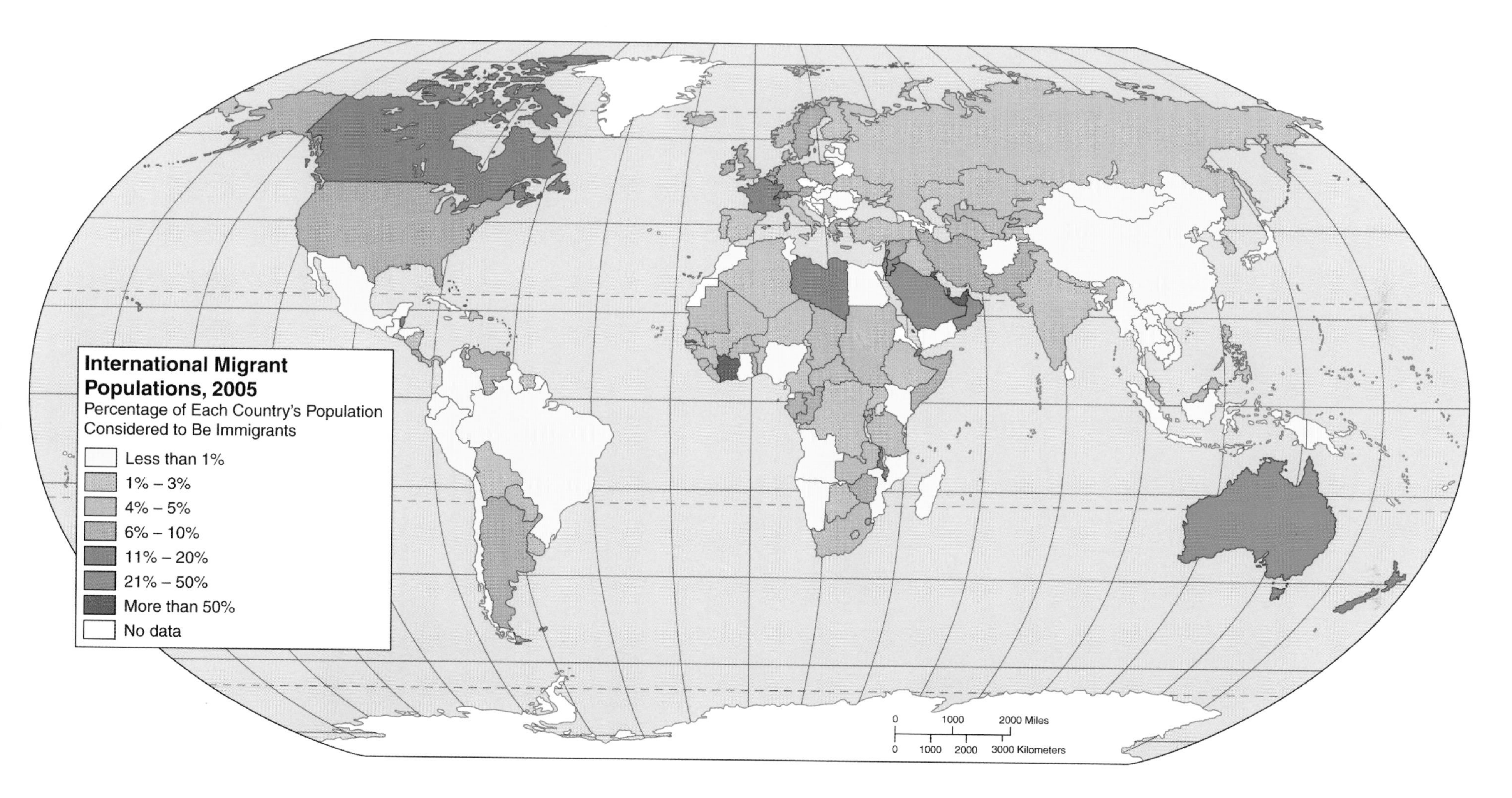

Migration—the movement from one place to another—takes a number of different forms. It may be a move within a country from an old job to a new one. It may also mean a migration, either forced or voluntary, from one country to another. The map here depicts international migration: migration between countries. Migration is distinguished from a refugee movement in that migrants are not defined as refugees granted a humanitarian and temporary protection status under international law. The nearly 2 million people who have fled Iraq for Iran are refugees who expect and hope to return to Iraq after the present civil war is ended. The Middle Americans who leave the Central American countries, Mexico, or the Caribbean for the United States plan to live in the United States permanently, although still retaining cultural and family ties to their native country. This map clearly shows that those countries viewed as having the most favorable opportunities for improvement in personal living conditions are those with the highest numbers of in-migrants; those countries that are overcrowded, with little economically upward mobility, or international conflict tend to be those with the greatest number of out-migrants.

Map 33 Average Daily Per Capita Supply of Calories (Kilocalories)

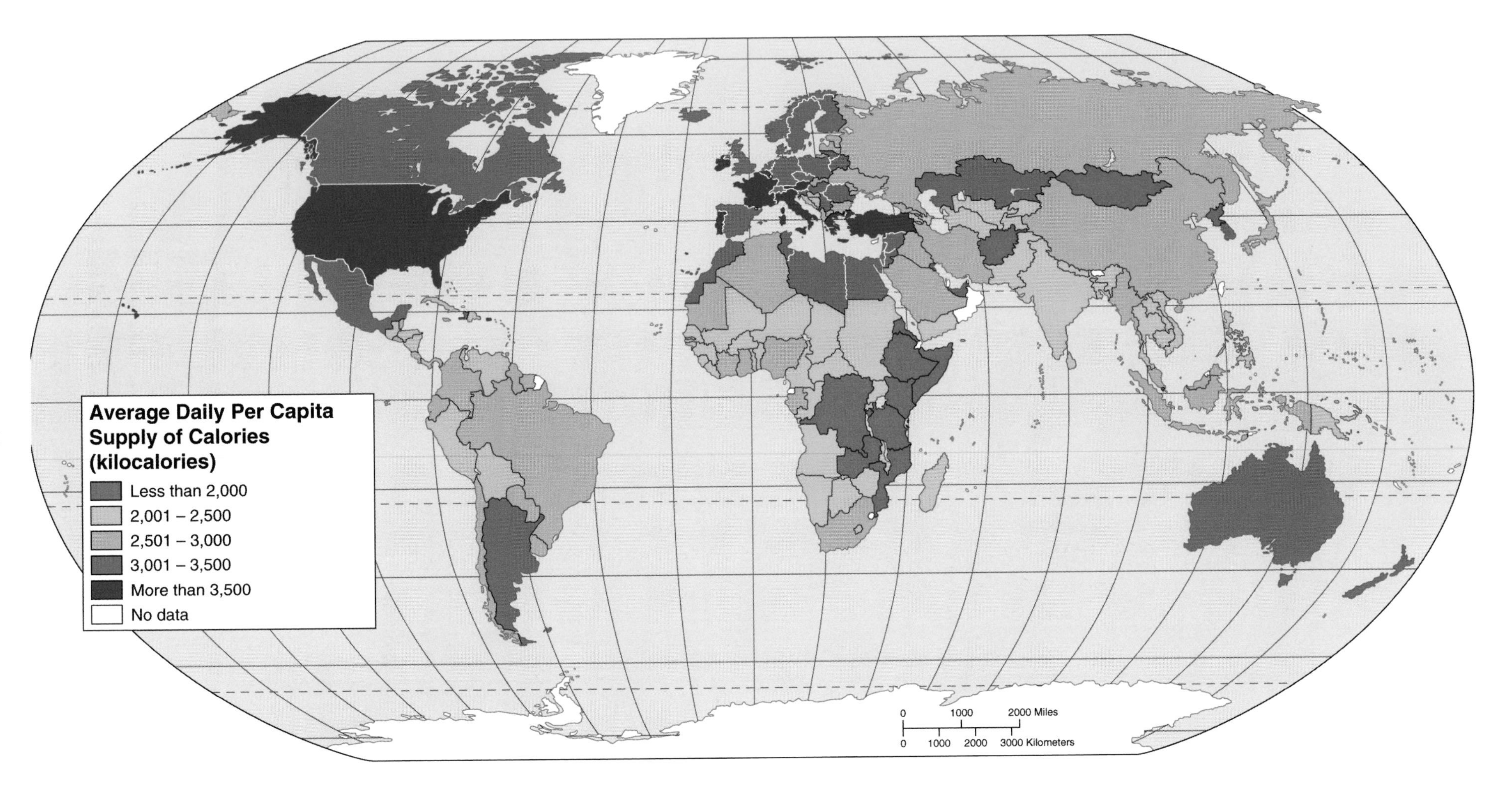

The data shown on this map, which indicate the presence or absence of critical food shortages, do not necessarily indicate the presence of starvation or famine. But they certainly do indicate potential problem areas for the next decade. The measurements are in calories from *all* food sources: domestic production, international trade, draw-down on stocks or food reserves, and direct foreign contributions or aid. The quantity of calories available is that amount, estimated by the UN's Food and Agriculture Organization (FAO), that reaches consumers. The calories actually consumed may be lower than the figures shown, depending on how much is lost in a variety of ways: in home storage (to pests such as rats and mice), in preparation and cooking, through consumption by pets and domestic animals, and as discarded foods, for example. The estimate of need is not a global uniform value but is calculated for each country on the basis of the age and sex distribution of the population and the estimated level of activity of the population. Compare this map with Map 59 for a good measure of potential problem areas for food shortages within the next decade.

Map 34 Proportion of Daily Food Intake from Animal Products

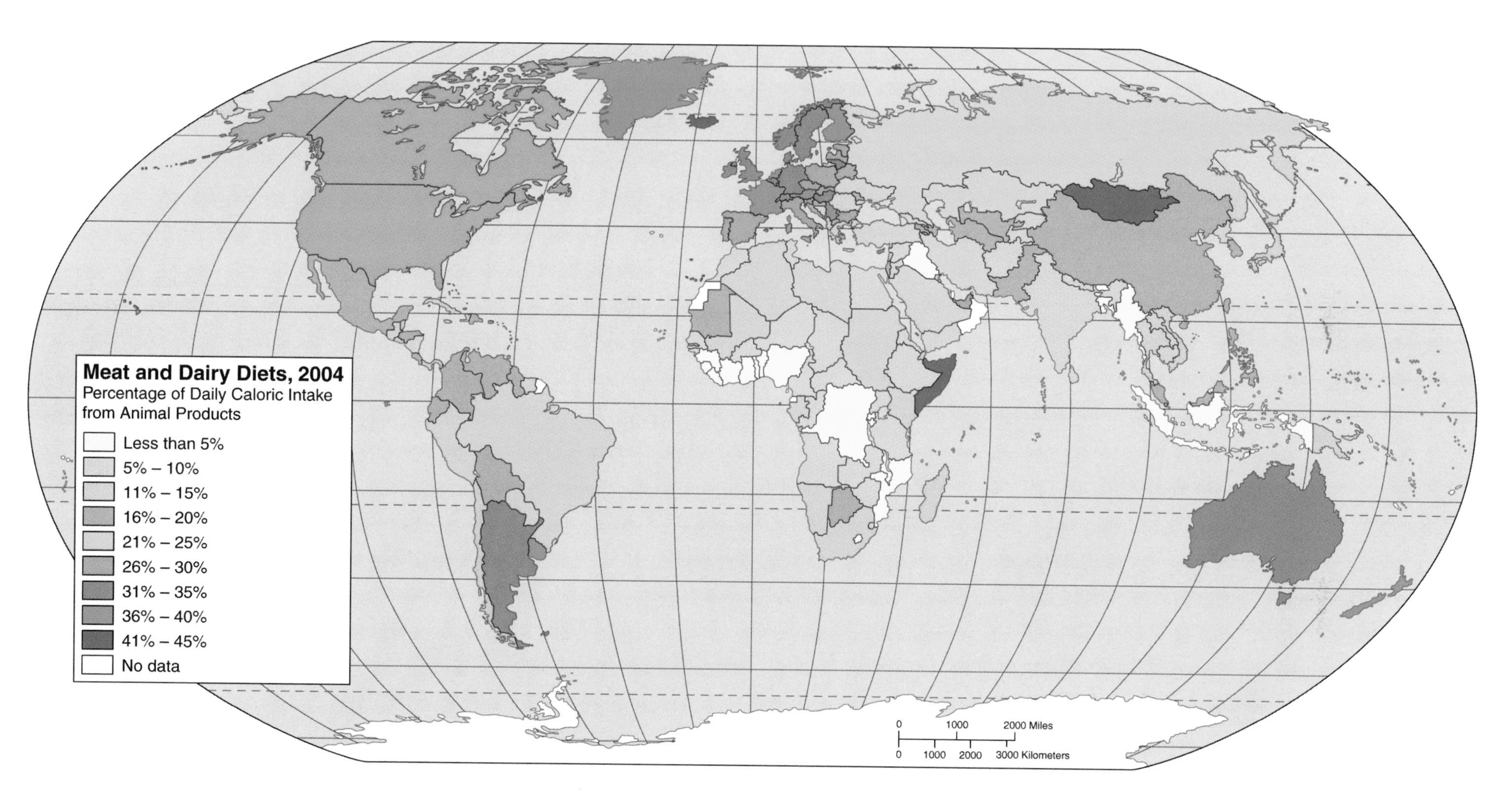

As the world grows proportionally wealthier, the percentage of the diet from meat and fish grows accordingly. While there is precious little scientific evidence to indicate that a vegetarian diet is, by definition, healthier than one richer in animal products, it is indisputable that more reliance on animal products produces a greater strain on environmental systems. Growing plants to feed animals to feed humans simply consumes a great deal more energy (in all forms) than growing plants to feed a human population directly. The long-term implications of altering land-use patterns to fit with greater wealth and, therefore, more meat consumption, are not generally favorable. And we are already beginning to notice the impact on fish population in the world's oceans as the result of an increasing demand for this food resource.

Map 35 Illiteracy Rates

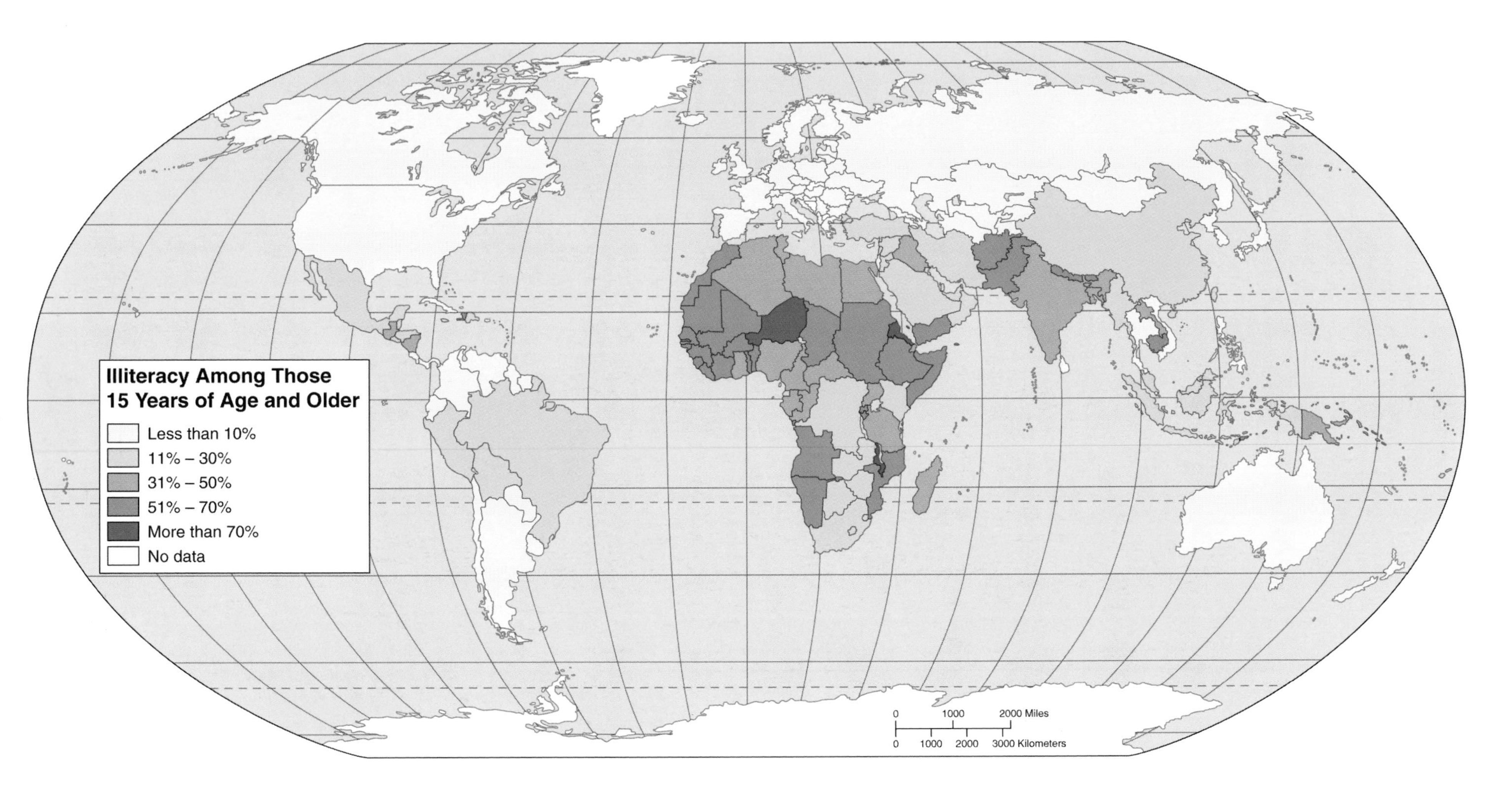

Illiteracy rates are based on the percentages of people age 15 or above (classed as adults in most countries) who are not able to write and read, with understanding, a brief, simple statement about everyday life written in their home- or official language. As might be expected, illiteracy rates tend to be higher in the less developed states, where educational systems are a low government priority. Rates of literacy or illiteracy also tend to be gender-differentiated, with women in many countries experiencing educational neglect or discrimination that makes it more likely they will be illiterate. In many developing countries, between five and ten times as many women will be illiterate as men, and the illiteracy rate for women may even exceed 90 percent. Both male and female illiteracy severely compromise economic development.

Map 36 Female/Male Inequality in Education and Employment

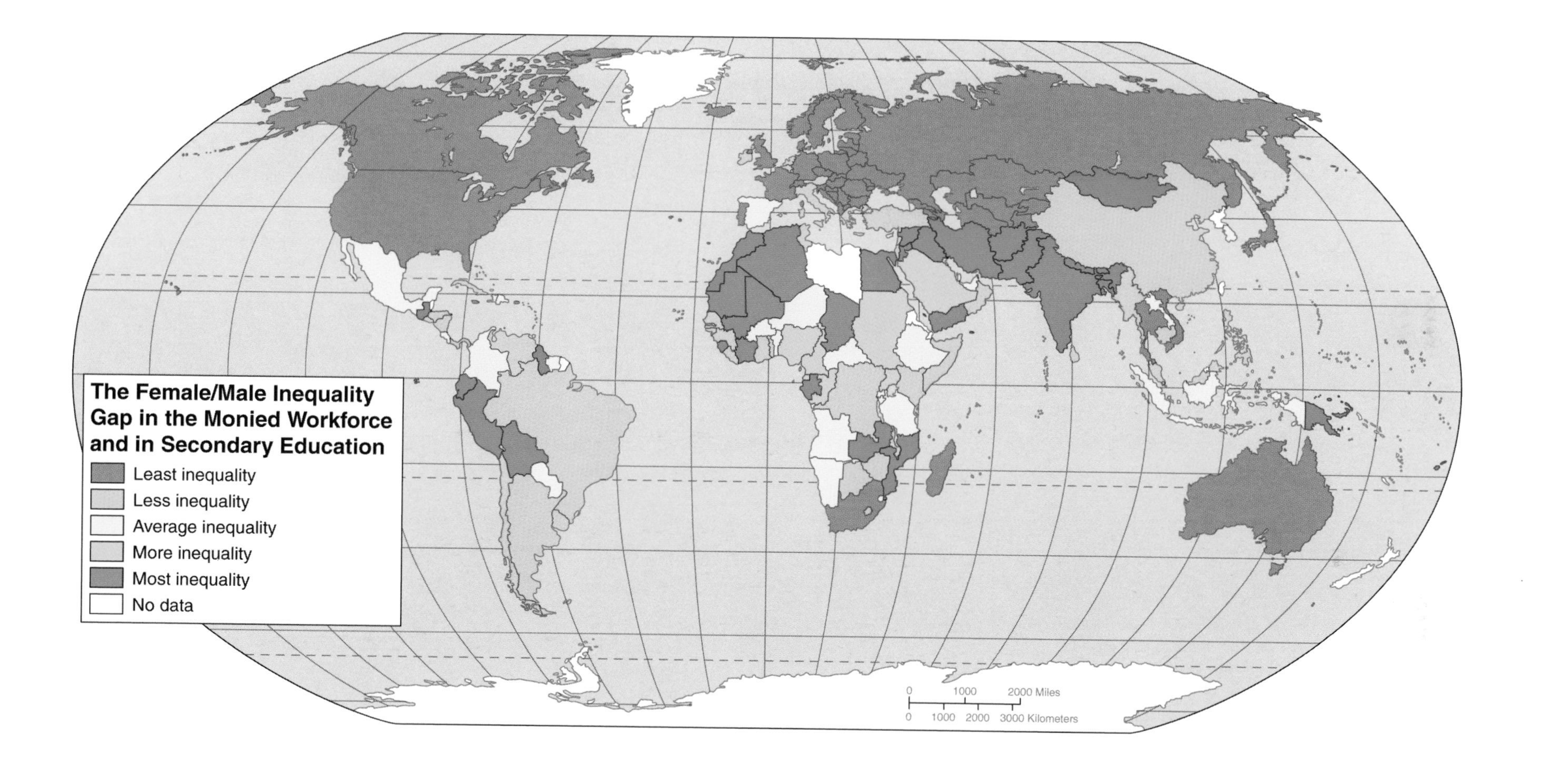

While women in developed countries, particularly in North America and Europe, have made significant advances in socioeconomic status in recent years, in most of the world they suffer from significant inequality when compared with their male counterparts. Although women have received the right to vote in most of the world's countries, in over 90 percent of these countries that right has only been granted in the last 50 years. In most regions, literacy rates for women still fall far short of those for men; in Africa and Asia, for example, only about half as many women are literate as are men. Women marry considerably younger than men and attend school for shorter periods of time.

Inequalities in education and employment are perhaps the most telling indicators of the unequal status of women in most of the world. Lack of secondary education in comparison with men prevents women from entering the workforce with equally high-paying jobs. Even where women are employed in positions similar to those held by men, they still tend to receive less compensation. The gap between rich and poor involves not only a clear geographic differentiation, but a clear gender differentiation as well.

Map 37 Global Scourges: Major Infectious Diseases

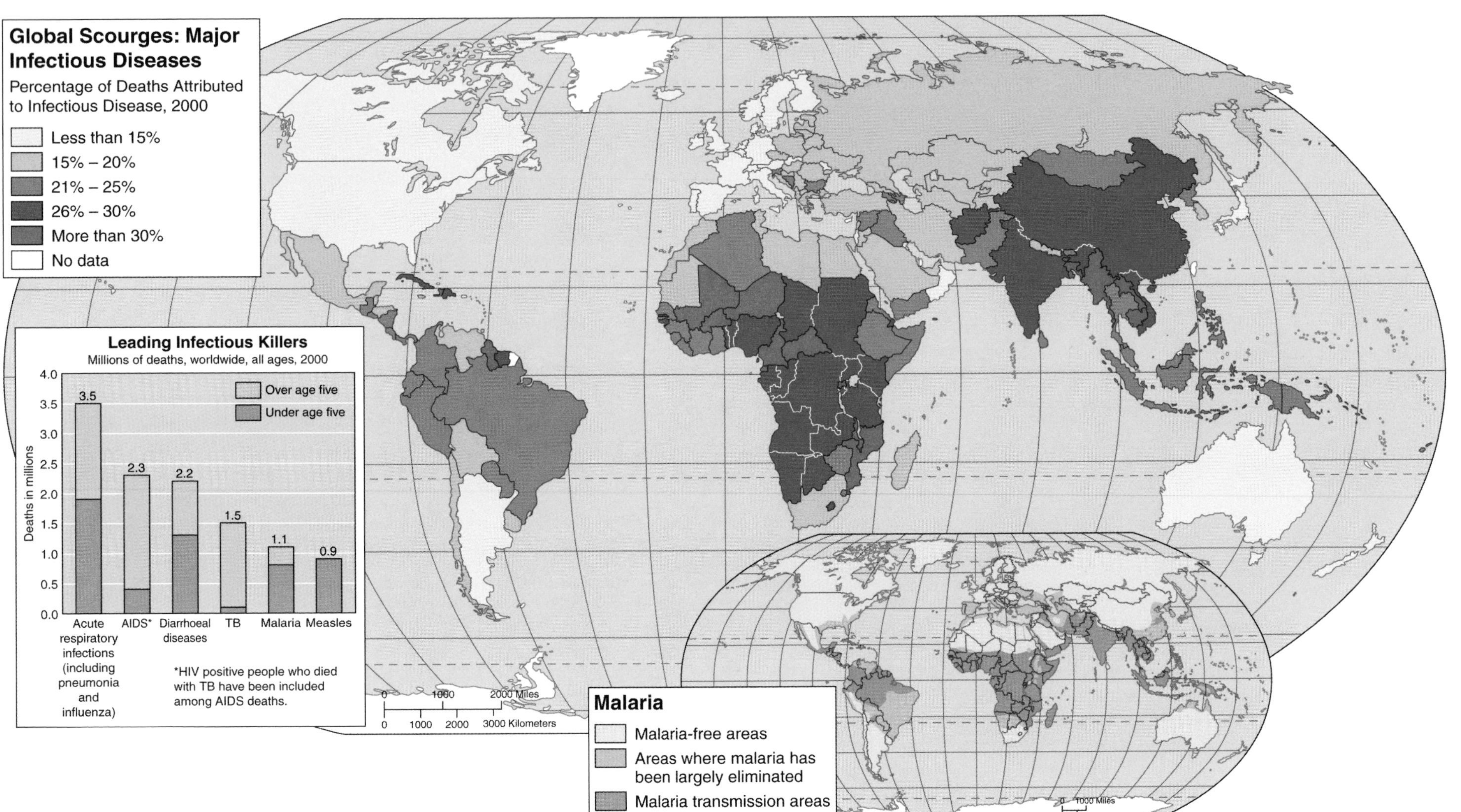

Infectious diseases are the world's leading cause of premature death and at least half of the world's population is, at any time, at risk of contracting an infectious disease. Although we often think of infectious diseases as being restricted to the tropical world (malaria, dengue fever), many if not most of them have attained global proportions. A major case in point is HIV/AIDS, which quite probably originated in Africa but has, over the last two decades, spread throughout the entire world. Major diseases of the nineteenth century, such as cholera and tuberculosis, are making a major comeback in many parts of the world, in spite of being preventable or treatable. Part of the problem with infectious diseases is that they tend to be associated with poverty (poor nutrition, poor sanitation, substandard housing, and so on) and, therefore, are seen as a problem of undeveloped countries, with the consequent lack of funding for prevention and treatment. Infectious diseases are also tending to increase because lifesaving drugs, such as antibiotics and others used in the fight against diseases, are losing their effectiveness as bacteria develop genetic resistance to them. The problem of global warming is also associated with a spread of infectious diseases as many disease vectors (certain species of mosquito, for example) are spreading into higher latitudes with increasingly warm temperatures and are spreading disease into areas where populations have no resistance to them. Infectious diseases have become something greater than simply a health issue of poor countries. They are now major social problems with potentially enormous consequences for the entire world.

Map 38 Adult HIV Prevalence 2004

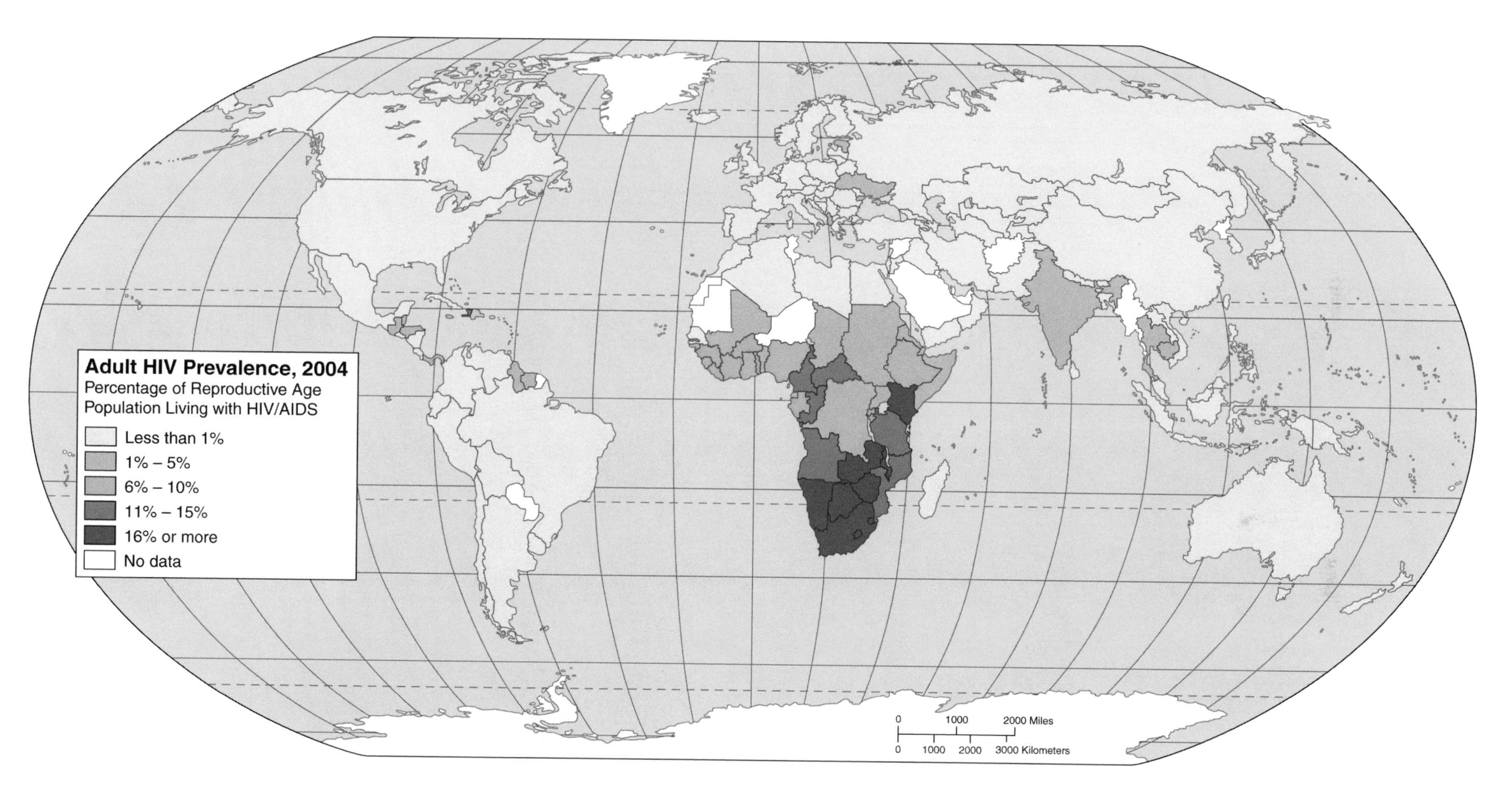

Of all the maps of the incidences of infectious disease, the global map of HIV/AIDS is the most dramatic as it impacts one continent so much more that it does other parts of the world. While there is no question that many African countries need to lower their growth rates, to do so by increasing death rates of children and young adults is dangerous. If the population profiles of countries like South Africa are plotted, they look like a bottle with a narrow bottom and a wide middle. In other words, the numbers of people who are going to be alive to be economically productive in the next generation are smaller than those in the older generations who may be forced to work well into what would be considered "old age" in order to support their sons and daughters. The stress placed by HIV/AIDS in some African countries in which more than 15% of the population are HIV positive is nearly incalculable. Try to imagine a United States with 20 million or more people between the ages of 15 and 45 dying of AIDS. And what becomes of the children orphaned by parents who have died of AIDS before seeing their offspring able to take care of themselves? In fact, many of these children are themselves born HIV positive. This is one of the world's most urgent health problems.

Map 39 The Index of Human Development

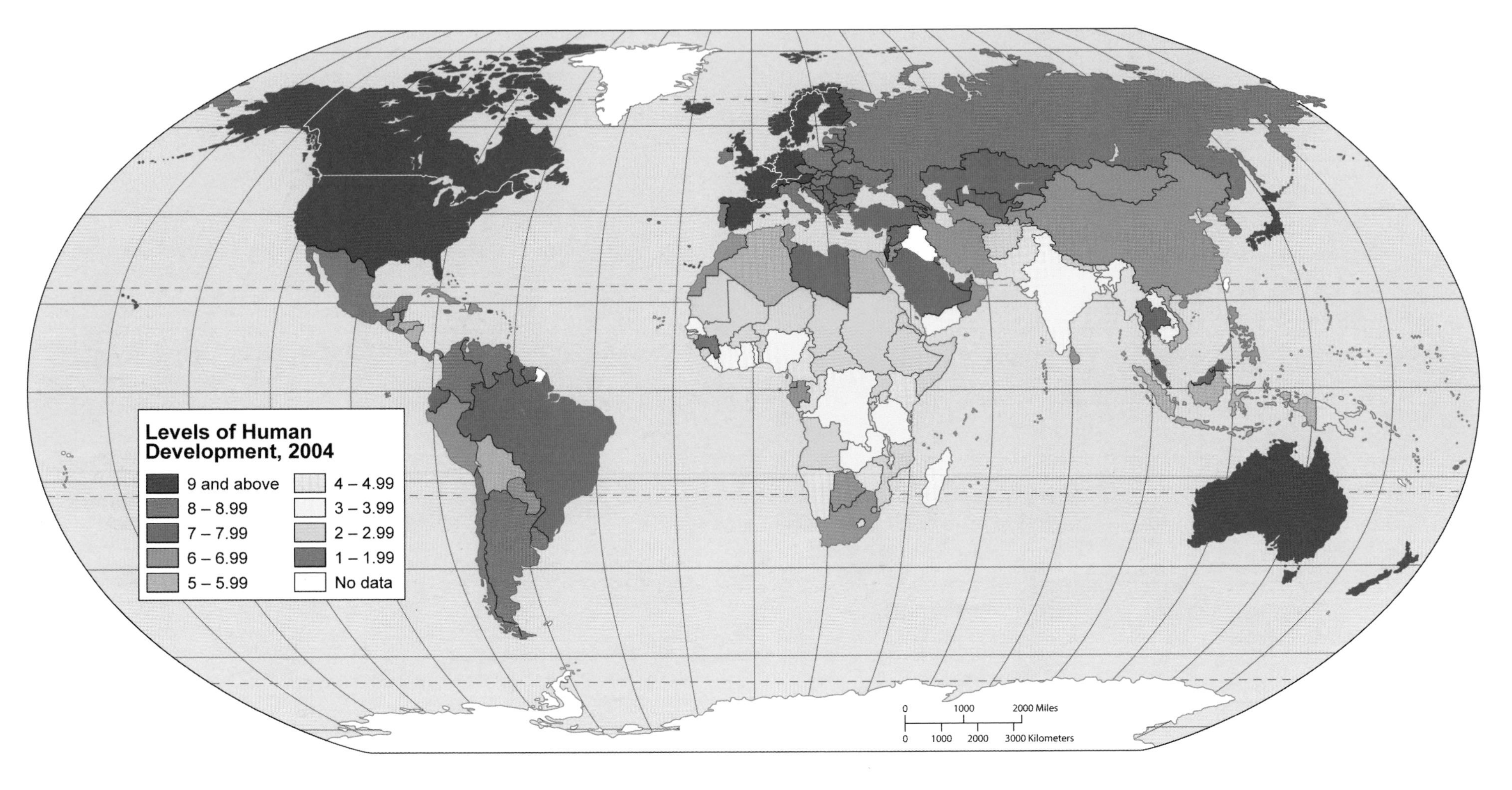

The development index upon which this map is based takes into account a wide variety of demographic, health, and educational data, including population growth, per capita gross domestic income, longevity, literacy, and years of schooling. The map reveals significant improvement in the quality of life in Middle and South America, although it is questionable whether the gains made in those regions can be maintained in the face of the dramatic population increases expected over the next 30 years. More clearly than anything else, the map illustrates the near-desperate situation in Africa and South Asia. In those regions, the unparalleled growth in population threatens to overwhelm all efforts to improve the quality of life. In Africa, for example, the population is increasing by 20 million persons per year. With nearly 45 percent of the continent's population aged 15 years or younger, this growth rate will accelerate as the women reach child-bearing age. Africa, along with South Asia, faces the very difficult challenge of providing basic access to health care, education, and jobs for a rapidly increasing population. The map also illustrates the striking difference in quality of life between those who inhabit the world's equatorial and tropical regions and those fortunate enough to live in the temperate zones, where the quality of life is significantly higher.

Map 40 Demographic Stress: The Youth Bulge

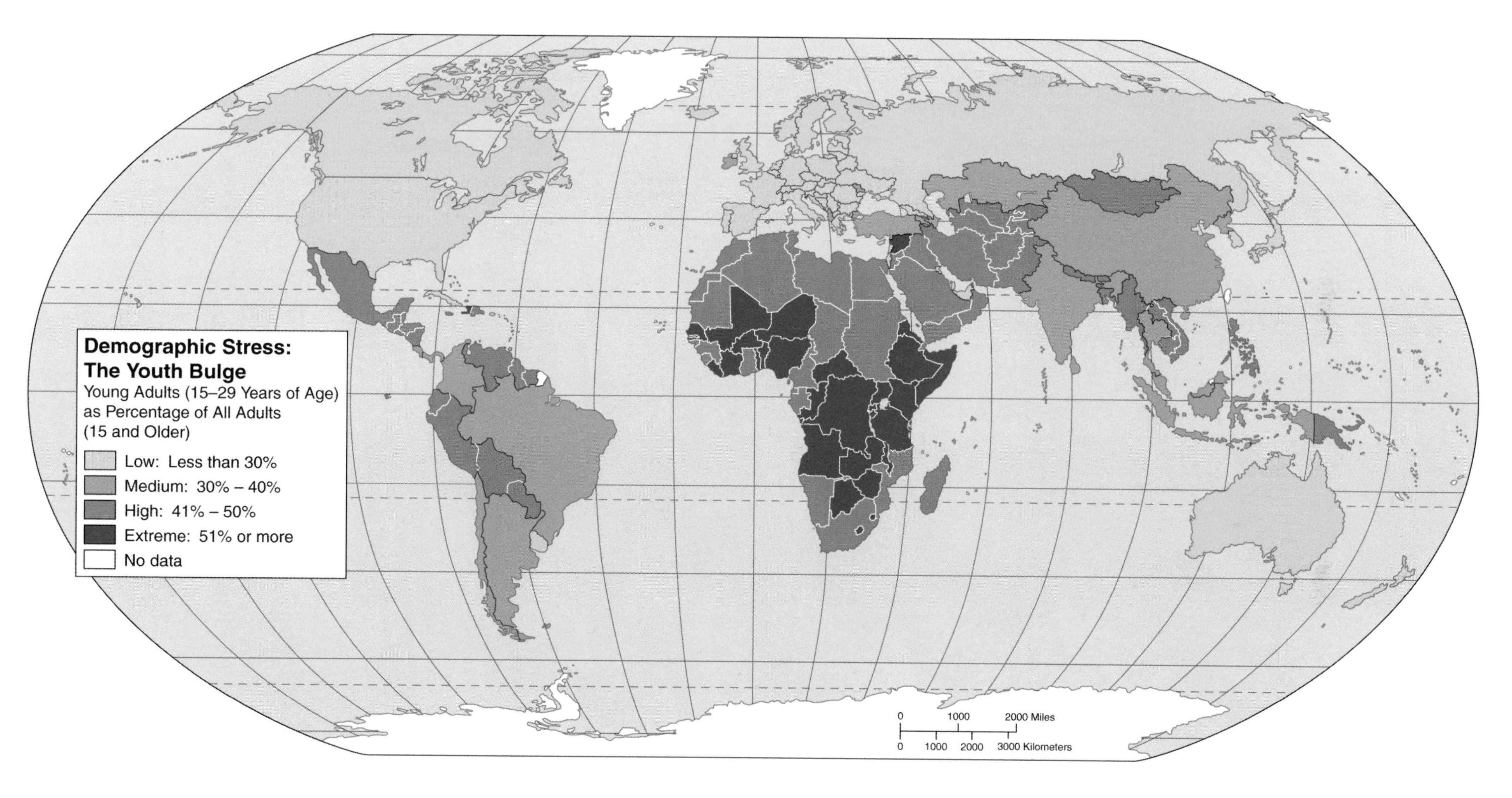

One of the greatest stresses of the demographic transition is when the death rate drops as the result of better public health and sanitation and the birth rate remains high for the same reasons it has always been high in traditional, agricultural societies: (1) the need for enough children to supply labor, which is one of the few ways to increase agricultural production in a non-mechanized agricultural system, and (2) the need for enough children to offset high infant and child mortality rates so that some children will survive to take care of parents in their old age. With declining death rates and steady birth rates, population growth skyrockets—particularly among the youngest cohorts of a population (between the ages of birth and 15). While this is, on the one hand, a demographic benefit since it increases the size of the labor force in the next generation and thereby helps to accelerate economic growth, it also means more people of child-bearing age in the next generation and, hence, greater numbers of births, which continue to swell the population in the youngest, most vulnerable, and most dependent portion of the population. The literature on population and conflict suggests that the larger the percentage of a population below the age of 25, the greater the chance for political violence and warfare. The suicide bombers of the modern Middle East are not oldsters.

Map 41 Demographic Stress: Rapid Urban Growth

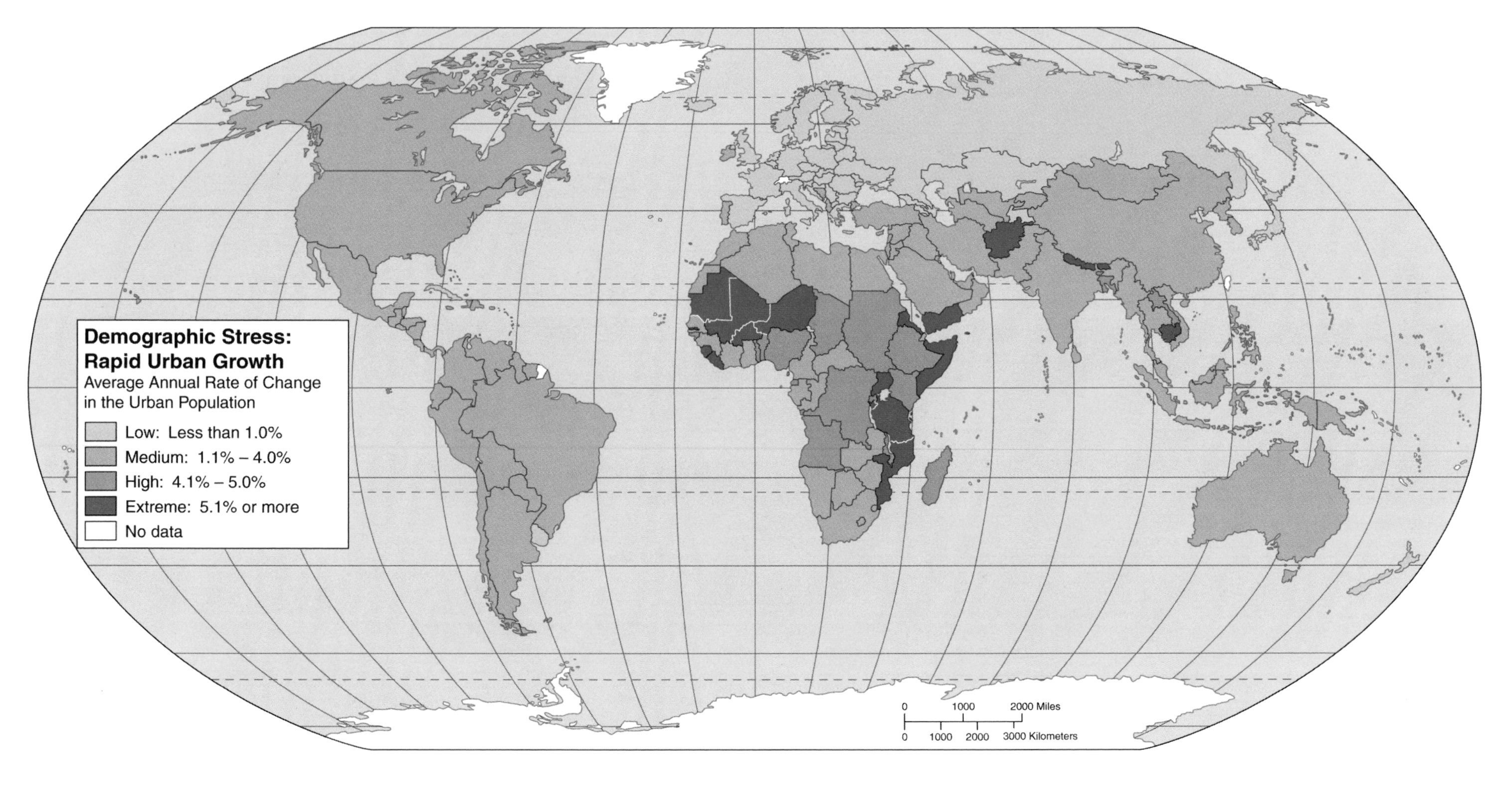

The trend toward urbanization—an increasing percentage of a country's population living in a city—is normally a healthy, modern trend. But in many of the world's developing countries, increasing urbanization is the consequence of high physiologic population densities (too many people for the available agricultural land) and the flight of poorly educated, untrained rural poor to the cities where they hope (often in vain) to find employment. High urbanization exists in many of the world's poorest countries where the urban poor live in conditions that make the worst living conditions in the inner cities of developed countries look positively luxurious. In the urban slums of South America, Africa, and South Asia, millions of people live in temporary housing of cardboard and flattened aluminum cans, with no public services such as water, sewage, or electricity. In Africa, where only 40% of the population is urbanized (in comparison with more than 90% in North America and Europe), the population of urban poor is greater than the total urban population of the United States and the European Union. This represents an increasingly destabilizing element of modern urban societies.

Map 42 Demographic Stress: Competition for Cropland

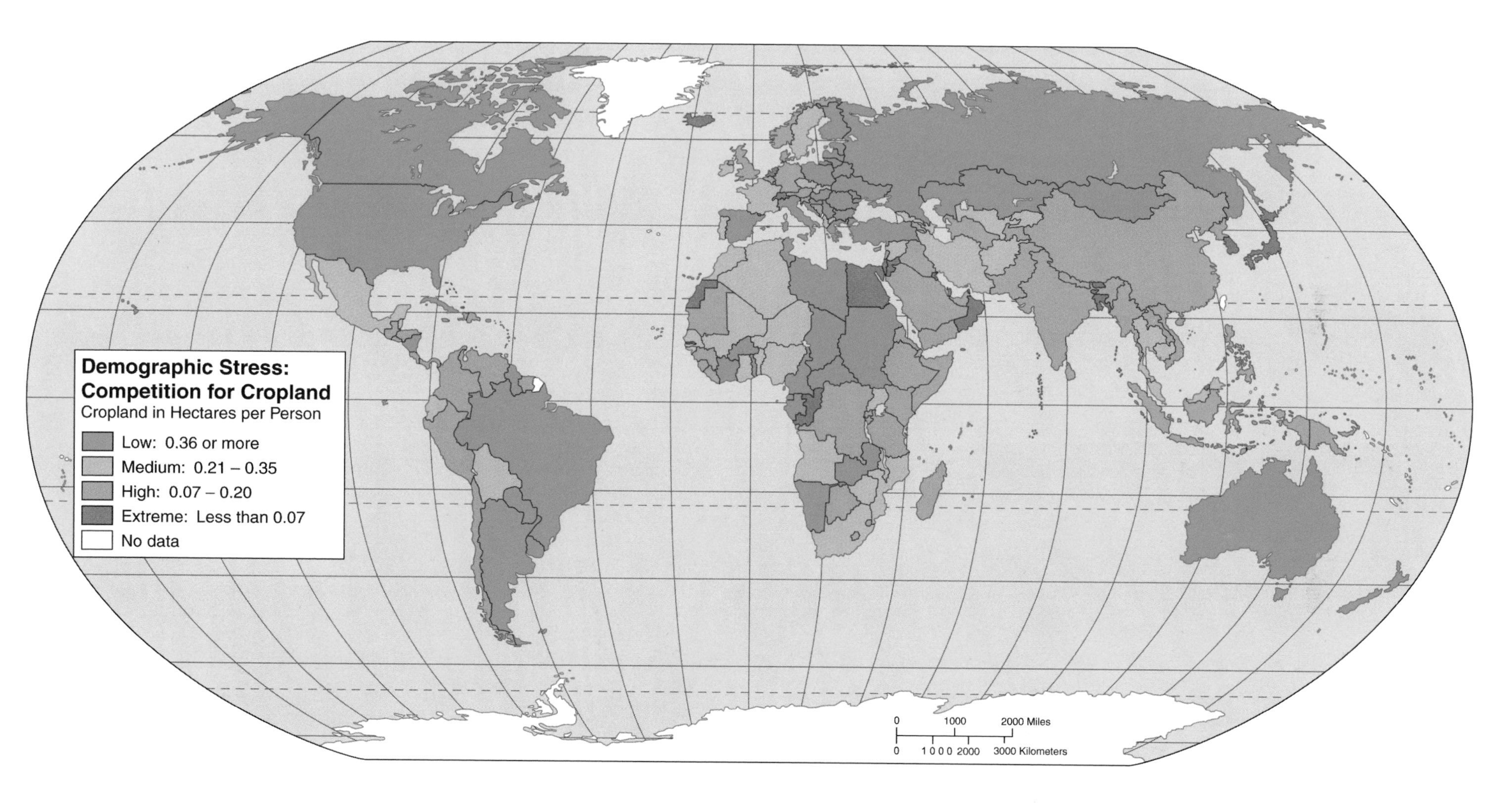

Many of the world's countries have reached the point where available acres or hectares in cropland are less than the numbers of people wishing to occupy them. For developed countries, who pay for agricultural imports with industrial exports, this is not an alarming trend and therefore countries like Germany or Italy (where the ratio between cropland and farmers is very low) have little to worry about. But in countries in South America, Africa, and South and East Asia, the trend toward more farmers and less available land *is* an alarming trend. In these developing regions, farmers depend on their crops for a relatively meager subsistence diet and—if they are lucky—a few bushels of rice or corn to take to the local market to sell or exchange for the small surpluses of other farmers. Despite the broad global trends toward urbanization and more productive agriculture (and this generally means mechanized agriculture), farm occupation and subsistence cropping remain mainstays of the economy in Africa south of the Sahara, in much of western South America, and in much of South and East Asia. Here, the increasingly small margin between the numbers of farmers and the amount of available farmland can lead to conflict among tribal communities or even among members of a single family.

Map 43 Demographic Stress: Competition for Fresh Water

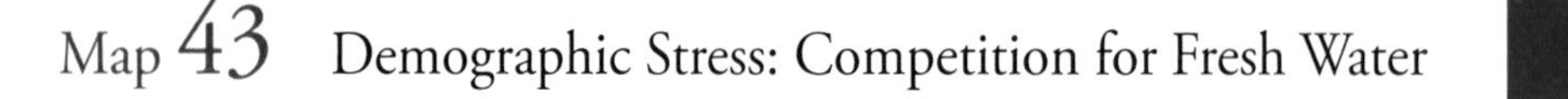

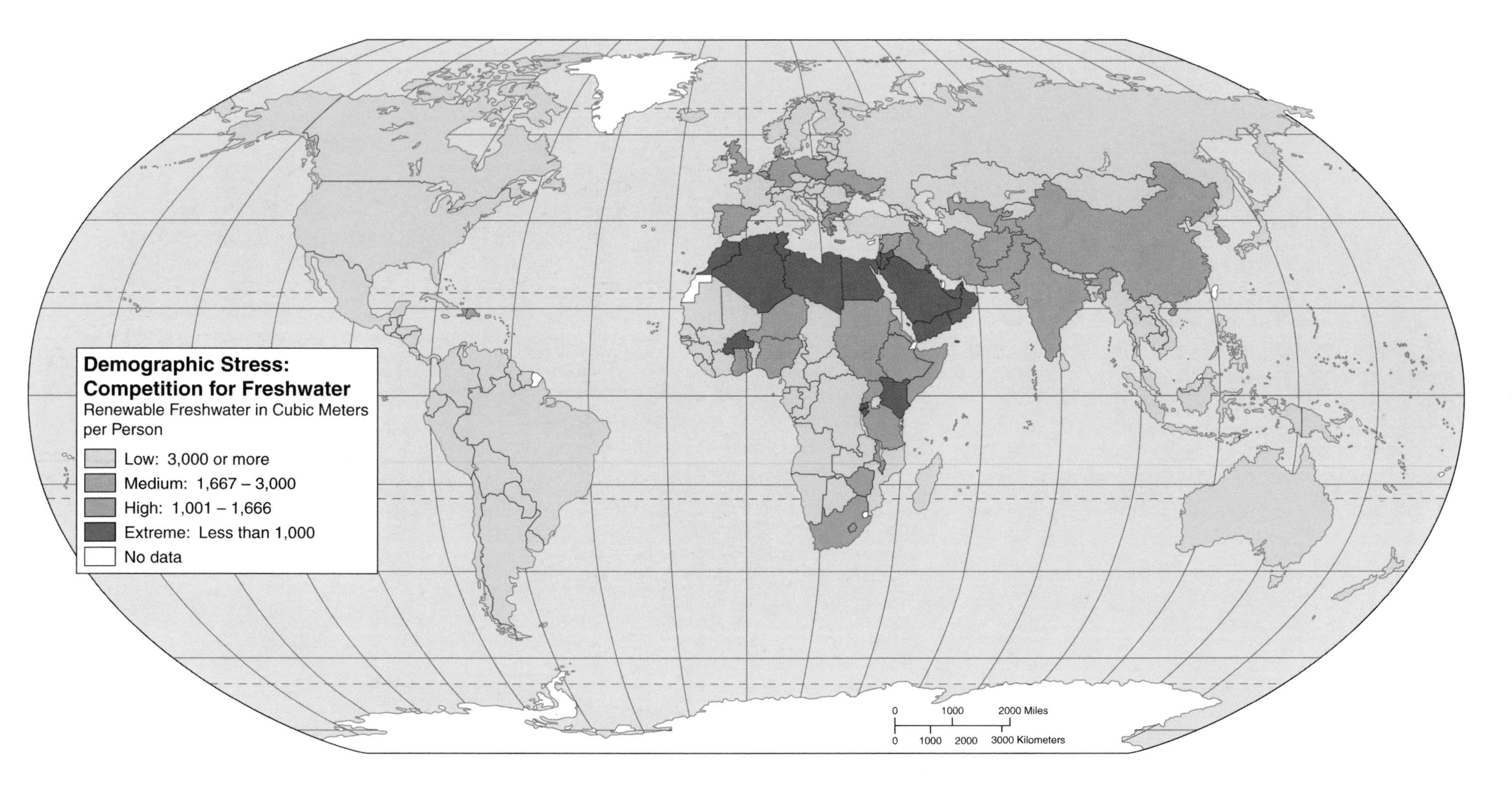

As rates of urbanization have increased, the competition between city dwellers and farmers for water has also increased. And as the population of farmers relative to the available surface or ground water supply has accelerated, so rural competition for access to fresh water for irrigation has increased. Obviously, this trend is most obvious in the world's drier regions. On this map, the greatest stress factors relating to water availability are in North and East Africa, the Middle East, and Central and East Asia. The current conflict between Israelis and Palestinians in the West Bank goes far beyond religion or politics: Some is the result of the more affluent Israeli farmers being able to drill wells to tap ground water, which lowers the water table and causes previously accessible surface wells in Palestinian villages to go dry. And part of the horrific Hutu-Tutsi civil war in Rwanda and the taking of European farmlands by the current government of Zimbabwe have more to do with the need for access to fresh water than to racial or ethnic differences. Given that fresh water is, for all practical purposes, a nonrenewable resource, more conflicts over access to water are going to erupt in the future.

Map 44 Demographic Stress: Death in the Prime of Life

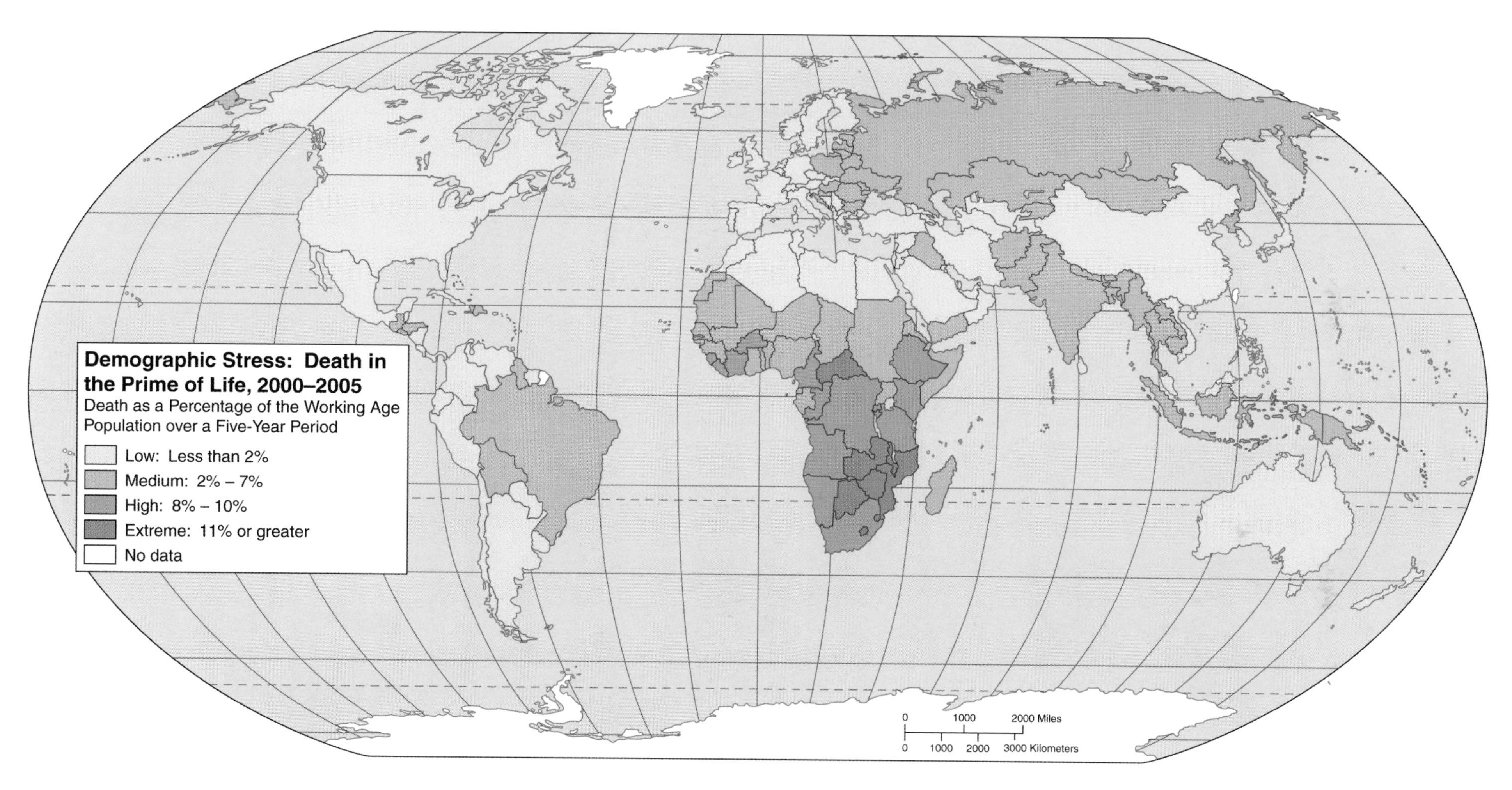

Most infectious diseases lay waste to the oldest and the youngest sectors of a population, leaving the healthier working-age population able to reproduce itself and to continue to grow an economy. Modern infectious diseases—most particularly HIV/AIDS—impacts not the youngest or oldest but those in the prime of life, in their most productive and reproductive years. Indeed, medical data from the United Nations show that 90% of HIV/AIDS-related deaths occur in people of working age. Some of the most severely impacted countries in southern Africa lose between 10% and 20% of their working age population every 5 years. Rates in the developed world are from 0.10% to 0.02% of this level. A prominent fact of life in much of the developing world—particularly in sub-Saharan Africa—is the long-distance labor migration. In this pattern, a man leaves his wife and family for extended periods of time (10 or more months of the year) to work in mines, on docks, as sailors. These men are highly vulnerable to the acquisition of HIV, which is then transmitted to wives when the husband returns home on his annual visit. Thus both the male and the female portion of the working-age population are impacted by early mortality from AIDS. None of this bodes well for economic development in those affected areas of the world.

Map 45 Demographic Stress: Interactions of Demographic Stress

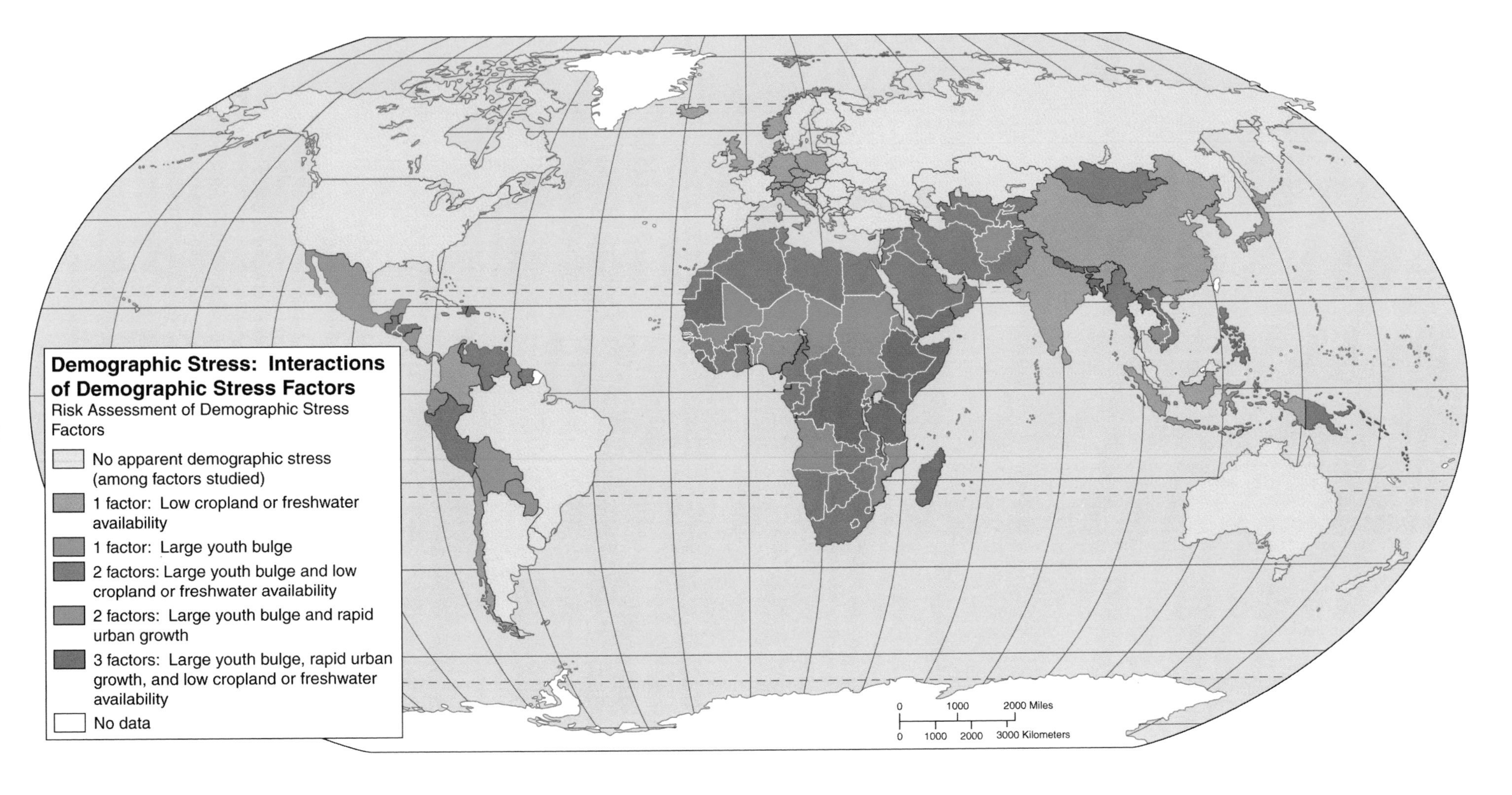

We have persisted in viewing internal and international conflicts as conflicts produced by religious differences, the acceptance or rejection of "freedom" or "democracy," or even, as argued by a prominent historian, a "Clash of Civilizations." While religious, political, or historical differences cannot be ignored in interpreting the causes of conflict, neither can simple demographic stresses produced by too many new urban dwellers without jobs or hopes of jobs, by too many farmers and too little farmland, by too many deaths among people who should be in the most productive portions of their lives. A massive study by Population Action International, from which these last half-dozen or so maps have been drawn, shows quite clearly that countries suffering from multiple demographic risk factors are, were, and will be much more prone to civil conflict than countries with only one or two demographic risk factors. A look at this map provides a predictor of future civil unrest, violence, and even civil war.

Part IV

Global Economic Patterns

Map 46 Rich and Poor Countries: Gross National Income
Map 47 Gross National Income Per Capita
Map 48 Economic Growth: GDP Change Per Capita
Map 49 Total Labor Force
Map 50 Employment by Economic Activity
Map 51 Economic Output Per Sector
Map 52 Agricultural Production Per Capita
Map 53 Exports of Primary Products
Map 54 Dependence on Trade
Map 55 The Indebtedness of States
Map 56 Global Flows of Investment Capital
Map 57 Central Government Expenditures Per Capita
Map 58 Purchasing Power Parity
Map 59 Private Per Capita Consumption 2003
Map 60 Energy Production Per Capita
Map 61 Energy Consumption Per Capita
Map 62 Energy Dependency
Map 63 Flows of Oil

Map 46 Rich and Poor Countries: Gross National Income

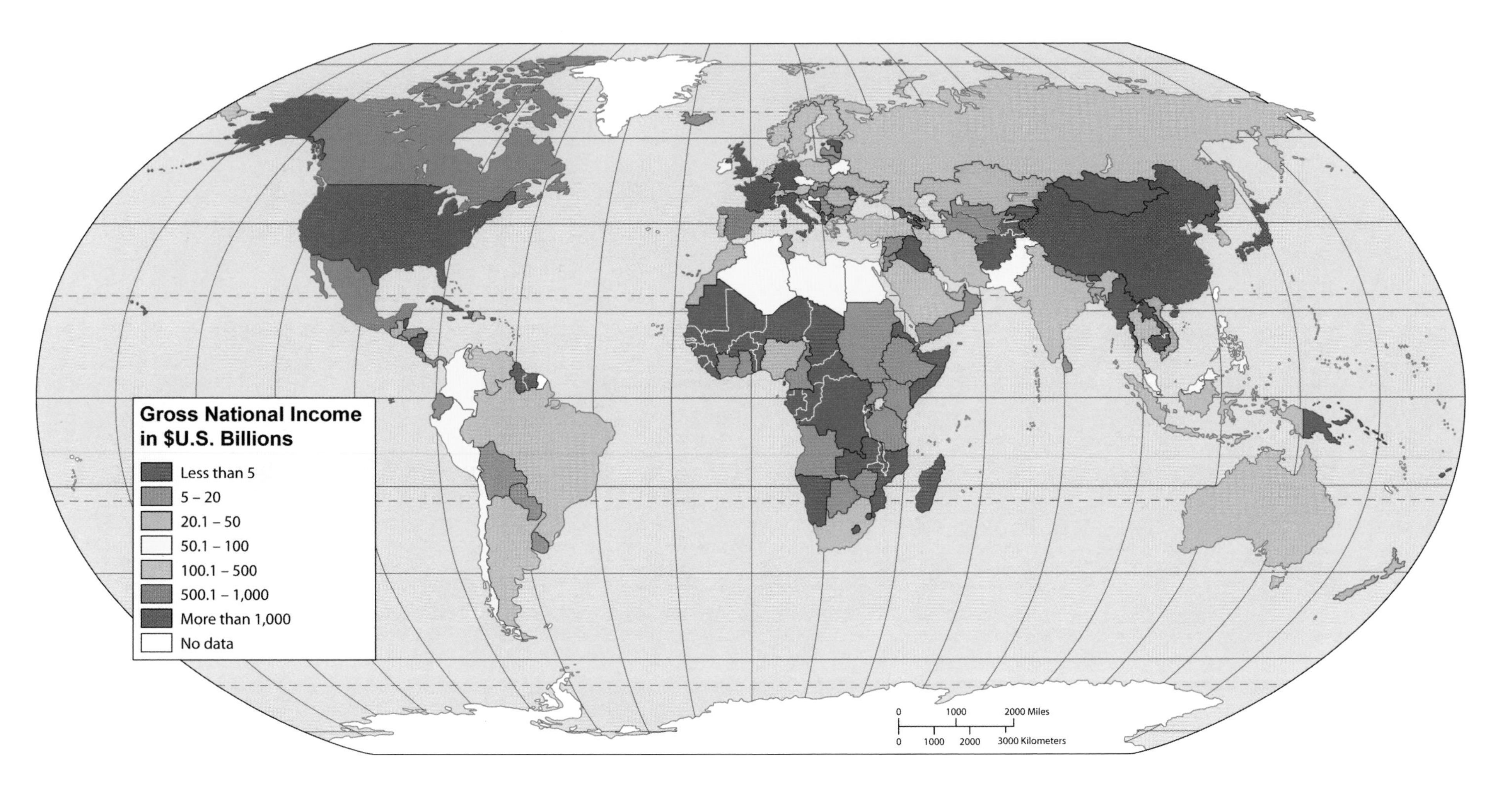

Gross National Income (GNI) is the broadest measure of national income and measures the total claims of a country's residents to all income from domestic and foreign products during a year. Although GNI is often misleading and commonly incomplete, it is often used by economists, geographers, political scientists, policy makers, development experts, and others not only as a measure of relative well-being but also as an instrument of assessing the effectiveness of economic and political policies. What is wrong with GNI? First of all, it does not take into account a number of real economic factors, such as environmental deterioration, the accumulation or degradation of human and social capital, or the value of household work. Yet in spite of these deficiencies, GNI is still a reasonable way to assess the relative wealth of nations: the vast differences in wealth that separate the poorest countries from the richest. One of the more striking features of the map is the evidence it presents that such a small number of countries possess so many of the world's riches (keeping in mind that GNI provides no measure of the distribution of wealth within a country).

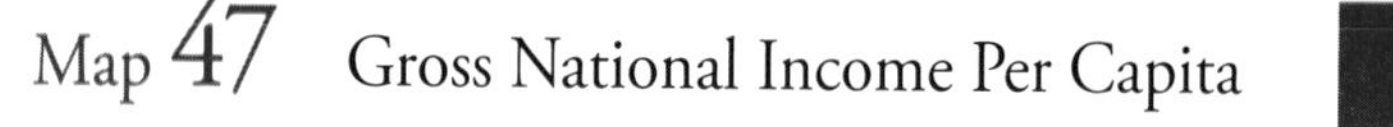

Map 47 Gross National Income Per Capita

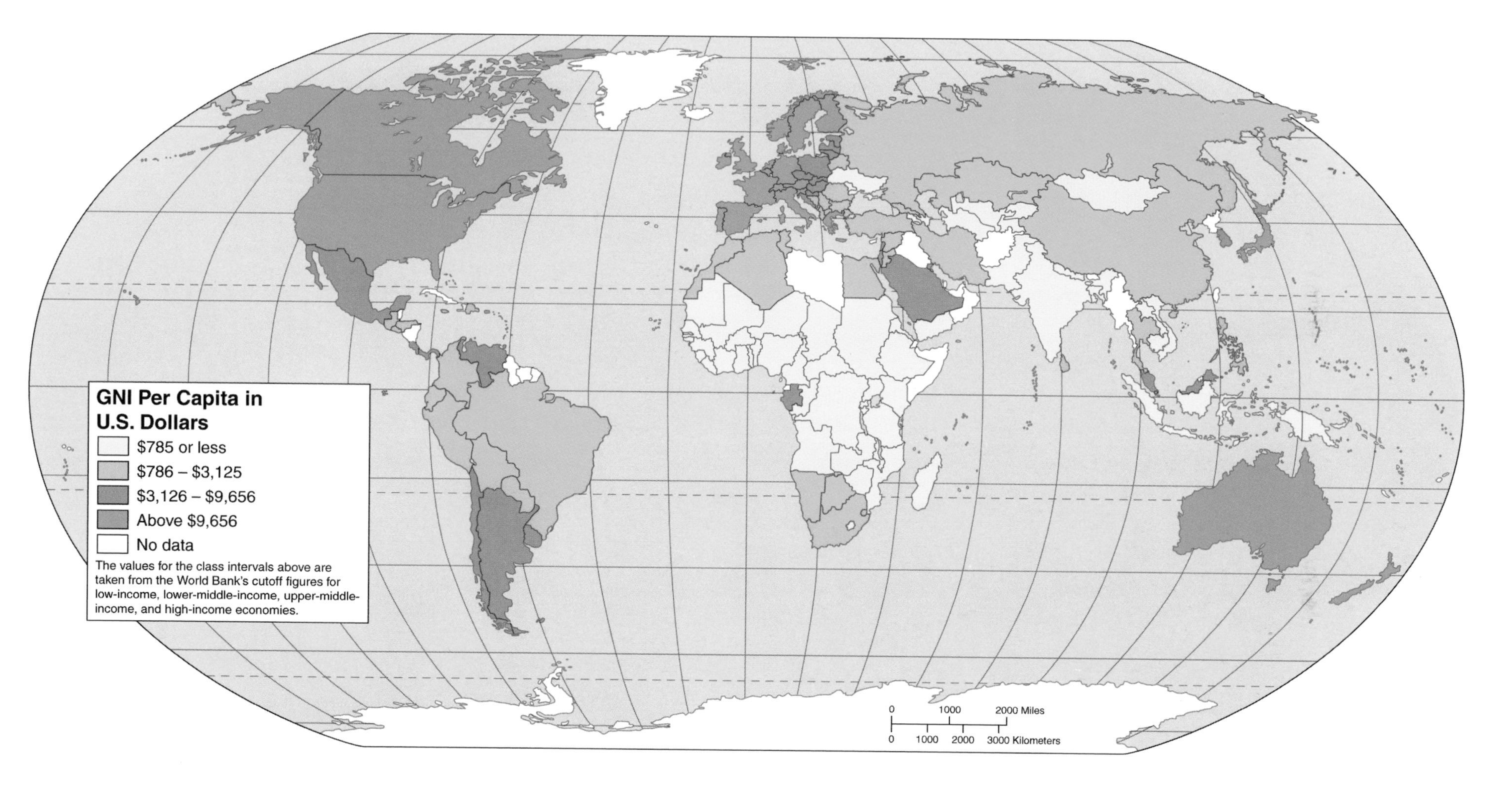

Gross National Income (GNI) in either absolute or per capita form should be used cautiously as a yardstick of economic strength because it does not measure the distribution of wealth among a population. There are countries (most notably, the oil-rich countries of the Middle East) where per capita GNI is high but where the bulk of the wealth is concentrated in the hands of a few individuals, leaving the remainder in poverty. Even within countries in which wealth is more evenly distributed (such as those in North America or Western Europe), there is a tendency for dollars or pounds sterling or euros to concentrate in the bank accounts of a relatively small percentage of the population. Yet the maldistribution of wealth tends to be greatest in the less developed countries, where the per capita GNI is far lower than in North America and Western Europe, and poverty is widespread. In fact, a map of GNI per capita offers a reasonably good picture of comparative economic well-being. It should be noted that a low per capita GNI does not automatically condemn a country to low levels of basic human needs and services. There are a few countries, such as Costa Rica and Sri Lanka, that have relatively low per capita GNI figures but rank comparatively high in other measures of human well-being, such as average life expectancy, access to medical care, and literacy.

Map 48 Economic Growth: GDP Change Per Capita

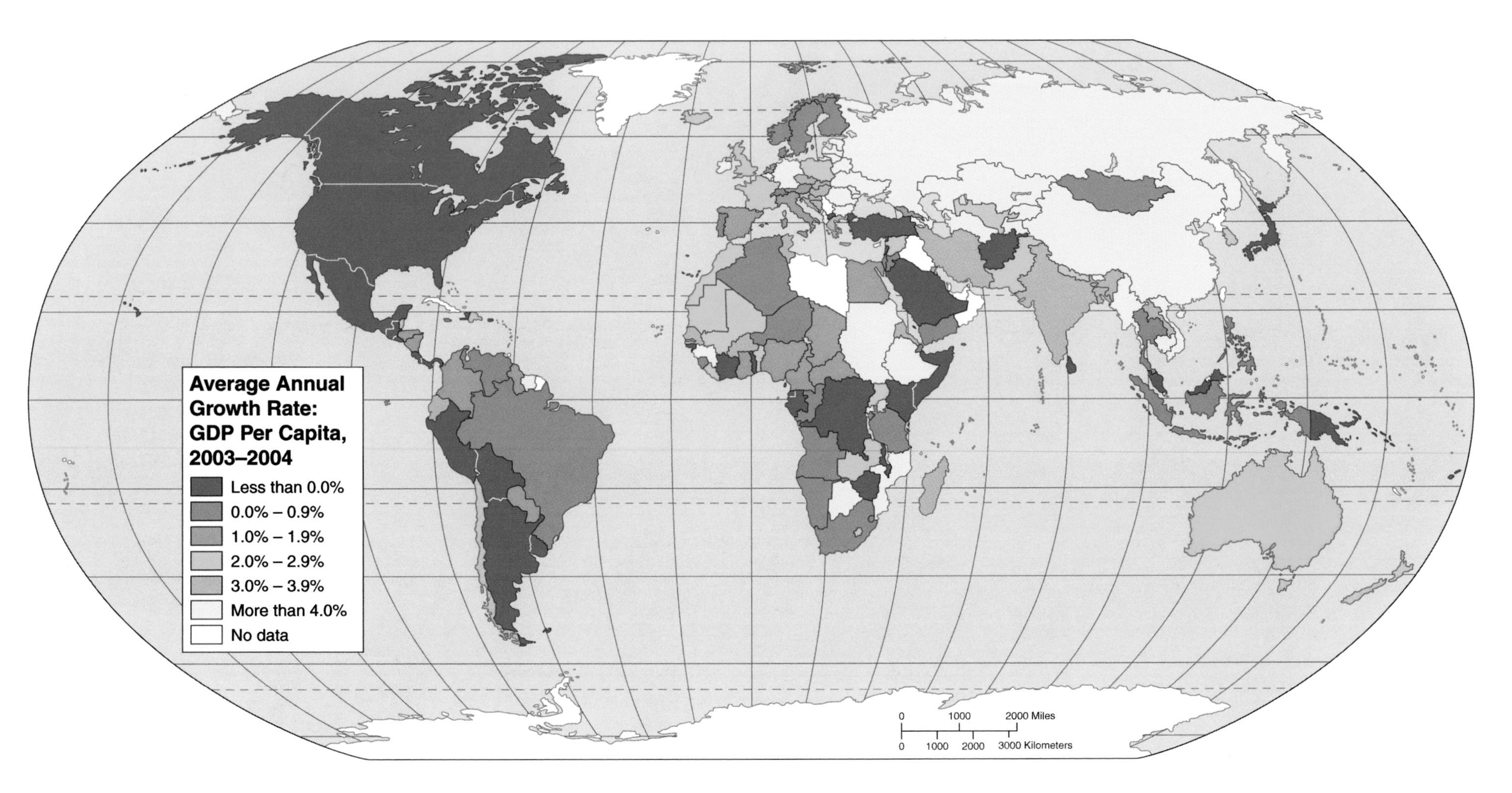

Gross Domestic Product or GDP is Gross National Income (GNI) less receipts of primary income from foreign sources. While the calculations of GDP growth per capita are complex, the growth rate is considered by the World Bank and international economists to be a particularly good measure of economic growth. One of the worldwide tendencies measured by GDP growth per capita is for continued economic development in Africa, and in South, Southeast, and East Asia where GDP grew at rates higher than the growth rates of "richer" countries in Europe and North and South America. This should not necessarily be viewed as a case of the poor catching up with the rich; in fact, it shows the huge impact that even relatively small production increases will have in countries with small GNIs and GDPs. Nevertheless, in spite of the continuing low economic growth through most of sub-Saharan Africa, the GDP growth rate of some of the world's poorer countries is an encouraging trend.

Map 49 Total Labor Force

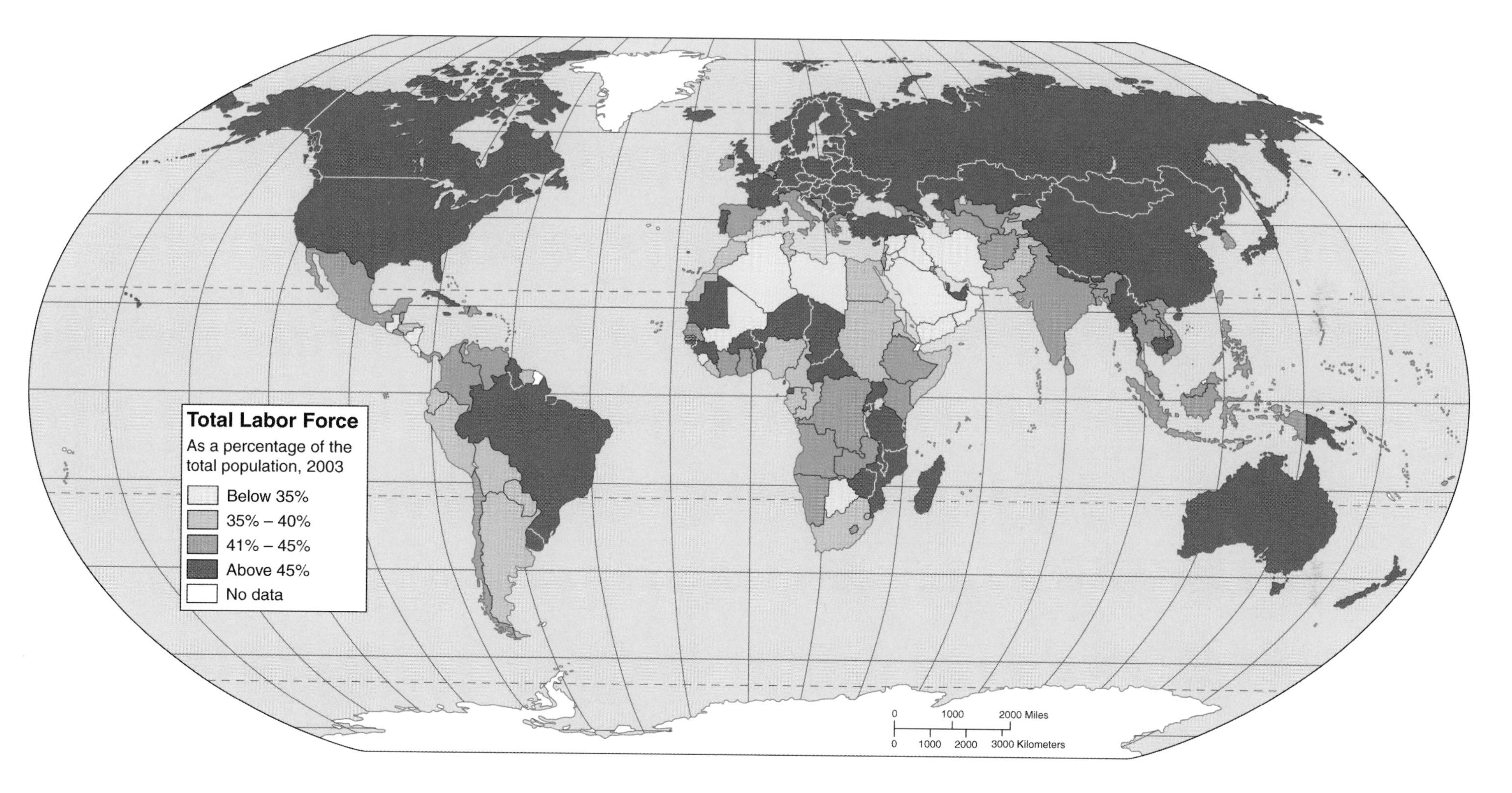

The term labor force refers to the economically active portion of a population, that is, all people who work or are without work but are available for and are seeking work to produce economic goods and services. The total labor force thus includes both the employed and the unemployed (as long as they are actively seeking employment). Labor force is considered a better indicator of economic potential than employment/unemployment figures, since unemployment figures will include experienced workers with considerable potential who are temporarily out of work. Unemployment figures will also incorporate persons seeking employment for the first time (many recent college graduates, for example). Generally, countries with higher percentages of total population within the labor force will be countries with higher levels of economic development. This is partly a function of levels of education and training and partly a function of the age distribution of populations. In developing countries, substantial percentages of the total population are too young to be part of the labor force. Also in developing countries a significant percentage of the population consist of women engaged in household activities or subsistence cultivation. These people seldom appear on lists of either employed or unemployed seeking employment and are the world's forgotten workers.

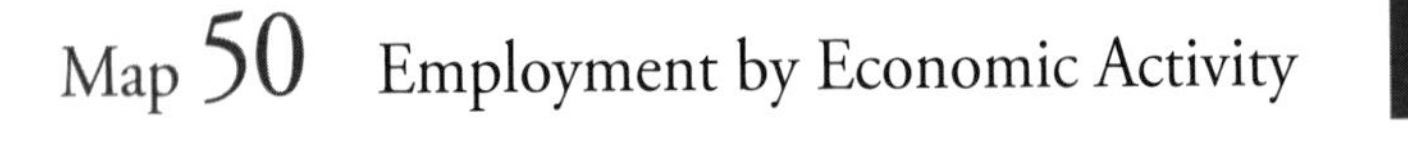

Map 50 Employment by Economic Activity

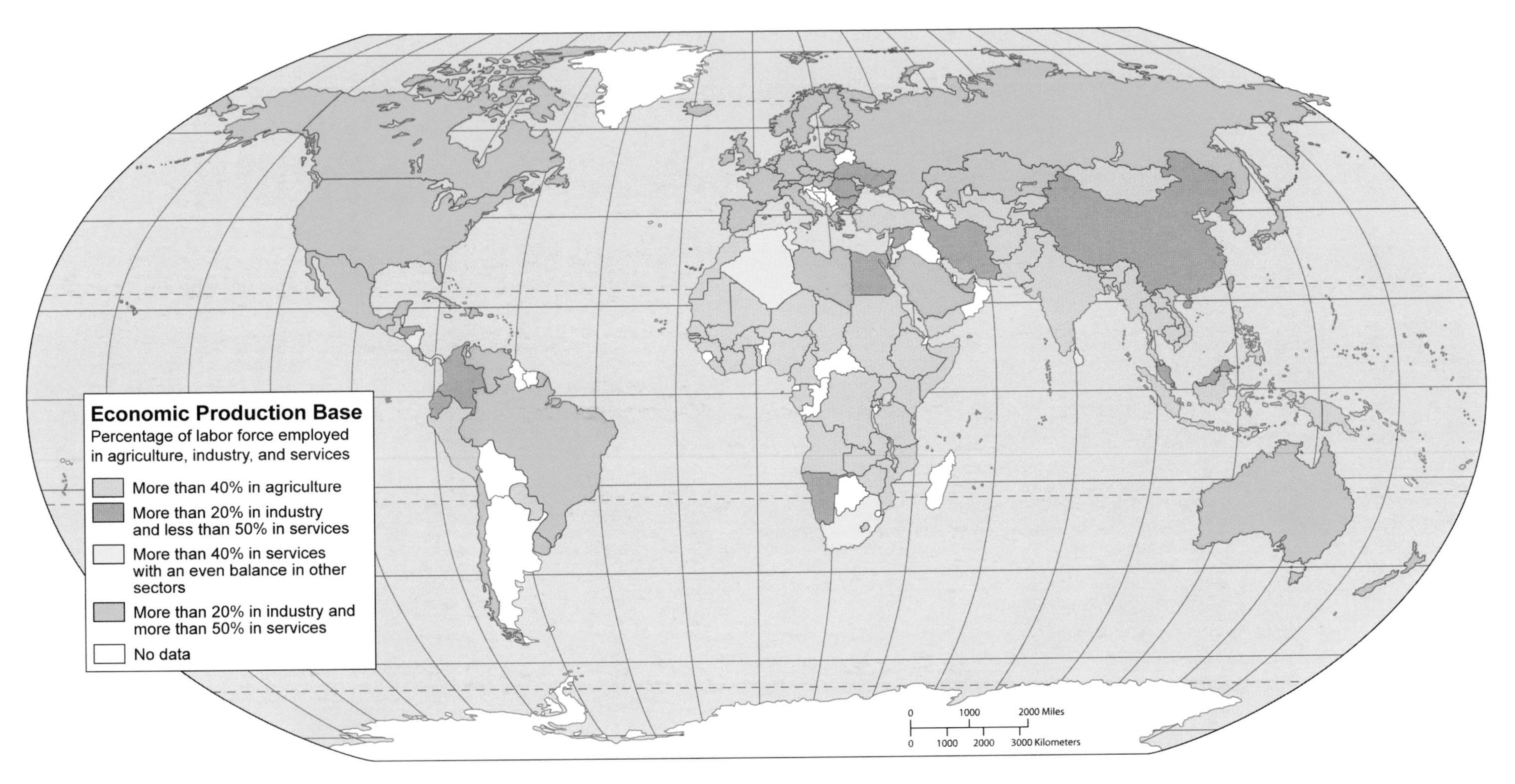

The employment structure of a country's population is one of the best indicators of the country's position on the scale of economic development. At one end of the scale are those countries with more than 40 percent of their labor force employed in agriculture. These are almost invariably the least developed, with high population growth rates, poor human services, significant environmental problems, and so on. In the middle of the scale are two types of countries: those with more than 20 percent of their labor force employed in industry and those with a fairly even balance among agricultural, industrial, and service employment but with at least 40 percent of their labor force employed in service activities. Generally, these countries have undergone the industrial revolution fairly recently and are still developing an industrial base while building up their service activities. This category also includes countries with a disproportionate share of their economies in service activities primarily related to resource extraction. On the other end of the scale from the agricultural economies are countries with more than 20 percent of their labor force employed in industry and more than 50 percent in service activities. These countries are, for the most part, those with a highly automated industrial base and a highly mechanized agricultural system (the "postindustrial," developed countries). They also include, particularly in Middle and South America and Africa, industrializing countries that are also heavily engaged in resource extraction as a service activity.

Map 51 Economic Output per Sector

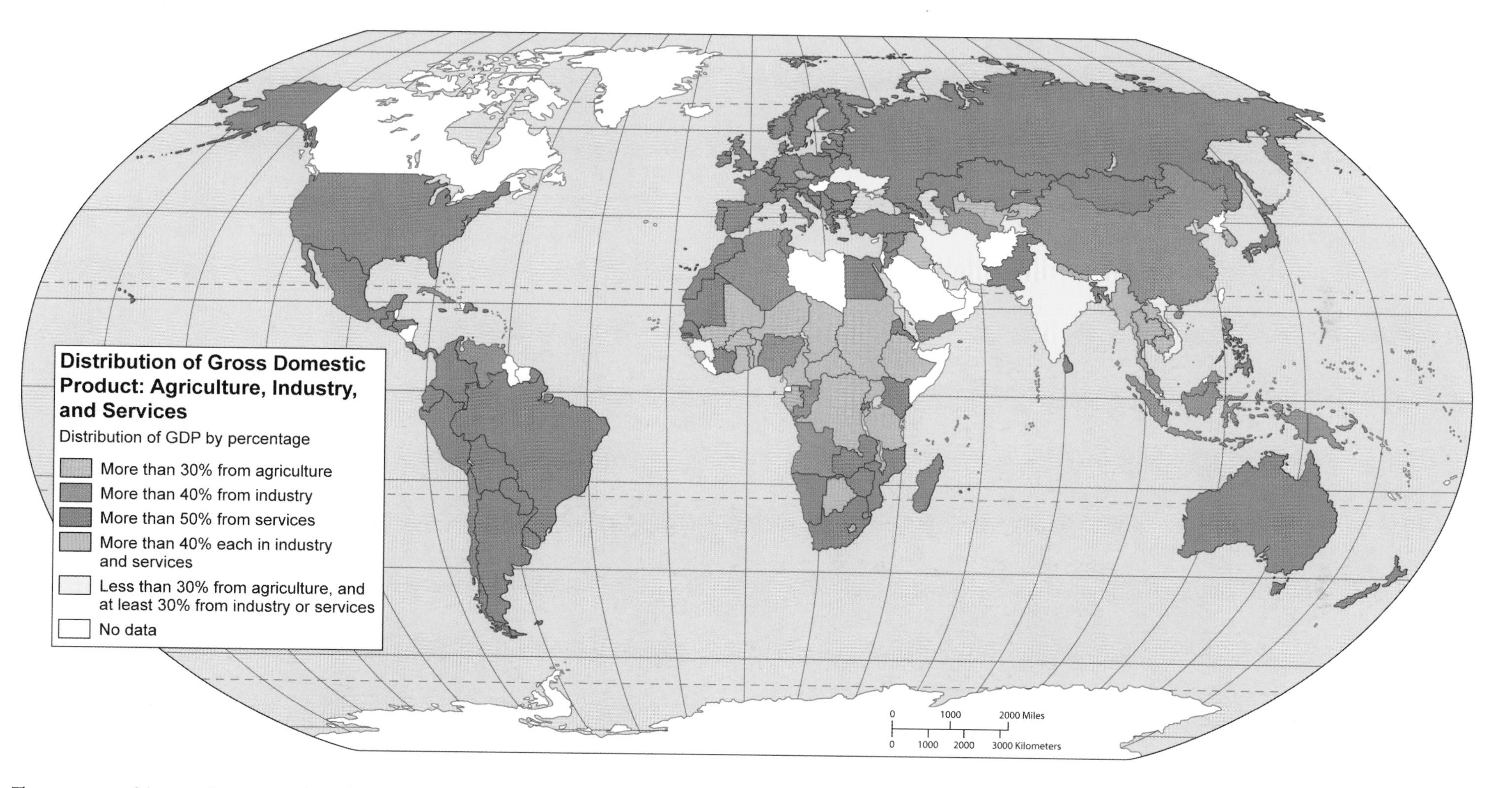

The percentage of the gross domestic product (the final output of goods and services produced by the domestic economy, including net exports of goods and nonfactor—nonlabor, noncapital—services) that is devoted to agricultural, industrial, and service activities is considered a good measure of the level of economic development. In general, countries with more than 40 percent of their GDP derived from agriculture are still in a *colonial dependency* economy—that is, raising agricultural goods primarily for the export market and dependent upon that market (usually the richer countries). Similarly, countries with more than 40 percent of GDP devoted to both agriculture and services often emphasize resource extractive (primarily mining and forestry) activities. These also tend to be *colonial dependency* countries, providing raw materials for foreign markets. Countries with more than 40 percent of their GDP obtained from industry are normally well along the path to economic development. Countries with more than half of their GDP based on service activities fall into two ends of the development spectrum. On the one hand are countries heavily dependent upon both extractive activities and tourism and other low-level service functions. On the other hand are countries that can properly be termed *postindustrial*: they have already passed through the industrial stage of their economic development and now rely less on the manufacture of products than on finance, research, communications, education, and other service-oriented activities.

Map 52 Agricultural Production Per Capita

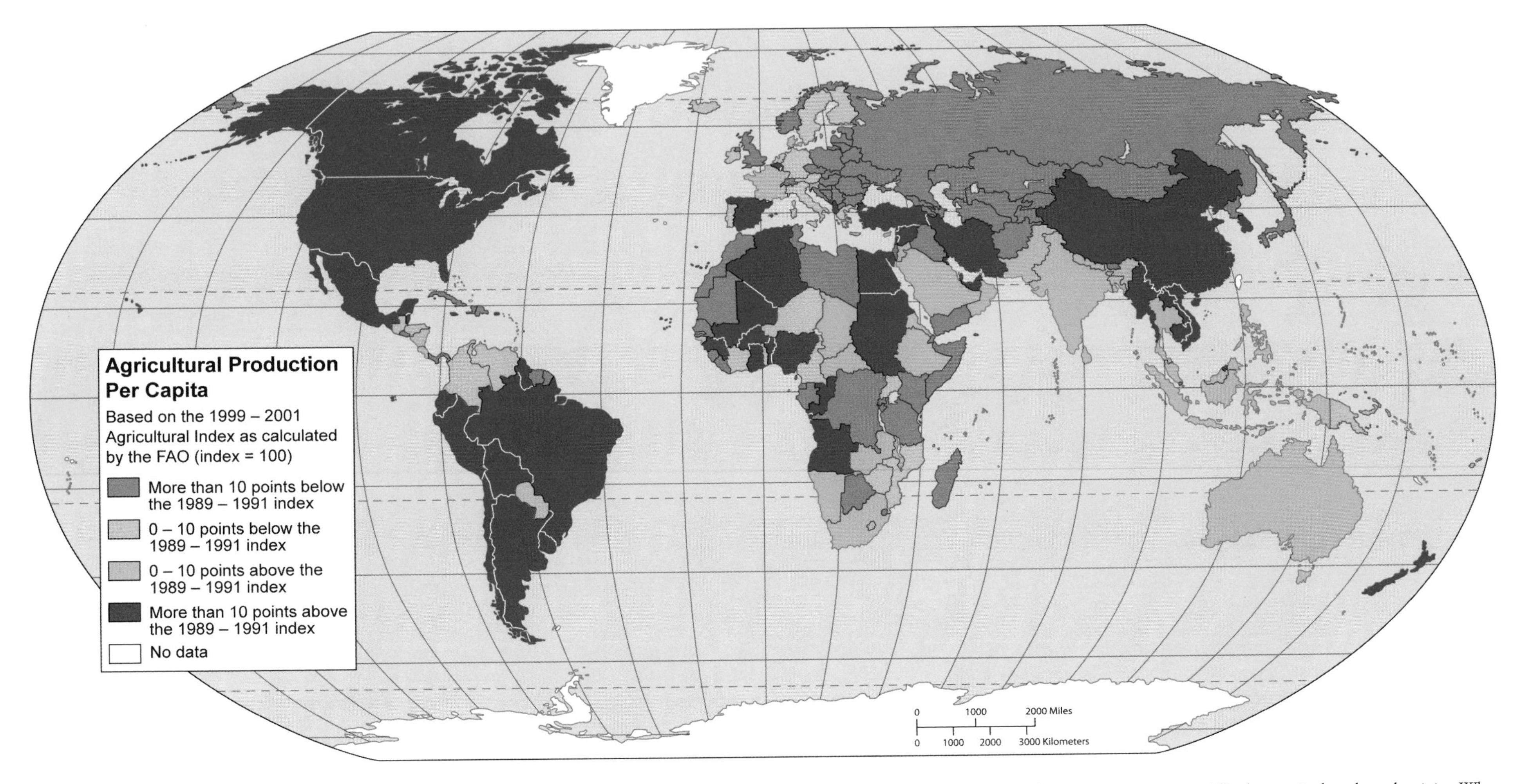

Agricultural production includes the value of all crops and livestock products originating within a country for the base year of 2002. The index value portrays the disposable output (after deductions for livestock feed and seed for planting) of a country's agriculture in comparison with the base period 1989–1991. Thus, the production values show not only the relative ability of countries to produce food but also show whether or not that ability has increased or decreased over a 10-year period. In general, global food production has kept up with or very slightly exceeded population growth. However, there are significant regional variations in the trend of food production keeping up with or surpassing population growth. For example, agricultural production in Africa and in Middle America has fallen, while production in South America, Asia, and Europe has risen. In the case of Africa, the drop in production reflects a population growing more rapidly than agricultural productivity. Where rapid increases in food production per capita exist (as in certain countries in South America, Asia, and Europe), most often the reason is the development of new agricultural technologies that have allowed food production to grow faster than population. In much of Asia, for example, the so-called Green Revolution of new, highly productive strains of wheat and rice made positive index values possible. Also in Asia, the cessation of major warfare allowed some countries (Cambodia, Laos, and Vietnam) to show substantial increases over the 1989–1991 index. In some cases, a drop in production per capita reflects government decisions to limit production in order to maintain higher prices for agricultural products. The United States and Japan fall into this category.

Map 53 Exports of Primary Products

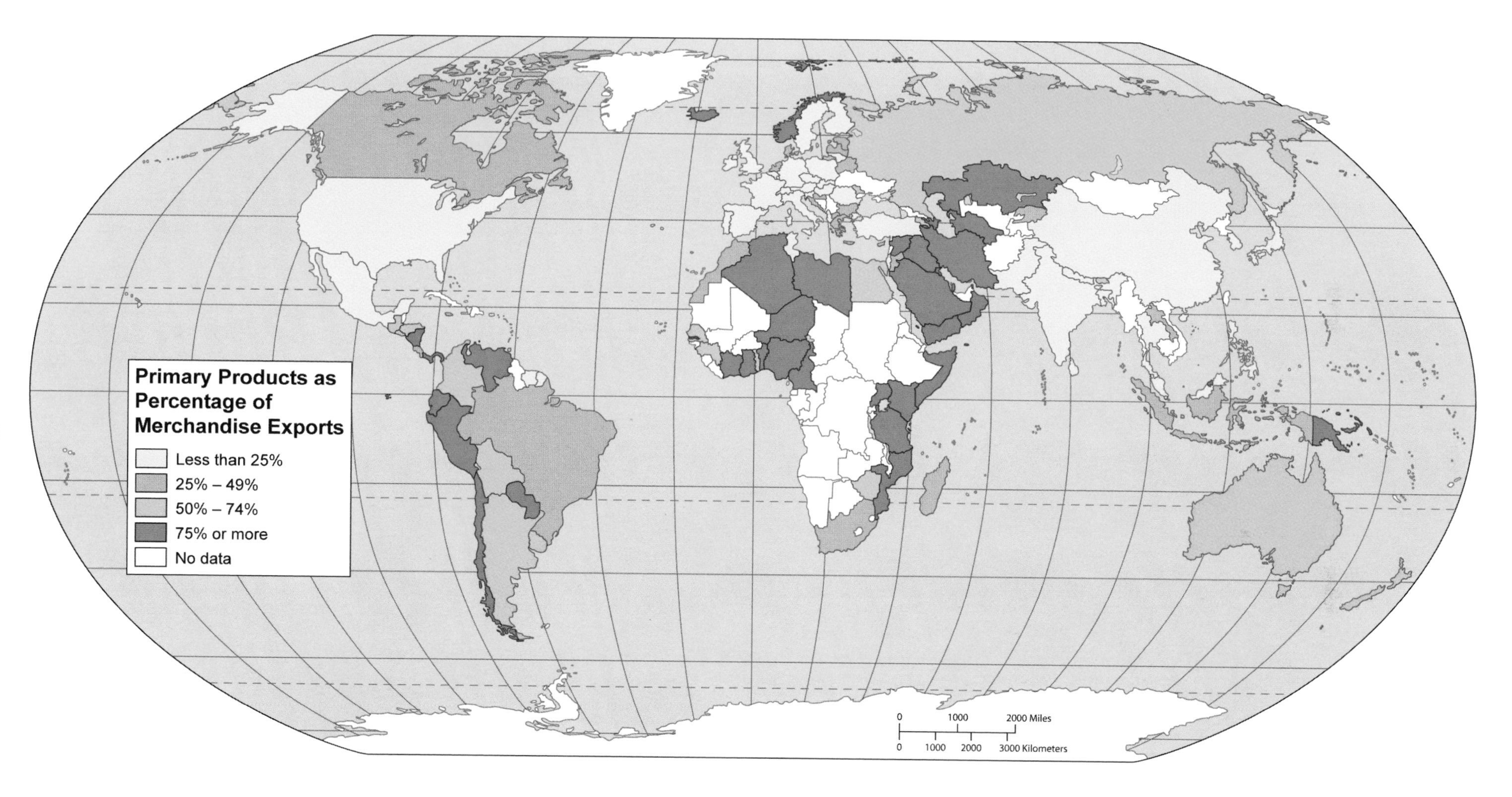

Primary products are those that require additional processing before they enter the consumer market: metallic ores that must be converted into metals and then into metal products such as automobiles or refrigerators; forest products such as timber that must be converted to lumber before they become suitable for construction purposes; and agricultural products that require further processing before being ready for human consumption. It is an axiom in international economics that the more a country relies on primary products for its export commodities, the more vulnerable its economy is to market fluctuations. Those countries with only primary products to export are hampered in their economic growth. A country dependent on only one or two products for export revenues is unprotected from economic shifts, particularly a changing market demand for its products. Imagine what would happen to the thriving economic status of the oil-exporting states of the Persian Gulf, for example, if an alternate source of cheap energy were found. A glance at this map shows that those countries with the lowest levels of economic development tend to be concentrated on primary products and, therefore, have economies that are especially vulnerable to economic instability.

Map 54 Dependence on Trade

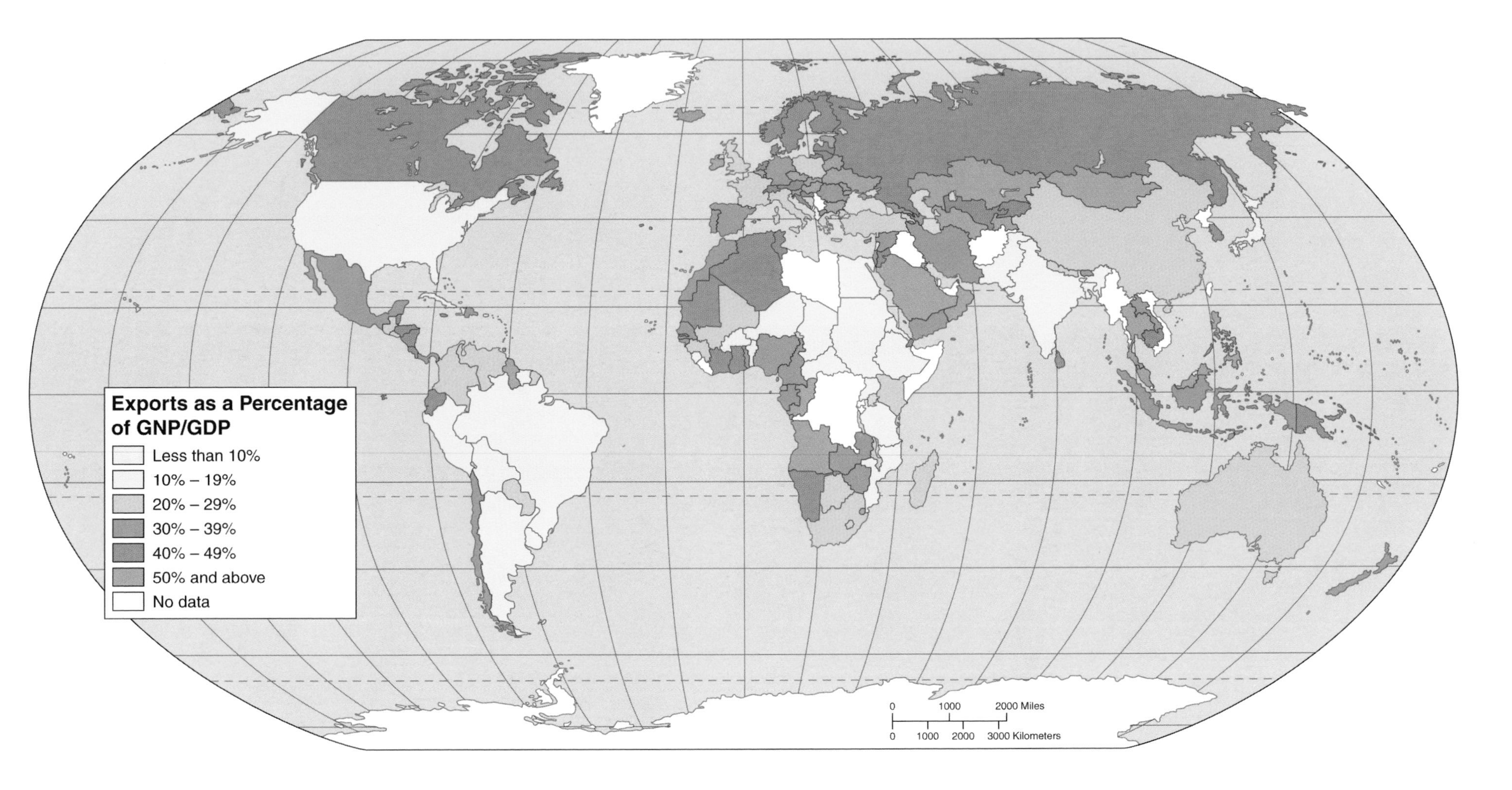

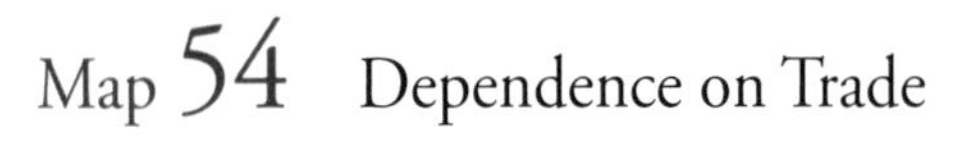

As the global economy becomes more and more a reality, the economic strength of virtually all countries is increasingly dependent upon trade. For many developing nations, with relatively abundant resources and limited industrial capacity, exports provide the primary base upon which their economies rest. Even countries like the United States, Japan, and Germany, with huge and diverse economies, depend on exports to generate a significant percentage of their employment and wealth. Without imports, many products that consumers want would be unavailable or more expensive; without exports, many jobs would be eliminated.

Map 55 The Indebtedness of States

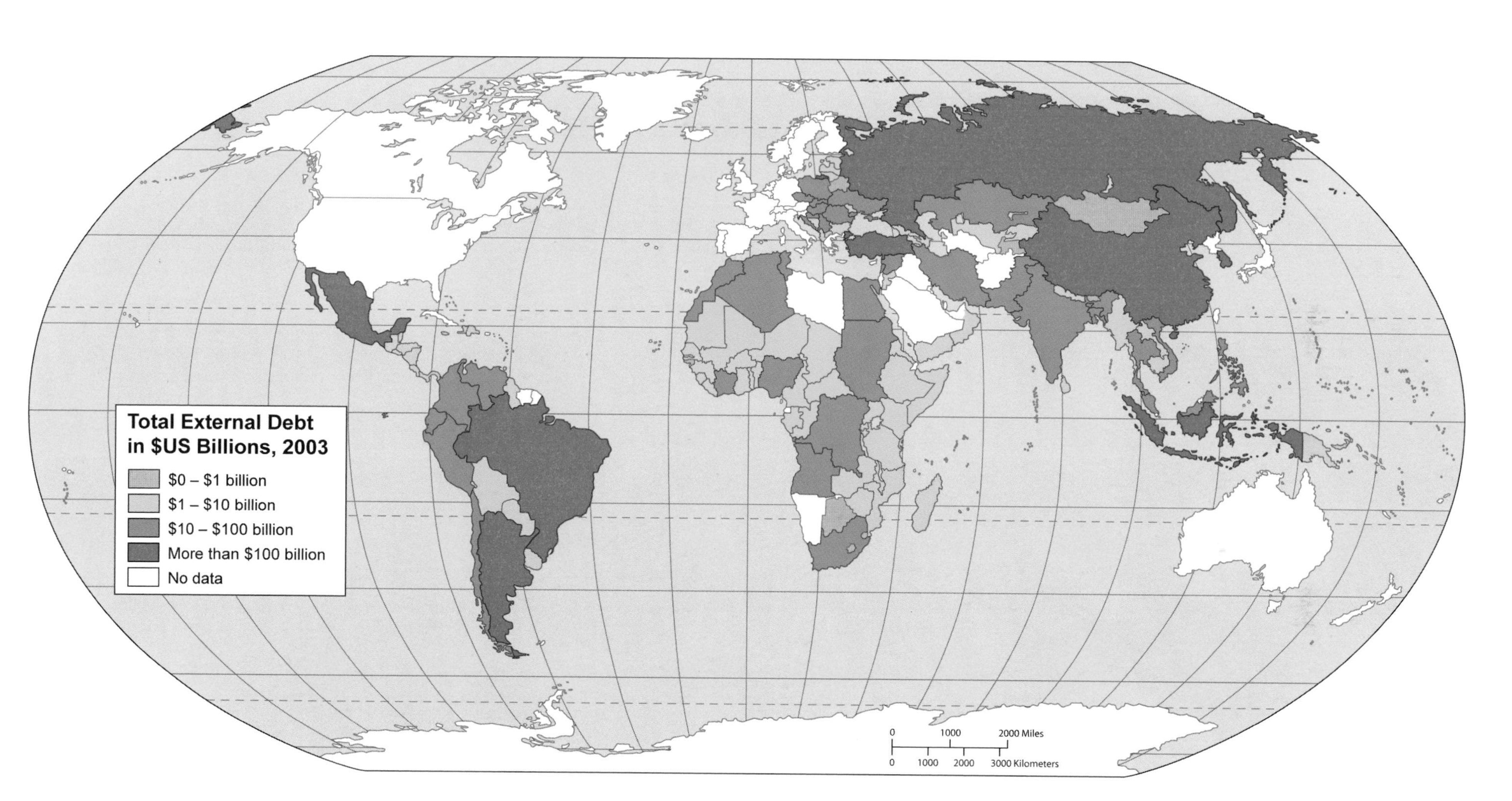

Many governments spend more on a wide variety of services and activities than they collect in taxes and other revenues. In order to finance this deficit spending, governments borrow money—often from banks or other investors outside their country. Repayment of these debts, or even meeting interest payments on them, often means expending a country's export income—in other words, exchanging a country's wealth in production or, more often, resources, for debt service. Where the debt is external, as it is in most developing countries, governments become more open to outside influence in political as well as economic terms. Even internal debt service or repayment of monies owed to investors within a country gives financial establishments a measure of influence over government decisions. The amounts of debt shown on the map indicate the total external indebtedness of states.

Map 56 Global Flows of Investment Capital

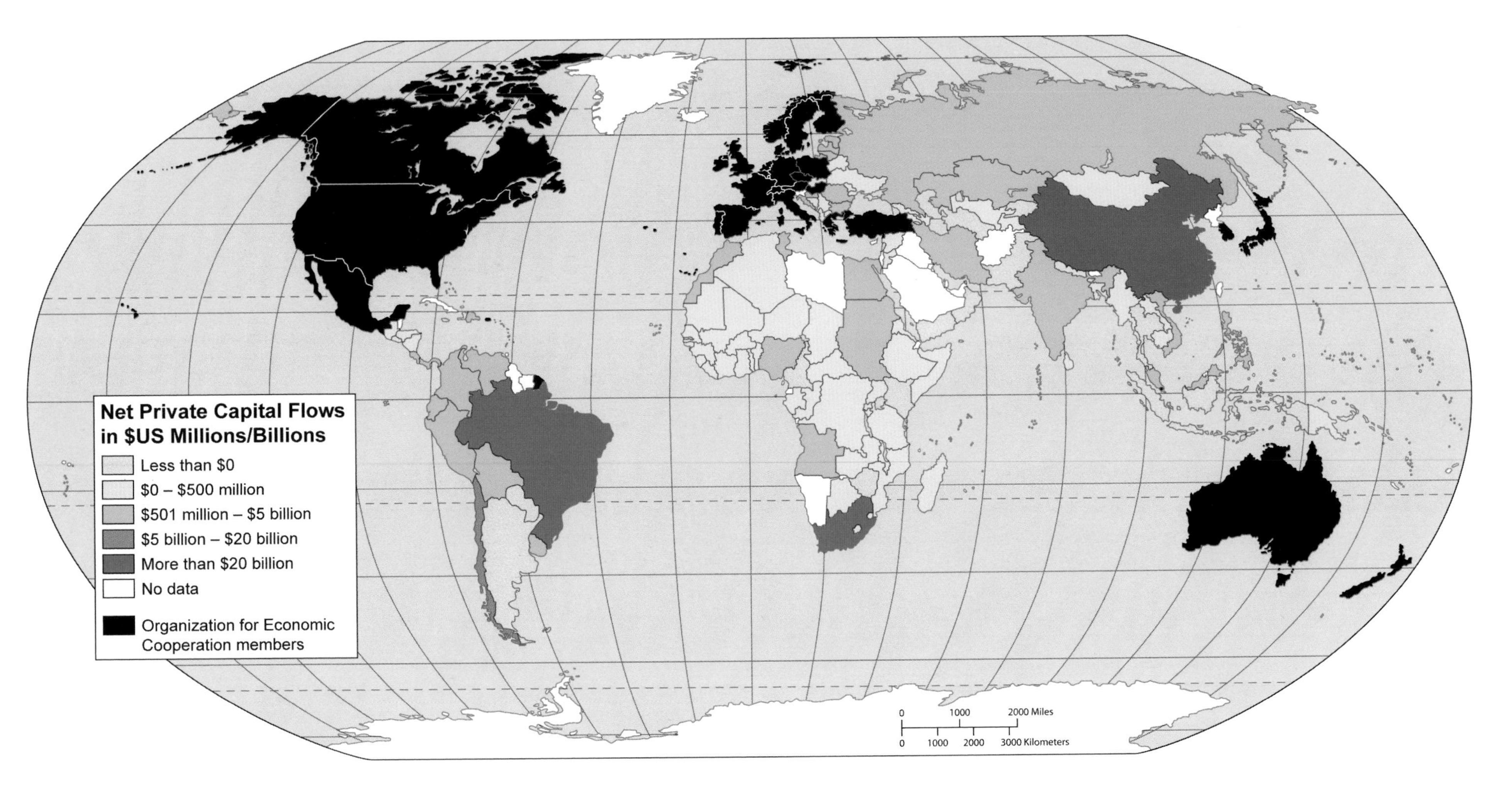

International capital flows include private debt and nondebt flows from one country to another, shown on the map as flows into a country. Nearly all of the capital comes from those countries that are members of the Organization for Economic Cooperation and Development (OECD), shown in black on the map. Capital flows include commercial bank lending, bonds, other private credits, foreign direct investment, and portfolio investment. Most of these flows are indicators of the increasing influence developed countries exert over the developing economies. Foreign direct investment or FDI, for example, is a measure of the net inflow of investment monies used to acquire long-term management interest in businesses located somewhere other than in the economy of the investor. Usually this means the acquisition of at least 10 percent of the stock of a company by a foreign investor and is, then, a measure of what might be termed "economic colonialism": control of a region's economy by foreign investors that could, in the world of the future, be as significant as colonial political control was in the past. International capital flows have increased greatly in the last decade as the result of the increasing liberalization of developing countries, the strong economic growth exhibited by many developing countries, and the falling costs and increased efficiency of communication and transportation services.

Map 57 Central Government Expenditures Per Capita

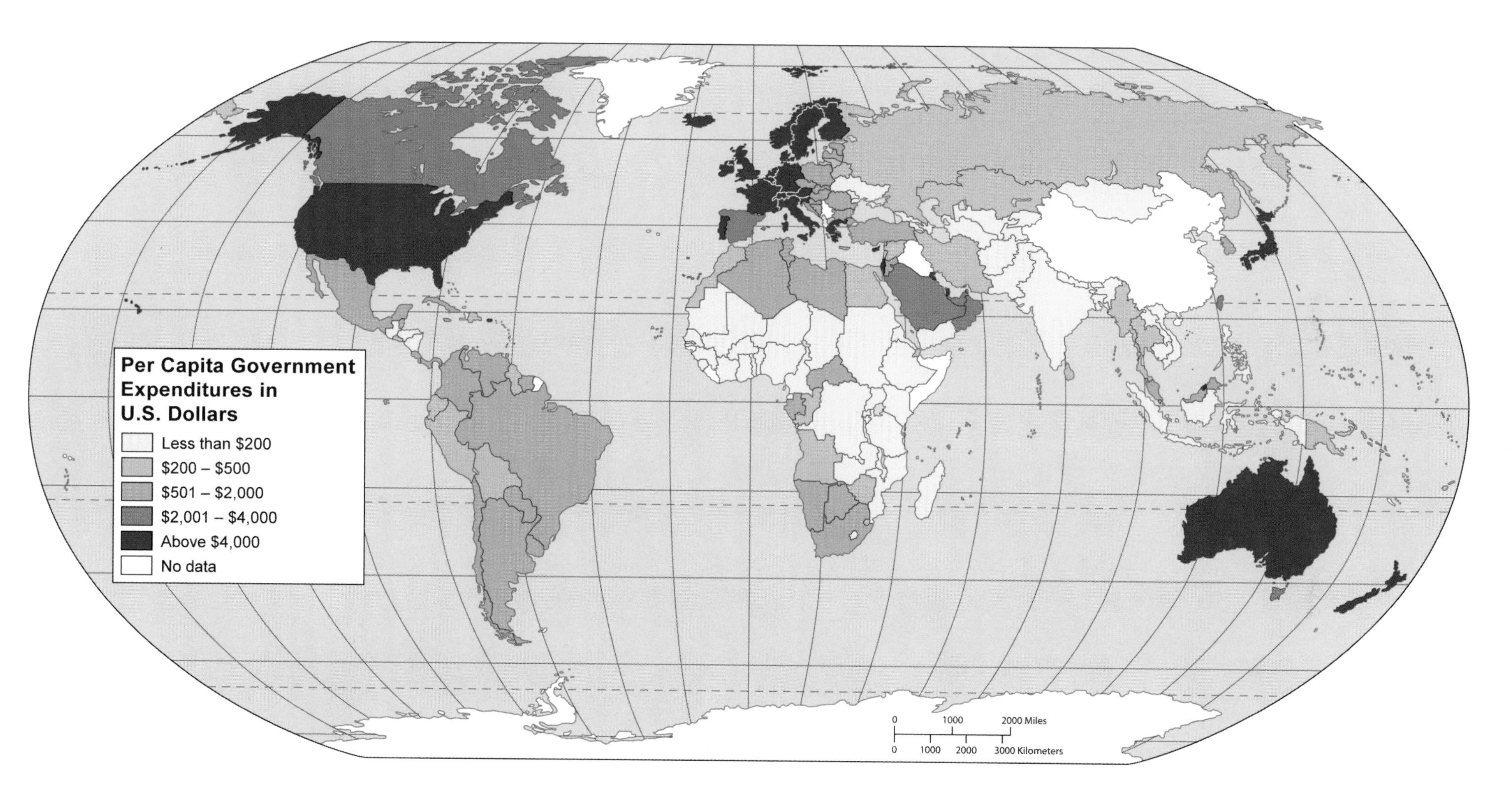

The amount of money that the central government of a country spends upon a variety of essential governmental functions is a measure of relative economic development, particularly when it is viewed on a per-person basis. These functions include such governmental responsibilities as agriculture, communications, culture, defense, education, fishing and hunting, health, housing, recreation, religion, social security, transportation, and welfare. Generally, the higher the level of economic development, the greater the per capita expenditures on these services. However, the data do mask some internal variations. For example, countries that spend 20 percent or more of their central government expenditures on defense will often show up in the more developed category when, in fact, all that the figures really show is that a disproportionate amount of the money available to the government is devoted to purchasing armaments and maintaining a large standing military force. Thus, the fact that Libya spends more than the average for Africa does not suggest that the average Libyan is much better off than the average Tanzanian. Nevertheless, this map—particularly when compared with Map 61, Energy Consumption Per Capita—does provide a reasonable approximation of economic development levels.

Map 58 Purchasing Power Parity

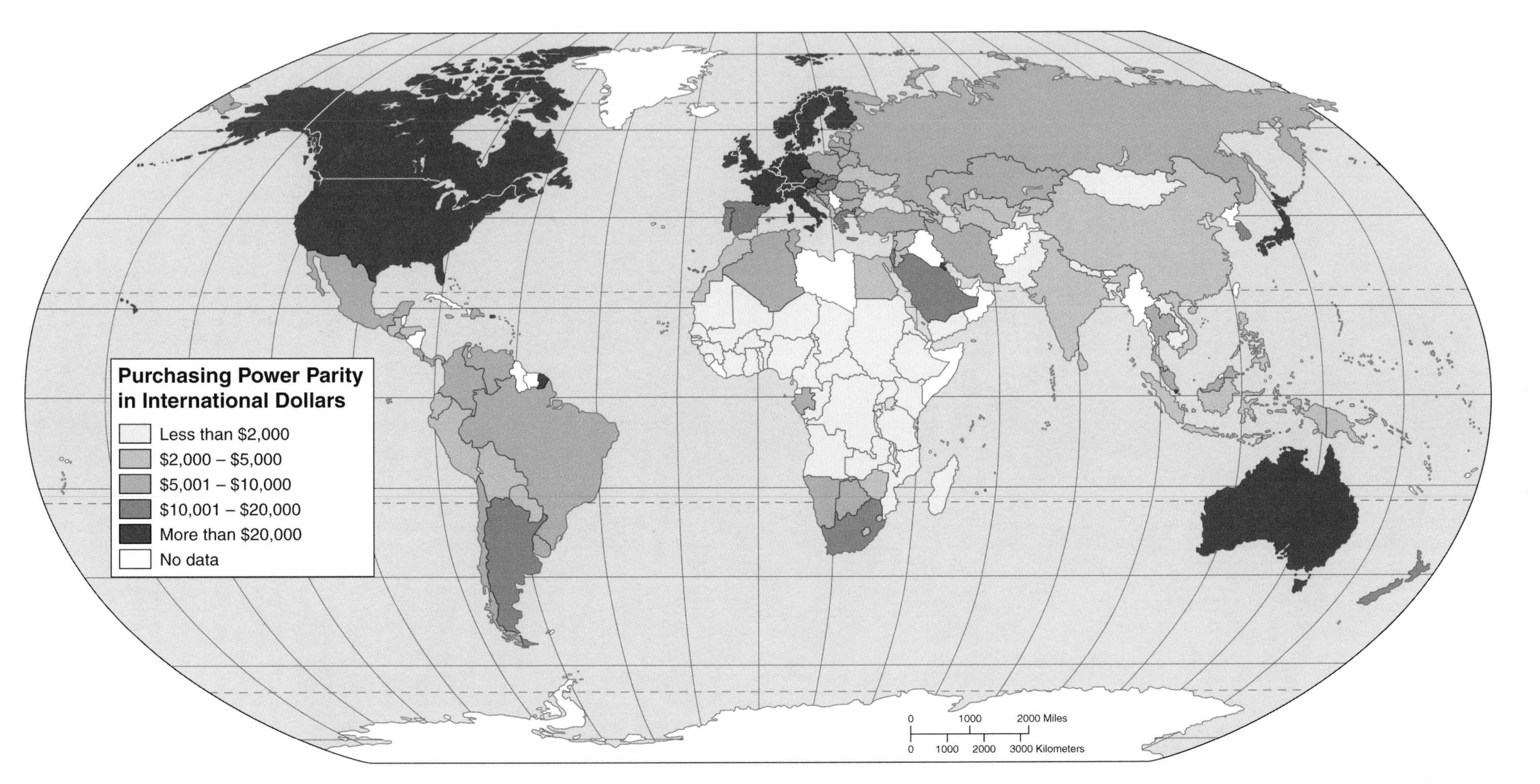

Of all the economic measures that separate the "haves" from the "have-nots," perhaps per capita Purchasing Power Parity (PPP) is the most meaningful. While per capita figures can mask significant uneven distributions within a country, they are generally useful for demonstrating important differences between countries. Per capita GNP and GDP (Gross Domestic Product) figures, and even per capita income, have the limitation of seldom reflecting the true purchasing power of a country's currency at home. In order to get around this limitation, international economists seeking to compare national currencies developed the PPP measure, which shows the level of goods and services that holders of a country's money can acquire locally. By converting all currencies to the "international dollar," the World Bank and other organizations using PPP can now show more truly comparative values, since the new currency value shows the number of units of a country's currency required to buy the same quantity of goods and services in the local market as one U.S. dollar would buy in an average country. The use of PPP currency values can alter the perceptions about a country's true comparative position in the world economy. More than per capita income figures, PPP provides a valid measurement of the ability of a country's population to provide for itself the things that people in the developed world take for granted: adequate food, shelter, clothing, education, and access to medical care. A glance at the map shows a clear-cut demarcation between temperate and tropical zones, with most of the countries with a PPP above $5,000 in the midlatitude zones and most of those with lower PPPs in the tropical and equatorial regions. Where exceptions to this pattern occur, they usually stem from a tremendous maldistribution of wealth among a country's population.

Map 59 Private Per Capita Consumption 2003

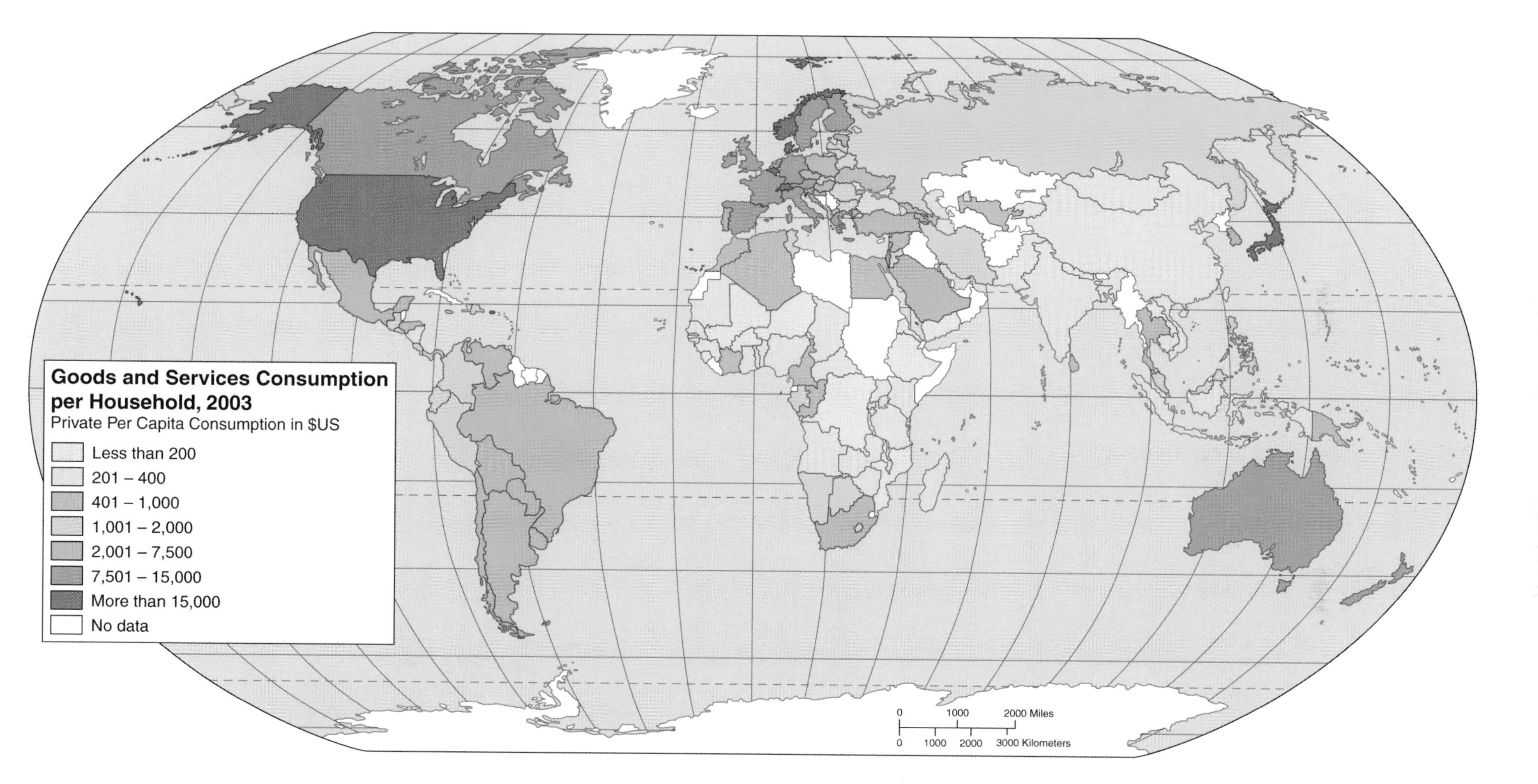

While we often examine production in analyzing global economic patterns, too seldom do we look at patterns of consumption—particularly on a per-person basis. When we do, the dramatic differences between the "rich" and "poor" (of which we are well aware) become even more dramatic. The consumption of key resources—minerals, agricultural products, harvests from the oceans, forestry products, water supplies—is among the most uneven distributions to be found among all the economic patterns of the world. Those who inhabit the wealthiest nations consume far more than their share of these key resources because they can afford to. And they produce far more than their share of the waste products, including atmospheric and oceanic pollution, of that consumption. Not only do they deplete their own resource bases but those of other countries as well. A trend that can be viewed as alarming by some and encouraging by others is that the number of "rich" or even "moderately well-off" countries is rising, and that number is rising among the world's most populous countries: India and China. What will happen to the global economy and the global environment (and we cannot separate the two) when India and China begin to consume and waste at the same levels as, say, the United States?

Map 60 Energy Production Per Capita

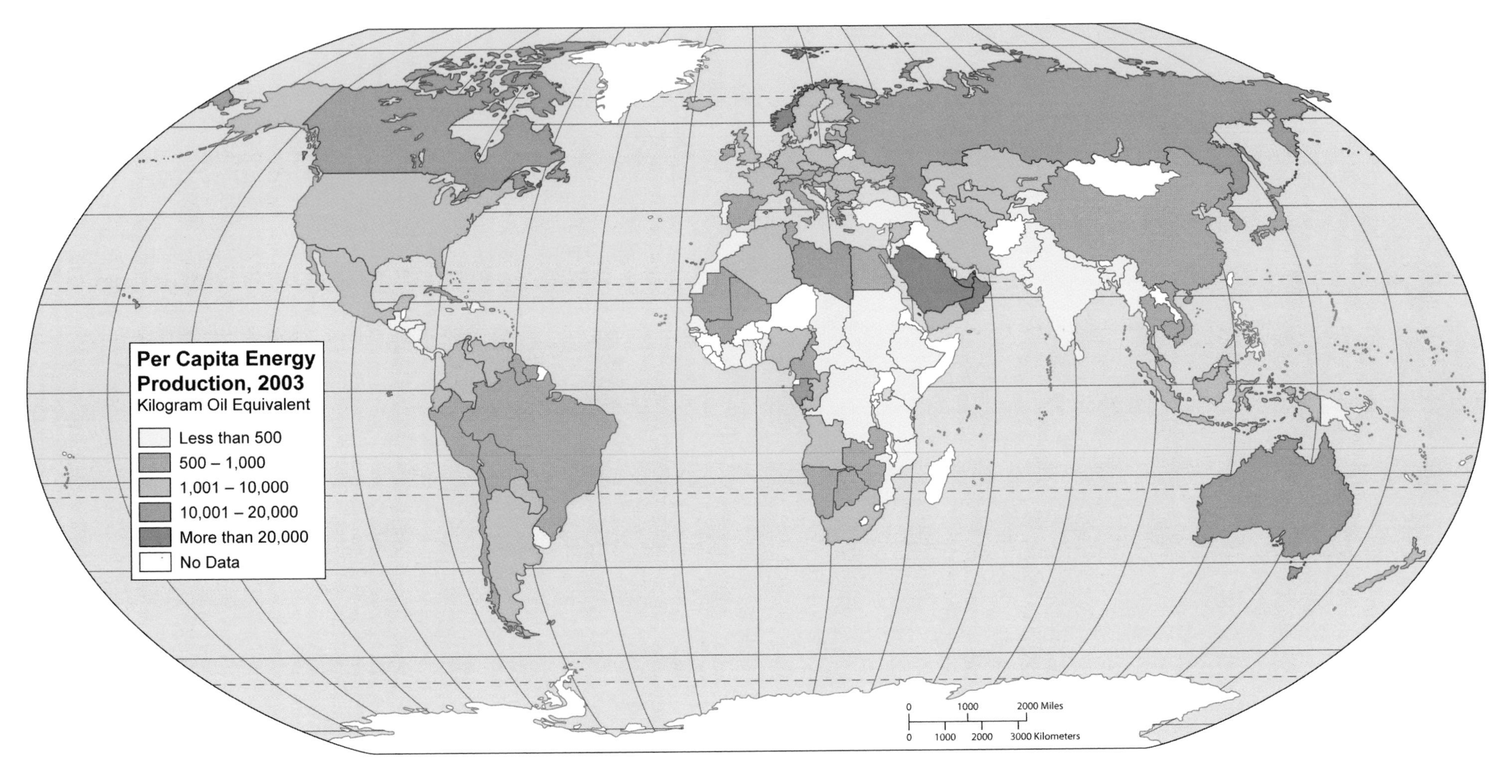

Energy production per capita is a measure of the availability of mechanical energy to assist people in their work. This map shows the amount of all kinds of energy—solid fuel (primarily coal), liquid fuel (primarily petroleum), natural gas, geothermal, wind, solar, hydroelectric, nuclear, waste recycling, and indigenous heat pumps—produced per person in each country. With some exceptions, wealthier countries produce more energy per capita than poor ones. Countries such as Japan and many European states rank among the world's wealthiest, but are energy-poor and produce relatively little of their own energy. They have the ability, however, to pay for imports. On the other hand, countries such as those of the Persian Gulf or the oil-producing states of Central and South America may rank relatively low on the scale of economic development but rank high as producers of energy. In many poor countries, especially in Central and South America, Africa, South Asia, and East Asia, large proportions of energy come from traditional fuels such as firewood and animal dung. Indeed for many in the developing world, the real energy crisis is a shortage of wood for cooking and heating.

Map 61 Energy Consumption Per Capita

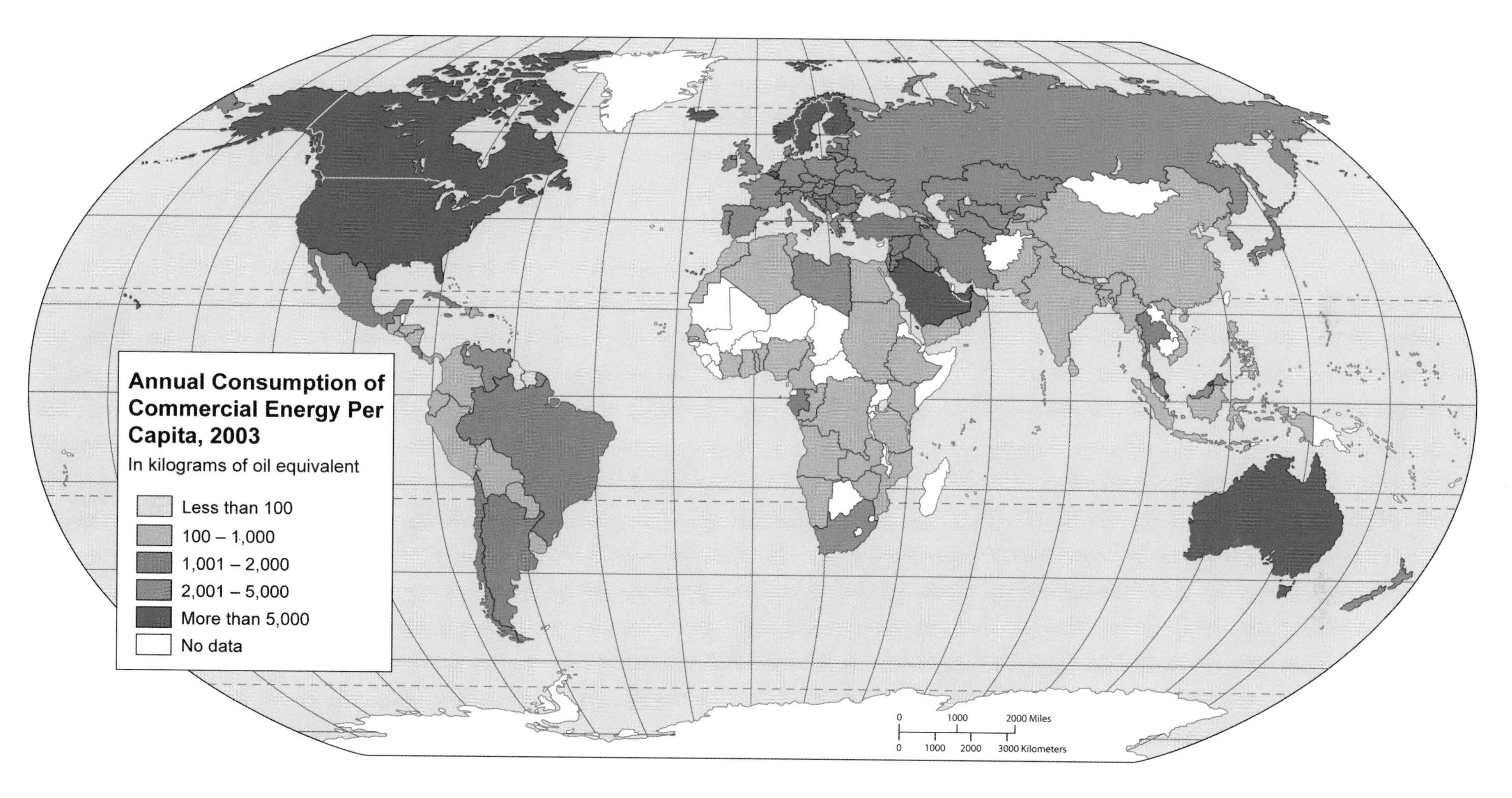

Of all the quantitative measures of economic well-being, energy consumption per capita may be the most expressive. All of the countries defined by the World Bank as having high incomes consume at least 100 gigajoules of commercial energy (the equivalent of about 3.5 metric tons of coal) per person per year, with some, such as the United States and Canada, having consumption rates in the 300 gigajoule range (the equivalent of more than 10 metric tons of coal per person per year). With the exception of the oil-rich Persian Gulf states, where consumption figures include the costly "burning off" of excess energy in the form of natural gas flares at wellheads, most of the highest-consuming countries are in the Northern Hemisphere, concentrated in North America and Western Europe. At the other end of the scale are low-income countries, whose consumption rates are often less than 1 percent of those of the United States and other high consumers. These figures do not, of course, include the consumption of noncommercial energy—the traditional fuels of firewood, animal dung, and other organic matter—widely used in the less developed parts of the world.

Map 62 Energy Dependency

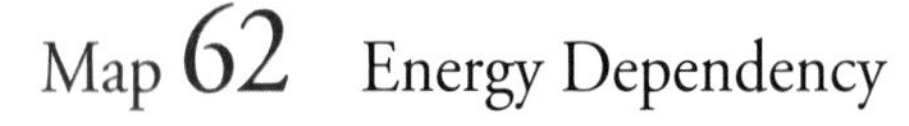

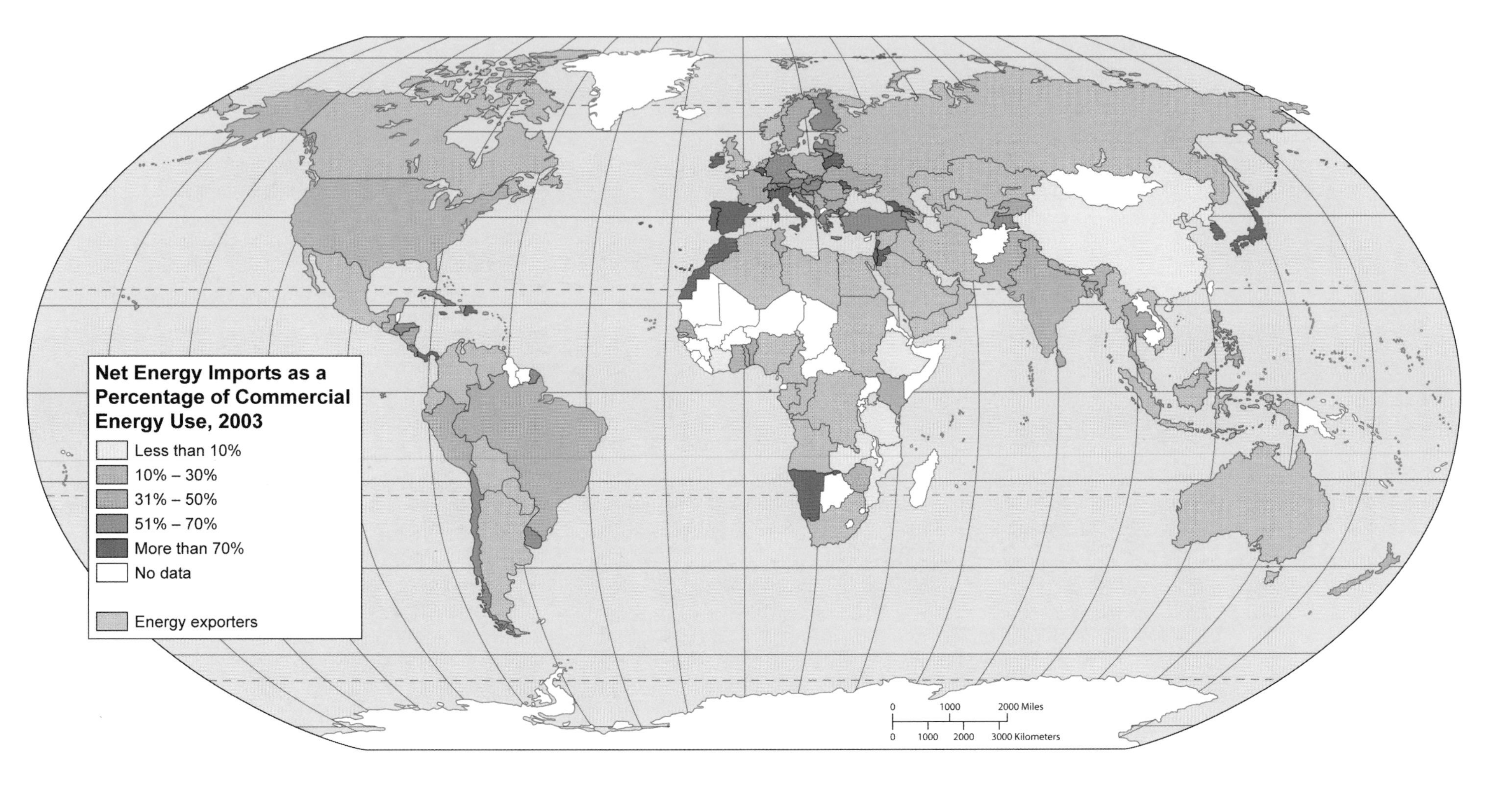

The patterns on the map show dependence on commercial energy before transformation to other end-use fuels such as electricity or refined petroleum products; energy from traditional sources such as fuelwood or dried animal dung is not included. Energy dependency is the difference between domestic consumption and domestic production of commercial energy and is most often expressed as a net energy import or export. A few of the world's countries are net exporters of energy: most are importers. The growth in global commercial energy use over the last decade indicates growth in the modern sectors of the economy—industry, transportation, and urbanization—particularly in the lesser developed countries. Still, the primary consumers of energy—and those having the greatest dependence on foreign sources of energy—are the more highly developed countries of Europe, North America, and Japan.

Map 63 Flows of Oil

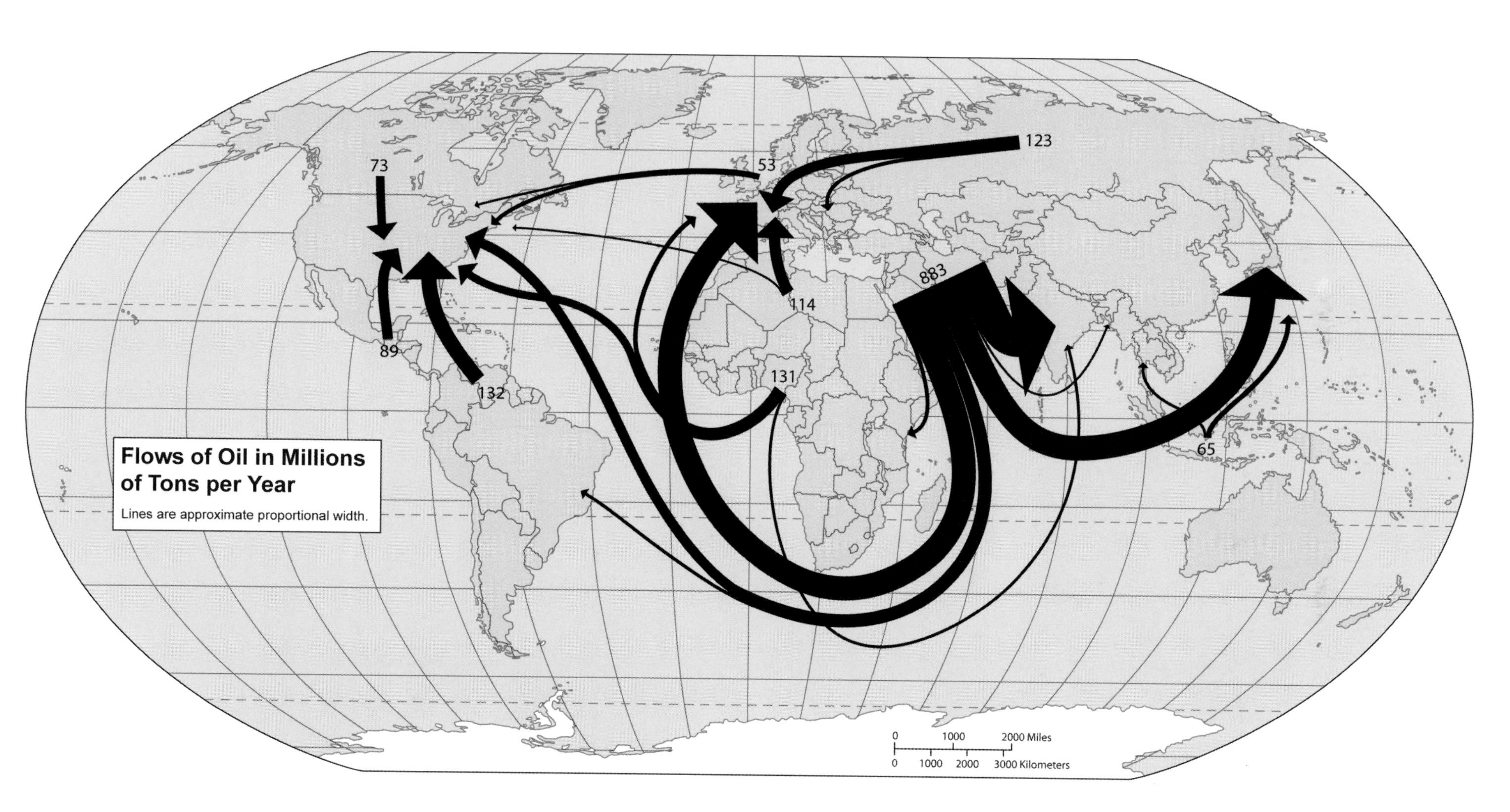

The pattern of oil movements from producing region to consuming region is one of the dominant facts of contemporary international maritime trade. Supertankers carry a million tons of crude oil and charge rates in excess of $0.10 per ton per mile, making the transportation of oil not only a necessity for the world's energy-hungry countries, but also an enormously profitable proposition. One of the major negatives of these massive oil flows is the damage done to the oceanic ecosystems—not just from the well-publicized and dramatic events like the wrecking of the *Exxon Valdez* but from the incalculable amounts of oil from leakage, scrubbings, purgings, and so on, which are a part of the oil transport technology. It is clear from the map that the primary recipients of these oil flows are the world's most highly developed economies.

Part V

Global Patterns of Environmental Disturbance

Map 64 Global Air Pollution: Sources and Wind Currents
Map 65 The Acid Deposition Problem: Air, Water, Soil
Map 66 Major Polluters and Common Pollutants
Map 67 Global Carbon Dioxide Emissions
Map 68 Potential Global Temperature Change
Map 69 Water Resources: Availability of Renewable Water Per Capita
Map 70 Water Resources: Annual Withdrawal Per Capita
Map 71 Pollution of the Oceans
Map 72 Food Supply from Marine and Freshwater Systems
Map 73 Cropland Per Capita: Changes 1996-2003
Map 74 World Pastureland 2005
Map 75 Fertilizer Use 2005
Map 76 Annual Deforestation Rates 1997-2003
Map 77 World Timber Production 2000
Map 78 The Loss of Biodiversity: Globally Threatened Animal Species
Map 79 The Loss of Biodiversity: Globally Threatened Plant Species
Map 80 Hotspots of Biodiversity
Map 81 The Risks of Desertification
Map 82 Soil Degradation
Map 83 The Degree of Human Disturbance

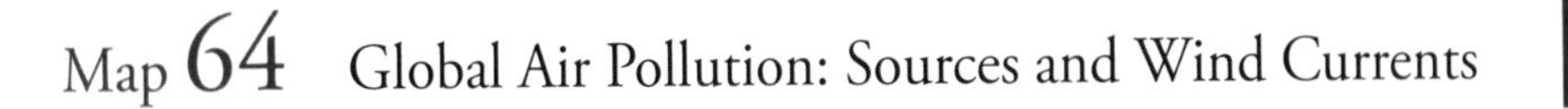

Map 64 Global Air Pollution: Sources and Wind Currents

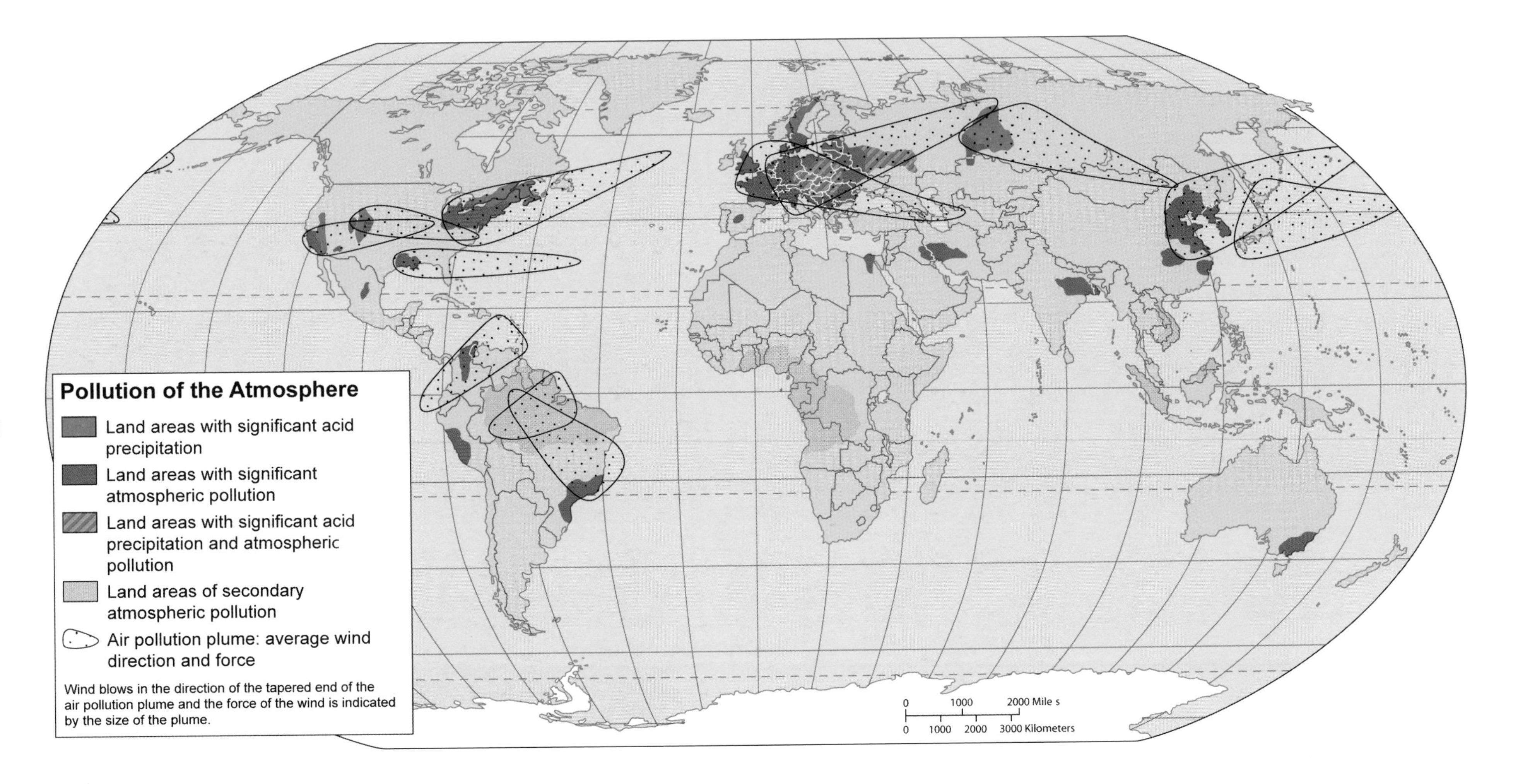

Almost all processes of physical geography begin and end with the flows of energy and matter among land, sea, and air. Because of the primacy of the atmosphere in this exchange system, air pollution is potentially one of the most dangerous human modifications in environmental systems. Pollutants such as various oxides of nitrogen or sulfur cause the development of acid precipitation, which damages soil, vegetation, and wildlife and fish. Air pollution in the form of smog is often dangerous for human health. And most atmospheric scientists believe that the efficiency of the atmosphere in retaining heat—the so-called greenhouse effect—is being enhanced by increased carbon dioxide, methane, and other gases produced by agricultural and industrial activities. The result, they fear, will be a period of global warming that will dramatically alter climates in all parts of the world.

Map 65 The Acid Deposition Problem: Air, Water, Soil

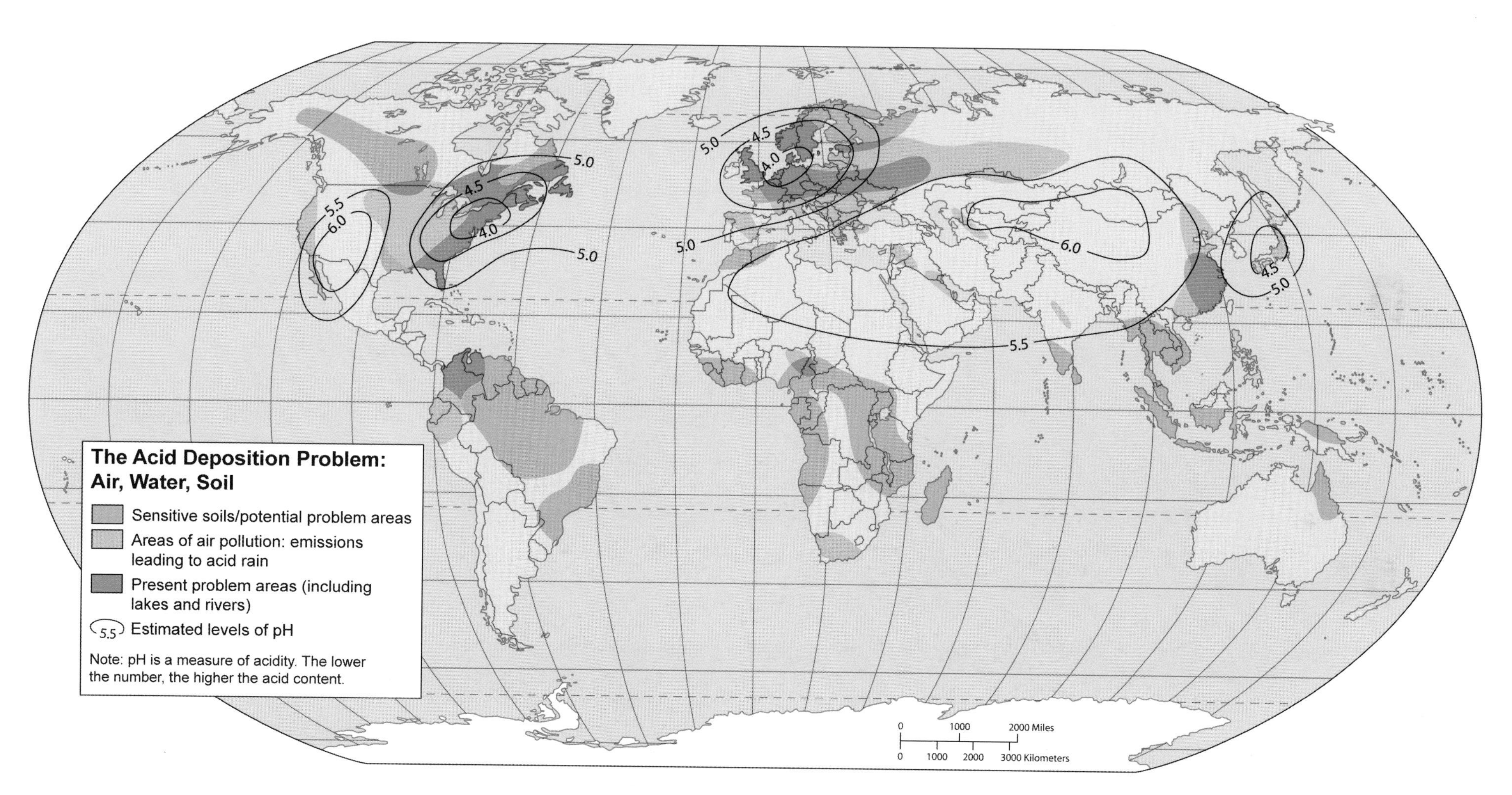

The term "acid precipitation" refers to increasing levels of acidity in snowfall and rainfall caused by atmospheric pollution. Oxides of nitrogen and sulfur resulting from incomplete combustion of fossil fuels (coal, oil, and natural gas) combine with water vapor in the atmosphere to produce weak acids that then "precipitate" or fall along with water or ice crystals. Some atmospheric acids formed by this process are known as "dry-acid" precipitates and they too will fall to earth, although not necessarily along with rain or snow. In some areas of the world, the increased acidity of streams and lakes stemming from high levels of acid precipitation or dry acid fallout has damaged or destroyed aquatic life. Acid precipitation and dry acid fallout also harm soil systems and vegetation, producing a characteristic burned appearance in forests that lends the same quality to landscapes that forest fires would. The region most dramatically impacted by acid precipitation is Central Europe where decades of destructive environmental practices, including the burning of high sulfur coal for commercial, industrial, and residential purposes, has produced the destruction of hundreds of thousands of acres of woodlands—a phenomenon described by the German foresters who began their study of the area following the lifting of the Iron Curtain as "Waldsterben": Forest Death.

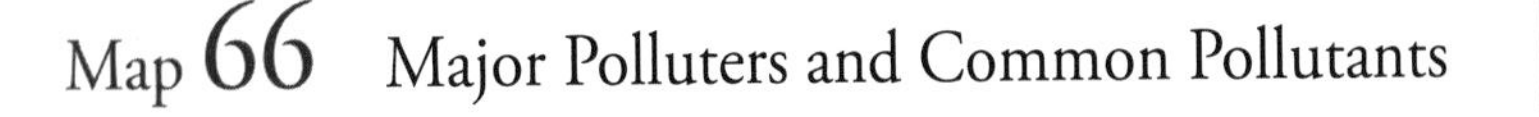

Map 66 Major Polluters and Common Pollutants

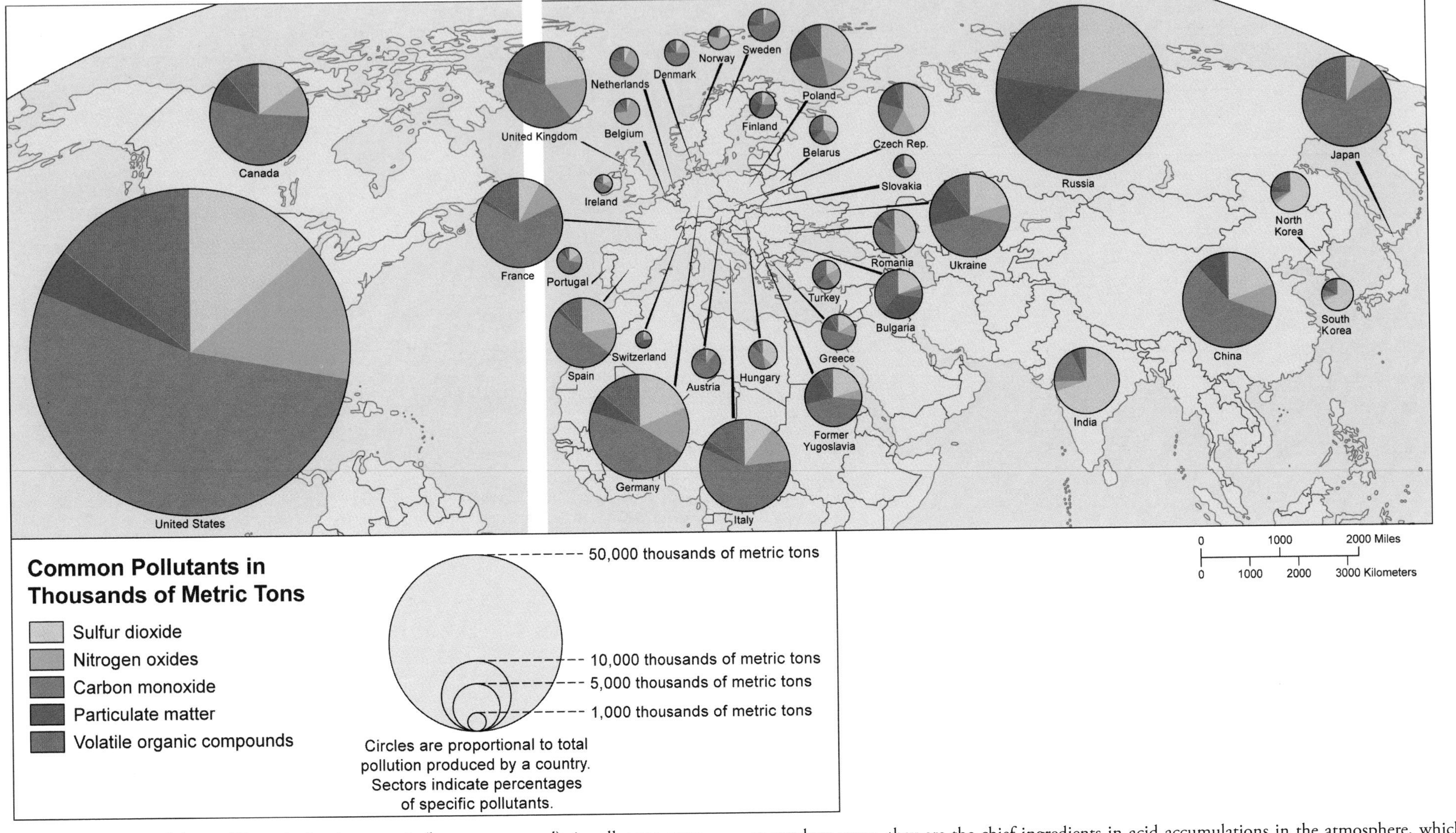

More than 90 percent of the world's total of anthropogenic (human-generated) air pollutants come from the heavily populated industrial regions of North America, Europe, South Asia (primarily in India), and East Asia (mainly in China, Japan, and the two Koreas). This map shows the origins of the five most common pollutants: sulfur dioxide, nitrogen oxide, carbon monoxide, particulate matter, and volatile organic compounds. These substances are produced both by industry and by the combustion of fossil fuels that generate electricity and power trains, planes, automobiles, buses, and trucks. In addition to combining with other components of the atmosphere and with one another to produce smog, they are the chief ingredients in acid accumulations in the atmosphere, which ultimately result in acid deposition, either as acid precipitation or dry acid fallout. Like other forms of pollutants, these air pollutants do not recognize political boundaries, and regions downwind of major polluters receive large quantities of pollutants from areas over which they often have no control.

Map 67 Global Carbon Dioxide Emissions

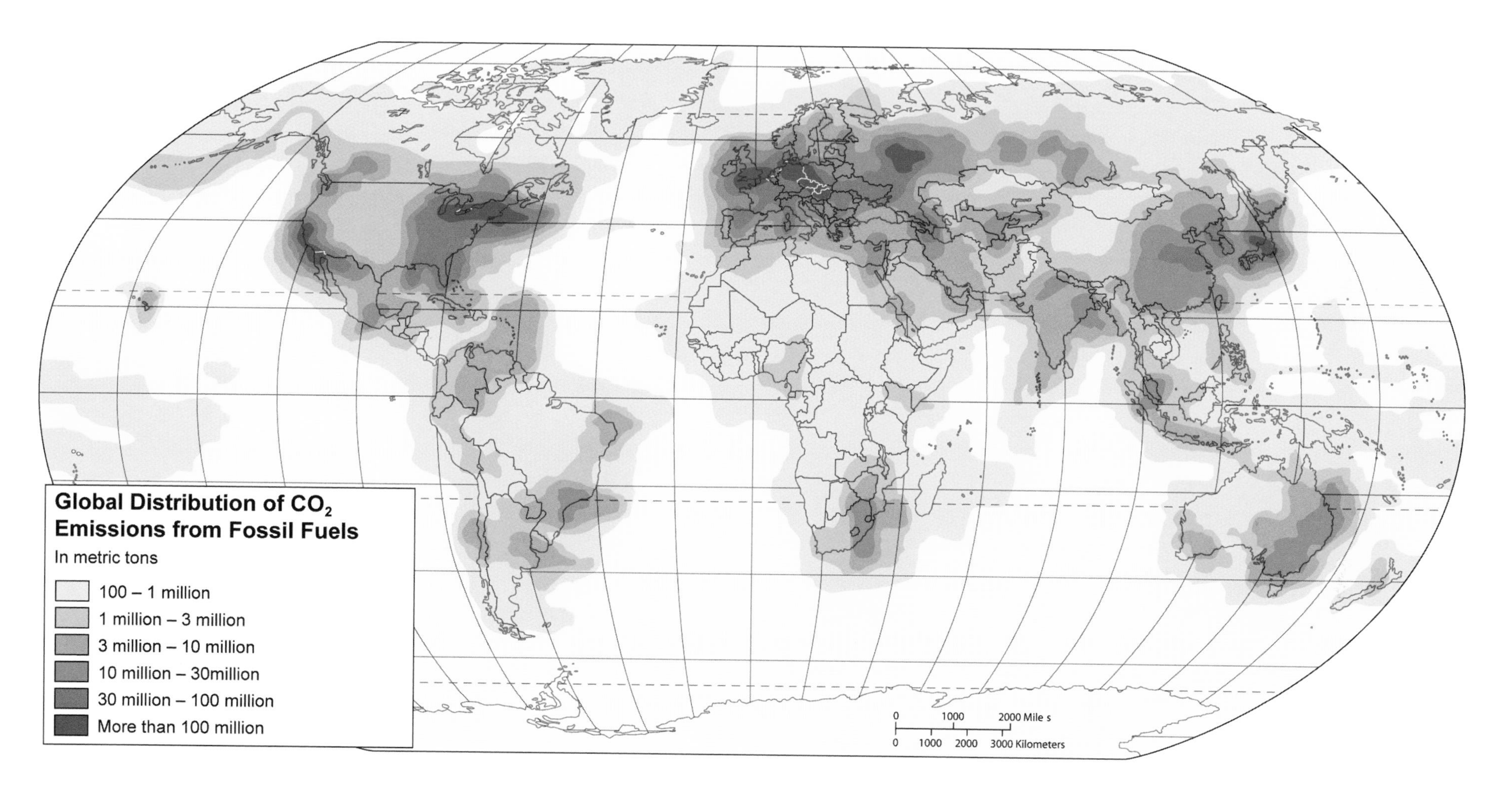

One of the most important components of the atmosphere is the gas carbon dioxide (CO_2), the byproduct of animal respiration, decomposition, and combustion. During the past 200 years, atmospheric CO_2 has risen dramatically, largely as the result of the tremendous increase in fossil fuel combustion brought on by the industrialization of the world's economy and the burning and clearing of forests by the expansion of farming. While CO_2 by itself is relatively harmless, it is an important "greenhouse gas." The gases in the atmosphere act like the panes of glass in a greenhouse roof, allowing light in but preventing heat from escaping. The greenhouse capacity of the atmosphere is crucial for organic life and is a purely natural component of the global energy cycle. But too much CO_2 and other greenhouse gases such as methane could cause the earth's atmosphere to warm up too much, producing the global warming that atmospheric scientists are concerned about. Researchers estimate that if greenhouse gases such as CO_2 continue to increase at their present rates, the earth's mean temperature could rise between 1.5 and 4.5 degrees Celsius by the middle of the next century. Such a rise in global temperatures would produce massive alterations in the world's climate patterns.

Map 68 Potential Global Temperature Change

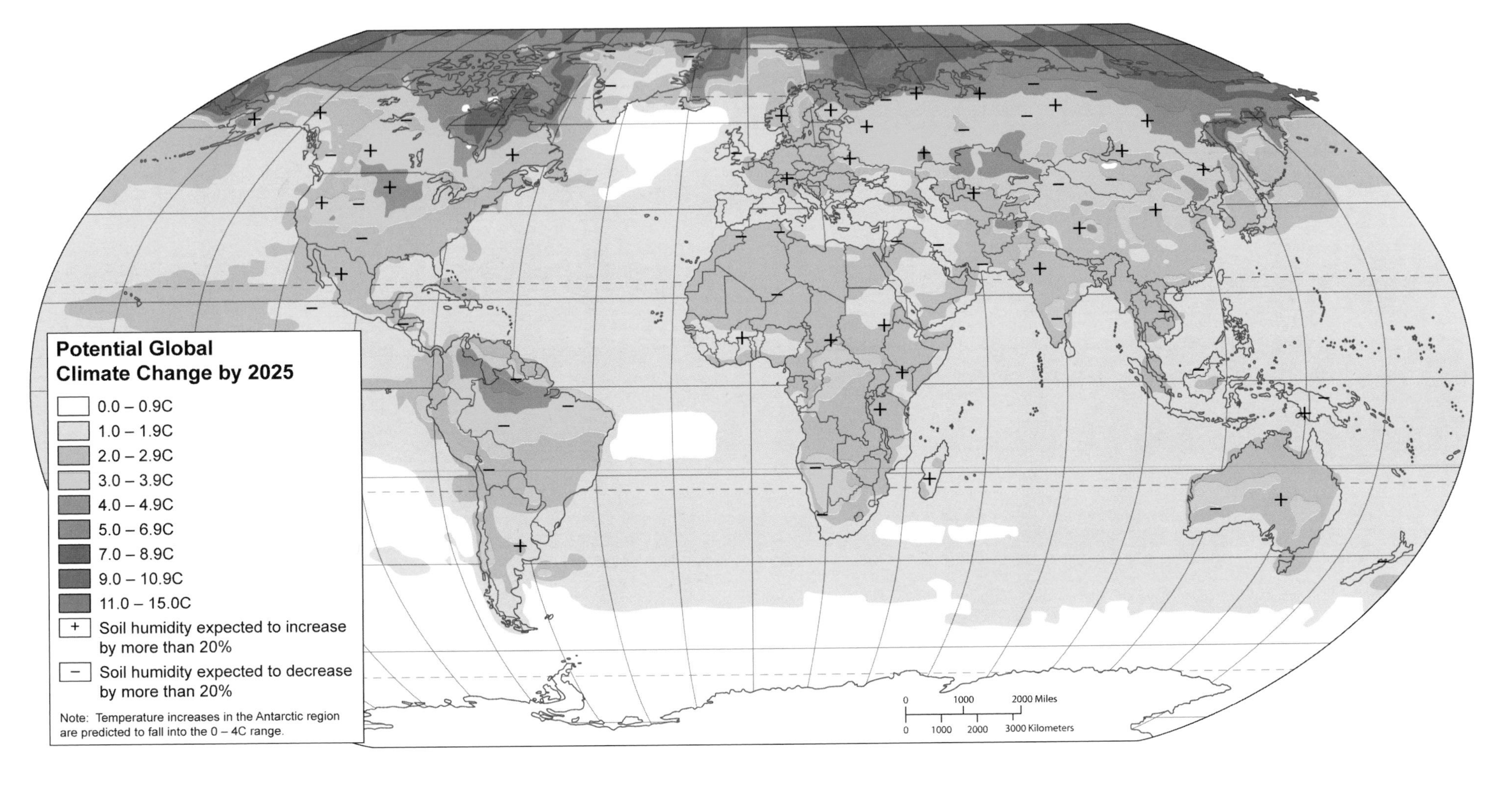

According to atmospheric scientists, one of the major problems of the twenty-first century will be "global warming," produced as the atmosphere's natural ability to trap and retain heat is enhanced by increased percentages of carbon dioxide, methane, chlorinated fluorocarbons or "CFCs," and other "greenhouse gases" in the earth's atmosphere. Computer models based on atmospheric percentages of carbon dioxide resulting from present use of fossil fuels show that warming is not just a possibility but a probability. Increased temperatures would cause precipitation patterns to alter significantly as well and would produce a number of other harmful effects, including a rise in the level of the world's oceans that could flood most coastal cities. International conferences on the topic of the enhanced greenhouse effect have resulted in several international agreements to reduce the emission of carbon dioxide or to maintain it at present levels. Unfortunately, the solution is not that simple since reduction of carbon dioxide emissions is, in the short run, expensive—particularly as long as the world's energy systems continue to be based on fossil fuels. Chief among the countries that could be hit by serious international mandates to reduce emissions are those highest on the development scale who use the highest levels of fossil fuels and, therefore, produce the highest emissions, and those on the lowest end of the development scale whose efforts to industrialize could be severely impeded by the more expensive energy systems that would replace fossil fuels.

Map 69 Water Resources: Availability of Renewable Water Per Capita

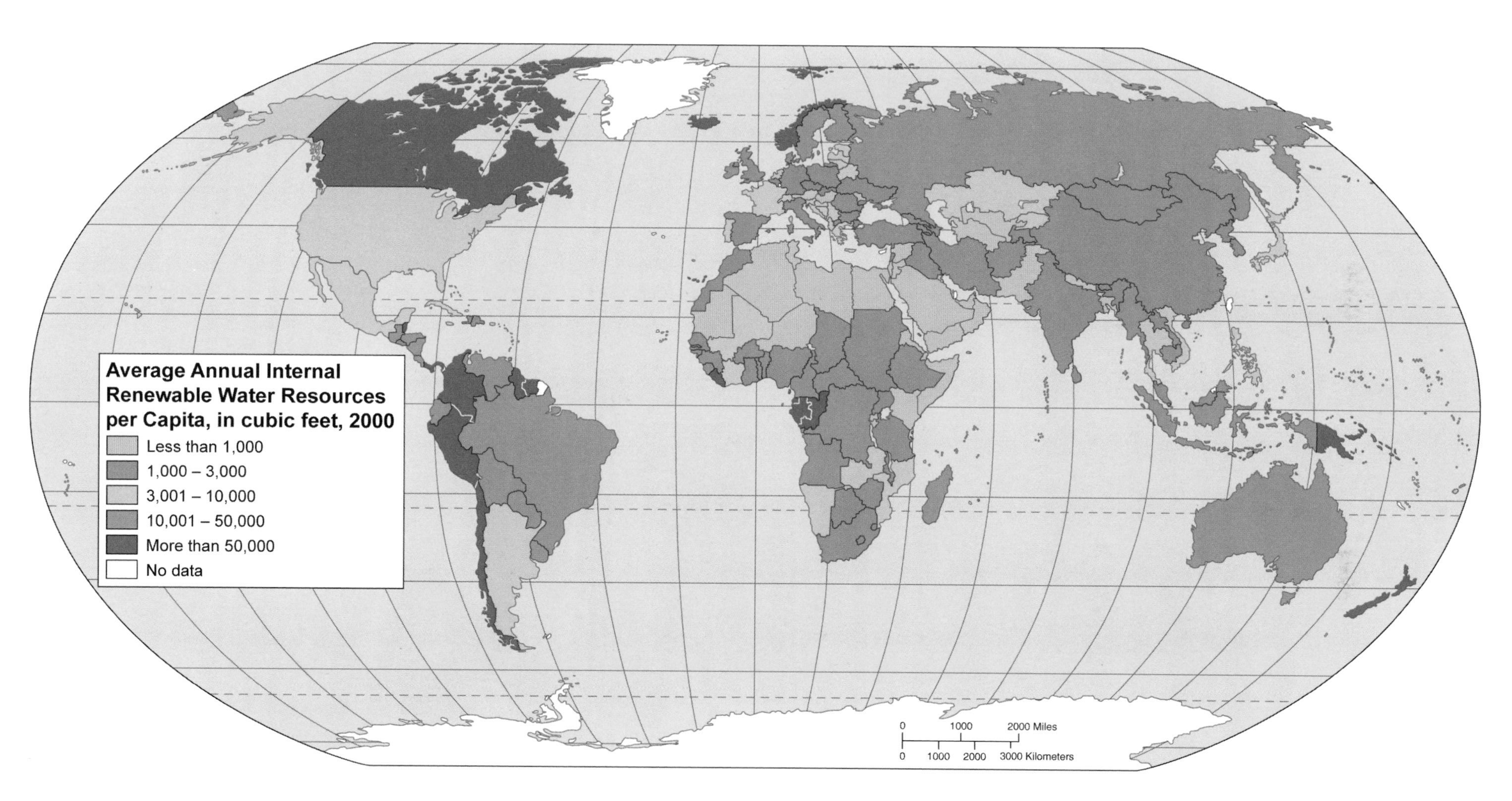

Renewable water resources are usually defined as the total water available from streams and rivers (including flows from other countries), ponds and lakes, and groundwater storage or aquifers. Not included in the total of renewable water would be water that comes from such nonrenewable sources as desalinization plants or melted icebergs. While the concept of renewable or flow resources is a traditional one in resource management, in fact, few resources, including water, are truly renewable when their use is excessive. The water resources shown here are indications of that principle. A country like the United States possesses truly enormous quantities of water. But the United States also uses enormous quantities of water. The result is that, largely because of excessive use, the availability of renewable water is much less than in many other parts of the world where the total supply of water is significantly less.

Map 70 Water Resources: Annual Withdrawal Per Capita

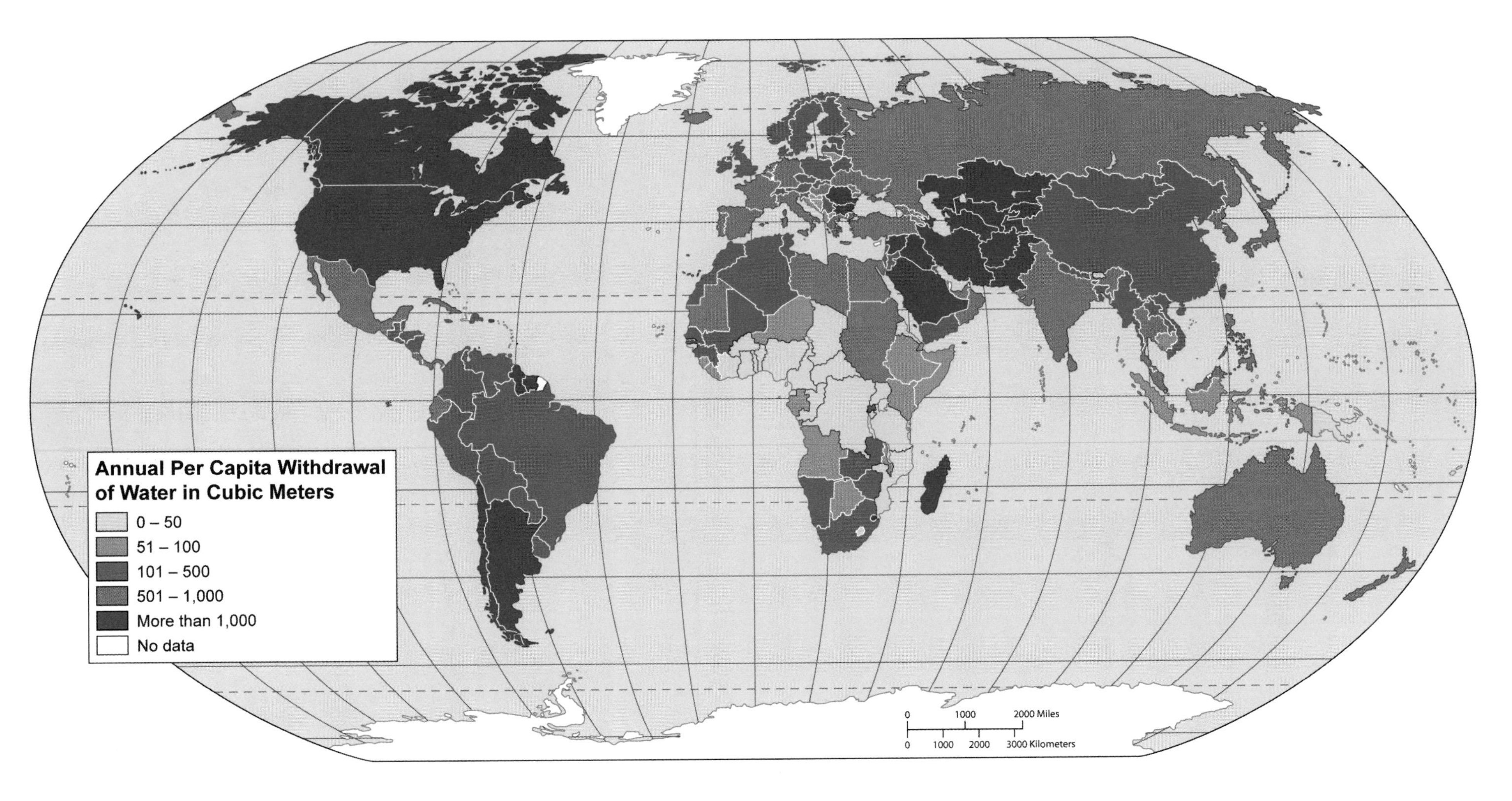

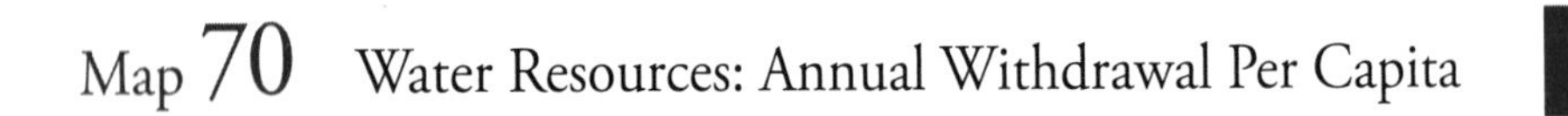

Water resources must be viewed like a bank account in which deposits and withdrawals are made. As long as the deposits are greater than the withdrawals, a positive balance remains. But when the withdrawals begin to exceed the deposits, sooner or later (depending on the relative sizes of the deposits and withdrawals) the account becomes overdrawn. For many of the world's countries, annual availability of water is insufficient to cover the demand. In these countries, reserves stored in groundwater are being tapped, resulting in depletion of the water supply (think of this as shifting money from a savings account to a checking account). The water supply can maintain its status as a renewable resource only if deposits continue to be greater than withdrawals, and that seldom happens. In general, countries with high levels of economic development and countries that rely on irrigation agriculture are the most spendthrift when it comes to their water supplies.

Map 71 Pollution of the Oceans

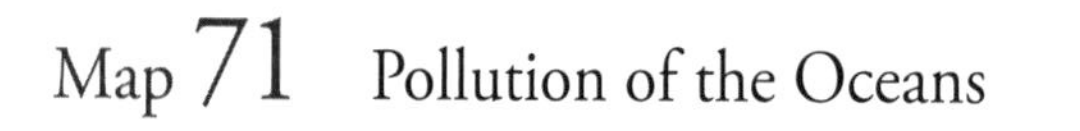

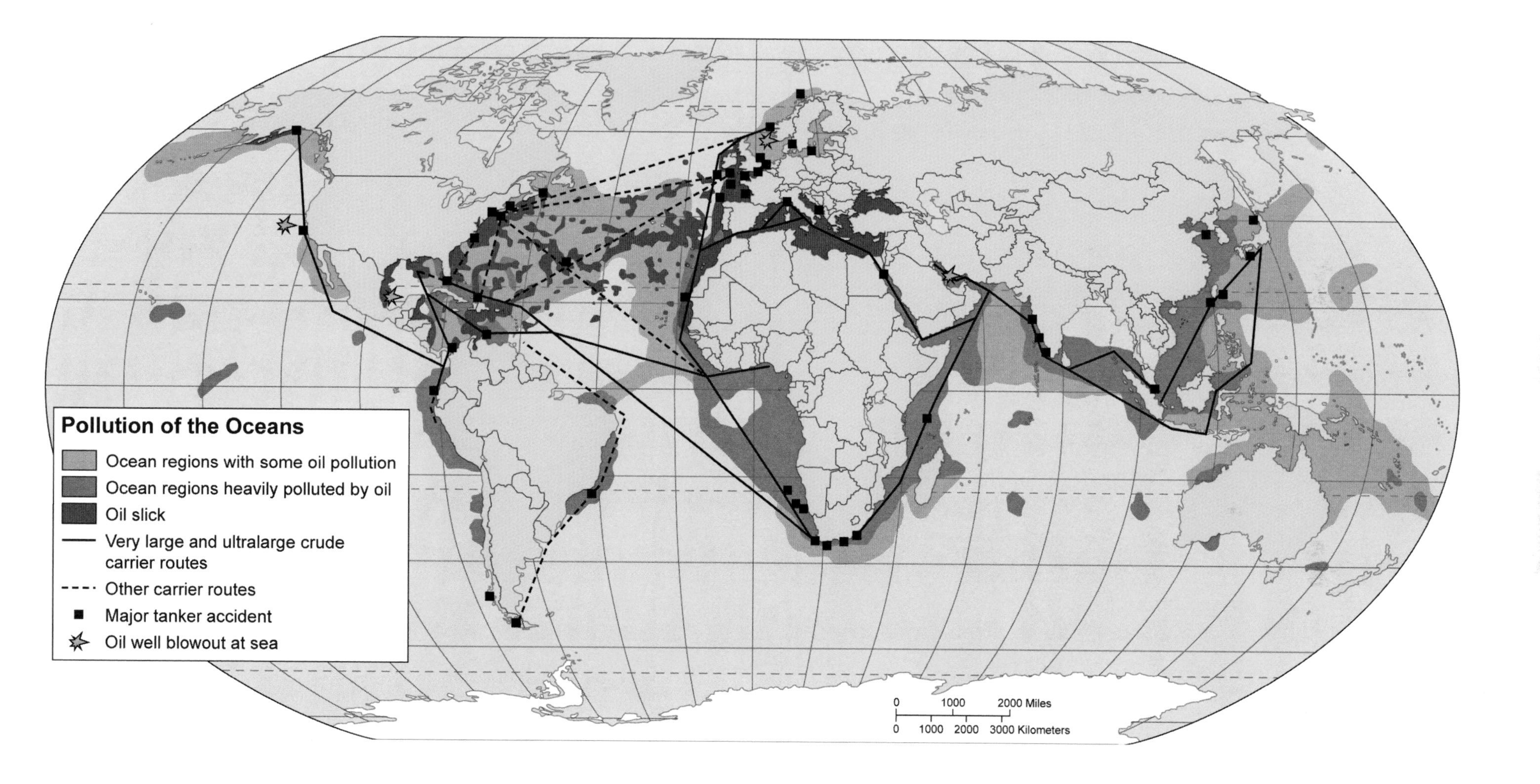

The pollution of the world's oceans has long been a matter of concern to physical geographers, oceanographers, and other environmental scientists. The great circulation systems of the ocean are one of the controlling factors of the earth's natural environment, and modifications to those systems have unknown consequences. This map is based on what we can measure: (1) areas of oceans where oil pollution has been proven to have inflicted significant damage to ocean ecosystems and life forms (including phytoplankton, the oceans' primary food producers, equivalent to land-based vegetation) and (2) areas of oceans where unusually high concentrations of hydrocarbons from oil spills may have inflicted some damage to the oceans' biota. A glance at the map shows that there are few areas of the world's oceans where some form of pollution is not a part of the environmental system. What the map does not show in detail, because of the scale, are the dramatic consequences of large individual pollution events: the wreck of the *Exxon Valdez* and the polluting of Prince William Sound, or the environmental devastation produced by the 1991 Gulf War in the Persian Gulf.

Map 72 Food Supply From Marine and Freshwater Systems

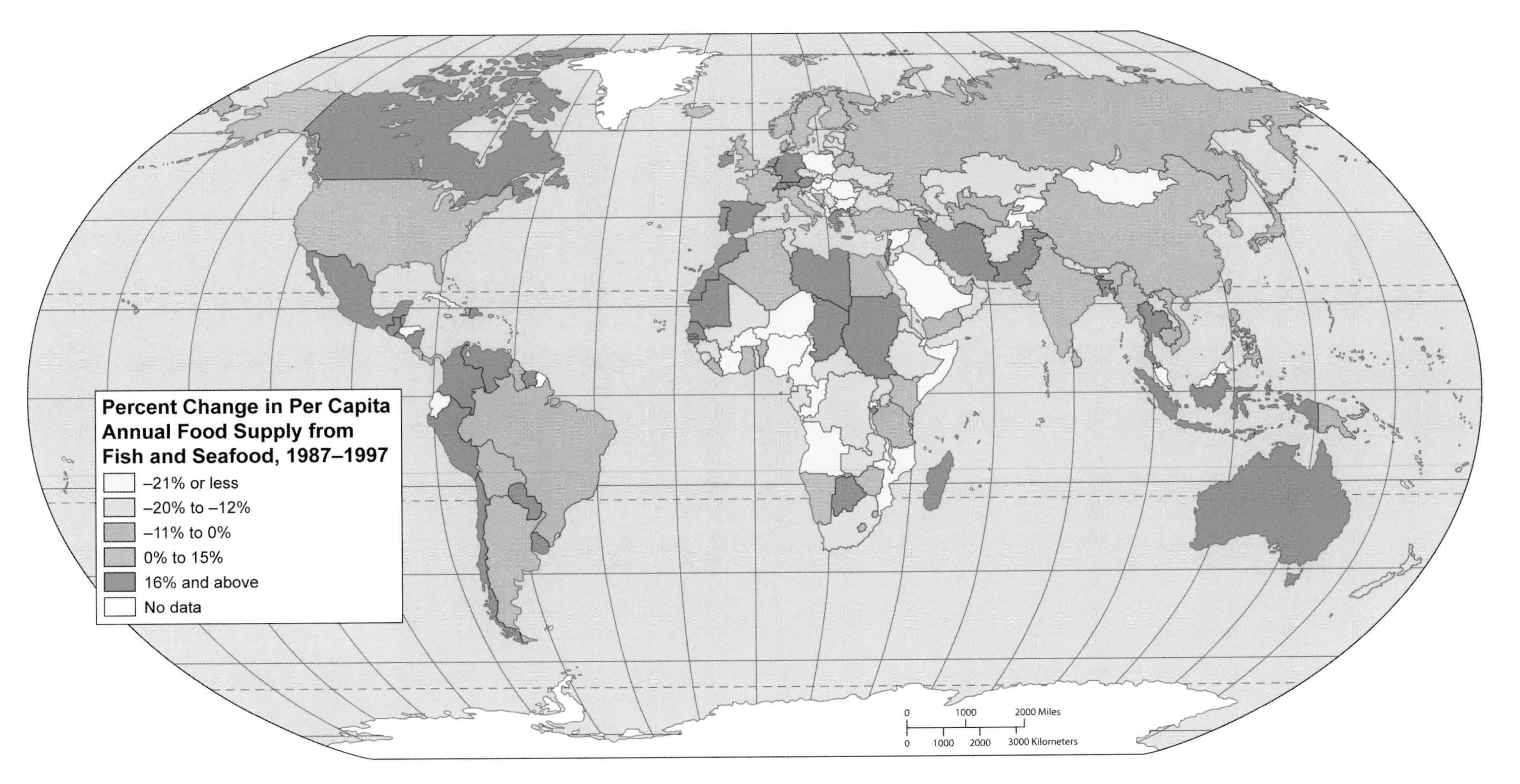

Not that many years ago, food supply experts were confidently predicting that the "starving millions" of the world of the future could be fed from the unending bounty of the world's oceans. While the annual catch from the sea helped to keep hunger at bay for a time, by the late 1980s it had become apparent that without serious human intervention in the form of aquaculture, the supply of fish would not be sufficient to offset the population/food imbalance that was beginning to affect so many of the world's regions. The development of factory-fishing with advanced equipment to locate fish and process them before they went to market increased the supply of food from the ocean, but in that increase was sown the seeds of future problems. The factory-fishing system, efficient in terms of economics, was costly in terms of fish populations. In some well-fished areas, the stock of fish that was viewed as near infinite just a few decades ago has dwindled nearly to the point of disappearance. This map shows both increases and decreases in the amount of individual countries' food supplies from the ocean. The increases are often the result of more technologically advanced fishing operations. The decreases are usually the result of the same thing: increased technology has brought increased harvests, which has reduced the supply of fish and shellfish and that, in turn, has increased prices. Most of the countries that have experienced sharp decreases in their supply of food from the world's oceans are simply no longer able to pay for an increasingly scarce commodity.

Map 73 Cropland Per Capita: Changes 1996-2003

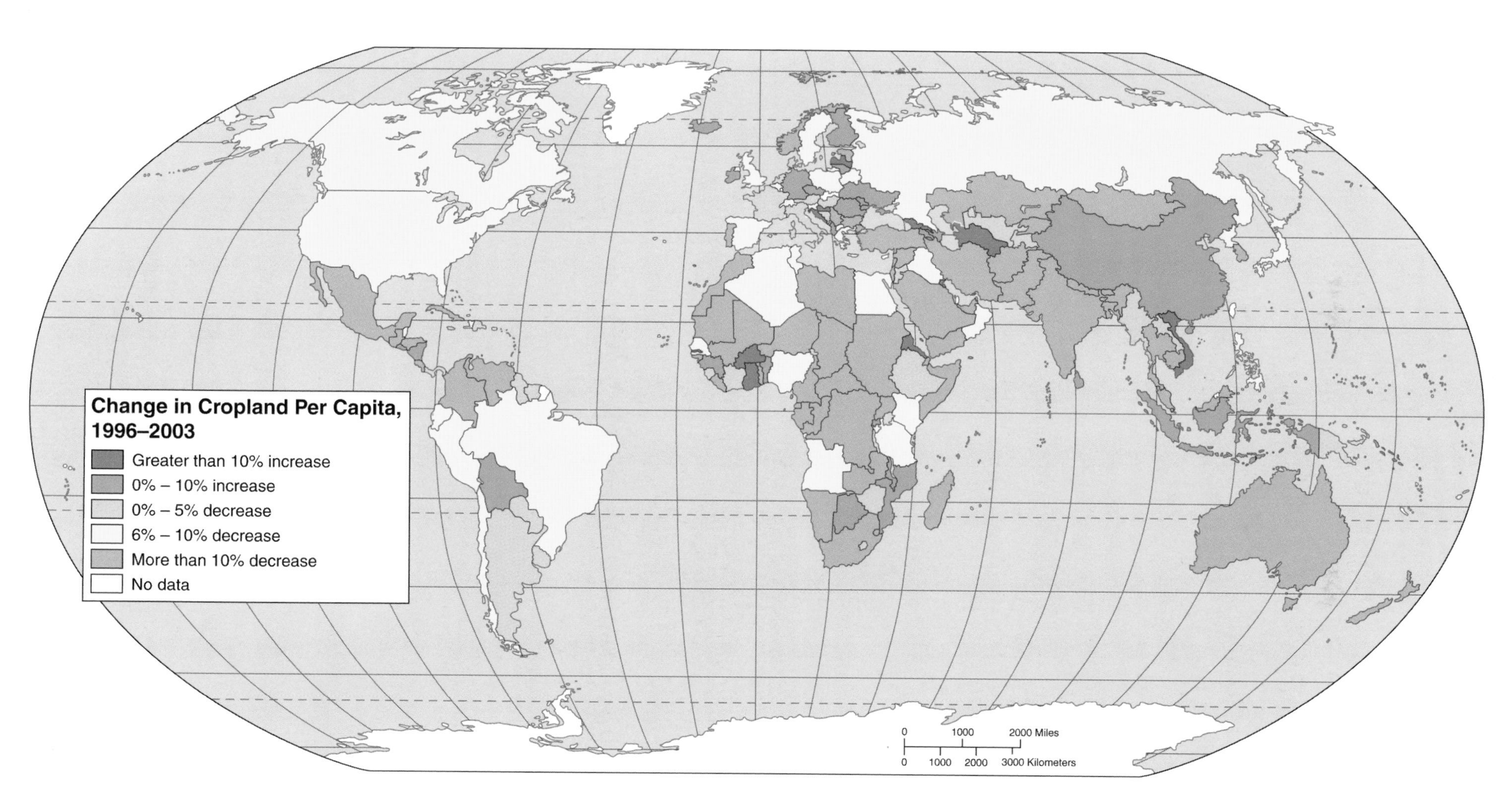

As population has increased rapidly throughout the world, the area of cultivated land has increased at the same time; in fact, the amount of farmland per person has gone up slightly. Unfortunately, the figures that show this also tell us that since most of the best (or even good) agricultural land in 1985 was already under cultivation, most of the agricultural area added since the early 1980s involves land that would have been viewed as marginal by the fathers and grandfathers of present farmers—marginal in that it was too dry, too wet, too steep to cultivate, too far from a market, and so on. The continued expansion of agricultural area is one reason that serious famine and starvation have struck only a few regions of the globe. But land, more than any other resource we deal with, is finite, and the expansion cannot continue indefinitely. Future gains in agricultural production are most probably going to come through more intensive use of existing cropland, heavier applications of fertilizers and other agricultural chemicals, and genetically engineered crops requiring heavier applications of energy and water, than from an increase in the amount of the world's cropland.

Map 74 World Pastureland 2005

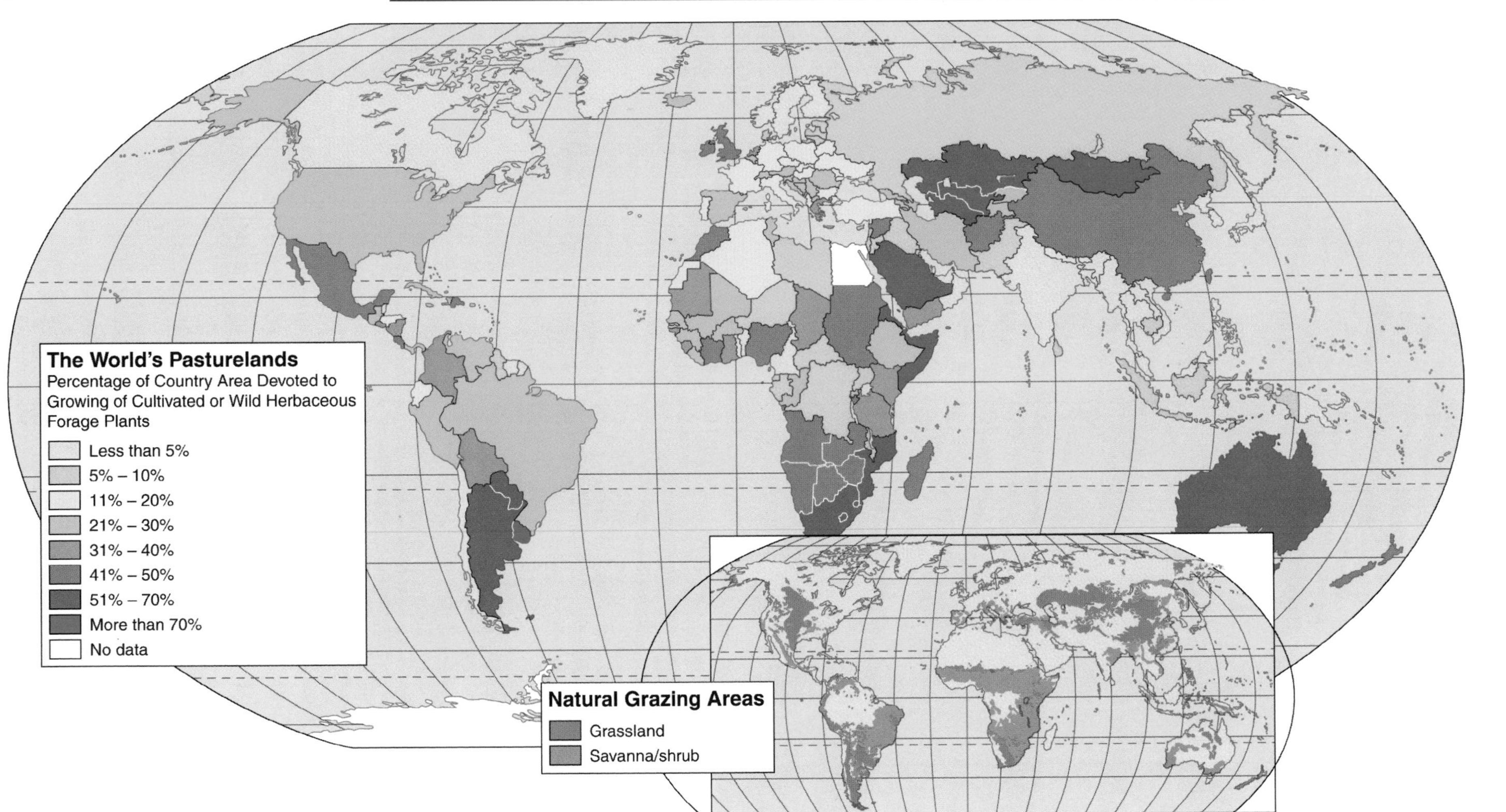

More than 25% of the world's surface is considered pastureland—either native wild grasses or human made pastures created by clearing forests and planting grass and other herbaceous forage plants for livestock feed. Two types of pasture predominate: the prairie and steppe grasses of the mid-latitudes and the savanna grasses of the sub-tropics and tropics. These two primary grassland biomes are shown in the inset map. The larger map depicts pastureland as a proportion of land area of individual countries. You will note that many of the countries with high levels of pastureland (Argentina and Australia, for example) are among the world's leading exporters of livestock products. Other countries with high levels of pastureland (Saudi Arabia and the Central Asian states) consume the bulk of their products domestically. The significance of the distribution and use of pastureland is that—whether it is in the African Sahel, China, Brazil, or the United States—the world's pasturelands are deteriorating rapidly under increasing demands to produce more animal products than a wealthier world can afford to pay for (see Map 34). Grassland degradation, particularly in areas where pastoral nomads use their animals as their chief source of food and income, creates woody scrublands where the carrying capacity for grazing animals is sharply diminished. In short, most of the world's pasturelands are overgrazed, and the world's supply of animal products is in jeopardy.

Map 75 Fertilizer Use 2005

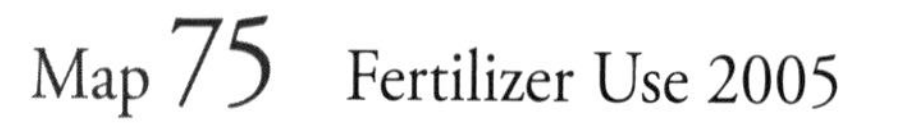

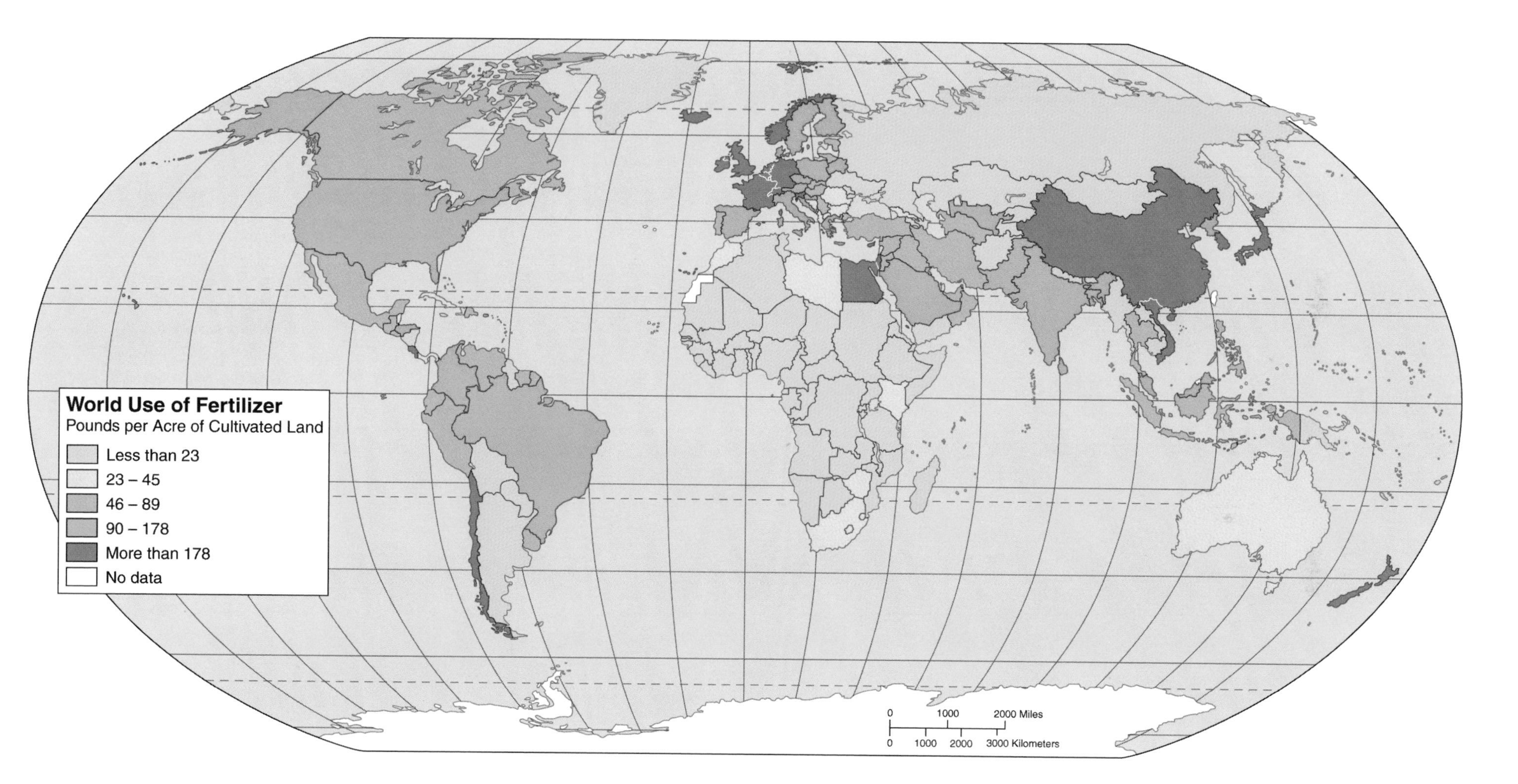

The use of fertilizer to maintain the productivity of agricultural lands is a wonderful agricultural invention—as long as the fertilizers used are natural rather than artificial. In most of the world's developed countries, such as those in Europe and North America, the use of animal manure to fertilize fields has decreased dramatically over the past century, in favor of artificial fertilizers that are cheaper and easier to use and—what is most important—increase crop yields more dramatically. The danger here is that artificial fertilizers normally have high concentrations of nitrates that tend to convert to nitrites in the soil, reducing the ability of soil bacteria to extract "free" nitrogen from the atmosphere. As more artificial fertilizer is used, natural soil fertility is decreased, creating the demand for more artificial fertilizers. In some areas, overuse of artificial fertilizers has actually created soils that are too "hot" chemically to produce crops. Countries with high fertilizer use in the developing world—northern South America, Southwest, South, and East Asia—still tend to use more natural fertilizers. But as farmers in those areas gain more ability to buy and use artificial fertilizers, their soils will also begin to suffer from overfertilization. Global agriculture needs to come to grips with the need to maintain productivity but to do so in a manner that is sustainable.

Map 76 Annual Deforestation Rates 1997-2003

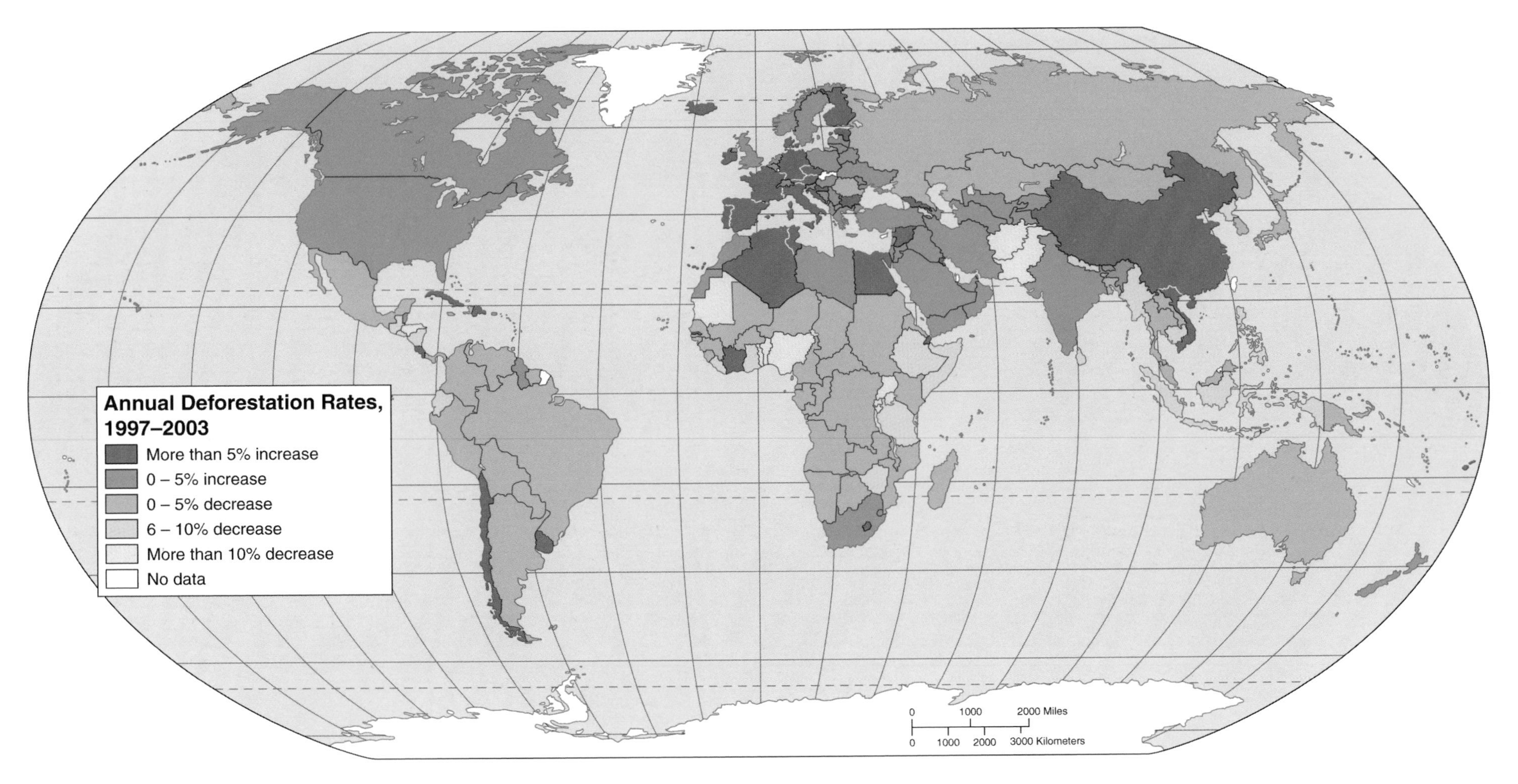

One of the most discussed environmental problems is that of deforestation. For most people, deforestation means clearing of tropical rain forests for agricultural purposes. Yet nearly as much forest land per year—much of it in North America, Europe, and Russia—is impacted by commercial lumbering as is cleared by tropical farmers and ranchers. Even in the tropics, much of the forest clearance is undertaken by large corporations producing high-value tropical hardwoods for the global market in furniture, ornaments, and other fine wood products. Still, it is the agriculturally driven clearing of the great rain forests of the Amazon Basin, west and central Africa, Middle America, and Southeast Asia that draws public attention. Although much concern over forest clearance focuses on the relationship between forest clearance and the reduction in the capacity of the world's vegetation system to absorb carbon dioxide (and thus delay global warming), of just as great concern are issues having to do with the loss of biodiversity (large numbers of plants and animals), the near-total destruction of soil systems, and disruptions in water supply that accompany clearing.

Map 77 World Timber Production 2000

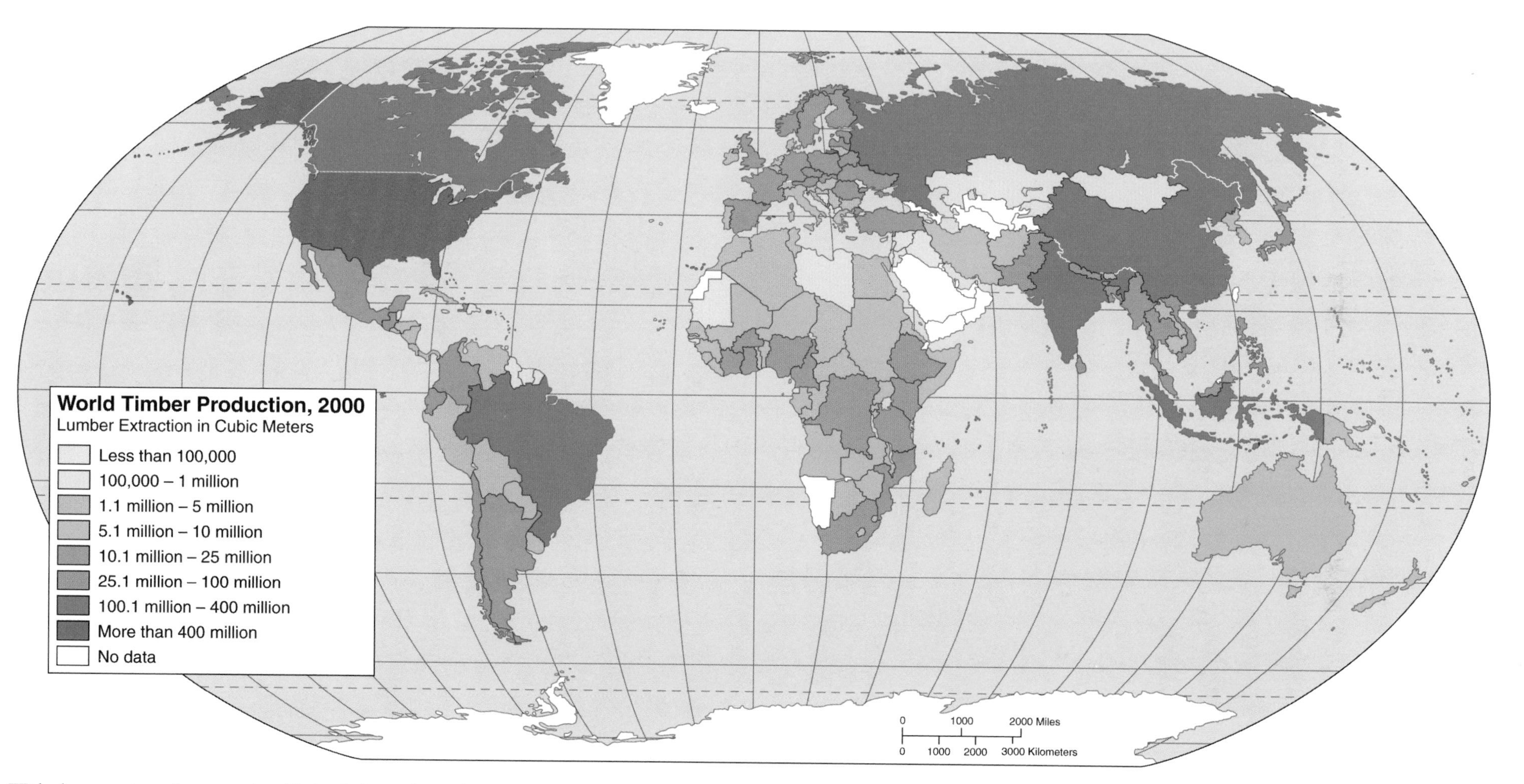

With the exception of water and arable land, forests have played a larger part in the transition from hunting-gathering to the emergence of modern industrial economies than any other natural resource. The chief uses of forests are for lumber, fuelwood, and pulp and paper products. The per capita use of wood does not tend to vary greatly between the developed and developing countries. In developed areas such as North America, Europe, and Russia, the bulk of forest products is used for lumber (much of it exported to forest-poor areas) and for pulp and paper products. In the developing countries in South America, Africa, and Asia (excluding Russia), more wood is used for fuel than for any other purpose. In those parts of the world where forests are used for lumber and pulp-paper products, more-or-less sustainable forest systems have been developed. These systems are often artificially maintained and are much simpler (and, hence, more fragile) ecosystems than the rich, complex natural ecosystems they have replaced. Where forests provide the chief source of fuel, the forest systems tend to remain healthier but are shrinking in size in proportion to the increase in human populations. Of the major forest ecosystems, those in greatest jeopardy are the tropical forests that are being cleared for agricultural land and the boreal or northern forests of Canada and Russia that are being simplified by overcutting.

Map 78 The Loss of Biodiversity: Globally Threatened Animal Species

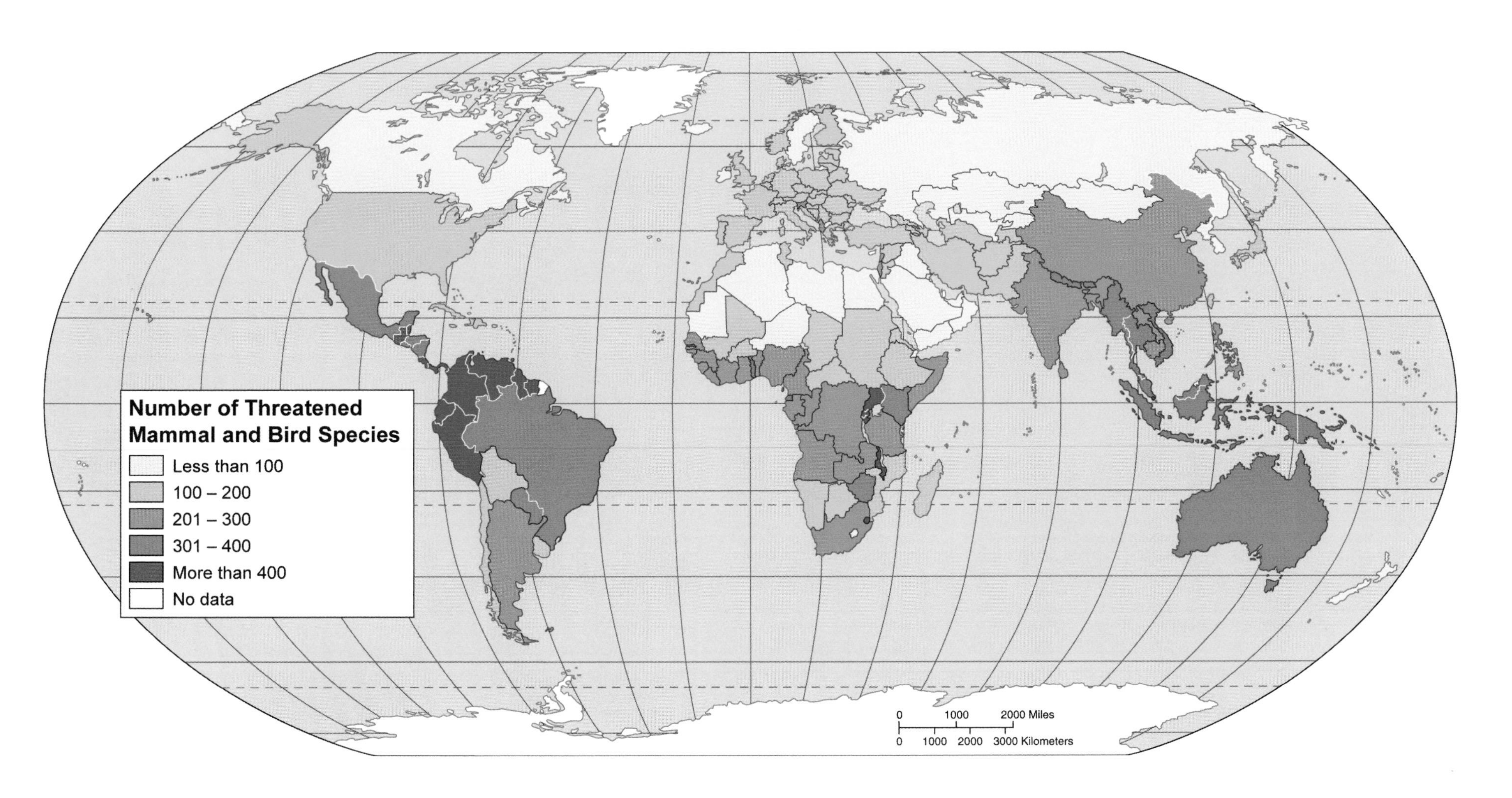

Threatened species are those in grave danger of going extinct. Their populations are becoming restricted in range, and the size of the populations required for sustained breeding is nearing a critical minimum. *Endangered species* are in immediate danger of becoming extinct. Their range is already so reduced that the animals may no longer be able to move freely within an ecozone, and their populations are at the level where the species may no longer be able to sustain breeding. Most species become threatened first and then endangered as their range and numbers continue to decrease. When people think of animal extinction, they think of large herbivorous species like the rhinoceros or fierce carnivores like lions, tigers, or grizzly bears. Certainly these animals make almost any list of endangered or threatened species. But there are literally hundreds of less conspicuous animals that are equally threatened. Extinction is normally nature's way of informing a species that it is inefficient. But conditions in the late twentieth century are controlled more by human activities than by natural evolutionary processes. Species that are endangered or threatened fall into that category because, somehow, they are competing with us or with our domesticated livestock for space and food. And in that competition the animals are always going to lose.

Map 79 The Loss of Biodiversity: Globally Threatened Plant Species

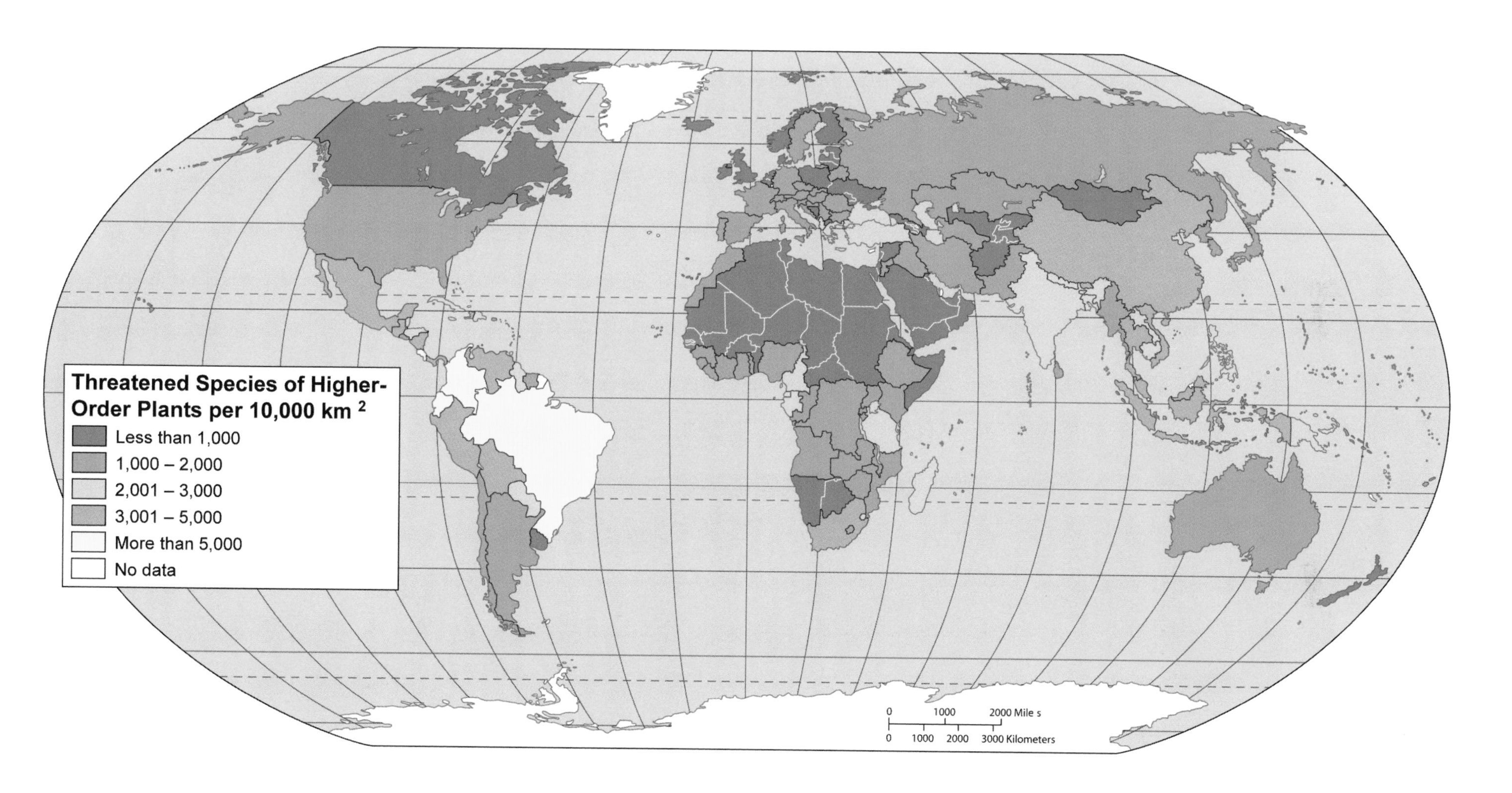

While most people tend to be more concerned about the animals on threatened and endangered species lists, the fact is that many more plants are in jeopardy, and the loss of plant life is, in all ecological regions, a more critical occurrence than the loss of animal populations. Plants are the primary producers in the ecosystem; that is, plants produce the food upon which all other species in the food web, including human beings, depend for sustenance. It is plants from which many of our critical medicines come, and it is plants that maintain the delicate balance between soil and water in most of the world's regions. When environmental scientists speak of a loss of biodiversity, what they are most often describing is a loss of the richness and complexity of plant life that lends stability to ecosystems. Systems with more plant life tend to be more stable than those with less. For these and other reasons, the scientific concern over extinction is greater when applied to plants than to animals. It is difficult for people to become as emotional over a teak tree as they would over an elephant. But as great a tragedy as the loss of the elephant would be, the loss of the teak would be greater.

Map 80 Hotspots of Biodiversity

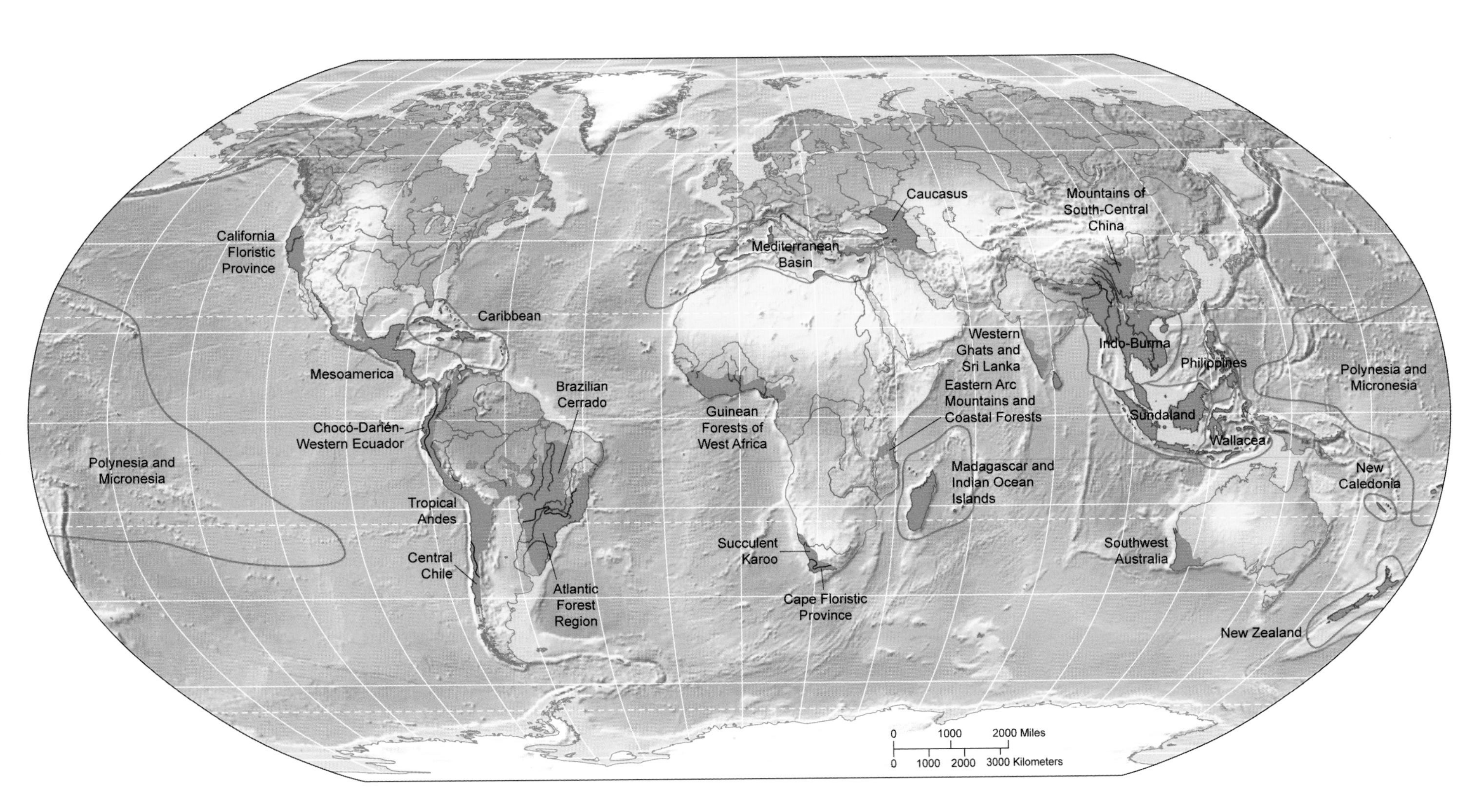

Where we have normally thought of tropical forest basins such as Amazonia as the world's most biologically diverse ecosystems, recent research has discovered the surprising fact that a number of hotspots of biological diversity exist outside the major tropical forest regions. These hotspot regions contain slightly less than 2 percent of the world's total land area but may contain up to 60 percent of the total world's terrestrial species of plants and animals. Geographically, the hotspot areas are characterized by vertical zonation (that is, they tend to be hilly to mountainous regions), long known to be a factor in biological complexity. They are also in coastal locations or near large bodies of water, locations that stimulate climatic variability and, hence, biological complexity. Although some of the hotspots are sparsely populated, others, such as Sundaland, are occupied by some of the world's densest populations. Protection of the rich biodiversity of these hotspots is, most biologists feel, of crucial importance to the preservation of the world's biological heritage.

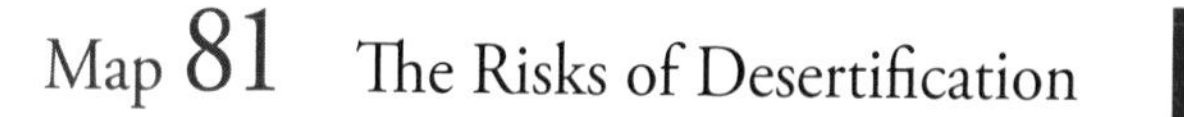

Map 81 The Risks of Desertification

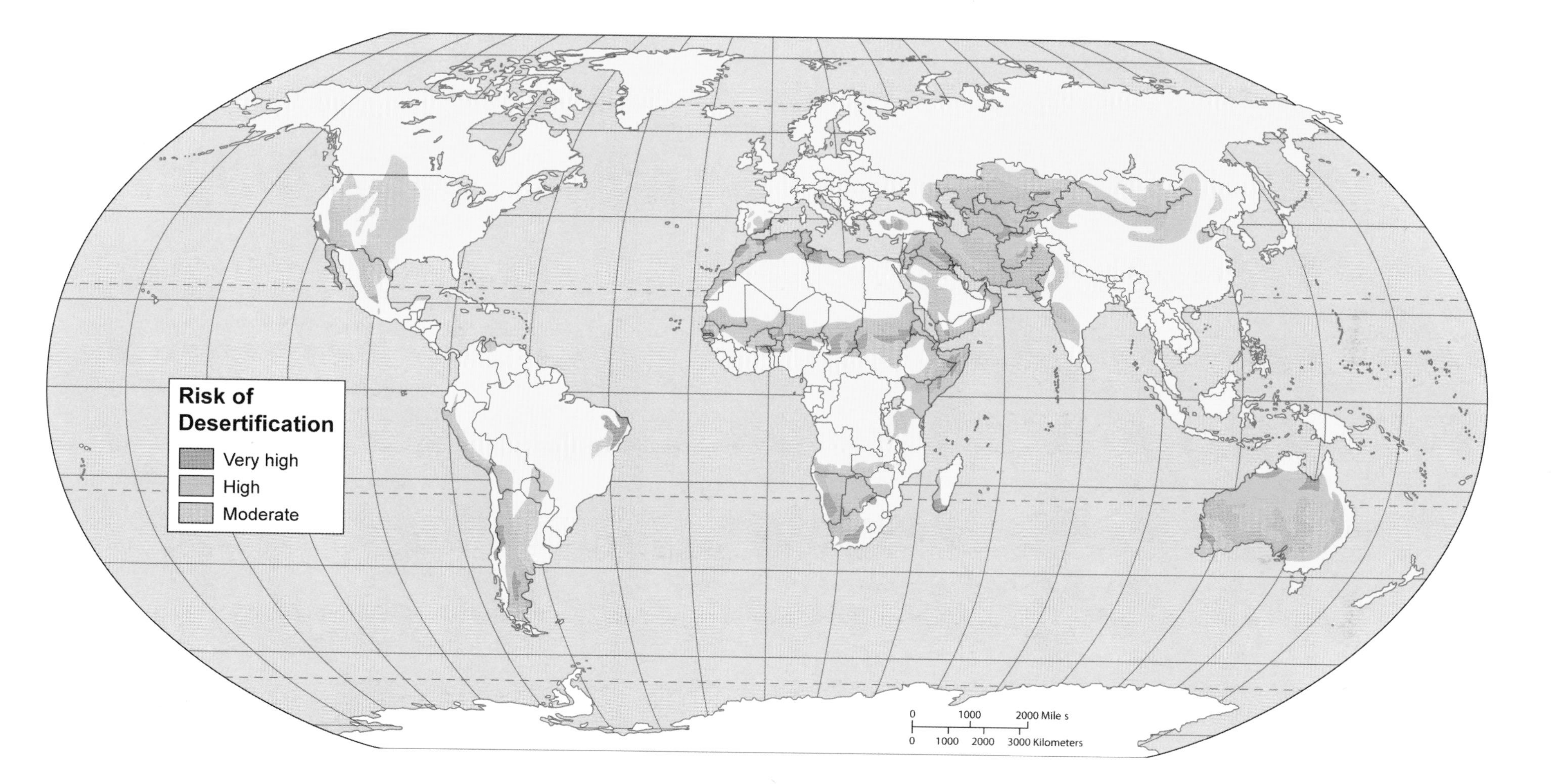

The awkward-sounding term *desertification* refers to a reduction in the food-producing capacity of drylands through vegetation, soil, and water changes that culminate in either a drier climate or in soil and plant systems that are less efficient in their use of water. Most of the world's existing drylands—the shortgrass steppes, the tropical savannas, the bunchgrass regions of the desert fringe—are fairly intensively used for agriculture and are, therefore, subject to the kinds of pressures that culminate in desertification. Most desertification is a natural process that occurs near the margins of desert regions. It is caused by dehydration of the soil's surface layers during periods of drought and by high water loss through evaporation in an environment of high temperature and high winds. This natural process is greatly enhanced by human agricultural activities that expose topsoil to wind and water erosion. Among the most important practices that cause desertification are (1) overgrazing of rangelands, resulting from too many livestock on too small an area of land; (2) improper management of soil and water resources in irrigation agriculture, leading to accelerated erosion and to salt buildup in the soil; (3) cultivation of marginal terrain with soils and slopes that are unsuitable for farming; (4) surface disturbances of vegetation (clearing of thorn scrub, mesquite, chaparral, and similar vegetation) without soil protection efforts being made or replanting being done; and (5) soil compaction by agricultural implements, domesticated livestock, and rain falling on an exposed surface.

Map 82 Soil Degradation

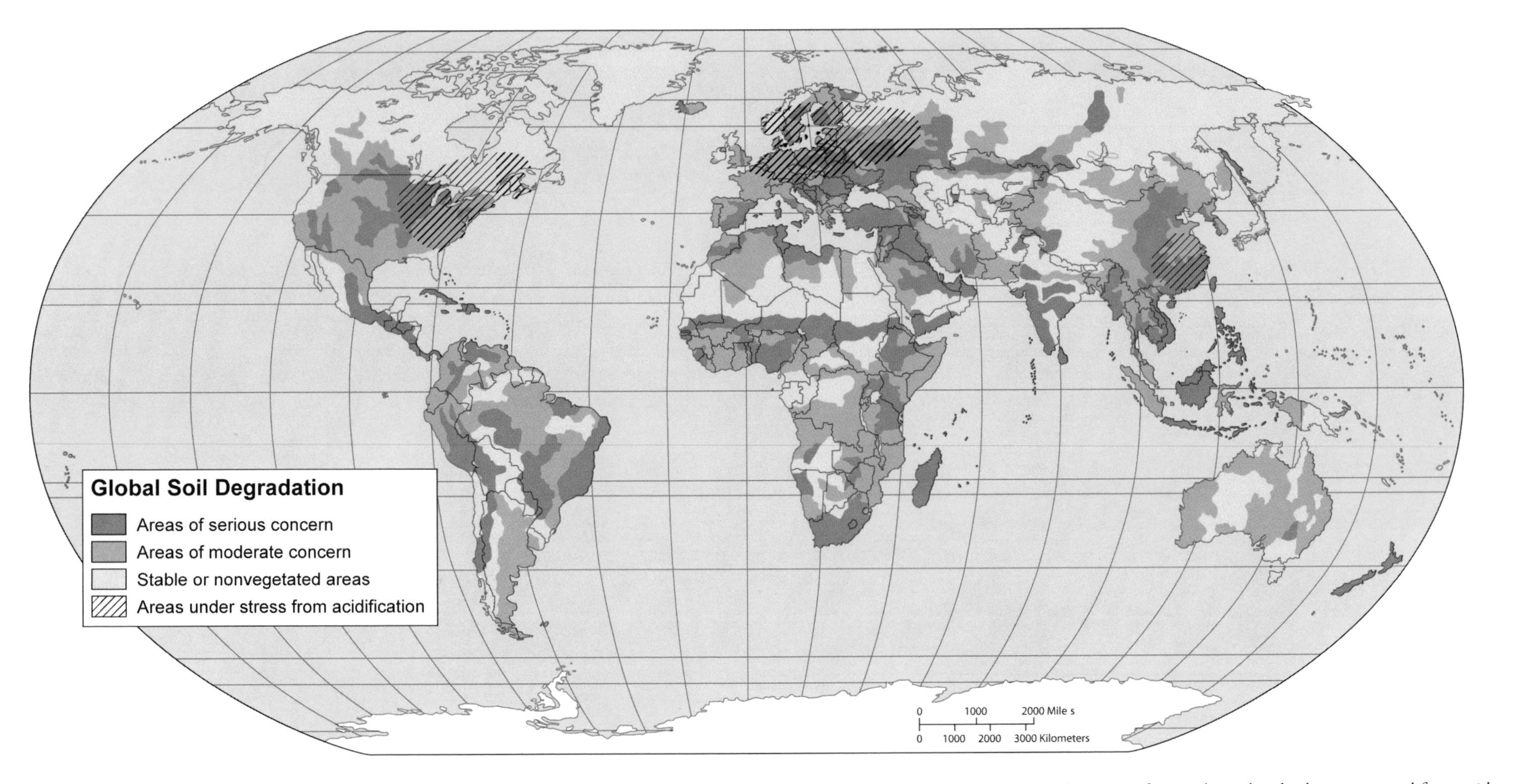

Recent research has shown that more than 3 billion acres of the world's surface suffer from serious soil degradation, with more than 22 million acres so severely eroded or poisoned with chemicals that they can no longer support productive crop agriculture. Most of this soil damage has been caused by poor farming practices, overgrazing of domestic livestock, and deforestation. These activities strip away the protective cover of natural vegetation forests and grasslands, allowing wind and water erosion to remove the topsoil that contains necessary nutrients and soil microbes for plant growth. But millions of acres of topsoil have been degraded by chemicals as well. In some instances these chemicals are the result of overapplication of fertilizers, herbicides, pesticides, and other agricultural chemicals. In other instances, chemical deposition from industrial and urban wastes and from acid precipitation has poisoned millions of acres of soil. As the map shows, soil erosion and pollution are not problems just in developing countries with high population densities and increasing use of marginal lands. They also afflict the more highly developed regions of mechanized, industrial agriculture. While many methods for preventing or reducing soil degradation exist, they are seldom used because of ignorance, cost, or perceived economic inefficiency.

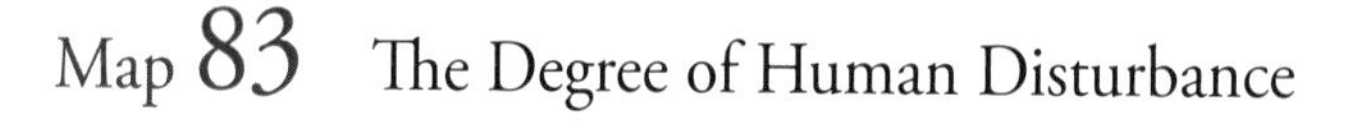

Map 83 The Degree of Human Disturbance

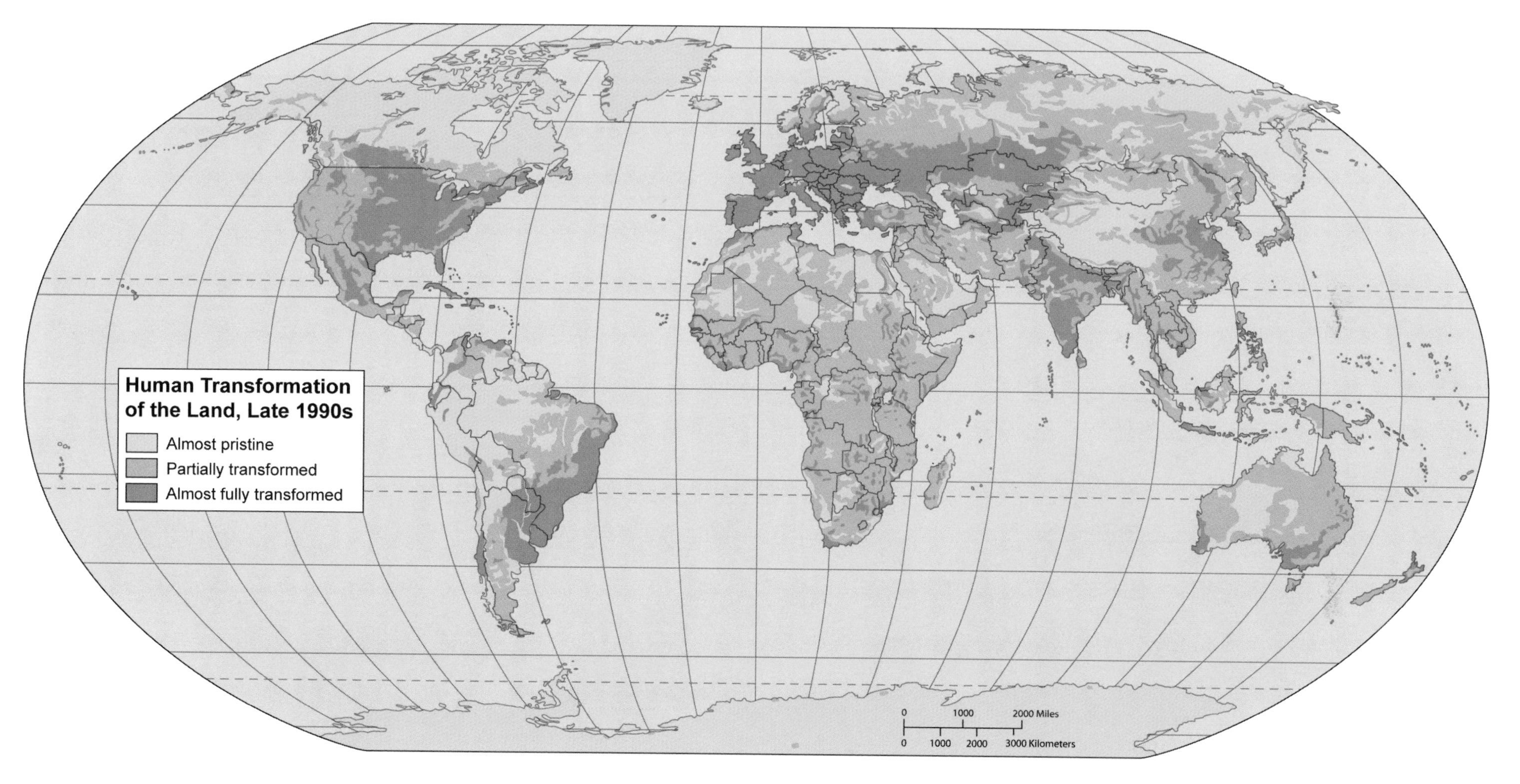

The data on human disturbance have been gathered from a wide variety of sources, some of them conflicting and not all of them reliable. Nevertheless, at a global scale this map fairly depicts the state of the world in terms of the degree to which humans have modified its surface. The almost pristine areas, covered with natural vegetation, generally have population densities under 10 persons per square mile. These areas are, for the most part, in the most inhospitable parts of the world: too high, too dry, too cold for permanent human habitation in large numbers. The partially transformed areas are normally agricultural areas, either subsistence (such as shifting cultivation) or extensive (such as livestock grazing). They often contain areas of secondary vegetation, regrown after removal of original vegetation by humans. They are also sometimes marked by a density of livestock in excess of carrying capacity, leading to overgrazing, which further alters the condition of the vegetation. The almost fully transformed areas are those of permanent and intensive agriculture and urban settlement. The primary vegetation of these regions has been removed, with no evidence of regrowth or with current vegetation that is quite different from natural (potential) vegetation. Soils are in a state of depletion and degradation, and, in drier lands, desertification is a factor of human occupation. The disturbed areas match closely those areas of the world with the densest human populations.

Part VI

Global Political Patterns

Map 84 Political Systems
Map 85 Sovereign States: Duration of Independence
Map 86 The Emergence of the State
Map 87 Organized States and Chiefdoms, A.D. 1500
Map 88 European Colonialism 1500-2000
Map 89 Post–Cold War International Alliances
Map 90 International Conflicts in the Post–World War II World
Map 91 Political Realms: Regional Changes, 1945-2003
Map 92 The Geopolitical World at the Beginning of the Twenty-First Century
Map 93 Flashpoints 2006
Map 94 Nations with Nuclear Weapons
Map 95 Military Power
Map 96 International Terrorist Incidents, 1998-2003
Map 97 Distribution of Minority Populations
Map 98 The Political Geography of a Global Religion: The Islamic World
Map 99 World Refugee Population
Map 100 Women's Rights
Map 101 Political and Civil Liberties 2004
Map 102 A Decade of Risk of Civil Conflict 2005-2015

Map 84 Political Systems

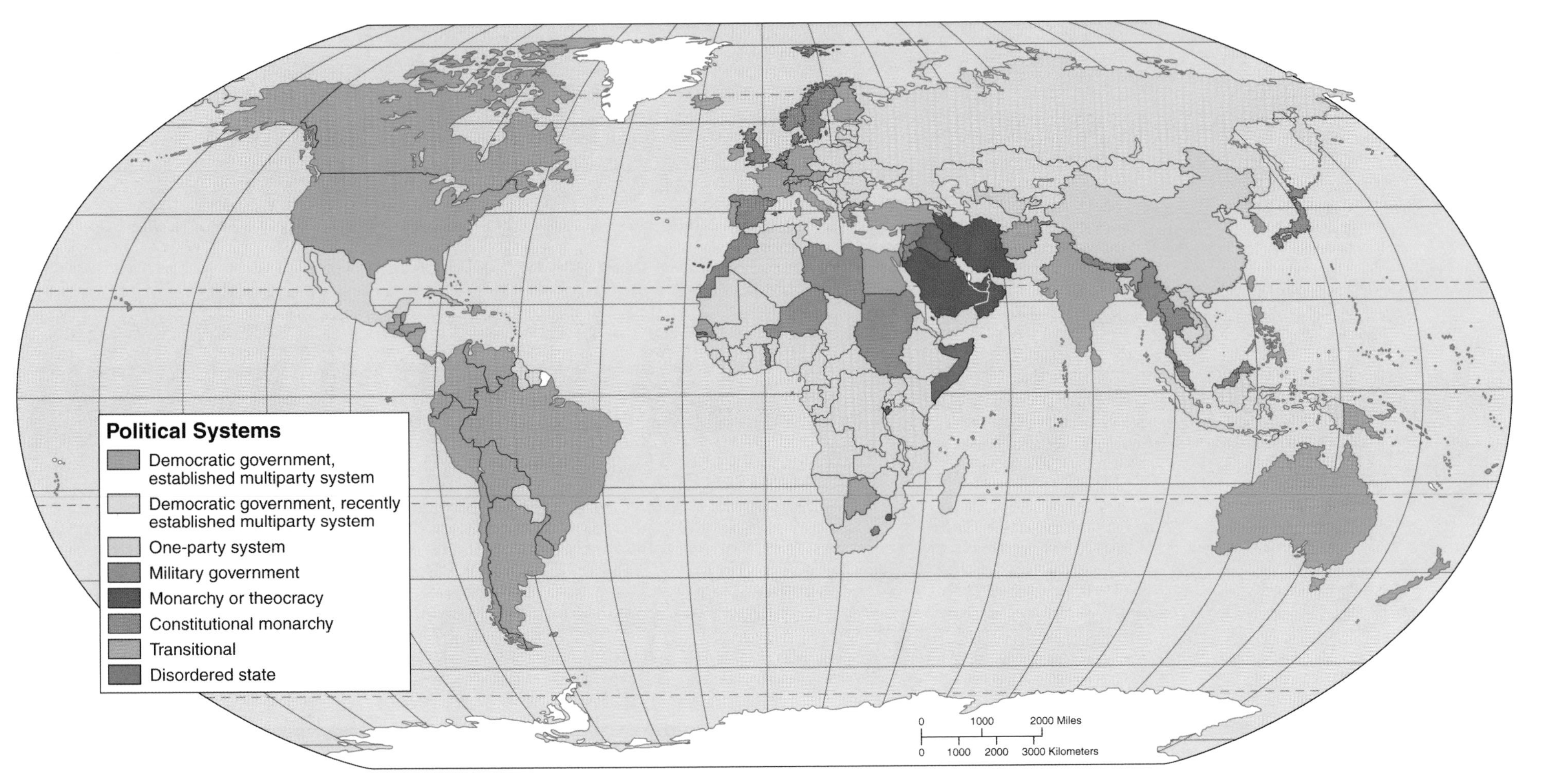

World political systems have changed dramatically during the last decade and may change even more in the future. The categories of political systems shown on the map are subject to some interpretation: established multiparty democracies are those in which elections by secret ballot with adult suffrage are and have been long-term features of the political landscape; recently established multiparty democracies are those in which the characteristic features of multiparty democracies have only recently emerged. The former Soviet satellites of eastern Europe and the republics that formerly constituted the USSR are in this category; so are states in emerging regions that are beginning to throw off the single-party rule that often followed the violent upheavals of the immediate postcolonial governmental transitions. The other categories are more or less obvious. One-party systems are states where single-party rule is constitutionally guaranteed or where a one-party regime is a fact of political life. Monarchies are countries with heads of state who are members of a royal family. In a constitutional monarchy, such as the U.K. and the Netherlands, the monarchs are titular heads of state only. Theoracies are countries in which rule is within the hands of a priestly or clerical class; today, this means primarily fundamentalist Islamic countries such as Iran. Military governments are frequently organized around a junta that has seized control of the government from civil authority; such states are often technically transitional, that is, the military claims that it will return the reins of government to civil authority when order is restored. Finally, disordered states are countries so beset by civil war or widespread ethnic that no organized government can be said to exist within them.

Map 85 Sovereign States: Duration of Independence

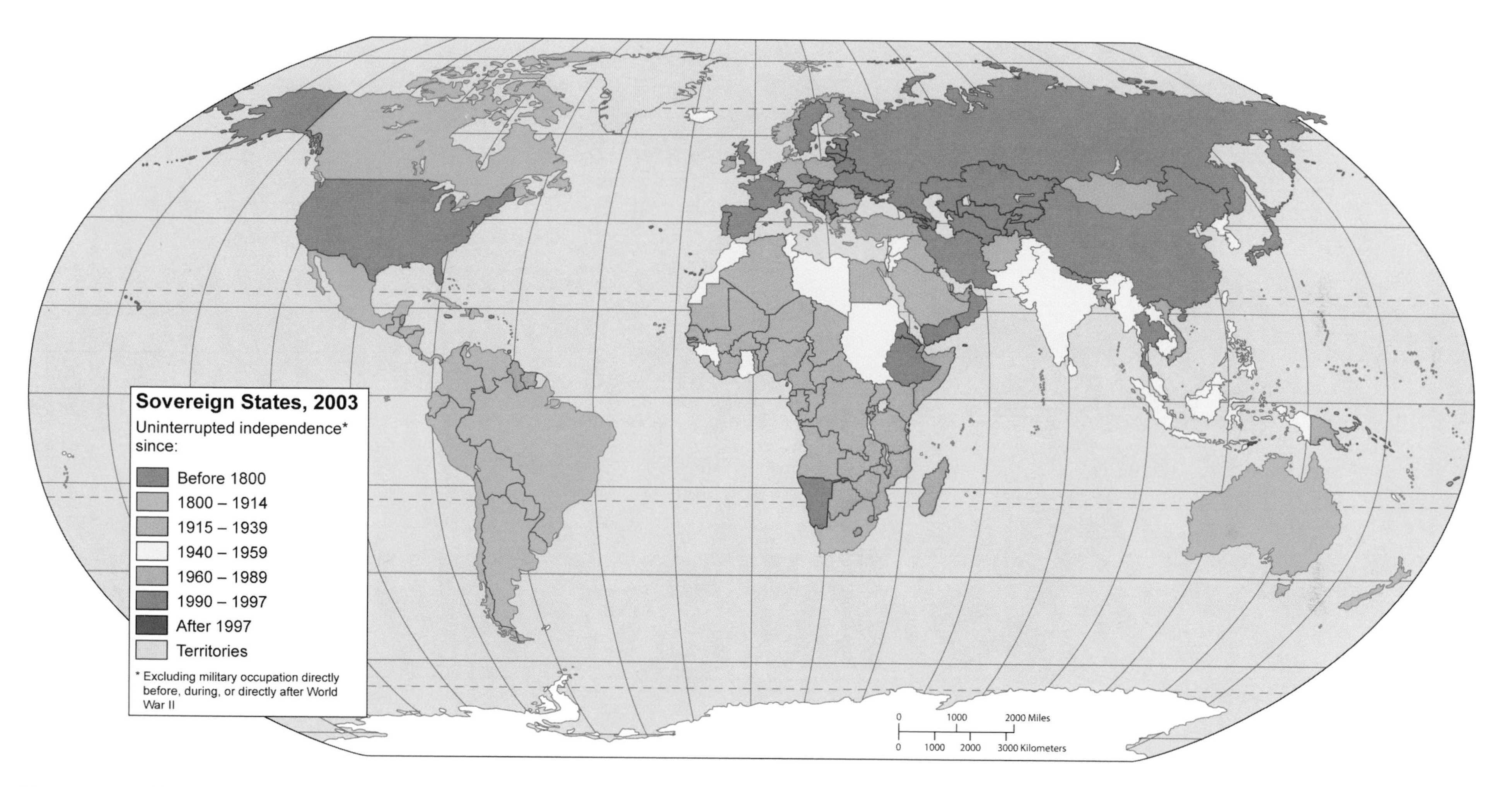

Most countries of the modern world, including such major states as Germany and Italy, became independent after the beginning of the nineteenth century. Of the world's current countries, only 27 were independent in 1800. (Ten of the 27 were in Europe; the others were Afghanistan, China, Colombia, Ethiopia, Haiti, Iran, Japan, Mexico, Nepal, Oman, Paraguay, Russia, Taiwan, Thailand, Turkey, the United States, and Venezuela). Following 1800, there have been five great periods of national independence. During the first of these (1800–1914), most of the mainland countries of the Americas achieved independence. During the second period (1915–1939), the countries of Eastern Europe emerged as independent entities. The third period (1940–1959) includes World War II and the years that followed, when independence for African and Asian nations that had been under control of colonial powers first began to occur. During the fourth period (1960–1989), independence came to the remainder of the colonial African and Asian nations, as well as to former colonies in the Caribbean and the South Pacific. More than half of the world's countries came into being as independent political entities during this period. Finally, in the last few years (1990–1997), the breakup of the existing states of the Soviet Union, Yugoslavia, and Czechoslovakia created 22 countries where only 3 had existed before.

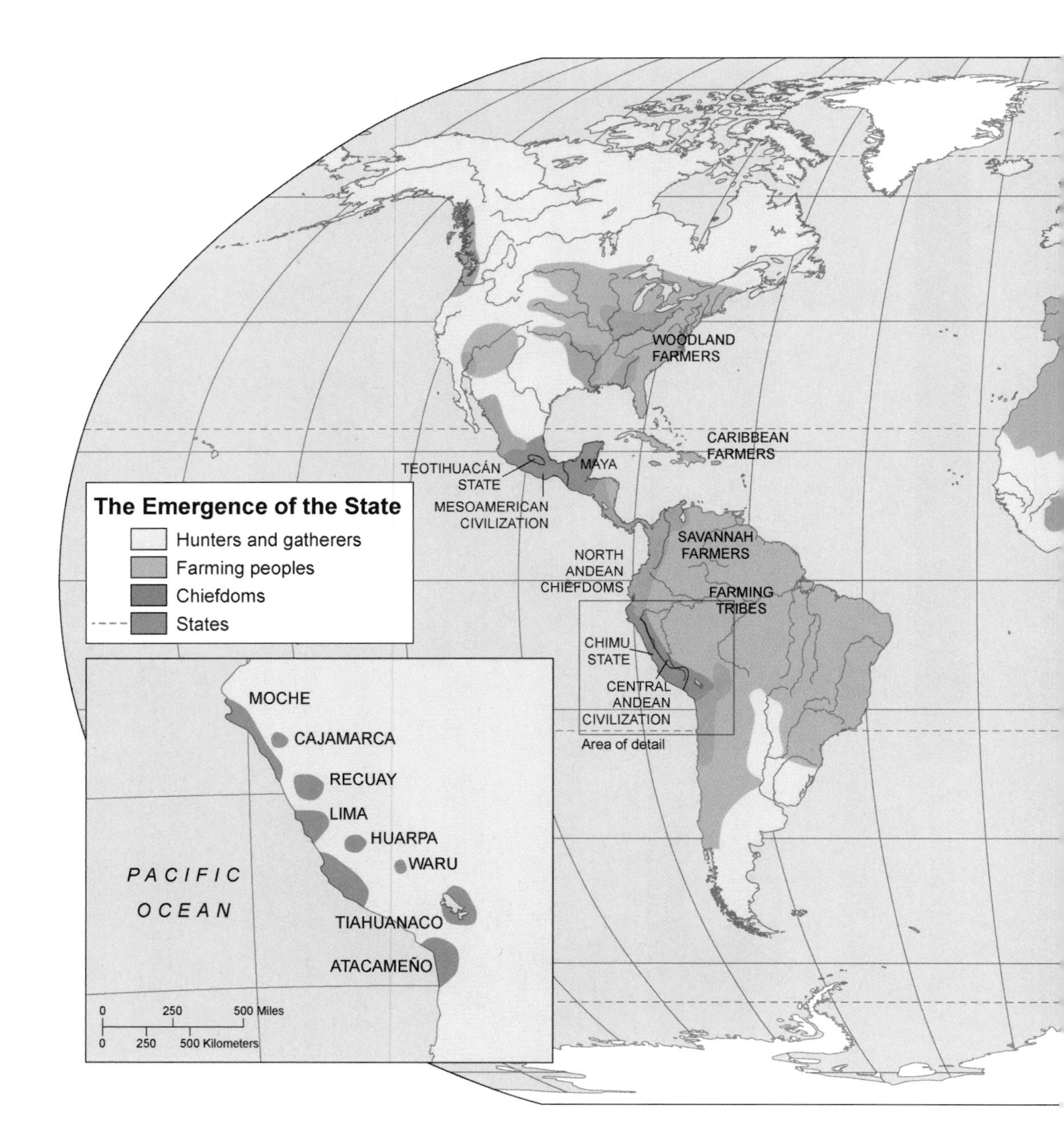

Agriculture is the basis of the development of the state, a form of complex political organization. Geographers and anthropologists believe that agriculture allowed for larger concentrations of population. Farmers do not need to be as mobile as hunters and gatherers to make a living, and people living sedentarily can have larger families than those constantly on the move. Ideas about access to land and ownership also change as people develop the social and political hierarchies that come with the transition from a hunting-gathering society to an agricultural one. Social stratification based on wealth and

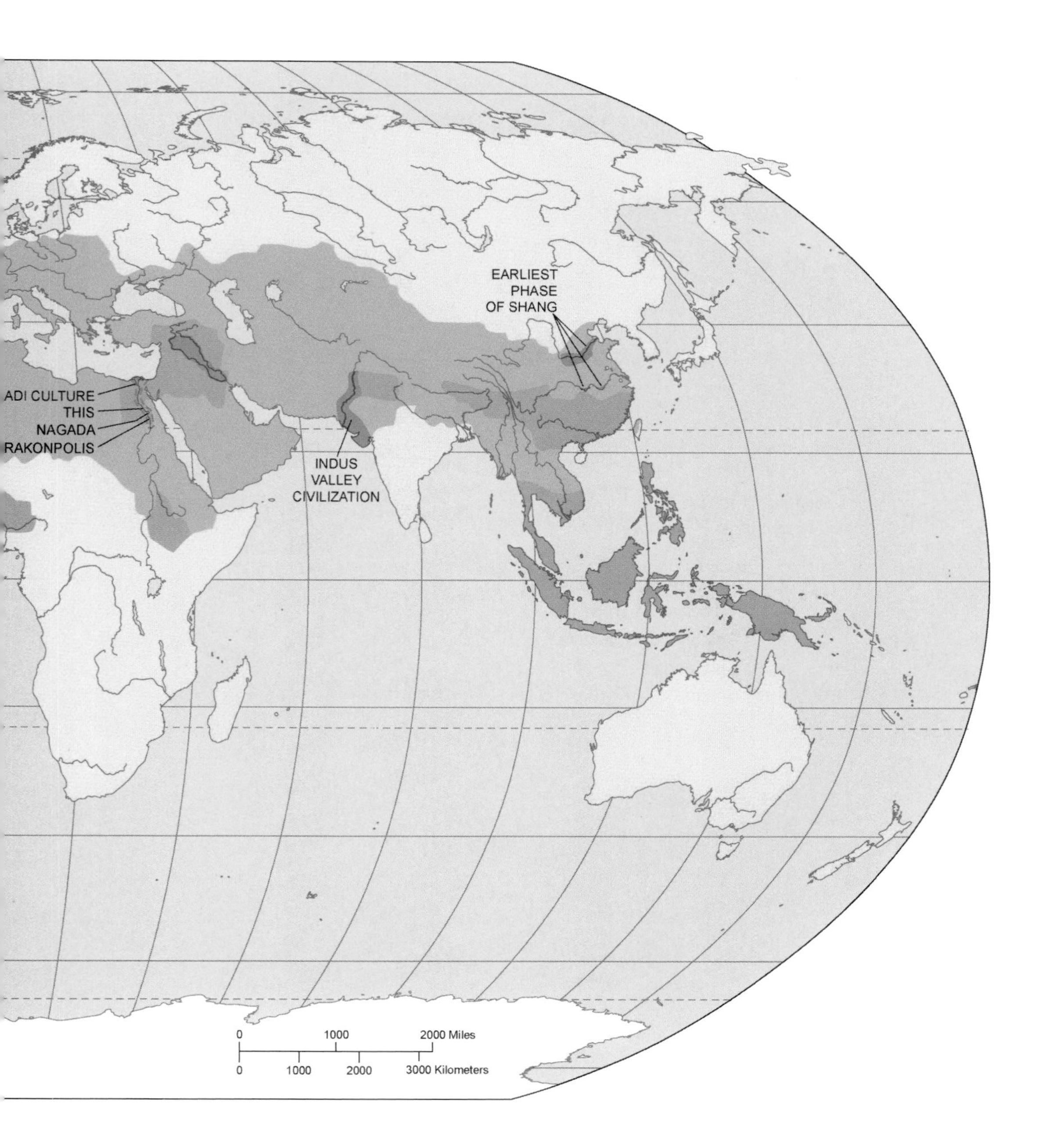

power creates different classes or groups, some of whom no longer work the land. An agricultural surplus supports those who perform other functions for society, such as artisans and craftspeople, soldiers and police, priests and kings. Thus, over time, egalitarian hunters and gatherers shifted to state-level societies in some parts of the world. The first true states are shown in the rust-colored areas of this map.

Map 87 Organized States and Chiefdoms, A.D. 1500

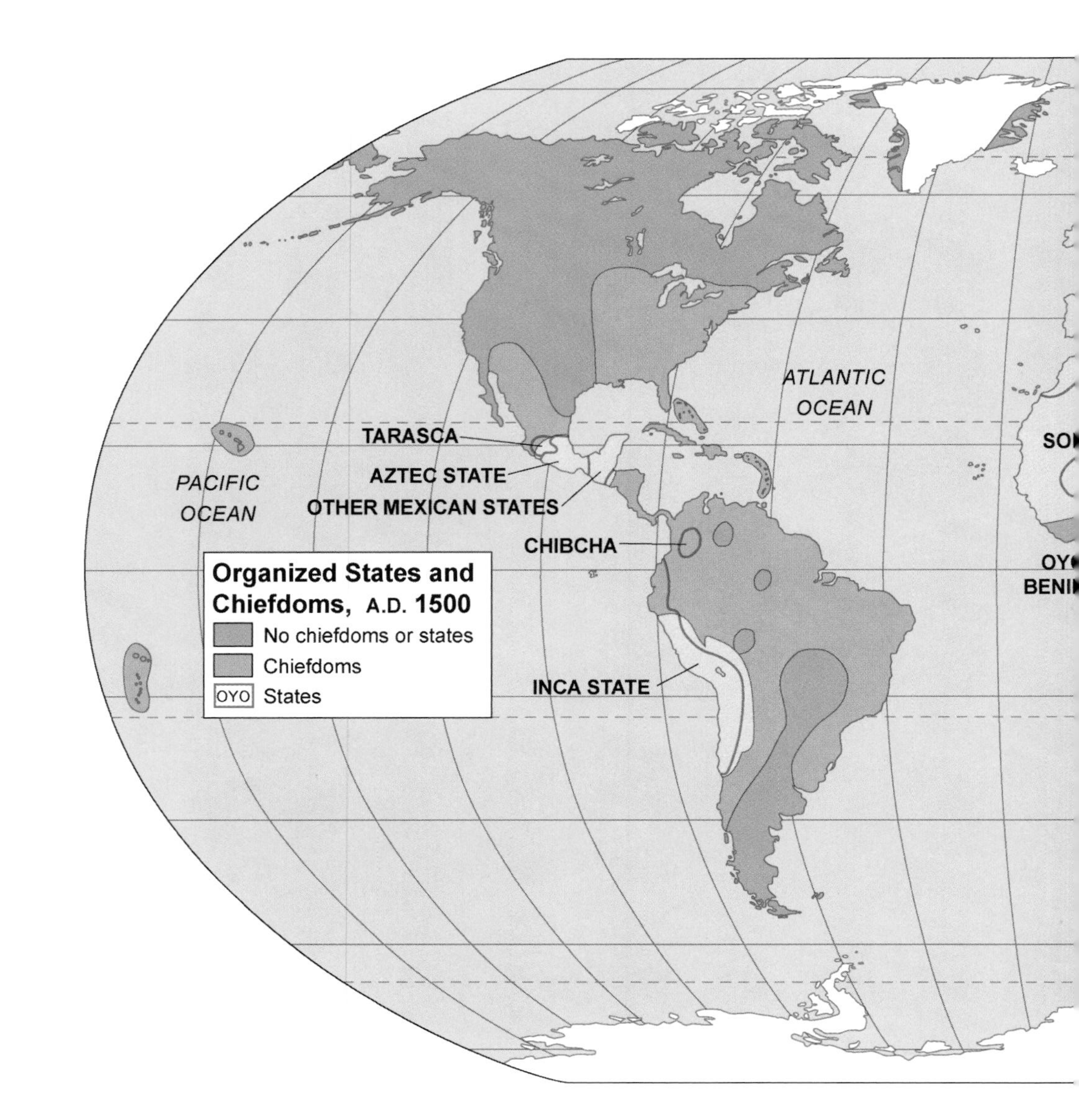

As Europeans began to expand outward through exploration and settlement from the late fifteenth to the eighteenth centuries, it was inevitable that they would find the complex political organizations of chiefdoms and states in many different parts of the world. Both chiefdoms and states are large-scale forms of political organization in which some people have privileged access to land, power, wealth, and prestige. Chiefdoms are kin-based societies in which wealth is distributed from upper to lower classes. States are organized on the basis of socioeconomic classes, headed by a

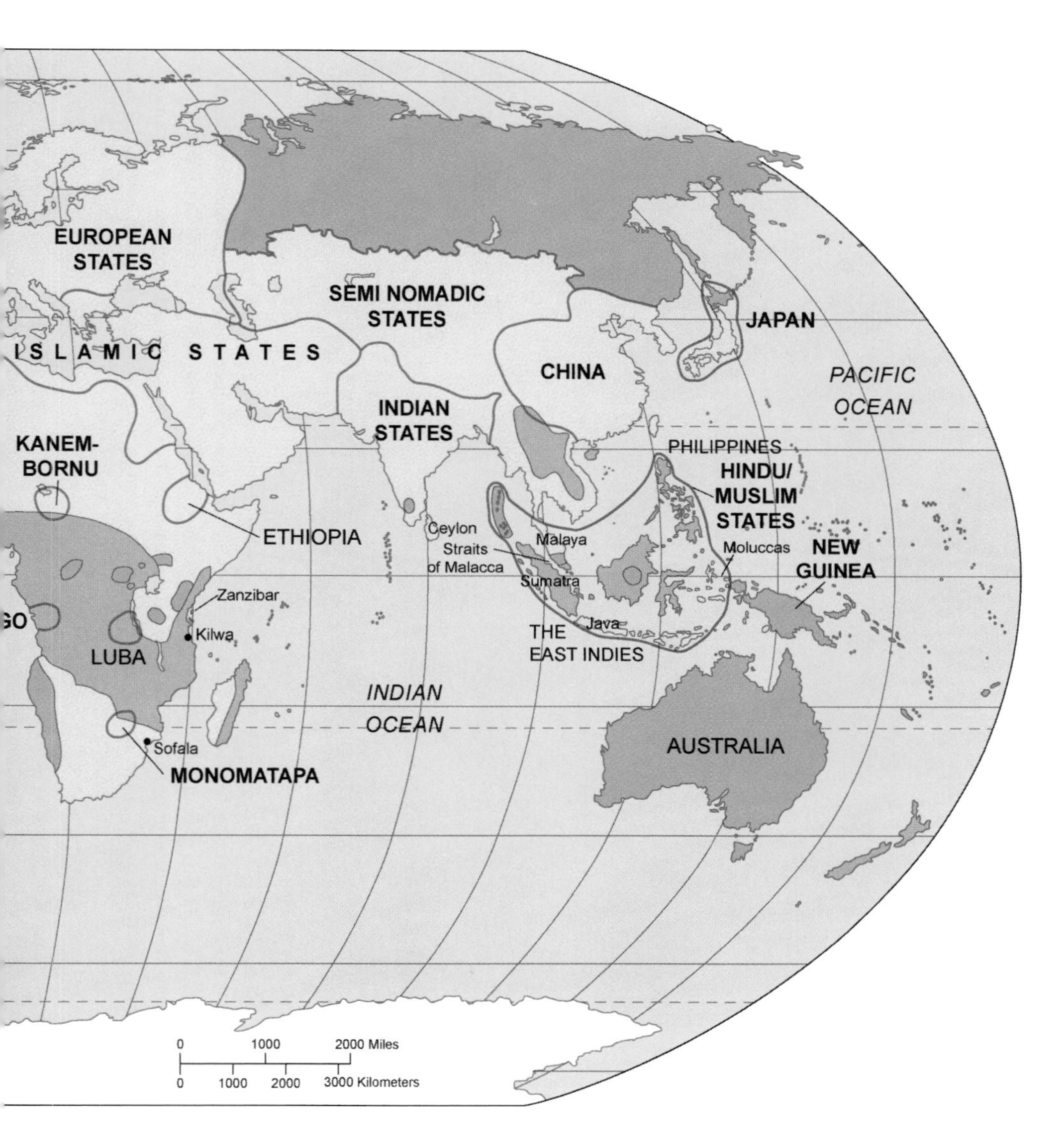

centralized government that is led by an elite. States in non-European areas included, just as they did in Europe, full-time bureaucracies and specialized subsystems for such activities as military action, taxation, operation of state religions, and social control. In many of the colonial regions of the world after the fifteenth century, Europeans actually found it easier to gain control of organized states and chiefdoms because those populations were already accustomed to some form of institutionalized central control.

Map 88 European Colonialism 1500–2000

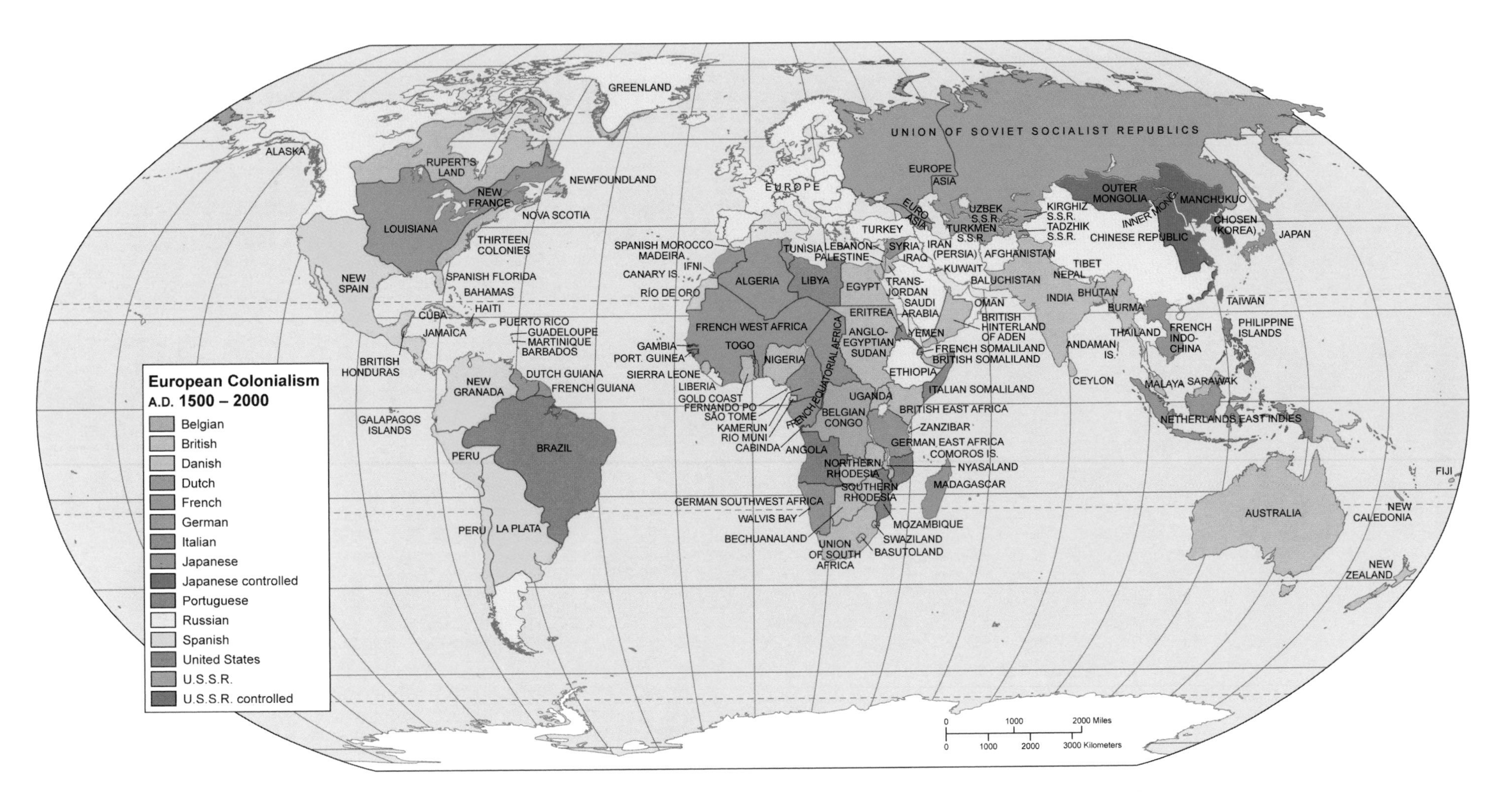

European nations have controlled many parts of the world during the last 500 years. The period of European expansion began when European explorers sailed the oceans in search of new trading routes and ended after World War II when many colonies in Africa and Asia gained independence. The process of colonization was very complex but normally involved the acquisition, extraction, or production of raw materials (including minerals, forest products, products from the sea, agricultural products, and animal furs/pelts) from the areas being controlled by the European colonial power in exchange for items of European manufacture. The concept of colonial dependency implied an economic structure in which the European country obtained raw materials from the colonial country in exchange for those manufactured items upon which populations in the colonial areas quickly came to depend. The colors on this map represent colonial control at its maximum extent and do not take into account shifting colonial control. In North America, for example, "New France" became British territory and "Louisiana" became Spanish territory after the Seven Years' (French and Indian) War.

Map 89 Post–Cold War International Alliances

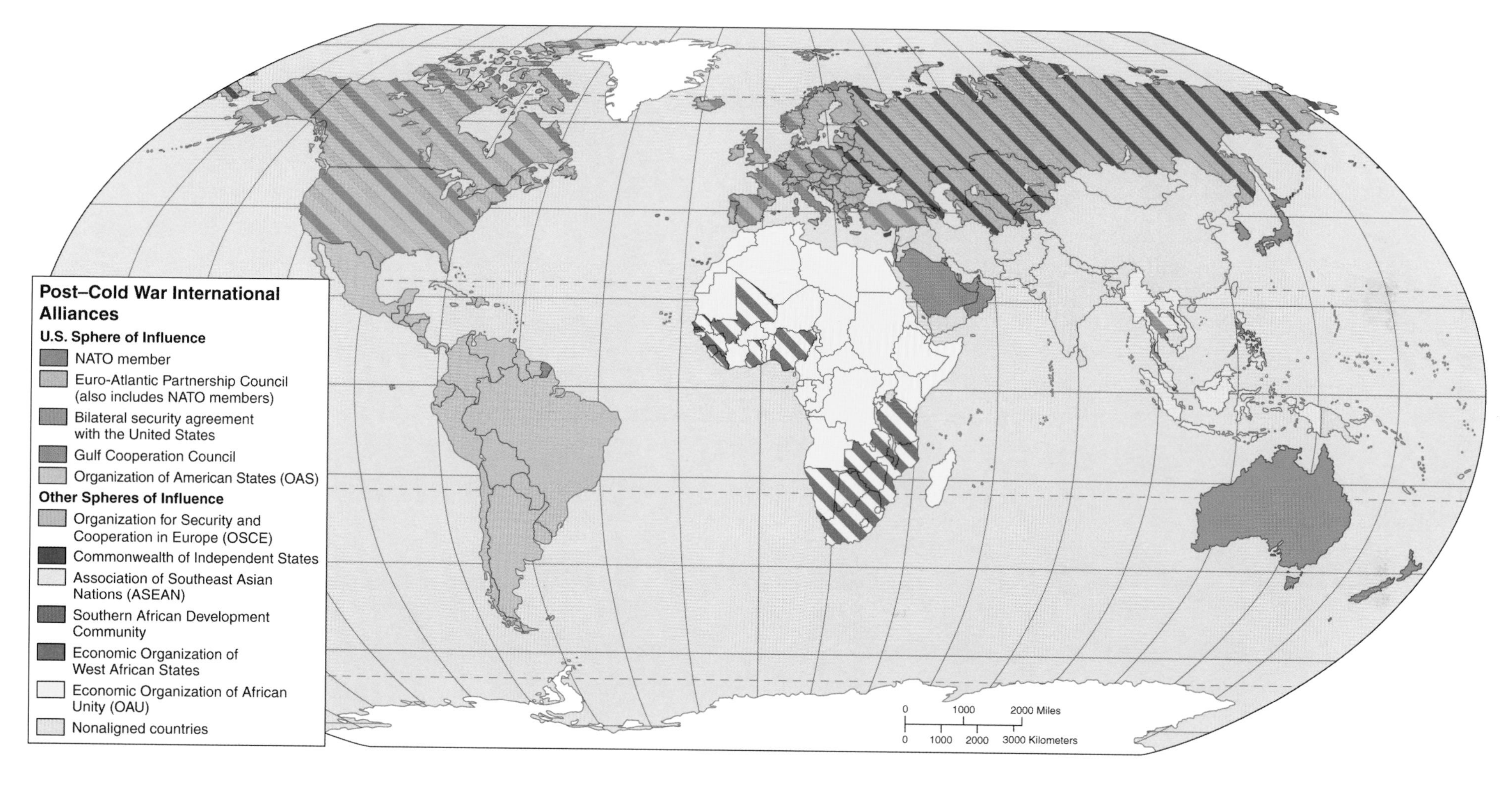

When the Warsaw Pact dissolved in 1992, the North Atlantic Treaty Organization (NATO) was left as the only major military alliance in the world. Some former Warsaw Pact members (Czech Republic, Hungary, and Poland) have joined NATO and others are petitioning for entry. The bipolar division of the world into two major military alliances is over, at least temporarily, leaving the United States alone as the world's dominant political and military power. But other international alliances, such as the Commonwealth of Independent States (including most of the former republics of the Soviet Union), will continue to be important. It may well be that during the first few decades of the twenty-first century economic alliances will begin to overshadow military ones in their relevance for the world's peoples.

Map 90 International Conflicts in the Post–World War II World

	Conflict[1]	Start Date	Major Belligerent Countries[2] (in alphabetical order)	
1	Palestine	1948	Egypt Iraq Israel	Jordan Lebanon Syria
2	Korean	1950	China North Korea South Korea United Nations: United States and 11 other countries	
3	Soviet-Hungarian	1956	Hungary	Soviet Union
4	Sinai	1956	Egypt France	Israel United Kingdom
5	Sino-Indian	1962	China	India
6	Kashmir	1965	India	Pakistan
7	Vietnam	1965	Australia North Vietnam South Korea	South Vietnam United States
8	Six-Day	1967	Egypt Israel	Jordan Syria
9	Soviet-Czech	1968	Czechoslovakia	Soviet Union
10	Football	1969	El Salvador	Honduras
11	Indo-Pakistani	1971	India	Pakistan
12	Yom Kippur	1973	Egypt Israel	Syria
13	Cyprus	1974	Cyprus	Turkey
14	Ogaden	1977	Ethiopia	Somalia
15	Cambodian-Vietnamese	1978	Cambodia China	Vietnam
16	Ugandan-Tanzanian	1978	Tanzania	Uganda
17	Afghanistan	1979	Afghanistan	Soviet Union
18	Persian Gulf	1980	Iran	Iraq
19	Angola	1981	Angola Cuba	South Africa
20	Falklands	1982	Argentina	United Kingdom
21	Saharan	1983	Chad	Libya
22	Lebanon	1987	France Israel Lebanon	Syria United States
23	Panama	1989	Panama	United States
24	Persian Gulf	1990	Iraq United Nations: United States and 7 other countries	
25	Yugoslavia	1990	Bosnia-Herzegovina	Croatia Serbia
26	Peruvian-Ecuadorian	1995	Ecuador	Peru
27	Albania	1995	Albania	Yugoslavia (Serbia-Montenegro)
28	Rwanda	1995	Burundi	Rwanda
29	East Timor	1995	Indonesia	New Guinea insurgency
30	Cameroon	1996	Cameroon	Nigeria
31	Northern Iraq	1996	Iraq	Kurdish insurgency
32	Eritrea	1997	Eritrea	Yemen
33	Iraq	1998	Great Britain Iraq	United States
34	Kosovo	1999	Albania NATO	Yugoslavia
35	Democratic Republic of the Congo	1998	Angola Chad Congo	Namibia Sudan Zimbabwe
36	Chechnya	1999	Chechnya	Russia
37	"War on Terrorism"	2001	Afghanistan (Taliban) al-Qaeda organization Great Britain	United States Others
38	Iraq	2003	Great Britain Iraq	United States

[1] "Conflict" implies at least 1,000 battle deaths.
[2] "Belligerent" implies country supplied at least 5% of the combat troops in the conflict.

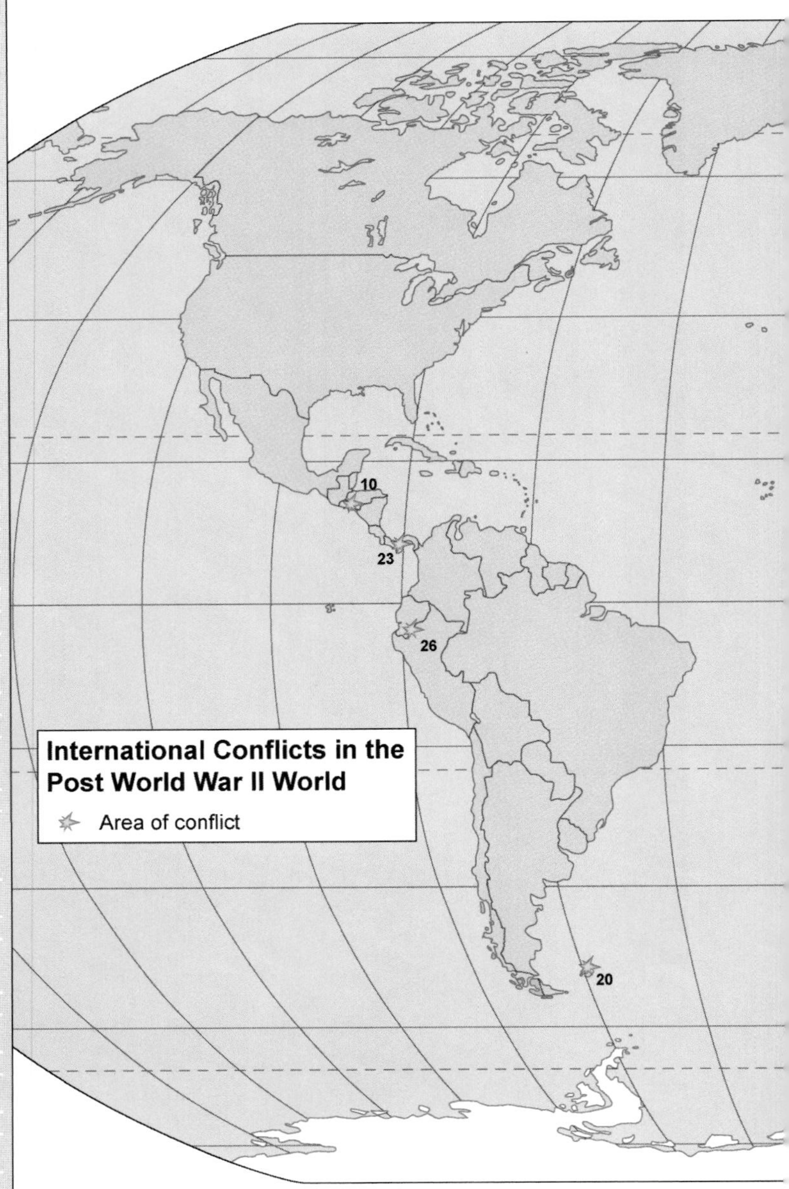

International Conflicts in the Post World War II World

Area of conflict

The Korean War and the Vietnam War dominated the post–World War II period in terms of international military conflict. But numerous smaller conflicts have taken place, with fewer numbers of belligerents and with fewer battle and related casualties. These smaller international conflicts have been mostly territorial conflicts, reflecting the continual readjustment of political boundaries and loyalties brought about by the end of colonial empires, and the dissolution of the Soviet Union. Many of these conflicts were not wars in the more traditional sense, in which two or more countries formally

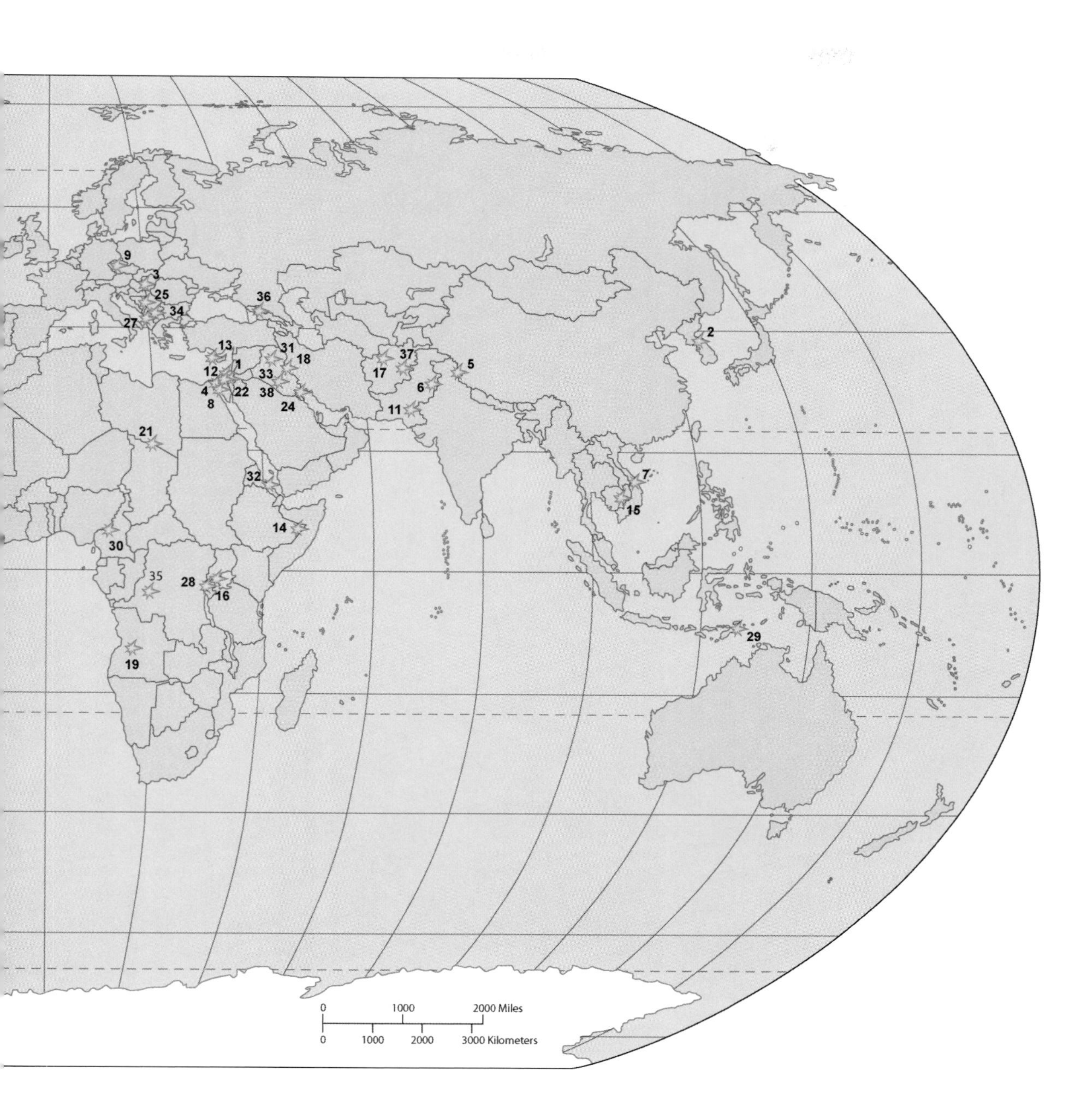

declare war on one another, severing diplomatic ties and devoting their entire national energies to the war effort. Rather, many of these conflicts were and are undeclared wars, sometimes fought between rival groups within the same country with outside support from other countries. The aftermath of the September 11, 2001, terrorist attacks on the United States indicate the dawn of yet another type of international conflict, namely a "war" fought between traditional nation-states and non-state actors.

Map 91 Political Realms: Regional Changes, 1945–2003

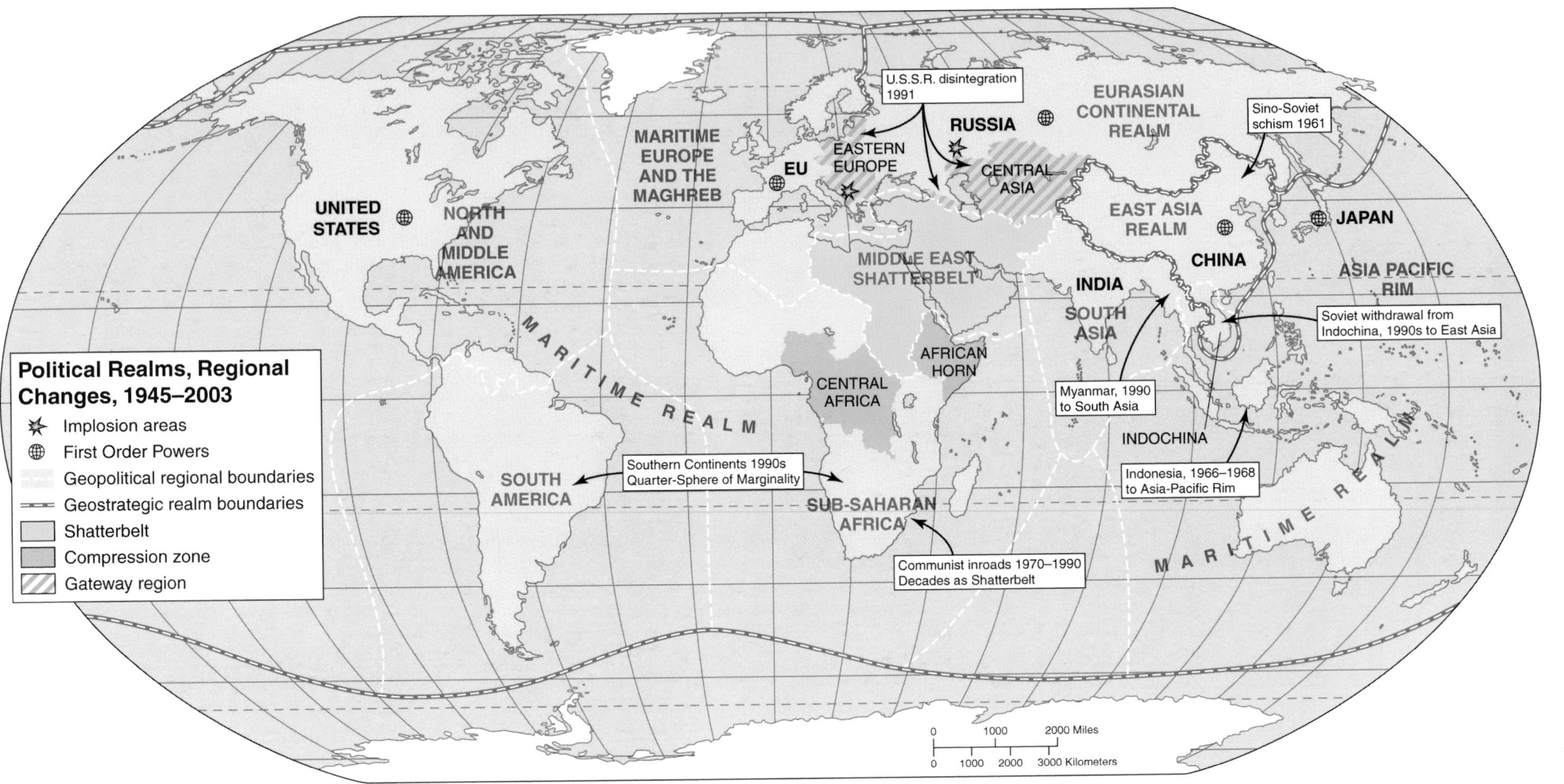

The Cold War following World War II shaped the major outlines of today's geopolitical relations. The Cold War included three phases. In the first, from 1945–1956, the Maritime Realm established a ring around the Continental Eurasian Realm in order to prevent its expansion. This phase included the Korean War (1950–1953), the Berlin Blockade (1948), the Truman doctrine and Marshall Plan (1947), and the founding of NATO (1949) and the Warsaw Pact (1955). Most of the world fell within one of the two realms: the Maritime (dominated by the United States) or the Eurasian Continental Realm (dominated by the Soviet Union). The Soviet Union sought to establish a ring of satellite states to protect it from a repeat of the invasions of World War II. The United States and other Maritime Realm states, in turn, sought to establish a ring of allies around the Continental Realm to prevent its expansion. South Asia was politically independent, but under pressure from both realms. During the second phase (1957–1979), Communist forces from the Continental Eurasian Realm penetrated deeply into the Maritime Realm. The Berlin Wall went up in 1961, Soviet missiles in Cuba ignited a crisis in 1962, and the United States became increasingly involved in the war in Vietnam (late 1960s). The Soviet Union sought increased political and military presence along important waterways including those in the Middle East, Southeast Asia, and the Caribbean. These regions became especially dangerous shatterbelts. The third phase (1980–1989) saw the retreat of Communist power from the Maritime Realm. China, after ten years of radical Communism and chaos of the Cultural Revolution (1966–1976), broke away from the Continental Eurasian Realm to establish a new East Asian realm. Soviet influence declined in the Middle East, Sub-Saharan Africa, and Latin America. In 1989 the Berlin Wall fell, and Eastern Europe began to establish democratic governments. In 1991 the Soviet Union broke apart as its constituent republics became independent states.

Map 92 The Geopolitical World at the Beginning of the Twenty-First Century

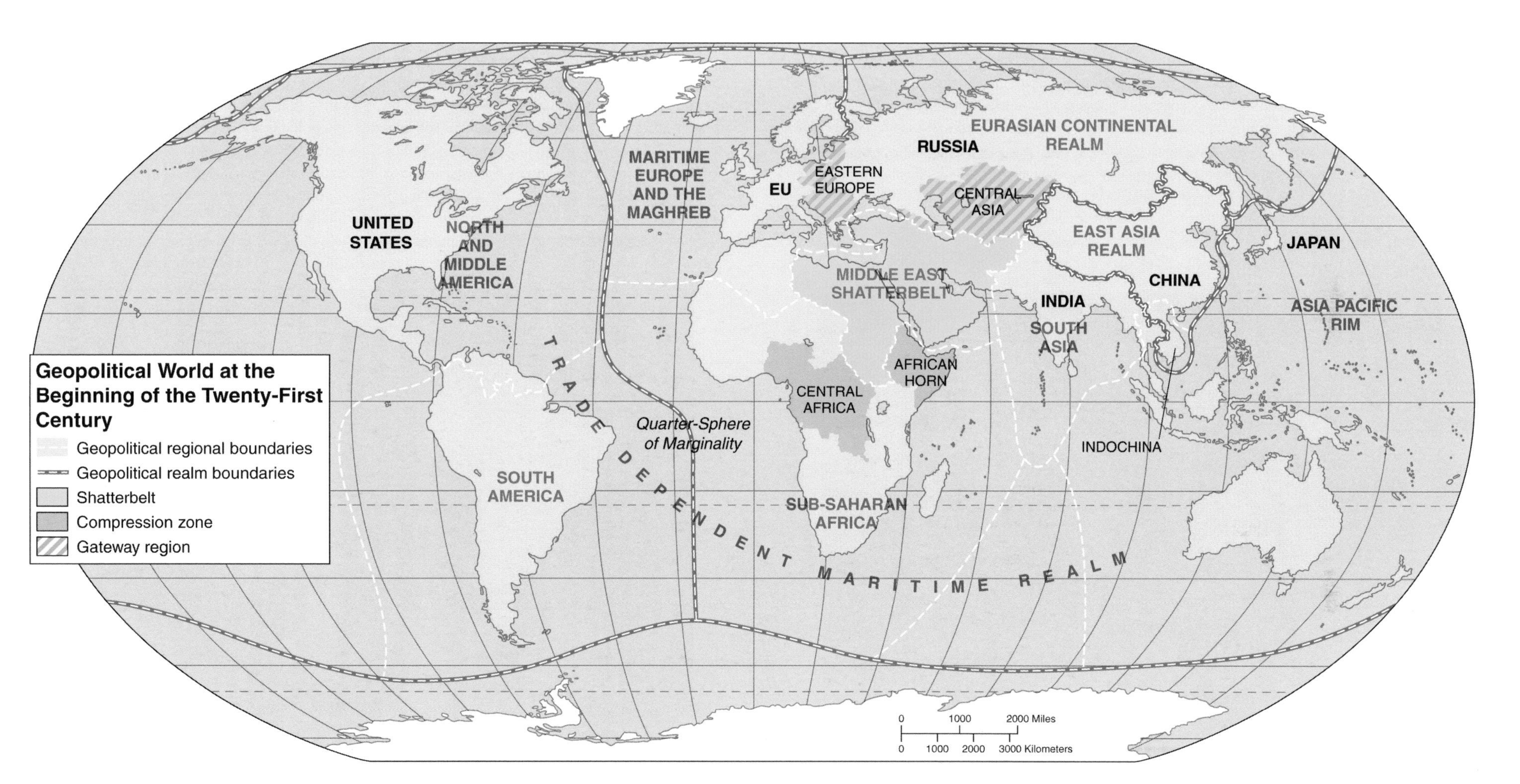

In the geostrategic structure of the world, the largest territorial units are realms. They are shaped by circulation patterns that link people, goods, and ideas. Realms are shaped by maritime and continental influences. Today's Atlantic and Pacific Trade-Dependent Maritime Realm has been shaped by international exchange over the oceans and their interior seas as mercantilism, capitalism, and industrialization gave rise to maritime-oriented states and to economic and political colonialism. The world's leading trading and economic powers are part of this realm. The Eurasian Continental Realm, centered around Russia, is inner-oriented, less influenced by outside economic or cultural forces, and politically closed, even after the fall of Communism. Expansion of NATO in Europe has increased its feeling of being "hemmed in." East Asia has mixed Maritime and Continental influences. China has traditionally been continental, but reforms that began in the late 1970s increased the importance of its maritime-oriented southern coasts. Even so, its trade volume is still low, and it maintains a hold on inland areas like Tibet and Xinjiang. Realms are subdivided into regions, some dependent on others, as South America is on North America. Regions located between powerful realms or regions may be shatterbelts (internally divided and caught up in competition between Great Powers) or gateways (facilitating the flow of ideas, goods, and people between regions). Compression zones are areas of conflict, but they are not contested by major powers.

Map 93 Flashpoints, 2006

Chechnya: The area in southern Russia known as the Caucasus Region is home to a large variety of non-Russian ethnic groups; many are Muslim and resent centuries of Russian domination and Soviet-era totalitarianism. After the Soviet Union disintegrated in 1991, several of these ethnic groups began agitating for more autonomy from Moscow or for outright independence. One of the more vocal groups with a history of opposition to Moscow's rule were the Chechens. The Chechens declared themselves a sovereign nation in 1991 and by 1994 relations between the breakaway government in Chechnya and the Russian government had drastically deteriorated. In December of that year, Russian forces attacked Chechnya, beginning the first of two (1994–96 and 1999–present) full-scale military conflicts that have also crept into the neighboring Russian autonomous area of Dagestan, itself largely Muslim. In the mid- and late 1990s Russia experienced several terrorist attacks in cities throughout the nation, which the Russian government attributed to Islamic extremists supporting Chechen independence. As a result, a second round of the conflict began in August 1999 with a full-scale Russian military assault on Dagestan and Chechnya. This assault is ongoing and continues to face intense resistance, with heavy casualties on both sides. In 2003 and 2004 Chechen rebels increased their pressure with urban terrorist activities in Russian cities, including Moscow.

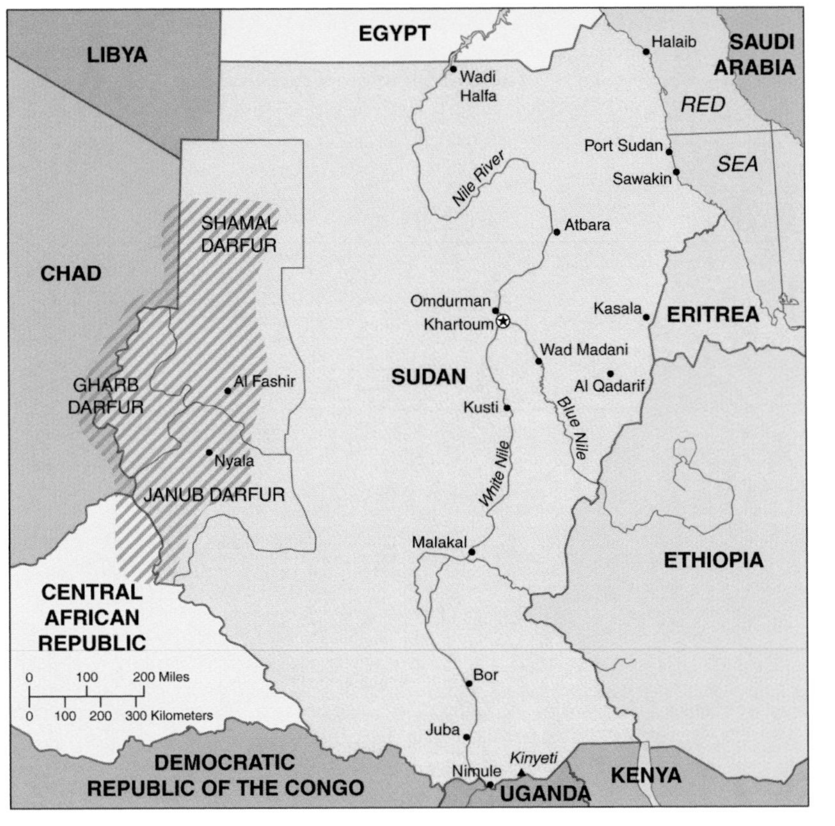

Sudan and Darfur Region: Since Sudan achieved independence from Great Britain in 1956, military regimes favoring Islamic-oriented governments have dominated national politics. These regimes have embroiled the country in a civil war for nearly all of the past half-century. These wars have been rooted in the attempts of northern economic, political, and social interests dominated by Muslims to control territories occupied by non-Muslim, non-Arab southern Sudanese such as the Dinka tribal groups. Since 1983, the war and war- and famine-related effects have resulted in more than 2 million deaths and over 4 million people displaced. The current ruling regime is a mixture of military elite and an Islamist party that came to power in a 1989 coup. Some northern opposition parties have made common cause with the southern rebels and entered the war as part of an anti-government alliance. Peace talks gained momentum in 2002-03 with the signing of several accords, including a cease-fire agreement. However, conflicts have continued to persist in the 3 Darfur provinces of western Sudan adjacent to the border with Chad and the Central African Republic where government-backed Muslim militia have attacked and killed tens of thousands of non-Muslim tribal peoples. In 2005 and 2006, areas of conflict spilled over the borders of Sudan to involve both Chad and the Central African Republic. The Darfur region is relatively water-rich and forested in a country that is chiefly desert and is therefore desired by Muslim pastoral groups from the north for settlement purposes. International organizations have labeled the conflict in Darfur as "genocide" against the non-Muslim populations.

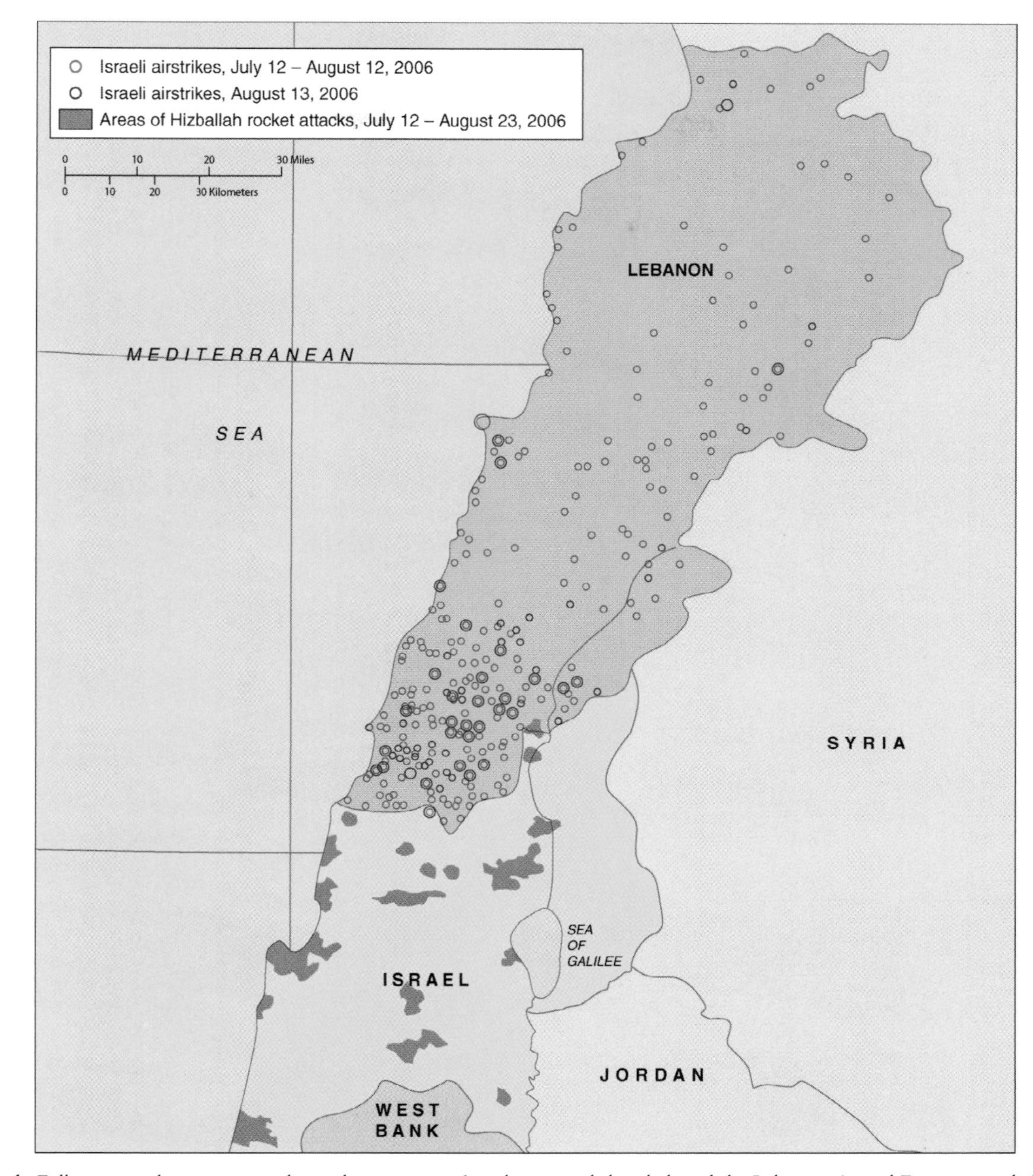

Lebanon-Israel: Following a devastating civil war between 1976 and 1991, Lebanon's Muslim majority (about 53%) and Christian minority (about 35%) agreed to rebuild its political institutions via a "blueprint for national reconciliation." In this, the Lebanese established a more equitable political system, particularly by giving Muslims a greater voice in the political process while institutionalizing sectarian divisions in the government. Lebanon's neighbor, Syria, intruded into Lebanese territory with military forces and, in response, Israel occupied a good part of southern Lebanon until 2000. Israeli withdrawal left many areas of Lebanon under Syrian control until the passage of a United Nations resolution in 2004, calling for Syria to withdraw from Lebanon and end its interference in Lebanese affairs. This encouraged the Lebanese government to oppose Syria's presence in Lebanon. In February 2005, the assassination of a popular former prime minister led to the "Cedar Revolution," massive demonstrations in Beirut against the Syrian presence. In April 2005, Syria finally withdrew the remainder of its military forces from Lebanon. In May–June 2005, Lebanon held its first legislative elections since the end of the civil war free of foreign interference, and the nation began rebuilding its former thriving economy. Most militias were disbanded, and the Lebanese Armed Forces extended authority over about two-thirds of the country. Unfortunately, Hizballah, a radical Shi'a organization listed by the U.S. State Department as a Foreign Terrorist Organization, retained its weapons and control over much of southern Lebanon. In July 2006, Hizballah militia kidnapped Israeli soldiers and launched rockets and mortars into northern Israel. Israel retaliated by blockading Lebanese ports and airports and by conducting airstrikes against known or suspected Hizballah locations throughout Lebanon. Since Hizballah militia are located throughout urban areas, Israeli retaliation resulted in many civilian deaths as well. And, of course, most of the Israelis killed by Hizballah rockets were civilians. Nine hundred thousand Lebanese and three hundred thousand Israelis were displaced by warfare, and normal life was disrupted across all of Lebanon and northern Israel. A UN resolution, which was approved by both Lebanese and Israeli governments in August 2006, called for the disarming of Hizballah, for Israel to withdraw, and for the deployment of the Lebanese Army and an enlarged UN peace-keeping force in southern Lebanon. The disarming of Hizballah has proven almost impossible, and Hizballah forces continue to occupy much of southern Lebanon. The situation remains tense.

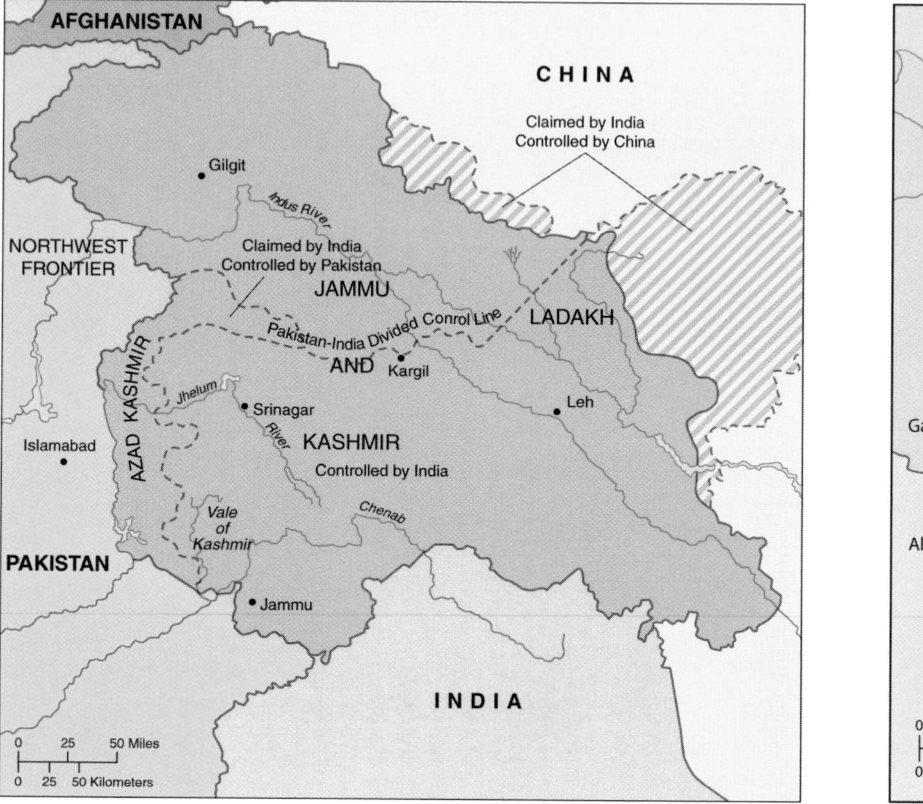

Jammu and Kashmir: When Britain withdrew from South Asia in 1947, the former states of British India were asked to decide whether they wanted to become part of a new Hindu India or a Muslim Pakistan. In the state of Jammu and Kashmir, the rulers were Hindu and the majority population was Muslim. The maharajah (prince) of Kashmir opted to join India, but an uprising of the Muslim majority precipitated a war between India and Pakistan over control of this high mountain region. In 1949 a cease-fire line was established by the UN, leaving most of the territory of Jammu and Kashmir in Indian hands. Since then Pakistan and India have waged intermittent skirmishes over the disputed territory that holds the headwaters of the Indus River, a life-giving stream to desert Pakistan. In 1999 extremist Muslim groups demanding independence escalated the periodic battles into a full-fledged, if small, war between two of Asia's major powers—both possessing nuclear weapons.

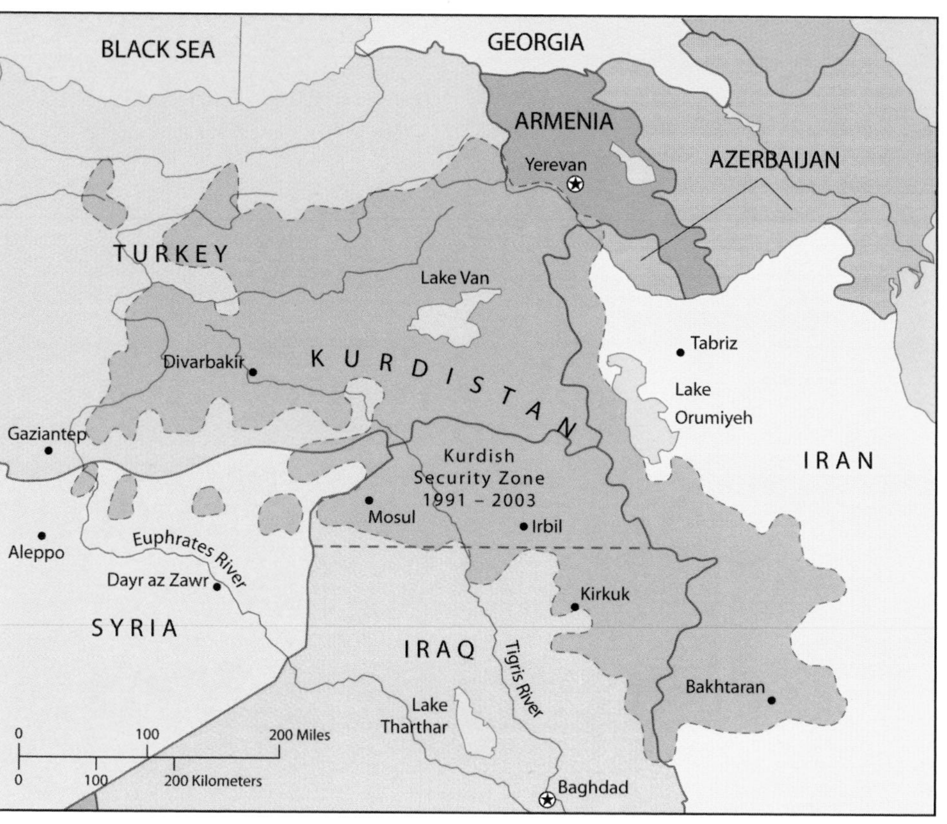

Kurdistan: Where Turkey, Iran, and Iraq meet in the high mountain region of the Tauros and Zagros mountains, a nation of 25 million people exists. This nation is "Kurdistan," but the Kurds, the occupants of this area for over 3,000 years, have no state, and receive much less attention than other stateless nations like the Palestinians. Following the 1991 Gulf War between Iraq and a U.S.-led coalition of European and Arabic states, the United Nations demarcated a Kurdish "security zone" in northern Iraq. From 1991 to 2003 the Security Zone was anything but secure as Iraqi militants from the south and Turks from the north infringed on Kurdish territory, and the internal militant extremist groups, such as the Kurdish Workers' Party, staged periodic attacks on rival villages. During the 2003 U.S.-led invasion of Iraq that eliminated the Baathist regime of Saddam Hussein, the Kurds played an important role in securing the northern portions of Iraq for the U.S.-British coalition and fought alongside American troops in expelling elements of the Iraqi army from cities like Mosul and Kirkuk. Rich in oil and history, Kurdistan will probably remain as a nation without a state, shared by Iraq, Turkey, and Iran—none of which is likely to give up substantial portions of territory for the establishment of a Kurdish state.

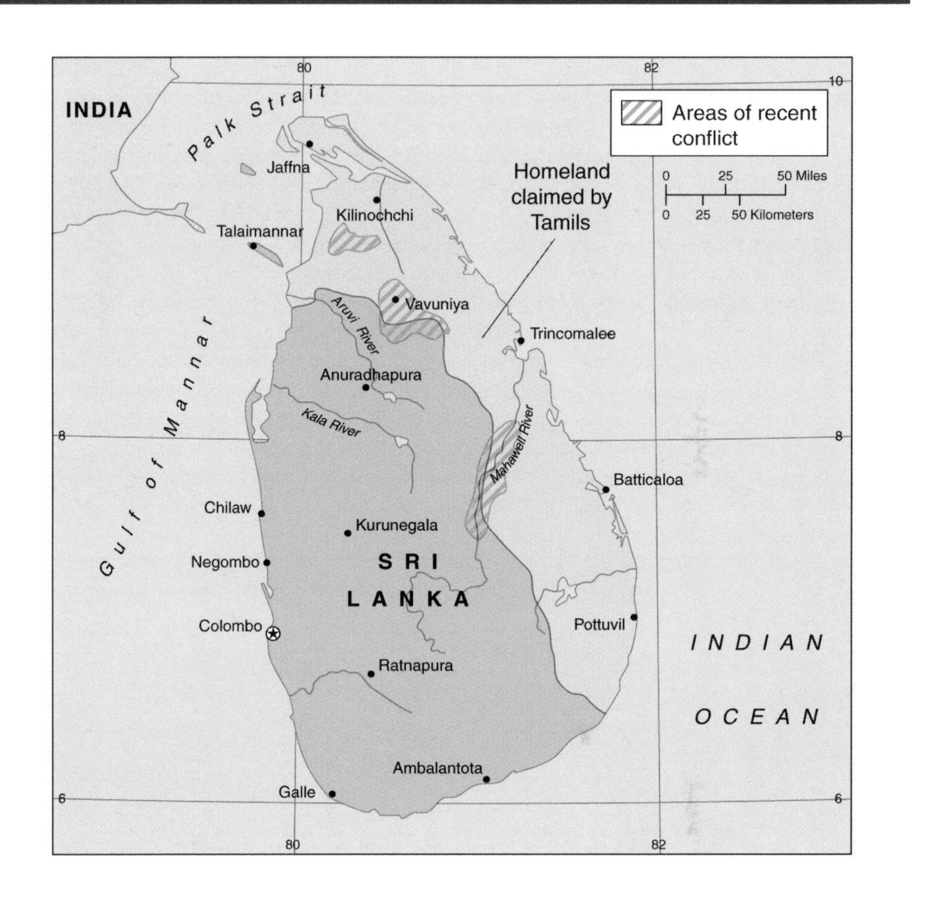

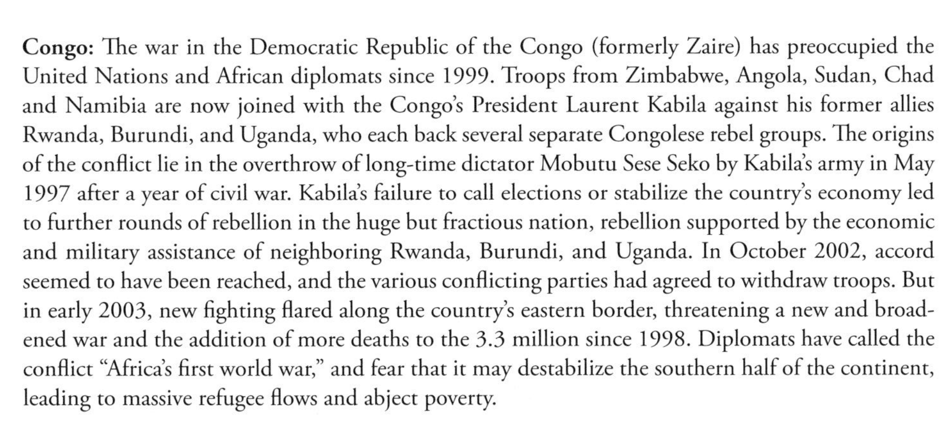

Congo: The war in the Democratic Republic of the Congo (formerly Zaire) has preoccupied the United Nations and African diplomats since 1999. Troops from Zimbabwe, Angola, Sudan, Chad and Namibia are now joined with the Congo's President Laurent Kabila against his former allies Rwanda, Burundi, and Uganda, who each back several separate Congolese rebel groups. The origins of the conflict lie in the overthrow of long-time dictator Mobutu Sese Seko by Kabila's army in May 1997 after a year of civil war. Kabila's failure to call elections or stabilize the country's economy led to further rounds of rebellion in the huge but fractious nation, rebellion supported by the economic and military assistance of neighboring Rwanda, Burundi, and Uganda. In October 2002, accord seemed to have been reached, and the various conflicting parties had agreed to withdraw troops. But in early 2003, new fighting flared along the country's eastern border, threatening a new and broadened war and the addition of more deaths to the 3.3 million since 1998. Diplomats have called the conflict "Africa's first world war," and fear that it may destabilize the southern half of the continent, leading to massive refugee flows and abject poverty.

Sri Lanka: The island state of Sri Lanka, historically known as Ceylon, is potentially one of the most agriculturally productive regions of Asia. Unfortunately for plans related to agricultural development, two quite different peoples have occupied the island country: The Buddhist Sinhalese originally from northern India and long the dominant population in Sri Lanka, and the minority Hindu Tamil, a Dravidian people from south India. Since independence from Britain, Sri Lankan governments have sought to "resettle" the Tamil population in south India, actions that finally precipitated an armed rebellion by Tamils against the Sinhalese-dominated government. The Tamils at present are demanding a complete separation of the state into two parts, with a Tamil homeland in the north and along the east coast. At one time viewed as an island paradise, Sri Lanka is now a troubled country with an uncertain future. A cease fire between Sinhalese and Tamil fighters was brokered in 2001 by Norway but fell apart in late 2003 with the resumption of violence. In May 2004, Norwegian diplomats again made an attempt to negotiate a cease fire but with questionable results.

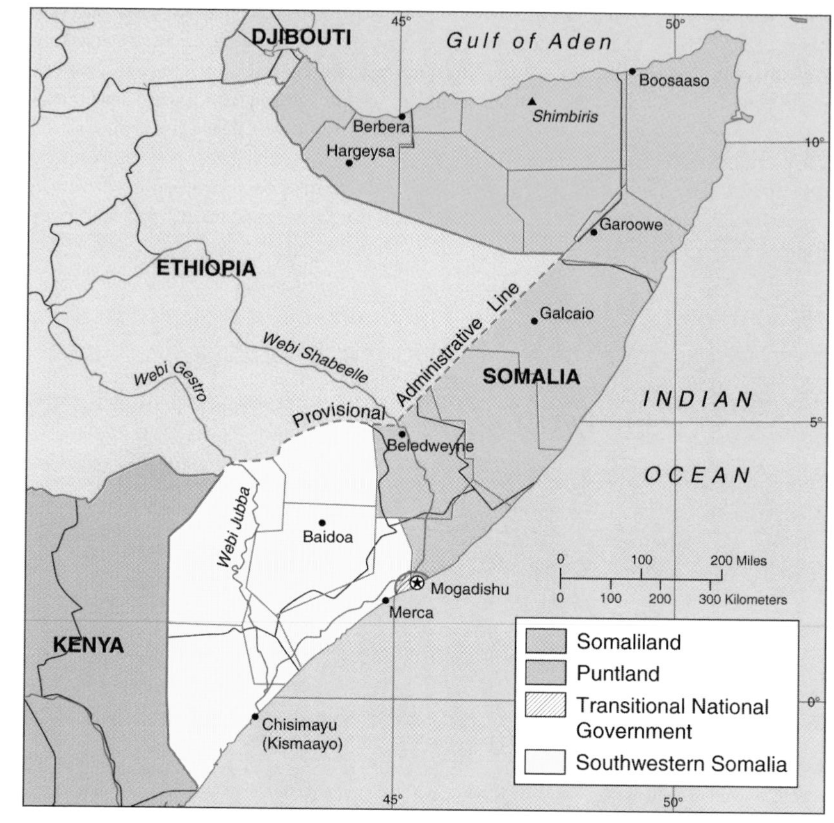

Somalia: With the ouster of the government led by Mohamed Said Barre in January 1991, turmoil, factional fighting, and anarchy have followed in Somalia, with several separate governments arising in different parts of the country. The northern clans declared an independent Republic of Somaliland. Which, although not recognized by any government, has maintained a stable existence, aided by the overwhelming dominance of a ruling clan and economic infrastucture left behind by British, Russian, and American military assistance programs. Puntland, the central portion of Somalia, from the Horn of Africa to the coast of the Indian Ocean and the border with Ethiopia, has been a self-governing autonomous state since 1998. The area of Southwestern Somalia is poorly organized and more conflict-ridden than any other part of the country, with much of that conflict centered on attempts to control the nominal Somali capital of Mogadishu and ongoing famine. UN humanitarian efforts were unable to either quell guerrilla activity or alleviate famine and the UN withdrew in 1995. In 2004 a new government, the Transitional Federal Government (TFG), was created for the entire country. The president and parliament of the new government have not yet moved to Mogadishu and discussions regarding the government's establishment are ongoing in Kenya. Numerous warlords and factions are still fighting for control of the capital city as well as for other southern regions.

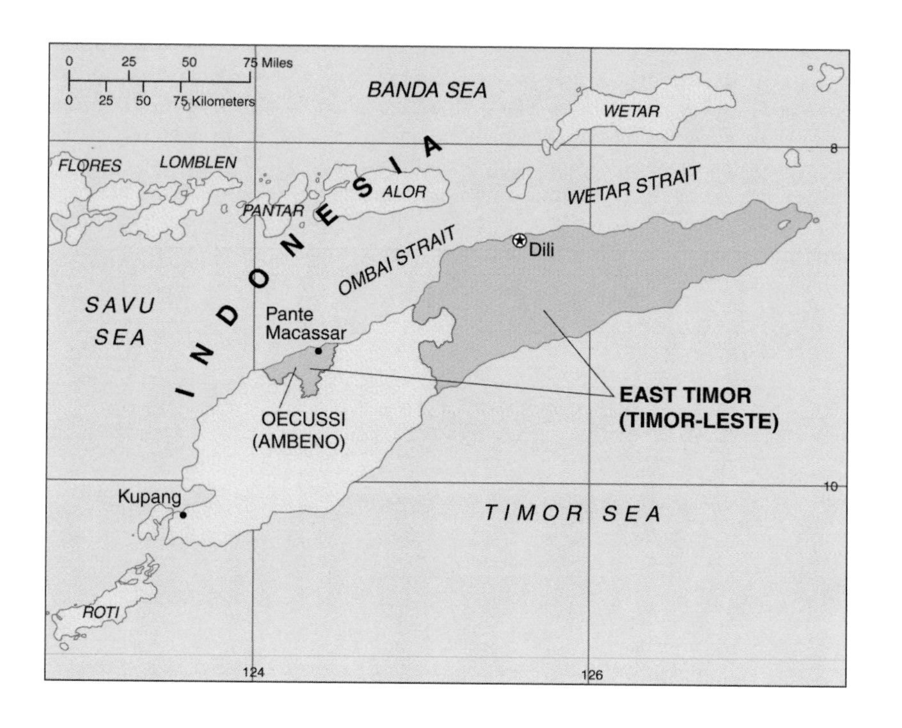

East Timor: The island of Timor was incorporated into Indonesia as a province in July 1976. For two decades, Indonesian troops attempted to pacify the natives of the island, at the cost of somewhere between 100,000 to 250,000 lives. In August 1999, in a UN-supervised popular referendum, an overwhelming majority of the people of the eastern portion of Timor voted for independence from Indonesia and incorporation as a separate state of East Timor. A multinational peacekeeping force was slated to arrive in September 1999 to oversee the transition. Prior to the arrival of that force, however, anti-independence Timorese militias—most of them linked directly to the Indonesian military—carried out a large-scale campaign of retribution against the supporters of independence, killing over 1,400 Timorese and forcing more than 300,000 to flee to West Timor as refugees. During the retribution insurgency, most of the country's infrastructure (homes, irrigation systems, schools, transportation lines, public utilities) was destroyed. The violence was brought to an end with the arrival of an Australian-led peacekeeping force at the end of September 1999. East Timor was officially recognized as an independent state in May 2002. The boundary between East Timor and Indonesia continues to be in dispute, however, and Indonesia and East Timor are contesting for control of coral islands north of the island, preventing delimitation of the northern maritime boundaries. Moreover, many refugees who left East Timor in 2003 still reside in Indonesia and refuse repatriation, placing pressures on both governments. Conflicts also exist with Australia over off-shore oil deposits to the south, hampering the creation of a southern maritime boundary with Indonesia. Any country with unresolved boundaries and large numbers of disaffected populations is, by definition, a flashpoint.

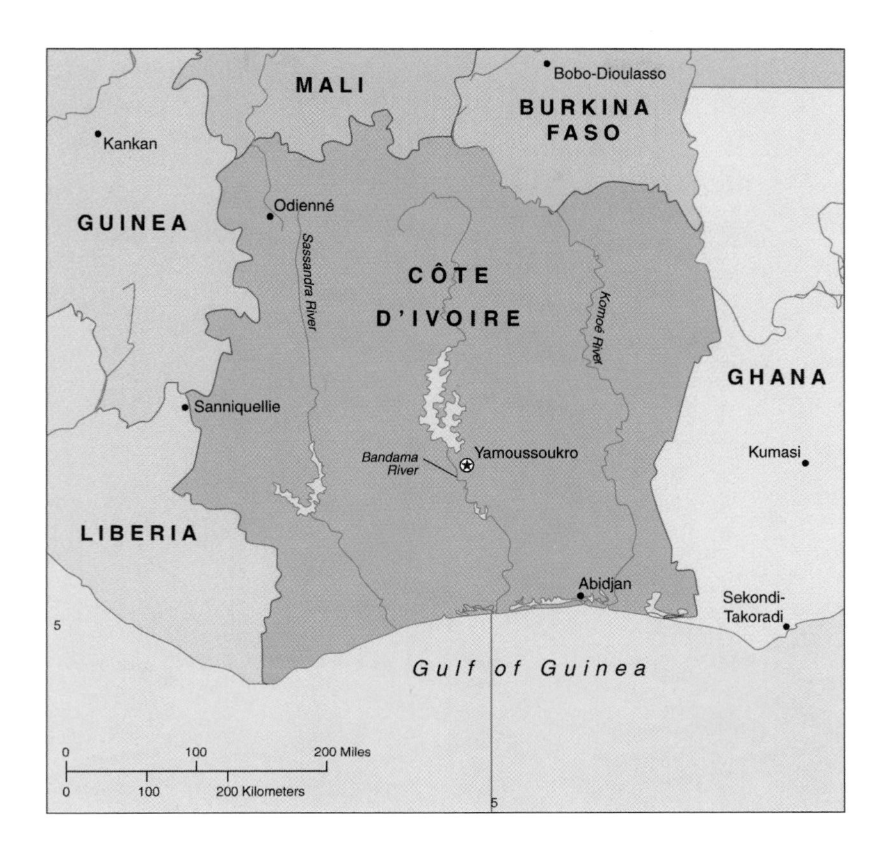

Côte d'Ivoire: Until relatively recently, its close ties to France following independence in 1960 and its development of cocoa production for export (along with significant foreign investment) made Côte d'Ivoire (Ivory Coast) one of the most prosperous of the tropical African states. Since 1999, however, political turmoil has disrupted economic development. In December 1999, the first military coup in Côte d'Ivoire's history overthrew the elected government. The leader of the coup rigged elections held in late 2000 and assumed "legitimate" power, but popular protest forced him to step aside in favor of the runner-up in the 2000 elections. Another coup, this one a failure, was launched by Ivorian dissidents and disaffected members of the military in September 2002. Rebel forces claimed the northern half of the country, and in January 2003 were granted ministerial positions in a unity government. For all practical purposes, Côte d'Ivoire has been since then in a state of civil war with the central government yet to exert control over the northern regions. In late 2006, anti-government protests intensified, leaving no apparent end in sight to a political impasse that has pitted the president against his prime minister. The international community and particularly the United Nations has been divided over how to deal with the continuing civil conflict in Côte d'Ivoire.

Colombia: Colombia was one of three countries (along with Ecuador and Venezuela) that emerged from the collapse of the newly independent country of Gran Colombia in 1830 (the former Spanish Viceroyalty of New Granada). Although possessing a rich natural resource base and agricultural abundance, Colombia has been torn apart internally for more than four decades as the result of conflict between government forces and antigovernment insurgent groups and illegal paramilitary groups. The two largest insurgent groups are the Revolutionary Armed Forces of Colombia (FARC) and the National Liberation Army (ELN). The largest illegal paramilitary group is a semi-organized array of disparate paramilitary forces called the United Self-Defense Groups of Colombia (AUC). Both insurgency and paramilitary organizations are heavily funded by the illegal trade in cocaine, the most important (but untaxed) commodity in the Colombian economy. The insurgents do not have the military or popular support necessary to overthrow the government but continue to engage in attacks against government forces and civilians, and many areas of the country are under the control of guerillas who double as drug traffickers. The insurgency groups, with a more political agenda, and the paramilitary groups, who have a fairly simple agenda of controlling the cocaine trade, not only battle the government forces but also one another. Large swaths of internal Colombia are under the control of either insurgents or paramilitary drug lords. Over the last few years, the Colombian government has increased efforts to regain control of these areas, with some mild success. However, neighboring countries remain concerned about the probability of the violence spilling into their borders. Colombian-organized illegal narcotics, guerrilla, and paramilitary activities penetrate all of its neighbors' borders and have created a serious refugee crisis with over 300,000 persons having fled the country, mostly into neighboring states.

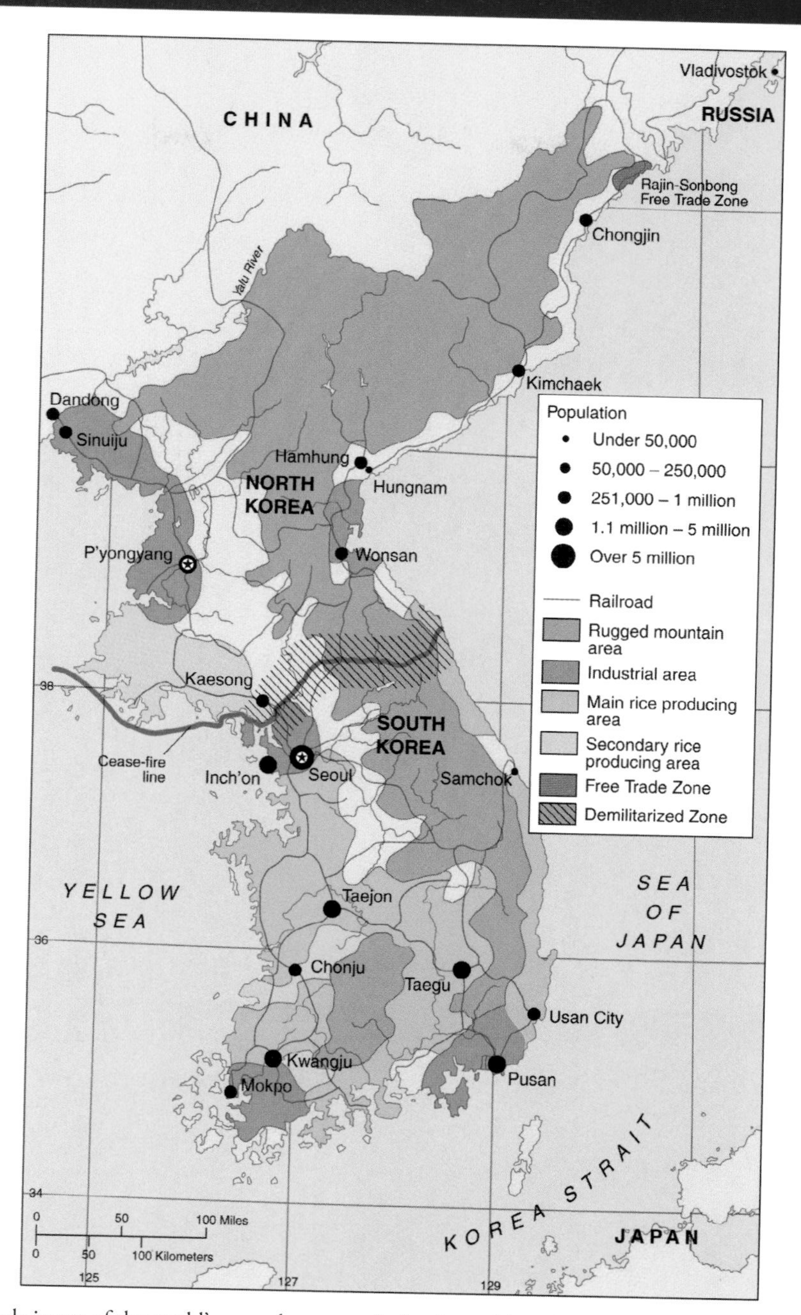

Korean Peninsula: The Korean peninsula is one of the world's most dangerous flashpoints, although active military conflict on even a minor scale has not existed since the 1950s. Throughout much of its history, Korea was an independent kingdom, although periodically a province of one of the Chinese dynasties. Beginning in the early twentieth century, Korea was annexed and occupied by Japan until the end of World War II. At that time, Korea was split into North Korea under Communist domination, and South Korea as an ostensible republic backed by the United States. Since the end of the Korean War of 1950–53—an attempt on the part of Communist North Korea, allied with China, to reunite the entire peninsula under Communist rule—South Korea has flourished economically and has become one of the most productive of so-called Pacific Tigers with strong and profitable trading links to many different countries. North Korea, on the other hand, adopted a policy of ostensible diplomatic and economic "self-reliance" and became one of the world's most reclusive, authoritarian, and isolated states. North Korea molded its political, economic, and military policies around the core ideological objective of eventual unification of Korea and still retains that objective. While South Korea was flourishing economically, North Korea so mismanaged and misallocated its resources that, since the mid-1990s, the country has not been able to feed itself. It continues to expend resources to maintain an army of 1 million—one of the world's largest. North Korea also has active military programs, including long-range missile development as well as chemical and biological weapons programs, massive conventional armed forces, and the development of nuclear weapons with long-range missile delivery systems. All of these programs are of major concern to the international community. In 2003, North Korea announced that it would violate a 1994 agreement with the United States to freeze and ultimately dismantle its existing plutonium-based program. It also expelled monitors from the International Atomic Energy Agency (IAEA) and withdrew completely from the international Non-Proliferation Treaty. By 2006, North Korea announced the development of nuclear weapons and delivery systems that were designed to "protect" North Korea against American aggression. Given that the United States still maintains a military presence in South Korea, the potential for destabilization of the entire peninsula exists up to and including the use of North Korean nuclear weapons against South Korea and the "American occupiers."

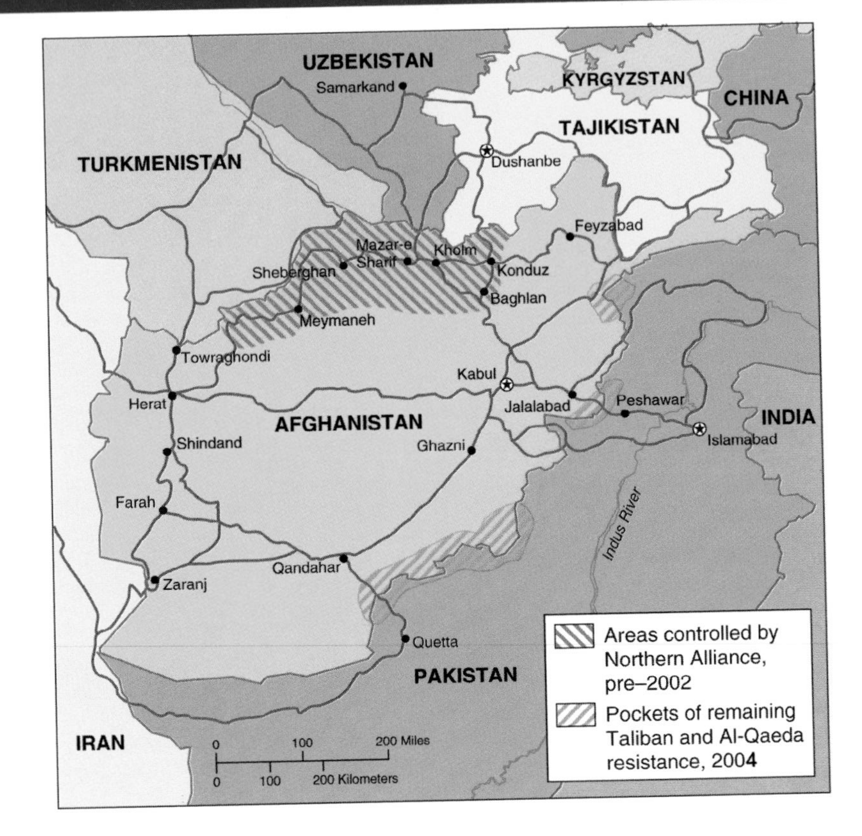

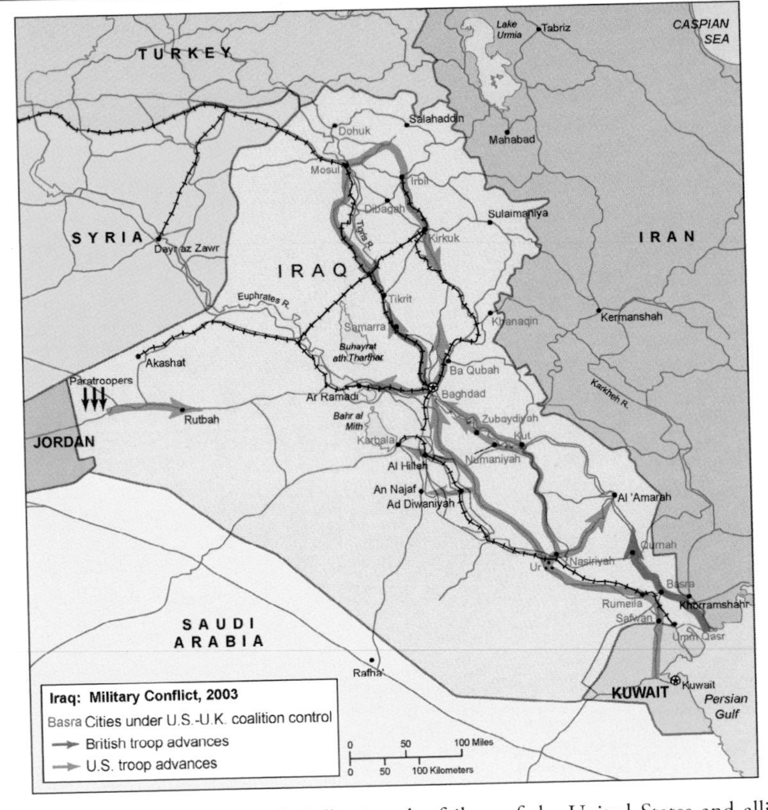

Afghanistan: In the aftermath of the tragic September 11, 2001, terrorist attacks on the World Trade Center and the Pentagon, the United States (backed to varying degrees by its allies) has declared a massive and global "war on terrorism" and any states that may provide "safe harbor" to terrorists. To date, the most prominent target of this U.S. declaration of war has been the Taliban regime of Islamic extremists who controlled about 95 percent of the territory of the beleaguered nation of Afghanistan. International observers believe that the Taliban regime has welcomed and provided a base for the al-Qaeda terrorist network dominated by Saudi expatriate and millionaire Osama bin Laden since the late 1990s. As a result of this intelligence, the U.S. and Britain pursued a daily bombardment of key al-Qaeda and Taliban installations inside Afghanistan for several months, with key logistical support in the form of air bases and supply depots provided by the government of Pakistan (in exchange for financial considerations and political support of the non-elected Pakistani government). U.S. and allied ground troops, aided by members of the Northern Alliance of Afghan rebels opposed to the Taliban regime, expelled the Taliban government in 2002. While now under home rule and with a duly elected government, Afghanistan still is plagued by warlords in remote areas of the country who refuse to recognize the legally constituted government. In addition, along the Afghanistan-Pakistan border, significant pockets of resistance from remnants of the former Taliban regime and from al-Qaeda forces are engaged in ongoing military conflict with American and Pakistani troops.

Iraq: Military Conflict, 2003–2004: Following the failure of the United States and allies Great Britain and Spain to secure approval from the United Nations to begin a UN-sponsored military conflict to "disarm" Iraq and effect "regime change" by removing the Baathist party dictator Saddam Hussein from power, the United States and its allies launched an independent military attack on Iraq in April 2003. The military campaign began with massive air and sea bombardments on government and military targets, then ground troops moved from Kuwait along highway routes to secure the oil fields and major urban areas. By early May 2003, virtually all of the country was under the control of the U.S.-led coalition of forces. The Iraqi military, for the most part, melted away into the general civilian population and there were no major pitched battles for territory in the short-lived war. Conflicts between British forces and paramilitary/political forces in the important southern city of Al Basrah produced casualties on both sides before the city was secured. And the U.S. troops met guerrilla resistance from political affiliates of Hussein's regime in cities such as An Najaf and Al Nasiriyah. In the northern parts of the country, U.S. troops inserted by parachute joined forces with Kurdish paramilitary groups to secure major cities like Mosul, Arbit, and Kirkuk. Although the U.S. and coalition forces controlled the country by June 2003, civil authority and infrastructure continued to be unavailable throughout large areas of Iraq by mid-2004. The spring of 2004 brought significant insurgency action against U.S. and coalition forces, who were increasingly viewed by the Iraqis as occupiers rather than liberators. A June 30, 2004, date was set to return Iraq to civilian Iraqi control.

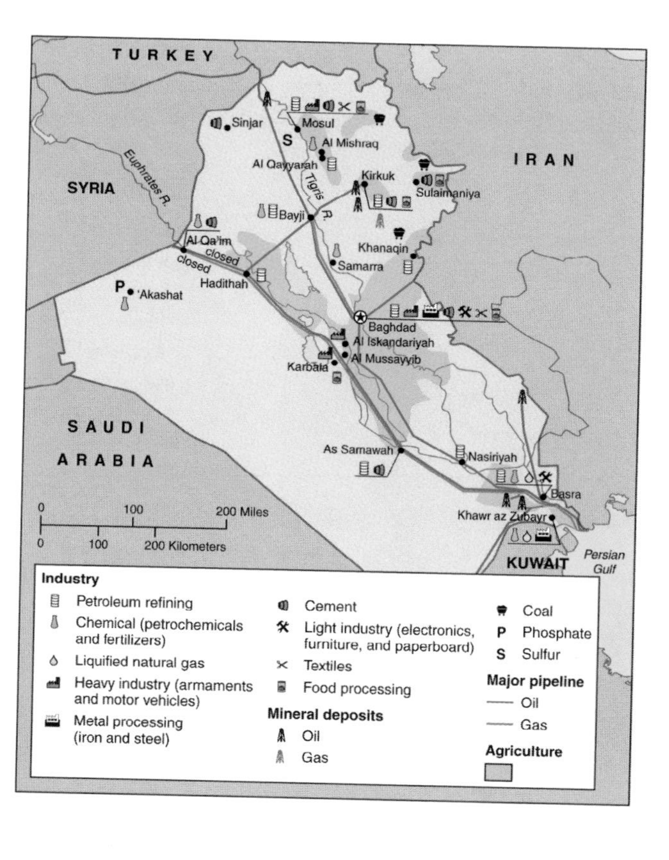

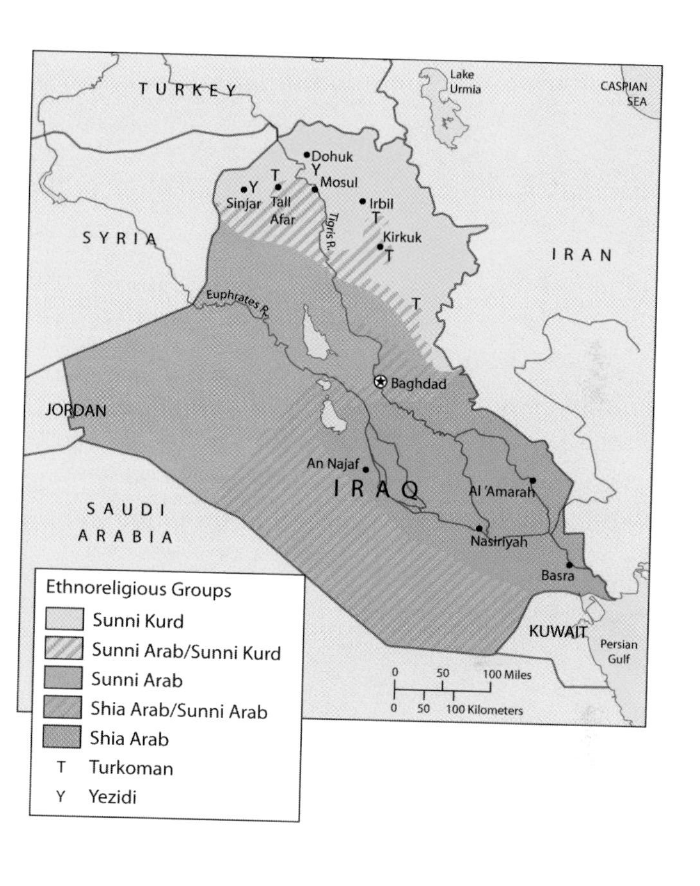

Iraq: Prior to the 1990–91 invasion of Kuwait by Iraq and the subsequent United Nations coalition's military expulsion of Iraq from its neighbor, Iraq was one of the most prosperous countries in the Middle East and the only one with full capacity to feed itself, even without the vast oil revenues generated by the country's immense reserves. Despite the inefficiencies of the Baathist dictatorship of Saddam Hussein, the country has a solid agricultural base and a burgeoning industry. The combination of military adventurism and conflict, in the form of a lengthy war with Iran and the ill-advised invasion of Kuwait, limited further economic development, however. Development was also problematic given the country's internal tensions between Arabic Sunni Muslims and Arabic Shiite Muslims, and between Arabs and Kurds and a few other minority populations in the northern parts of the country.

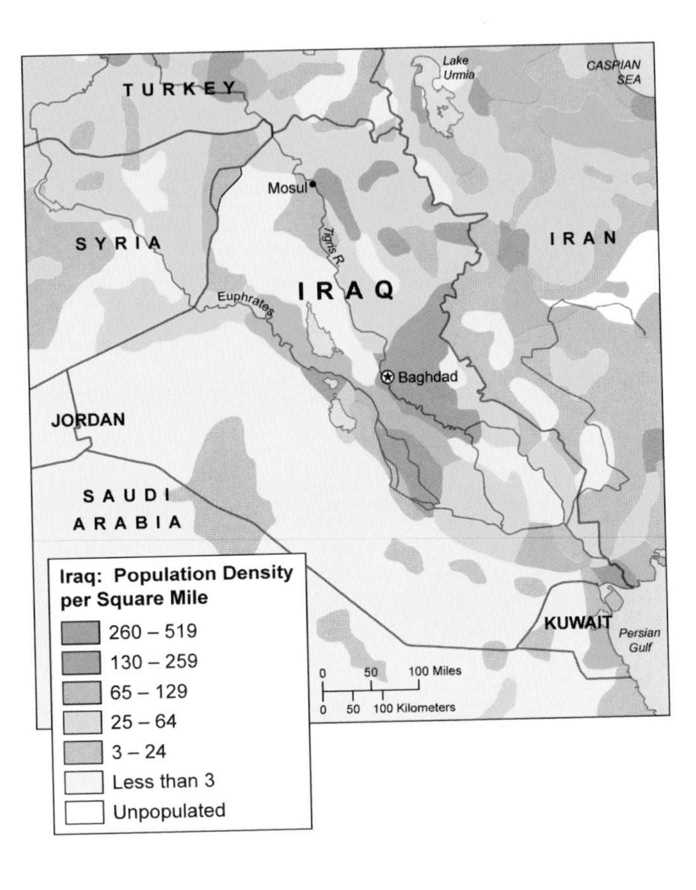

TURKEY
Lake Urmia
CASPIAN SEA
Mosul
Tigris R.
SYRIA
IRAN
IRAQ
Euphrates
Baghdad
JORDAN
SAUDI ARABIA
KUWAIT
Persian Gulf
0 50 100 Miles
0 50 100 Kilometers
Iraq: Population Density per Square Mile
260 – 519
130 – 259
65 – 129
25 – 64
3 – 24
Less than 3
Unpopulated

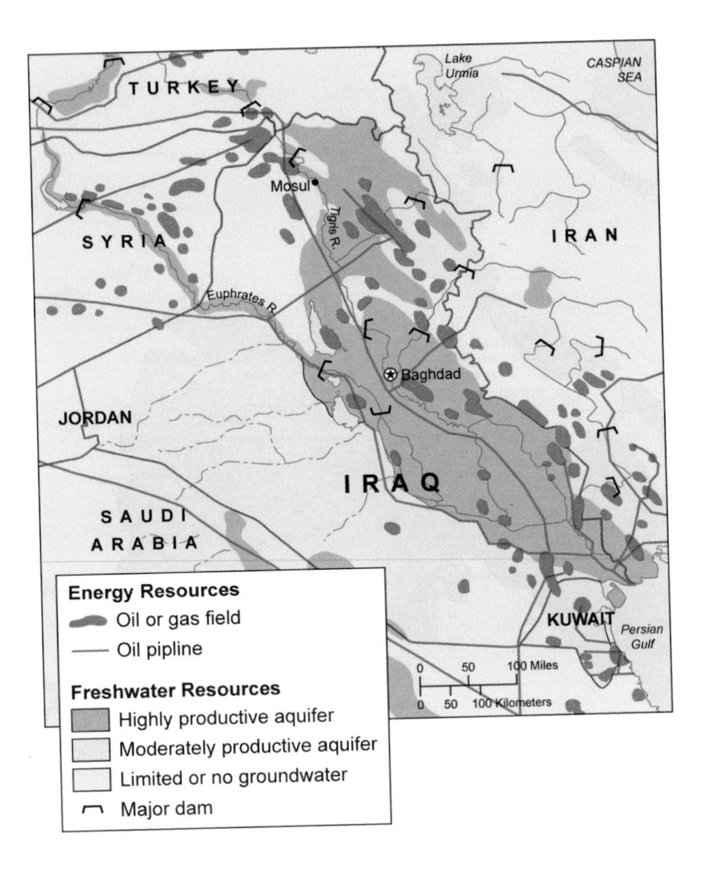

Lake Urmia
CASPIAN SEA
TURKEY
Mosul
Tigris R.
SYRIA
IRAN
Euphrates R.
Baghdad
JORDAN
IRAQ
SAUDI ARABIA
KUWAIT
Persian Gulf
0 50 100 Miles
0 50 100 Kilometers
Energy Resources
Oil or gas field
Oil pipline
Freshwater Resources
Highly productive aquifer
Moderately productive aquifer
Limited or no groundwater
Major dam

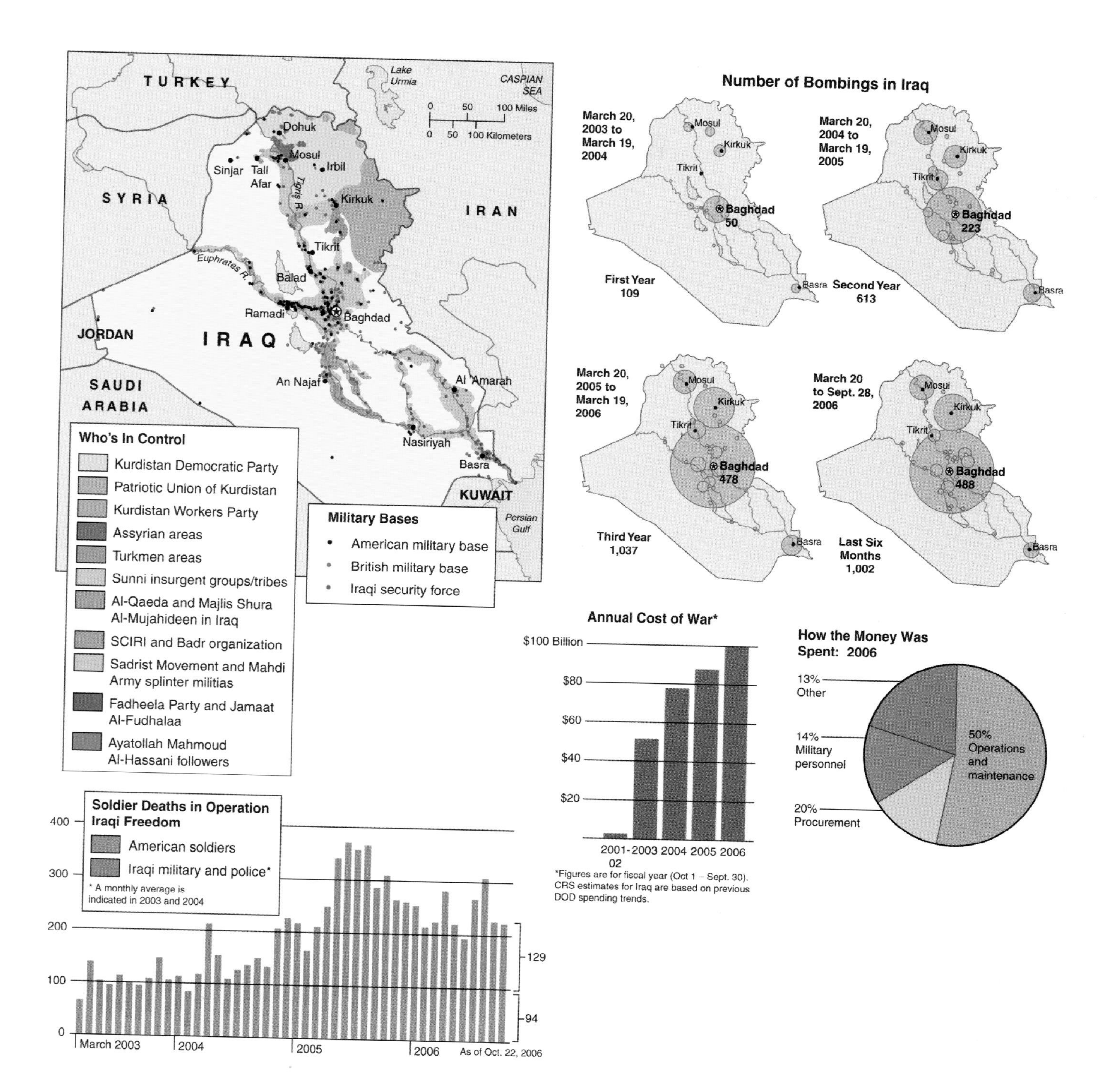

More than 3 years after the U.S.-led invasion, Iraq is in critical condition. Sectarian violence is rampant, death tolls continue to rise, and the price of Operation Iraqi Freedom has reached a staggering $318 billion. This is a look at the current state of Iraq.

Map 94 Nations With Nuclear Weapons

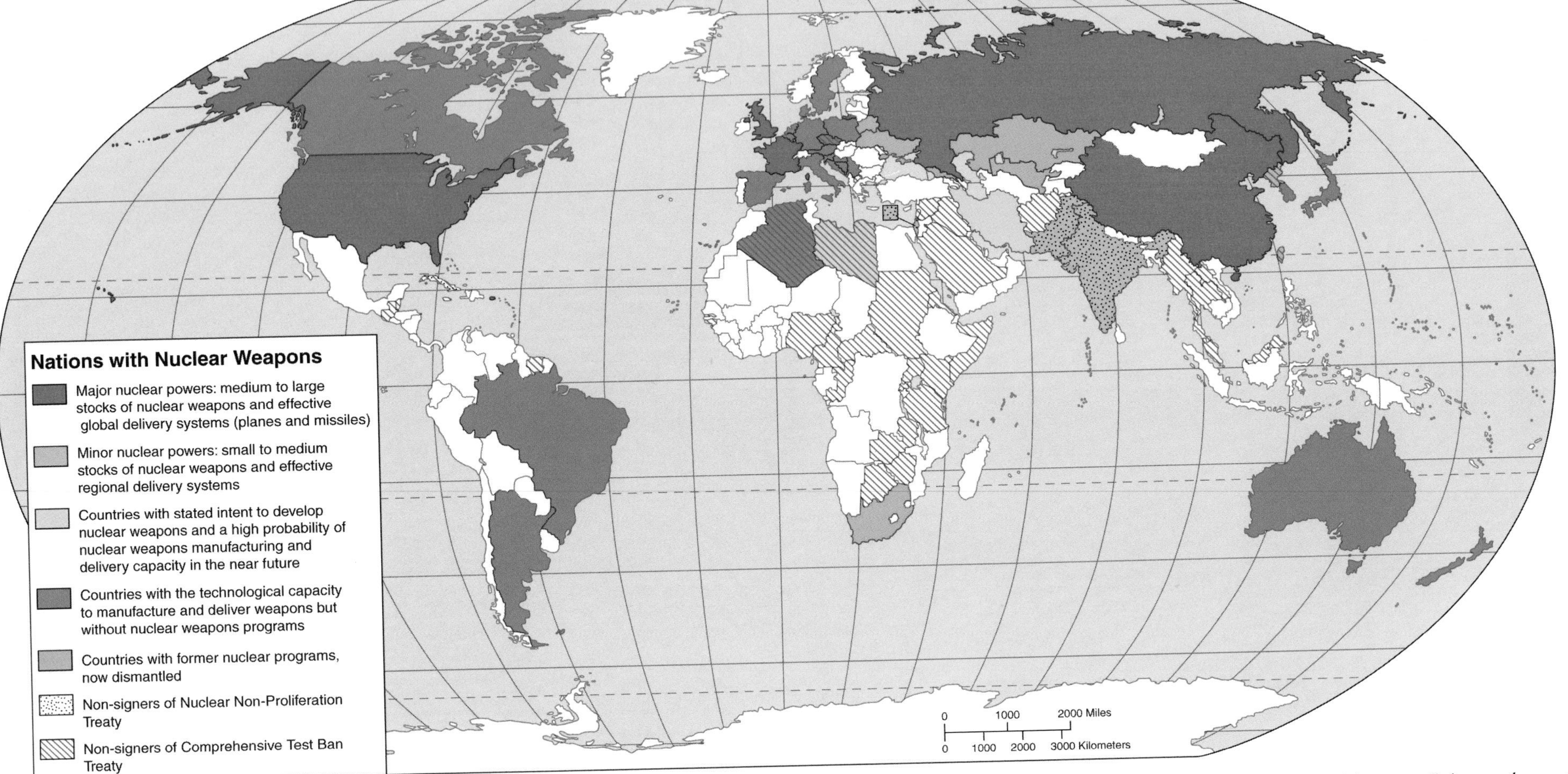

Since 1980, the number of countries possessing the capacity to manufacture and deliver nuclear weapons has grown dramatically, increasing the chances of accidental or intentional nuclear exchanges. In addition to the traditional nuclear powers of the United States, Russia, China, the United Kingdom, and France, must now be added Israel, India, and Pakistan as countries that, without possessing the large stocks of weapons of the major powers, nor the extensive delivery systems of the United States and Russia, still have effective regional (and possibly global) delivery systems and medium stocks of warheads. Countries such as Kazakhstan, Ukraine, Georgia, and Belarus that were created out of what had been the Soviet Union did have some nuclear capacity in the 1991–1995 period but have since had all nuclear weapons removed from their territories. However, North Korea has recently announced the re-establishment of its suspended nuclear weapons programs and may possess a small stock of nuclear warheads, along with the capacity to deliver those weapons regionally. Both Iran and Libya have nuclear ambitions, as did Iraq until the overthrow of the Baathist regime of Saddam Hussein by a U.S.-led military coalition in 2003. The proliferation of nuclear states threatens global security, and the objective of the Nuclear Non-Proliferation Treaty was to reduce the chances for expanding nuclear arsenals worldwide. This treaty has been partially successful in that a number of countries in the developed world certainly have the capacity to manufacture and deliver nuclear weapons but have chosen not to do so. These countries include Canada, European countries other than the United Kingdom and France, South Korea, Japan, Australia, and New Zealand, and Brazil and Argentina in South America. On the other side of the coin, the intent of North Korea to emerge as a nuclear power may force countries such as South Korea and Japan to re-think their positions as non-nuclear countries.

Map 95 Military Power

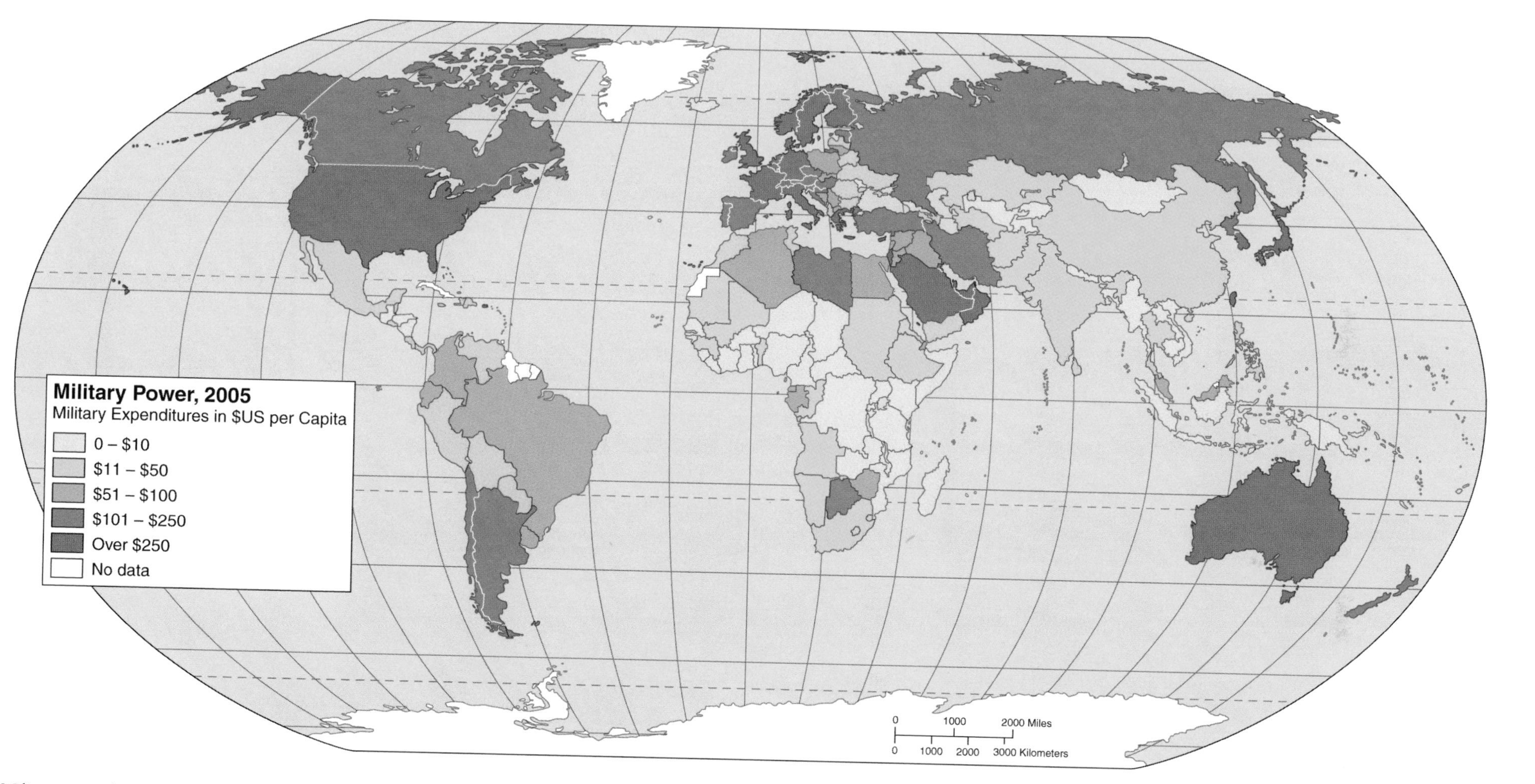

Military expenditures, measured in $US per capita, is often a good indicator of two things: a strong propensity toward defending "the Homeland" (few countries in the post–Cold War world actually think in terms of military aggression toward their neighbors), and an extensive network of political and military alliances that require significant military expenditures in the name of collective self-defense. The enormous per capita military expenditures of Saudi Arabia are explained in terms of the first category. Saudi Arabia has a limited ground force and almost no navy but one of the world's most advanced, well-trained, and well-equipped air forces. They view this as necessary to defend the homeland against attack from a neighboring country with which they might share political or, more likely, religious difference. Falling into the second category are those countries, such as the United States, and the other NATO allies that have obligation not only to themselves but to many others, including countries that are not members of NATO. The example of NATO troops and materiel being sent to Bosnia comes immediately to mind. Beyond these "normal" military expenditures are those of countries wishing to become regional powers through a large military force: Libya and Iran are excellent examples, although just during the last year, the expenditures by Libya have decreased and those of Iran have increased (2006 data were not available for the map). Finally, there are large per-capita military expeditures in what might be categorized as "rogue" states like North Korea for which the money spent on one of the world's largest standing armies and the development of a nuclear weapons system seem to make little sense in light of the population living at the starvation level.

Map 96 International Terrorist Incidents, 1998–2003

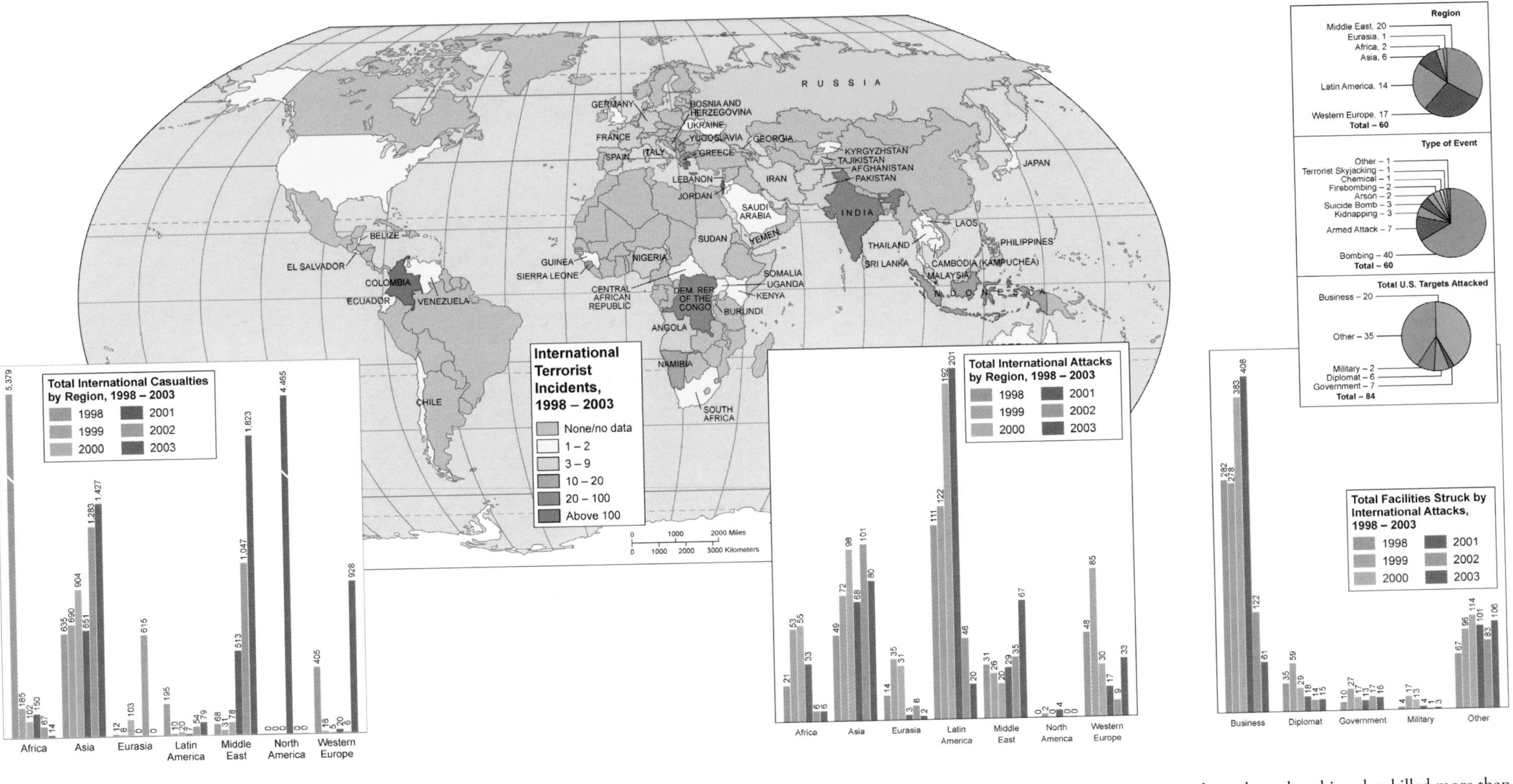

Americans have made a mantra of the saying "the world has changed" as a consequence of the terrorist attacks on the World Trade Center and the Pentagon on September 11, 2001. As the map above and the accompanying graphs of terrorist activities before and after 9/11/01 indicate, however, the world did not change, although the focus of a major terrorist attack shifted from Africa, Asia, and the Middle East to North America. Many other areas of the world have lived with terrorism and terrorist activity for years. In 2000 and 2001, despite the enormous losses in the United States in the 9/11 attacks, more lives were lost in Asia and Africa as a result of terrorism than were lost in North America. The world did not change, but Americans' perception of that world and their place in it has certainly changed. Events subsequent to 9/11 indicate a shift back to more "normal" patterns in North America from 2002 to the present, a time during which terrorist activities have been virtually absent. This is in sharp contrast to other parts of the world where terrorist activities have been on the upswing, the Madrid train bombing that killed nearly 200 people, and the London subway bombing that killed more than 50. Two qualifiers need to be added regarding the data presented: the term "casualties" refers to dead and wounded (the number of the dead represents a relatively small percentage of total casualties by region); the term "terrorist attacks" does not include insurgency events in areas of military conflict or occupation and, therefore, does not include attacks carried out against U.S. forces or others in places like Afghanistan or Iraq where military action continued throughout the time frame of the data. And, as terrible as they are, terrorist attacks do not include the casualties of sectarian violence such as that between Iraqi Shiites and Sunnis or between paramilitary organizations such as Hizballah and the legal state of Israel. In late 2003, the manner of collecting data on terrorist incidents and casualties therefrom changed, as did the definition of what constituted an international terrorist event. Hence, data for 2004 and 2005 are not comparable to data from 2003 and earlier.

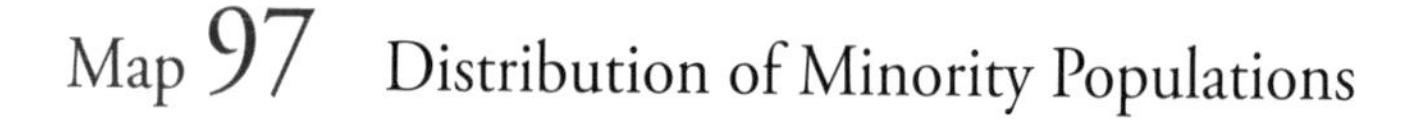

Map 97 Distribution of Minority Populations

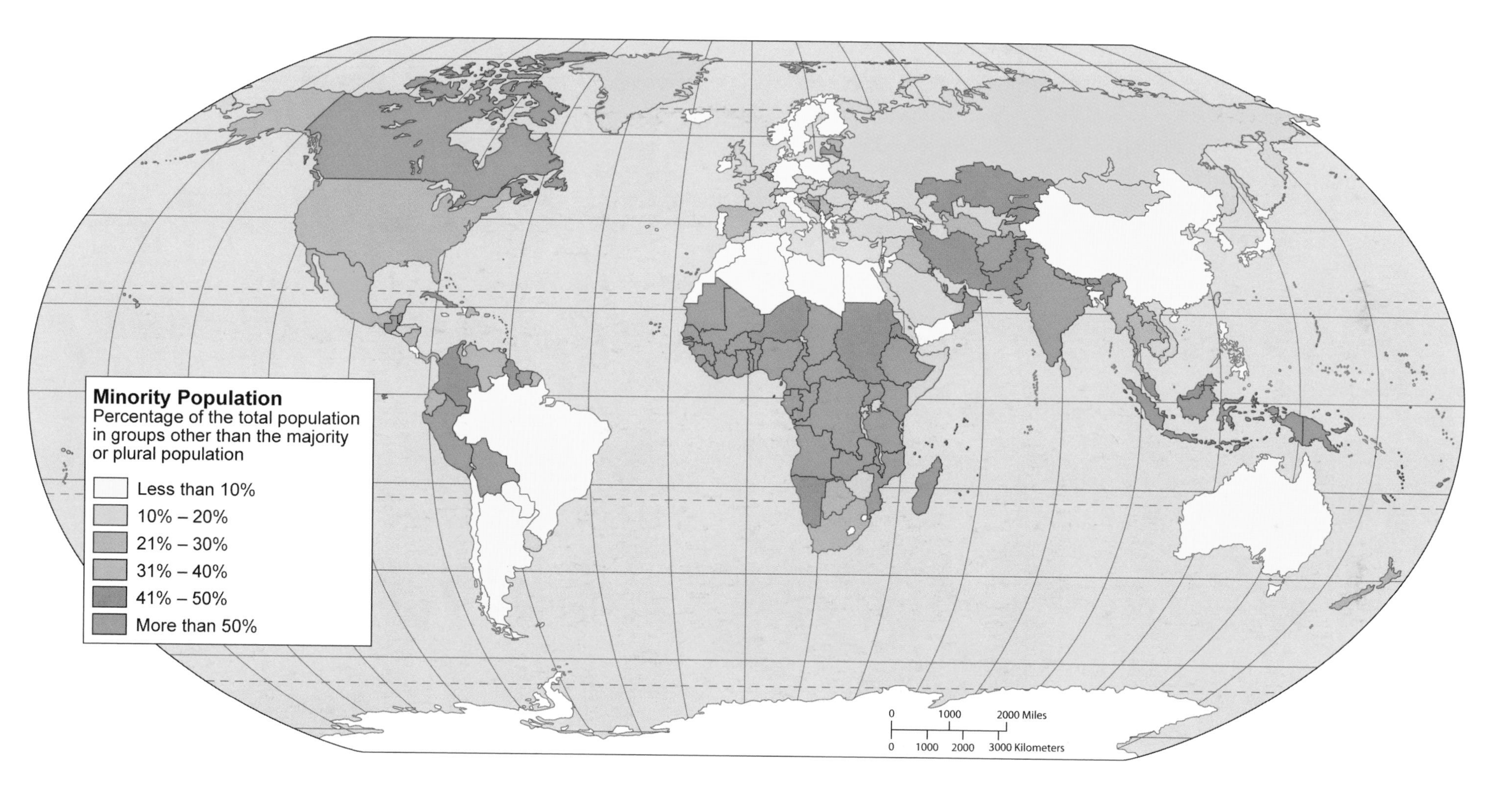

The presence of minority ethnic, national, or racial groups within a country's population can add a vibrant and dynamic mix to the whole. Plural societies with a high degree of cultural and ethnic diversity should, according to some social theorists, be among the world's most healthy. Unfortunately, the reality of the situation is quite different from theory or expectation. The presence of significant minority populations played an important role in the disintegration of the Soviet Union; the continuing existence of minority populations within the new states formed from former Soviet republics threatens the viability and stability of those young political units. In Africa, national boundaries were drawn by colonial powers without regard for the geographical distribution of ethnic groups, and the continuing tribal conflicts that have resulted hamper both economic and political development. Even in the most highly developed regions of the world, the presence of minority ethnic populations poses significant problems: witness the separatist movement in Canada, driven by the desire of some French-Canadians to be independent of the English majority, and the continuing ethnic conflict between Flemish-speaking and Walloon-speaking Belgians. This map, by arraying states on a scale of homogeneity to heterogeneity, indicates areas of existing and potential social and political strife.

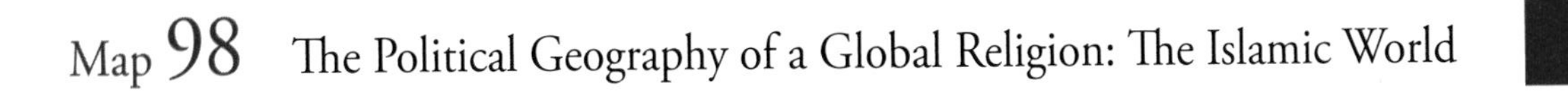

Map 98 The Political Geography of a Global Religion: The Islamic World

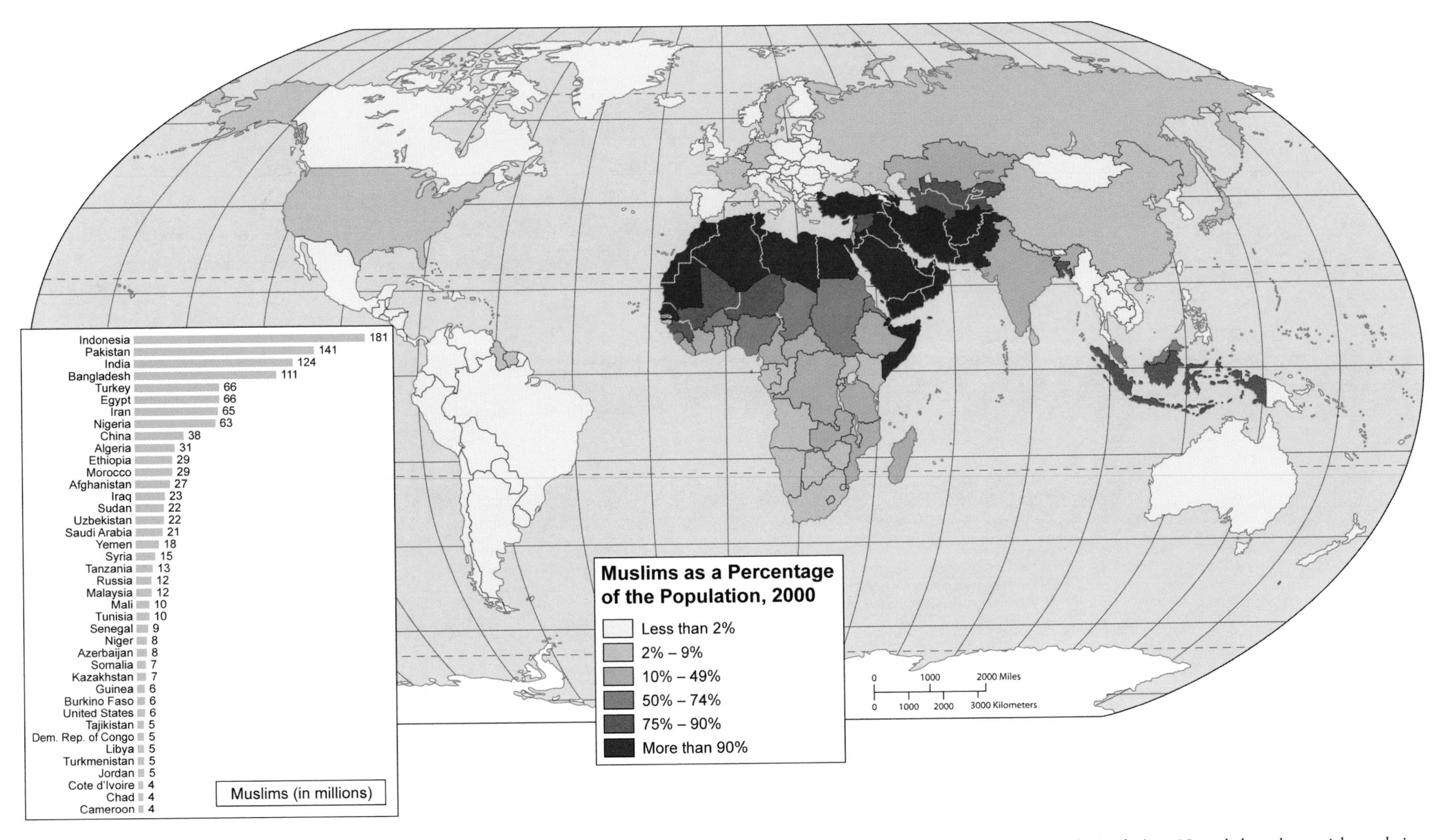

Islam, as a religion, does not promote conflict. The term **jihad**, often mistranslated to mean "holy war," in fact refers to the struggle to find God and to promote the faith. In spite of the beneficent nature of Islamic teachings, the tensions between Muslims and adherents of other faiths often flare into warfare. A comparison of this map with the map of international conflict will show a disproportionate number of wars in that portion of the world where Muslims are either majority or significant minority populations. The reasons for this are based more in the nature of government, cultures, and social structure, than in the tenets of the faith of Islam. Nevertheless, the spatial correlations cannot be ignored. Similarly, terrorist incidents falling considerably short of open armed warfare are spatially consistent with the distribution of Islam and even more consistent with the presence of Islamic fundamentalism or "Islamism," which tends to be less tolerant and more aggressive than the mainstream of the religion. Terrorism is also consistent with those areas where the legacy of colonialism or the persistent presence of non-Islamic cultures intrude into the Islamic world.

Map 99 World Refugee Population

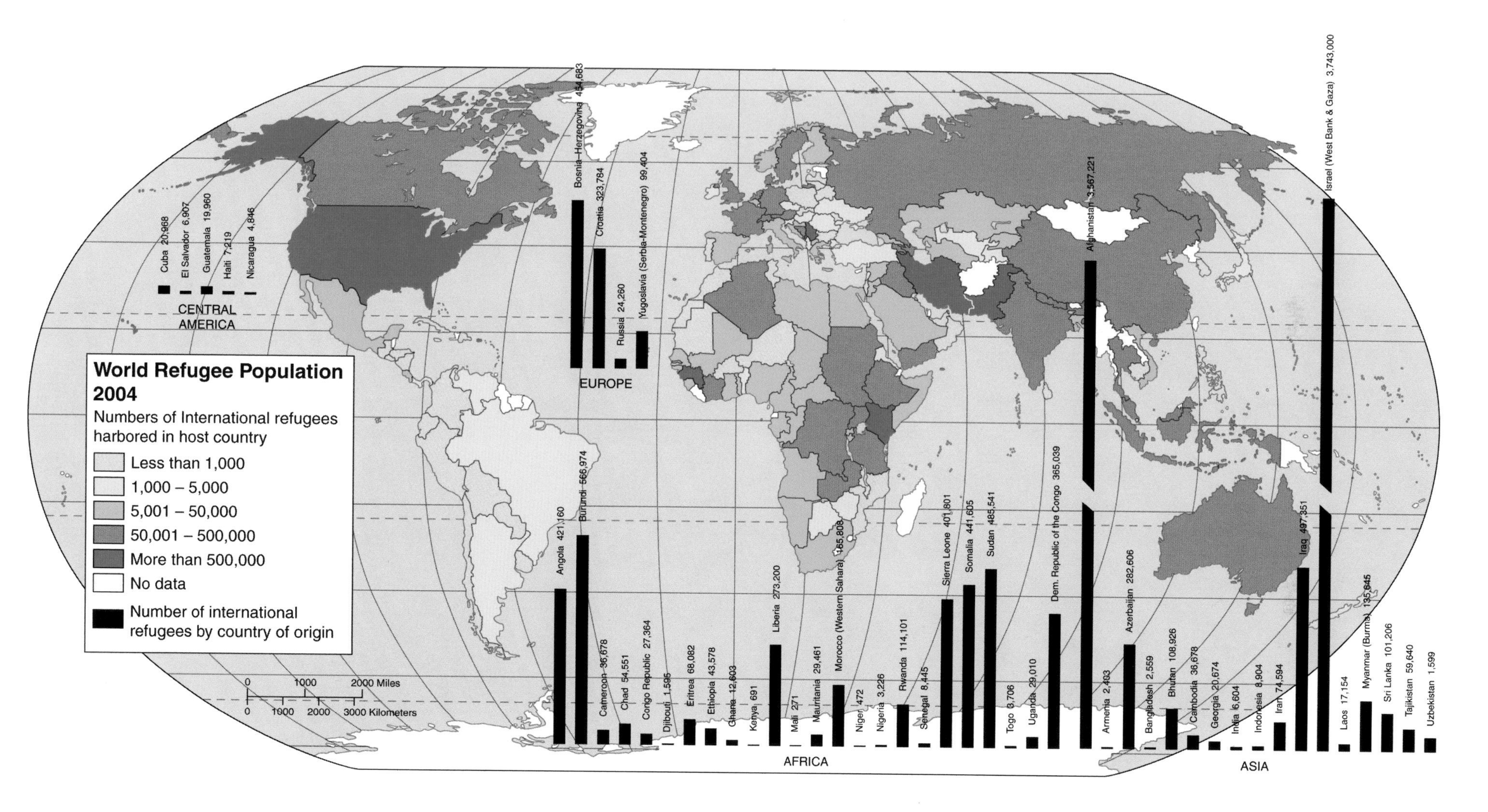

Refugees are persons who have been driven from their homes, normally by armed conflict, and have sought refuge by relocating. The most numerous refugees have traditionally been international refugees, who have crossed the political boundaries of their homelands into other countries. This refugee population is recognized by international agencies, and the countries of refuge are often rewarded financially by those agencies for their willingness to take in externally displaced persons. In recent years, largely because of an increase in civil wars, there have been growing numbers of internally displaced persons—those who leave their homes but stay within their country of origin. There are no rewards for harboring such internal refugee populations.

Map 100 Women's Rights

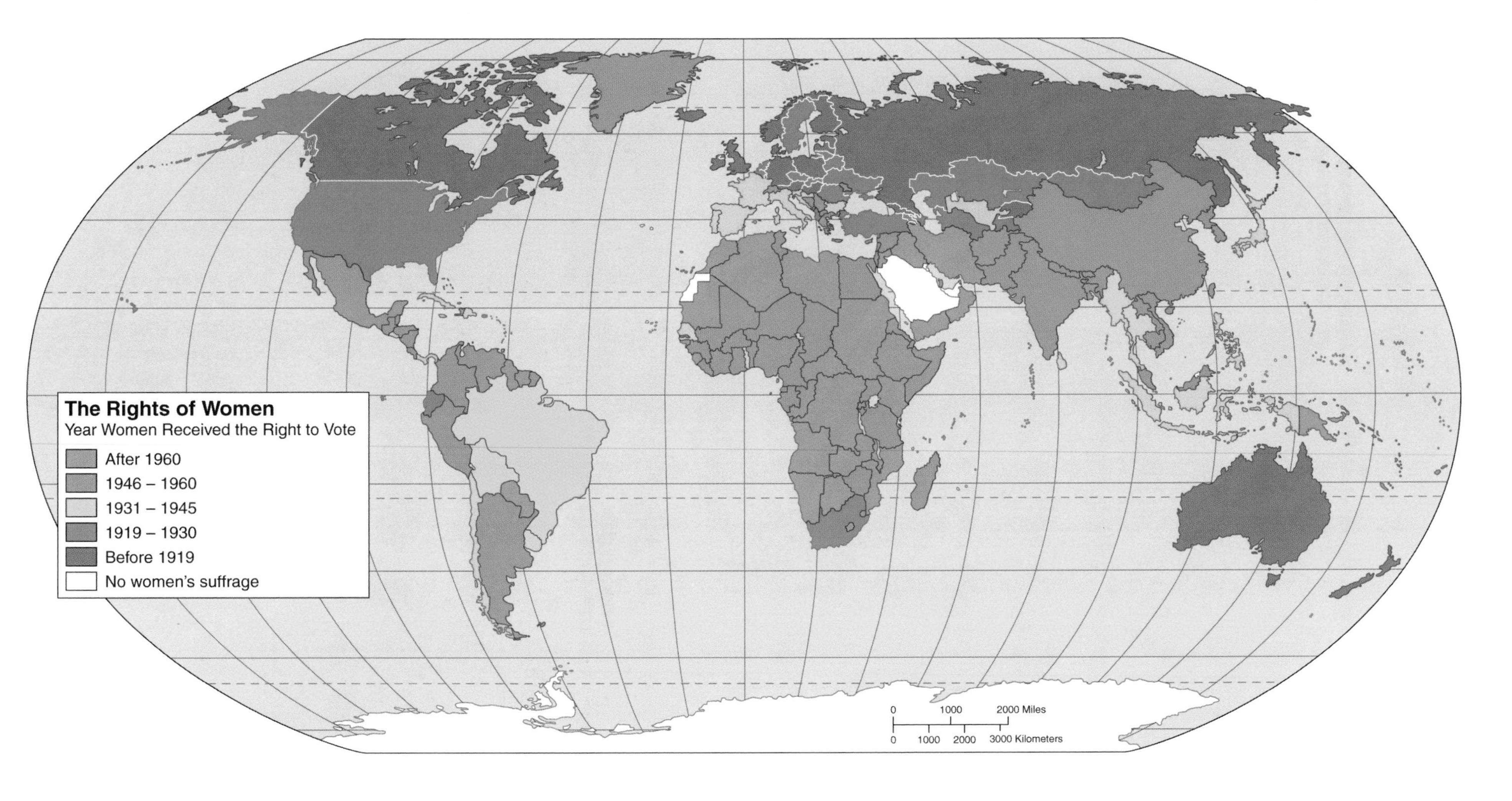

The "rights" referred to in this map refer primarily to the right to vote. But where women have the right to vote in free elections, the other fundamental rights tend to become available as well: the right to own property, the right to an education, the right to leave a domestic alliance without fear of retribution, or the right to be treated as a human being rather than property. But the time lag between women receiving the right to vote and their attainment of other fundamental human rights does not occur immediately or, in many cases, even relatively quickly. On the map, the most recent countries to grant suffrage to women are in Africa and Southwest Asia. In these regions, women still do not have access to many of the basic rights of what we would consider to be a civilized life. And, of course, there are still areas where women cannot vote: Kuwait, for example. Neither men nor women are allowed to vote in Brunei, Saudi Arabia, United Arab Emirates, or Western Sahara.

Map 101 Political and Civil Liberties, 2004

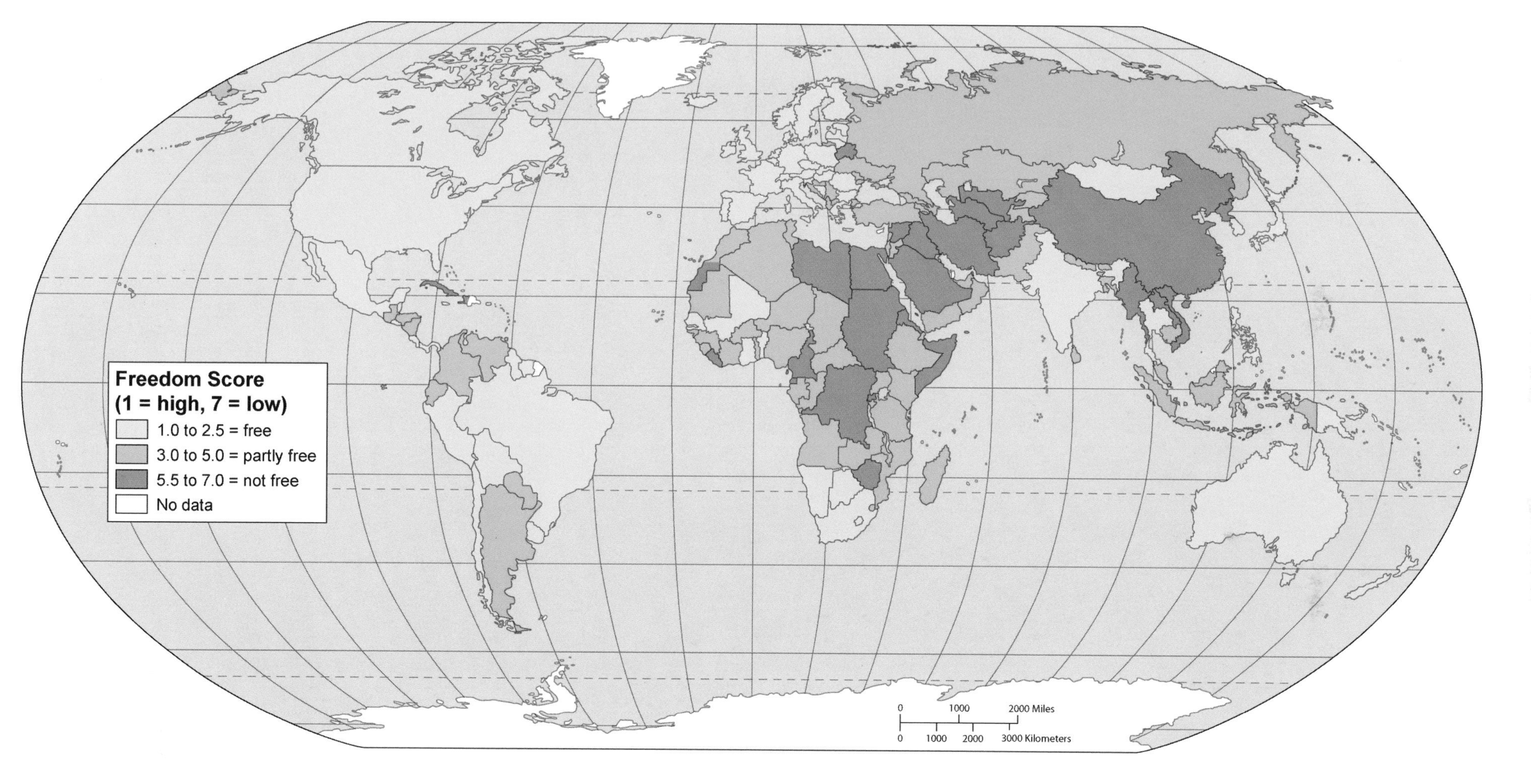

Although measures of political and civil liberty are somewhat difficult to obtain and assess, there are some generally accepted standards that can be evaluated: open elections and competitive political parties, the rule of law, freedoms of speech and press, judicial systems separate from other branches of government, and limits on the power of elected or appointed governmental officials. Interstingly, there appear to be correlations between "degrees of freedom" and such other characteristics of a state as per capita wealth, environmental quality, and healthy economic growth—characteristics that may be mutually contradictory. There is no empirical evidence of a causal link between democratic institutions and consumption; on the other hand, there is clear evidence of a positive relationship between wealth and consumption. Therefore, the three variables are closely correlated and should be used in assessing the nature of the state in any part of the world.

Map 102 A Decade of Risk of Civil Conflict 2005–2015

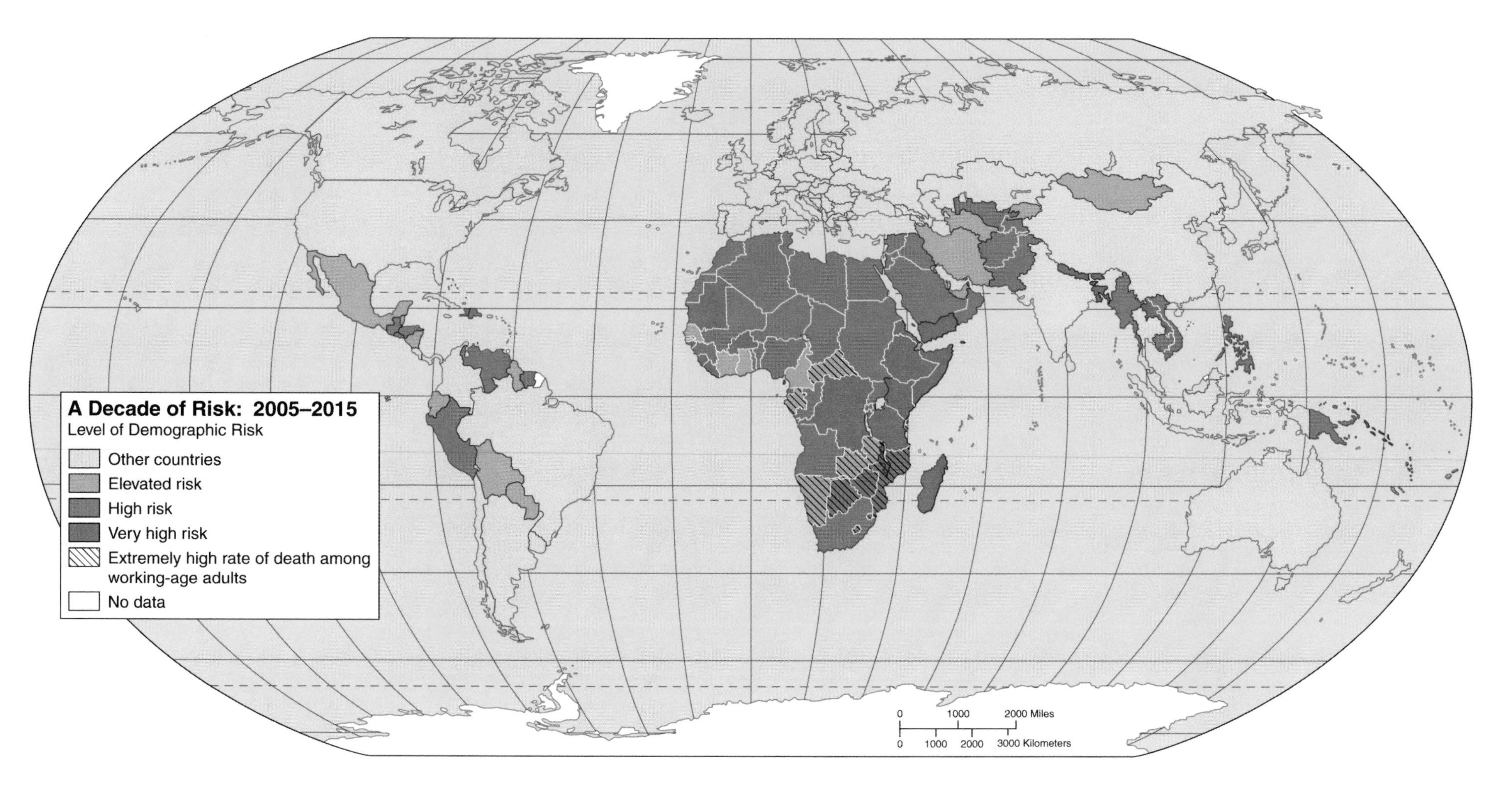

The point has been made earlier in this atlas (see maps 40–45) that demographic stresses enhance the risk of civil conflict, as opposed to international conflict. This map uses three demographic factors—percentage of a country's populations between the ages of 15 and 29, rate of urban population growth, and per capita availability of cropland and fresh water—to assess the risk of domestic violence and civil conflict. It needs to be noted that the data are available on a national level and that geographically large countries such as Brazil, Russia, China, or India may well have areas of demographic stress within them that are masked by the portrayal of something other than subnational data. Nevertheless, the implications of the map are striking: Those parts of the world where civil conflict is in the highest risk category are almost all in Africa and Southwest Asia. We have already seen evidences of such civil conflict: the Hutu-Tutsi civil war of Rwanda and the Shiite-Sunni chaos of Iraq. We describe those as "tribal" or "sectarian" civil conflicts. But how much of the actual conflict is the result of a combination of demographic stresses?

Part VII

World Regions

Map 103 North America: Political Divisions
Map 104 North America: Physical Features
Map 105 North America: Thematic
Map 105a Environment and Economy: The Use of Land
Map 105b Population Density
Map 105c Population Distribution: Clustering
Map 105d Deforestation in the Americas
Map 106 Canada
Map 107 United States
Map 108 Middle America
Map 109 The Caribbean
Map 110 South America: Political Divisions
Map 111 South America: Physical Features
Map 112 South America: Thematic
Map 112a Environment and Economy
Map 112b Population Density
Map 112c Environmental Problems
Map 112d Protection of Natural Areas
Map 113 Northern South America
Map 114 Southern South America
Map 115 Europe: Political Divisions
Map 116 Europe: Physical Features
Map 117 Europe: Thematic
Map 117a Environment and Economy: The Use of Land
Map 117b Population Density
Map 117c European Political Boundaries, 1914-1948
Map 117d Political Changes 1989-2004
Map 118 Western Europe
Map 119 Eastern Europe

Map 120 Northern Europe
Map 121 Africa: Political Divisions
Map 122 Africa: Physical Features
Map 123 Africa: Thematic
Map 123a Environment and Economy
Map 123b Population Density
Map 123c Colonial Patterns
Map 123d African Cropland and Dryland Degradation
Map 124 Northern Africa
Map 125 Southern Africa
Map 126 Asia: Political Divisions
Map 127 Asia: Physical Features
Map 128 Asia: Thematic
Map 128a Environment and Economy
Map 128b Population Density
Map 128c Industrialization: Manufacturing and Resources
Map 128d South, South-East, and East Asian Deforestation
Map 129 Northern Asia
Map 130 Southwestern Asia
Map 130a The Middle East: Territorial Changes, 1918-2004
Map 130b South-Central Eurasia: An Ethnolinguistic Crazy Quilt
Map 131 East Asia
Map 132 Southeast Asia
Map 133 Australasia and Oceania: Political Divisions
Map 134 Australasia and Oceania: Physical Features
Map 135 Australasia: Thematic
Map 135a Environment and Economy
Map 135b Population Density
Map 135c Climatic Patterns
Map 136 The Pacific Rim
Map 137 The Atlantic Ocean
Map 138 The Indian Ocean
Map 139 The Arctic
Map 140 Antarctica

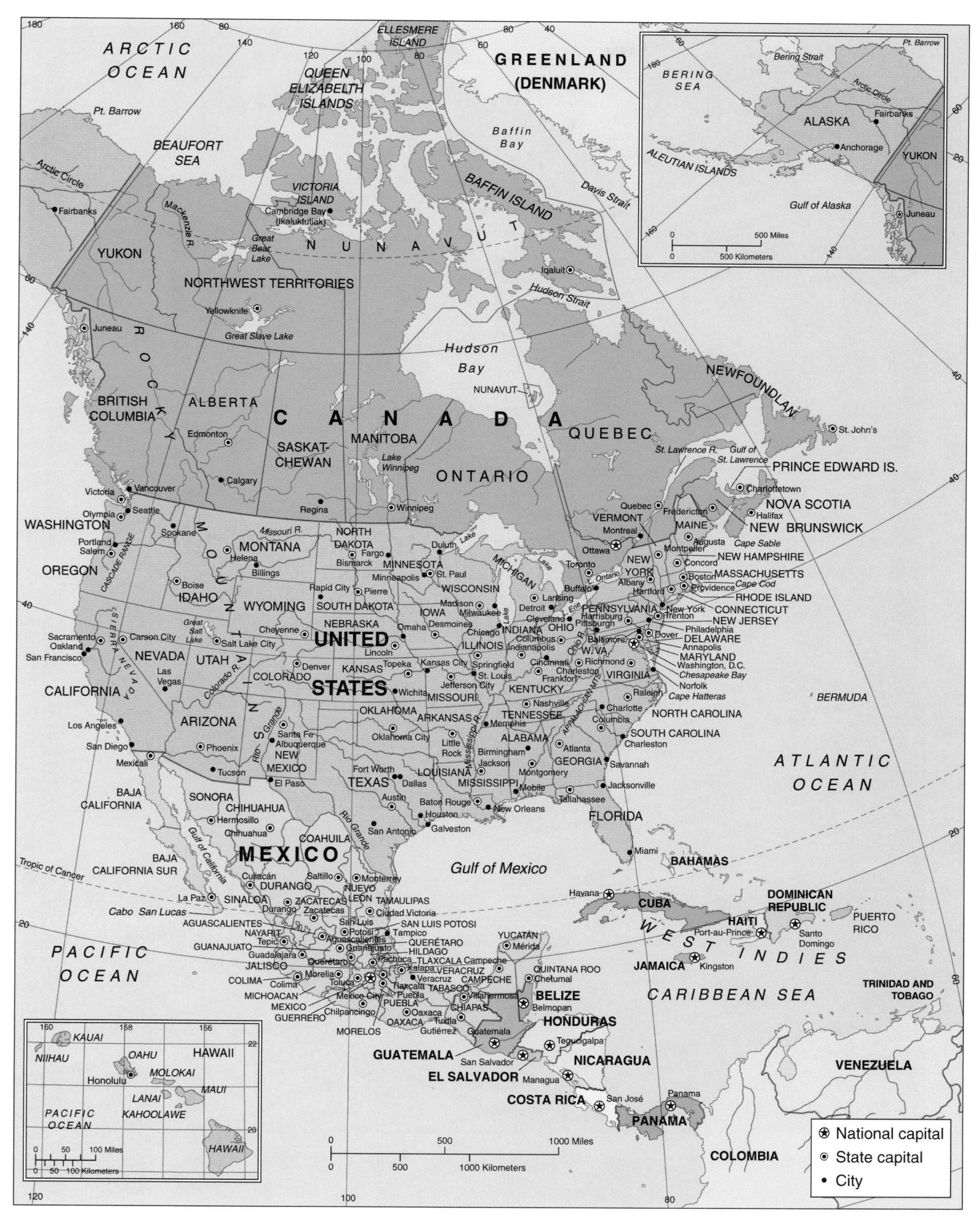
ARCTIC OCEAN
ELLESMERE ISLAND
QUEEN ELIZABETH ISLANDS
GREENLAND (DENMARK)
Pt. Barrow
BEAUFORT SEA
Baffin Bay
Arctic Circle
VICTORIA ISLAND
Cambridge Bay (Ikaluktutiak)
BAFFIN ISLAND
Davis Strait
Fairbanks
Mackenzie R.
Great Bear Lake
NUNAVUT
YUKON
NORTHWEST TERRITORIES
Iqaluit
Hudson Strait
Yellowknife
Great Slave Lake
Juneau
ROCKY MOUNTAINS
Hudson Bay
NUNAVUT
NEWFOUNDLAND
BRITISH COLUMBIA
ALBERTA
CANADA
SASKATCHEWAN
MANITOBA
QUEBEC
St. John's
Edmonton
Lake Winnipeg
ONTARIO
St. Lawrence R.
Gulf of St. Lawrence
PRINCE EDWARD IS.
Calgary
Victoria
Vancouver
Charlottetown
Olympia
Seattle
Regina
Winnipeg
Quebec
Fredericton
NOVA SCOTIA
Halifax
WASHINGTON
VERMONT
MAINE
NEW BRUNSWICK
Spokane
Missouri R.
NORTH DAKOTA
Montreal
Augusta
Cape Sable
Portland
Salem
MONTANA
Fargo
Duluth
Lake Superior
Ottawa
Montpelier
Helena
Bismarck
MINNESOTA
NEW HAMPSHIRE
OREGON
CASCADE RANGE
Billings
St. Paul
Lake Michigan
Lake Huron
Toronto
NEW YORK
Concord
Minneapolis
Boston
MASSACHUSETTS
Boise
Rapid City
Pierre
WISCONSIN
Lake Ontario
Albany
Cape Cod
IDAHO
Buffalo
Providence
WYOMING
SOUTH DAKOTA
Hartford
RHODE ISLAND
Madison
Milwaukee
Lansing
Lake Erie
IOWA
Detroit
PENNSYLVANIA
New York
CONNECTICUT
Great Salt Lake
NEBRASKA
Des Moines
Cleveland
Harrisburg
Trenton
NEW JERSEY
Sacramento
Carson City
Cheyenne
Omaha
Chicago
INDIANA
OHIO
Pittsburgh
Philadelphia
Oakland
San Francisco
Salt Lake City
UNITED STATES
Columbus
Dover
DELAWARE
SIERRA NEVADA
NEVADA
UTAH
Lincoln
ILLINOIS
Indianapolis
Baltimore
Annapolis
W. VA.
MARYLAND
Las Vegas
Colorado R.
Denver
KANSAS
Topeka
Kansas City
Springfield
Cincinnati
Richmond
Washington, D.C.
COLORADO
St. Louis
Charleston
Frankfort
VIRGINIA
Chesapeake Bay
CALIFORNIA
Wichita
Jefferson City
KENTUCKY
Norfolk
MISSOURI
Raleigh
Cape Hatteras
BERMUDA
Nashville
APPALACHIAN MTS.
OKLAHOMA
Charlotte
NORTH CAROLINA
ARIZONA
Rio Grande
Santa Fe
ARKANSAS
TENNESSEE
Memphis
Columbia
Los Angeles
Albuquerque
Oklahoma City
Little Rock
ALABAMA
Atlanta
SOUTH CAROLINA
Charleston
San Diego
Phoenix
NEW MEXICO
Mississippi R.
Birmingham
ATLANTIC OCEAN
Mexicali
Fort Worth
Jackson
GEORGIA
Savannah
Tucson
TEXAS
Dallas
LOUISIANA
Montgomery
El Paso
MISSISSIPPI
Jacksonville
BAJA CALIFORNIA
SONORA
Austin
Mobile
Tallahassee
CHIHUAHUA
Baton Rouge
New Orleans
FLORIDA
Hermosillo
Houston
Chihuahua
Galveston
San Antonio
COAHUILA
Gulf of California
MEXICO
Miami
BAJA CALIFORNIA SUR
BAHAMAS
Tropic of Cancer
Culiacán
Saltillo
Monterrey
Gulf of Mexico
DURANGO
NUEVO LEÓN
La Paz
SINALOA
ZACATECAS
TAMAULIPAS
Havana
DOMINICAN REPUBLIC
Cabo San Lucas
Durango
Zacatecas
Ciudad Victoria
CUBA
AGUASCALIENTES
San Luis
SAN LUIS POTOSÍ
YUCATÁN
HAITI
PUERTO RICO
NAYARIT
Potosí
Tampico
WEST INDIES
Port-au-Prince
Santo Domingo
Tepic
Aguascalientes
Mérida
GUANAJUATO
Guanajuato
QUERÉTARO
PACIFIC OCEAN
Guadalajara
HILDAGO
Kingston
JALISCO
Querétaro
Pachuca
TLAXCALA
Campeche
JAMAICA
Xalapa
VERACRUZ
QUINTANA ROO
COLIMA
Morelia
Veracruz
CAMPECHE
Chetumal
TRINIDAD AND TOBAGO
Colima
Toluca
Tlaxcala
TABASCO
MICHOACÁN
México City
Puebla
Villahermosa
BELIZE
CARIBBEAN SEA
MÉXICO
PUEBLA
CHIAPAS
Belmopan
GUERRERO
Chilpancingo
Oaxaca
OAXACA
Tuxtla Gutiérrez
HONDURAS
MORELOS
Guatemala
Tegucigalpa
GUATEMALA
San Salvador
NICARAGUA
VENEZUELA
EL SALVADOR
Managua
COSTA RICA
San José
Panama
PANAMA
COLOMBIA
0 500 1000 Miles
0 500 1000 Kilometers
National capital
State capital
City
BERING SEA
Bering Strait
Arctic Circle
ALASKA
Fairbanks
Anchorage
YUKON
ALEUTIAN ISLANDS
Gulf of Alaska
Juneau
0 500 Miles
0 500 Kilometers
KAUAI
NIIHAU
OAHU
HAWAII
MOLOKAI
Honolulu
MAUI
LANAI
KAHOOLAWE
PACIFIC OCEAN
HAWAII
0 50 100 Miles
0 50 100 Kilometers

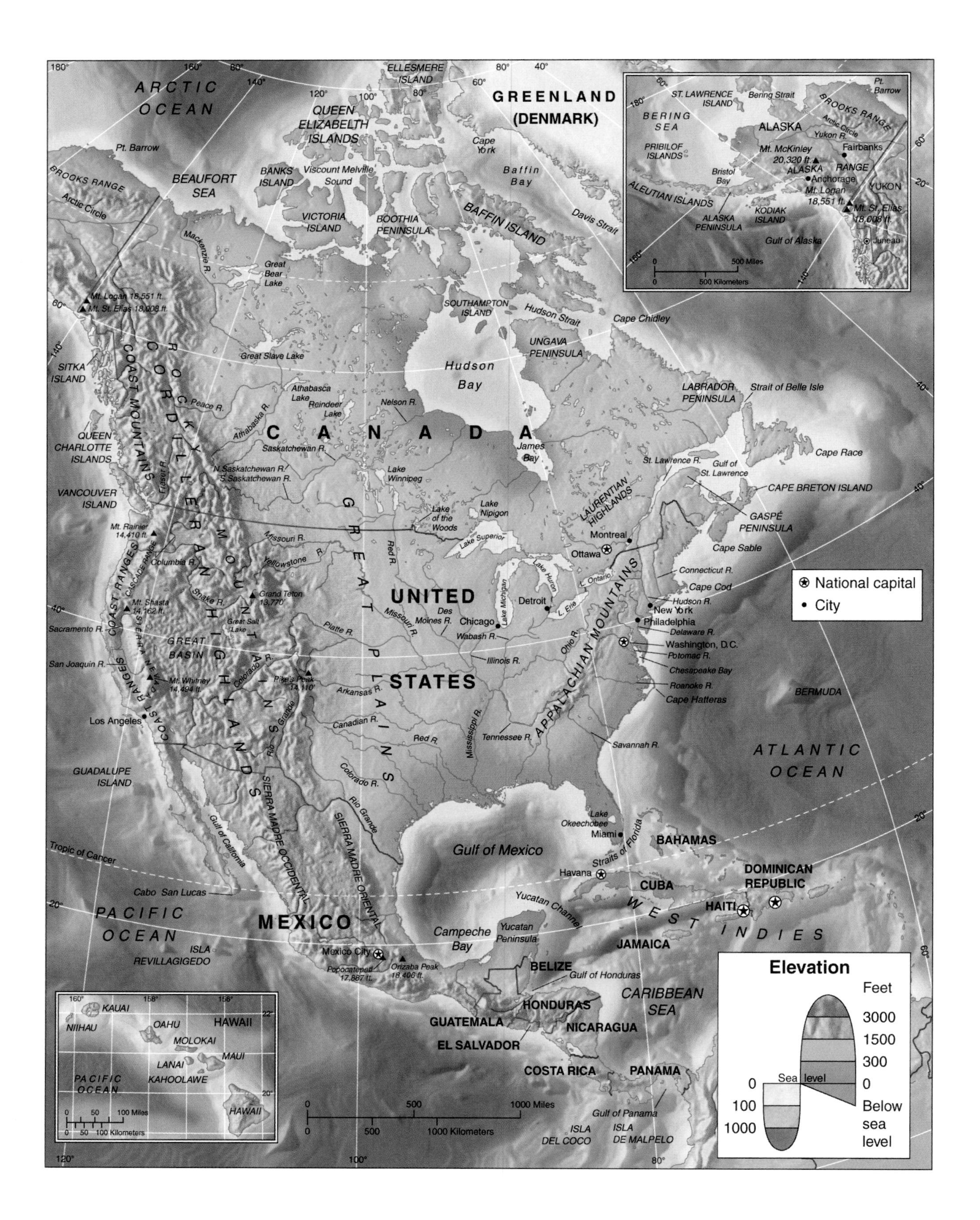
ARCTIC OCEAN
ELLESMERE ISLAND
QUEEN ELIZABELTH ISLANDS
GREENLAND (DENMARK)
Cape York
Pt. Barrow
BEAUFORT SEA
BANKS ISLAND
Viscount Melville Sound
Baffin Bay
BROOKS RANGE
Arctic Circle
VICTORIA ISLAND
BOOTHIA PENINSULA
BAFFIN ISLAND
Davis Strait
Mackenzie R.
Great Bear Lake
Mt. Logan 18,551 ft.
Mt. St. Elias 18,008 ft.
SOUTHAMPTON ISLAND
Hudson Strait
Cape Chidley
UNGAVA PENINSULA
Great Slave Lake
Hudson Bay
SITKA ISLAND
COAST MOUNTAINS
ROCKY CORDILLERAN
Peace R.
Athabasca Lake
Reindeer Lake
Athabasca R.
Nelson R.
LABRADOR PENINSULA
Strait of Belle Isle
QUEEN CHARLOTTE ISLANDS
CANADA
Saskatchewan R.
James Bay
Cape Race
N. Saskatchewan R.
S. Saskatchewan R.
Lake Winnipeg
St. Lawrence R.
Gulf of St. Lawrence
CAPE BRETON ISLAND
VANCOUVER ISLAND
LAURENTIAN HIGHLANDS
Lake of the Woods
Lake Nipigon
GASPÉ PENINSULA
Mt. Rainier 14,410 ft.
Missouri R.
Lake Superior
Montreal
Ottawa
Cape Sable
GREAT PLAINS
Red R.
Yellowstone R.
Columbia R.
CASCADE RANGE
COAST RANGES
Lake Huron
L. Ontario
Connecticut R.
Cape Cod
National capital
City
Grand Teton 13,770'
Snake R.
MOUNTAIN HIGHLANDS
UNITED STATES
Lake Michigan
Detroit
L. Erie
APPALACHIAN MOUNTAINS
Hudson R.
New York
Philadelphia
Mt. Shasta 14,162 ft.
Great Salt Lake
Des Moines R.
Chicago
Delaware R.
Sacramento R.
SIERRA NEVADA
GREAT BASIN
Platte R.
Wabash R.
Ohio R.
Washington, D.C.
Potomac R.
San Joaquin R.
Colorado R.
Illinois R.
Chesapeake Bay
Mt. Whitney 14,494 ft.
Pike's Peak 14,110'
Arkansas R.
Roanoke R.
Cape Hatteras
BERMUDA
Los Angeles
Canadian R.
Mississippi R.
Tennessee R.
Red R.
Savannah R.
Rio Grande
ATLANTIC OCEAN
GUADALUPE ISLAND
Colorado R.
SIERRA MADRE OCCIDENTAL
SIERRA MADRE ORIENTAL
Lake Okeechobee
Miami
Straits of Florida
BAHAMAS
Gulf of California
Gulf of Mexico
Tropic of Cancer
Havana
CUBA
DOMINICAN REPUBLIC
Cabo San Lucas
Yucatan Channel
WEST INDIES
HAITI
PACIFIC OCEAN
MEXICO
Campeche Bay
Yucatan Peninsula
JAMAICA
ISLA REVILLAGIGEDO
Mexico City
Popocatepetl 17,887 ft.
Orizaba Peak 18,406 ft.
BELIZE
Gulf of Honduras
CARIBBEAN SEA
HONDURAS
GUATEMALA
NICARAGUA
EL SALVADOR
COSTA RICA
PANAMA
Gulf of Panama
ISLA DEL COCO
ISLA DE MALPELO
0 500 1000 Miles
0 500 1000 Kilometers
ST. LAWRENCE ISLAND
Bering Strait
BERING SEA
ALASKA
PRIBILOF ISLANDS
Mt. McKinley 20,320 ft.
Fairbanks
Yukon R.
Bristol Bay
ALASKA RANGE
Anchorage
Mt. Logan 18,551 ft.
YUKON
ALEUTIAN ISLANDS
ALASKA PENINSULA
KODIAK ISLAND
Mt. St. Elias 18,008 ft.
Gulf of Alaska
Juneau
0 500 Miles
0 500 Kilometers
KAUAI
NIIHAU
OAHU
HAWAII
MOLOKAI
MAUI
LANAI
KAHOOLAWE
PACIFIC OCEAN
HAWAII
0 50 100 Miles
0 50 100 Kilometers
Elevation
Feet
3000
1500
300
0
Sea level
100
1000
Below sea level

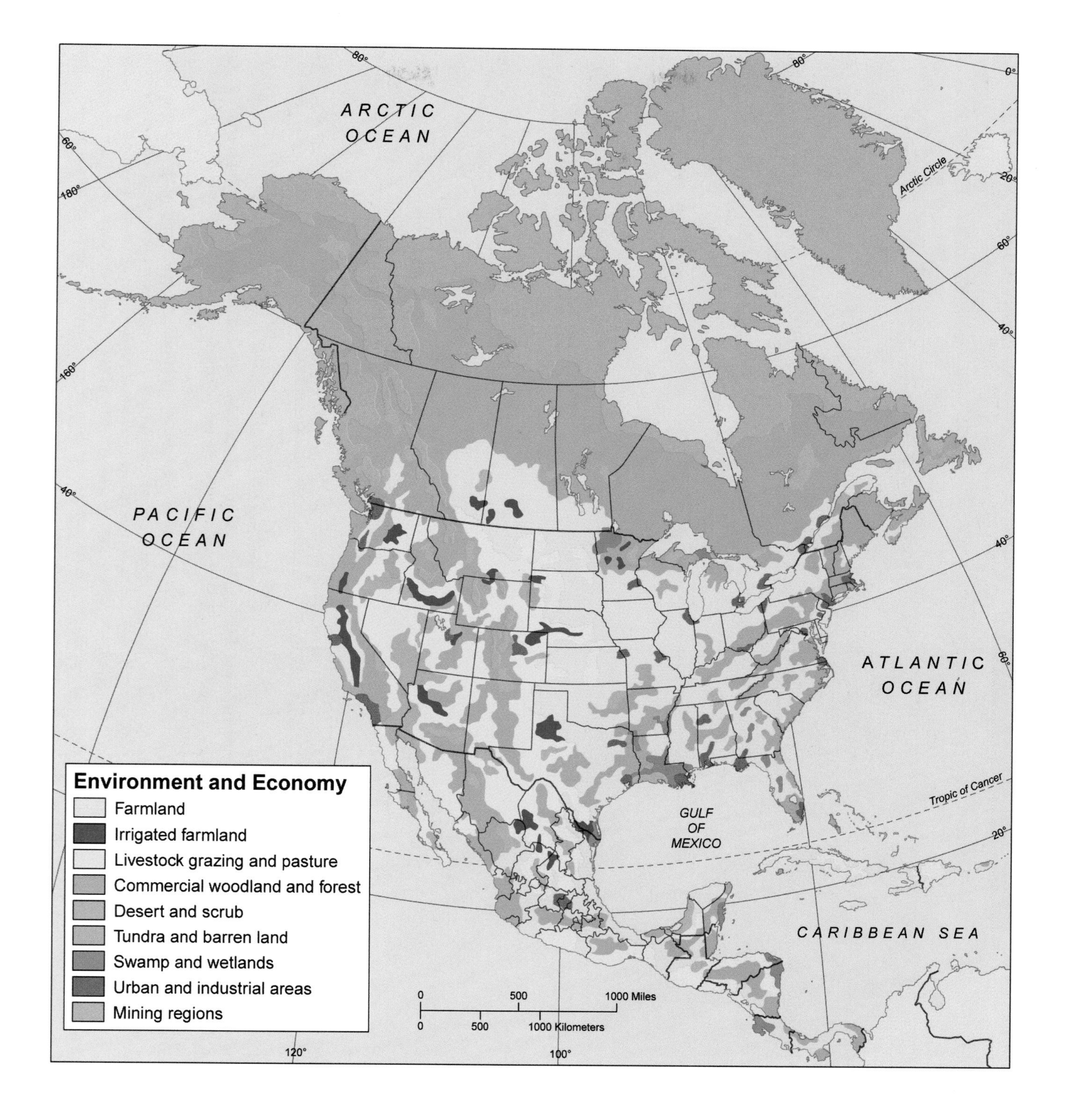

The use of land in North America represents a balance between agriculture, resource extraction, and manufacturing that is unmatched. The United States, as the world's leading industrial power, is also the world's leader in commercial agricultural production. Canada, despite its small population, is a ranking producer of both agricultural and industrial products and Mexico has begun to emerge from its developing nation status to become an important industrial and agricultural nation as well. The countries of Middle America and the Caribbean are just beginning the transition from agriculture to modern industrial economies. Part of the basis for the high levels of economic productivity in North America is environmental: a superb blend of soil, climate, and raw materials. But just as important is the cultural and social mix of the plural societies of North America, a mix that historically aided the growth of the economic diversity necessary for developed economies.

Map 105b Population Density

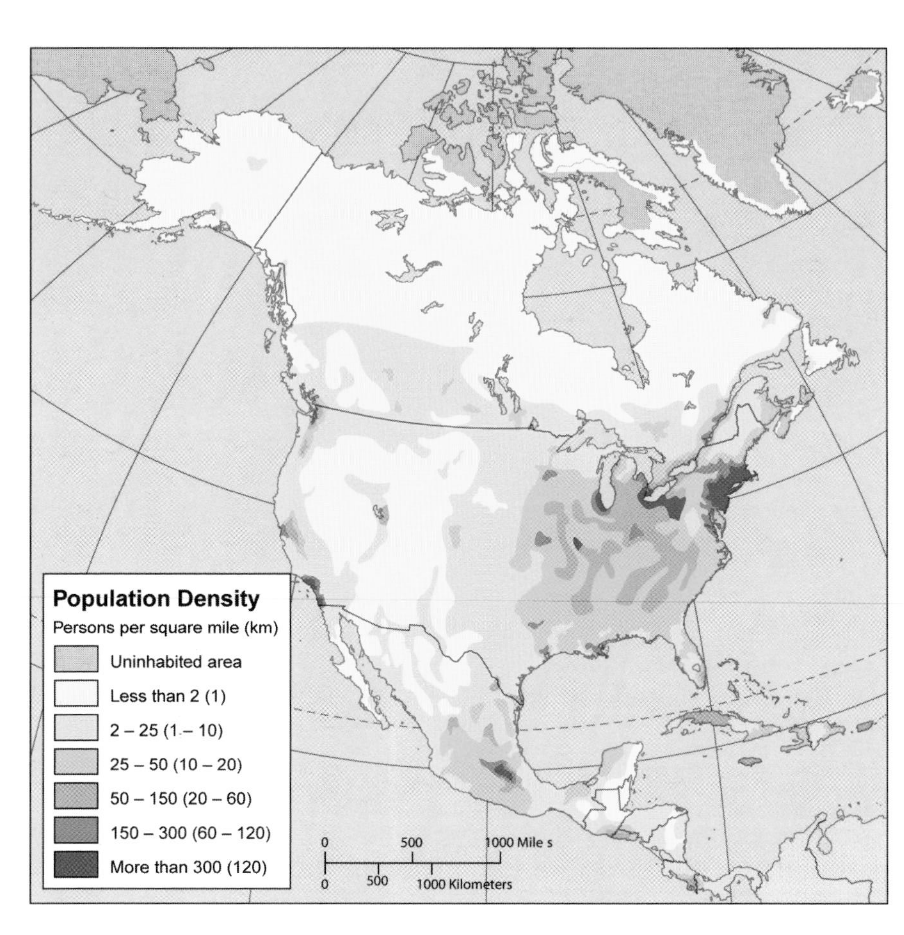

North America contains nearly 500 million people and the United States, with over 250 million inhabitants, is the third most populous country in the world, after China and India. Most of the present North American population has roots in the Old World. The native populations of the Americas had little or no resistance to Old World diseases in 1500 and within a couple of centuries of first contact with Europeans, most of the native peoples had either died out or preserved their genetic heritage by mixing with disease-resistant Old World populations. This left a North American population that is largely European, but with significant minorities resulting from the slave laborers imported from Africa, and from the mixture of native Americans with Europeans and/or Africans. The density of that population is largely the consequence of environmental factors (good soil, the availability of water, the presence of other resources) and cultural/economic ones (agricultural production, urbanization, industrialization).

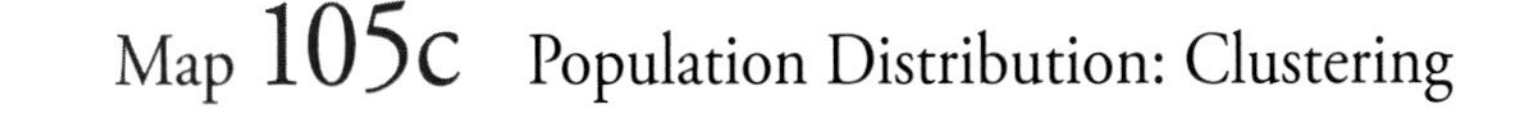

Map 105c Population Distribution: Clustering

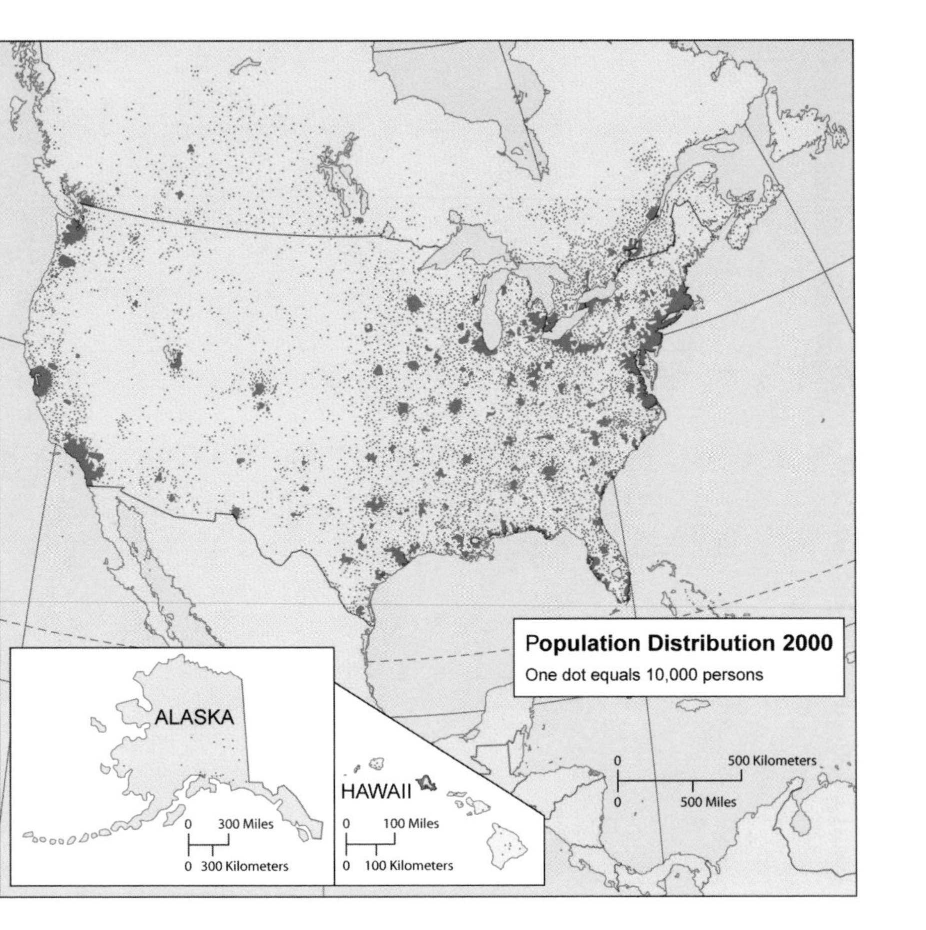

Although population clustering is characteristic of highly economically developed regions, the Anglo-American population exhibits a remarkably clustered pattern. The primary reasons for this remarkable development are agricultural technology and affluence. A highly developed agricultural technology allows a small number of farmers, using sophisticated machinery, to grow and harvest enormous quantities of agricultural produce on large farms. In the United States, the world's leader in commercial agricultural production, only 2 percent of the population are farmers and that population is thinly distributed over wide areas. The vast majority of Americans live and work in city-suburb systems in which the widespread availability of private automobiles allows people to live considerable distances from where they work, keeping overall urban population densities relatively low but allowing for extensive urbanization—with cities large enough in area to be visible as population clusters on maps at this scale.

Map 105d Deforestation in the Americas

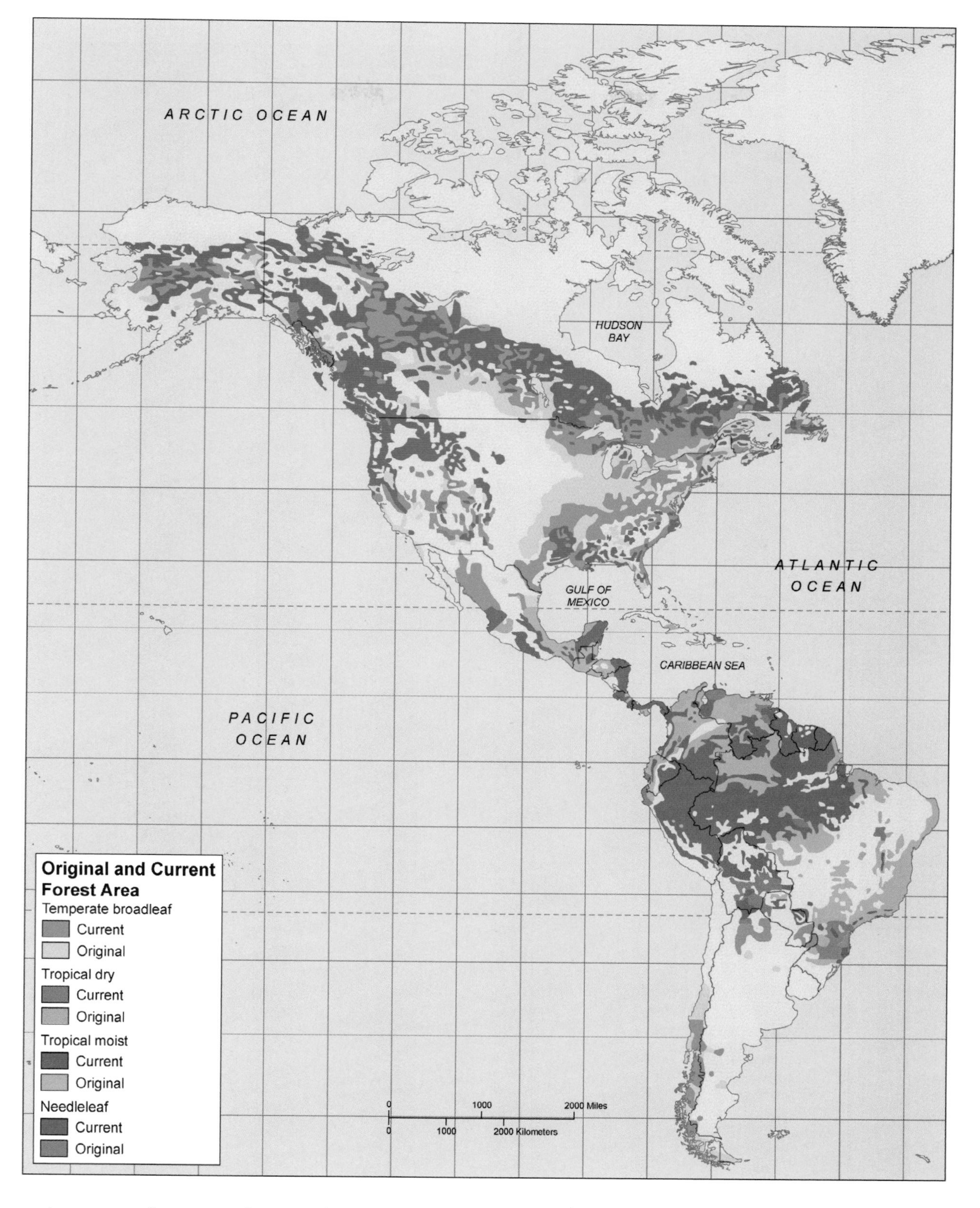

This map shows current forest cover adjacent to the estimated forest cover that would be present had there been no human intervention over the past several thousand years since the end of the last ice age, and assuming that climatic conditions would have been pretty much the same as they have been during that time. While in the popular imagination, it is the tropical forest regions of South America and Middle America that have been "deforested," the map shows otherwise. The greatest losses of forest cover have been in the now-agricultural regions of North America and the more temperate regions of South America (most in southern Brazil). Beside these forest losses in the temperate zones, the forest losses of the Amazon Basin seem relatively minor. This does not suggest that the current deforestation of Amazonia and other tropical forest regions is something about which we should not be concerned, but it does put that current deforestation in a more accurate historical context.

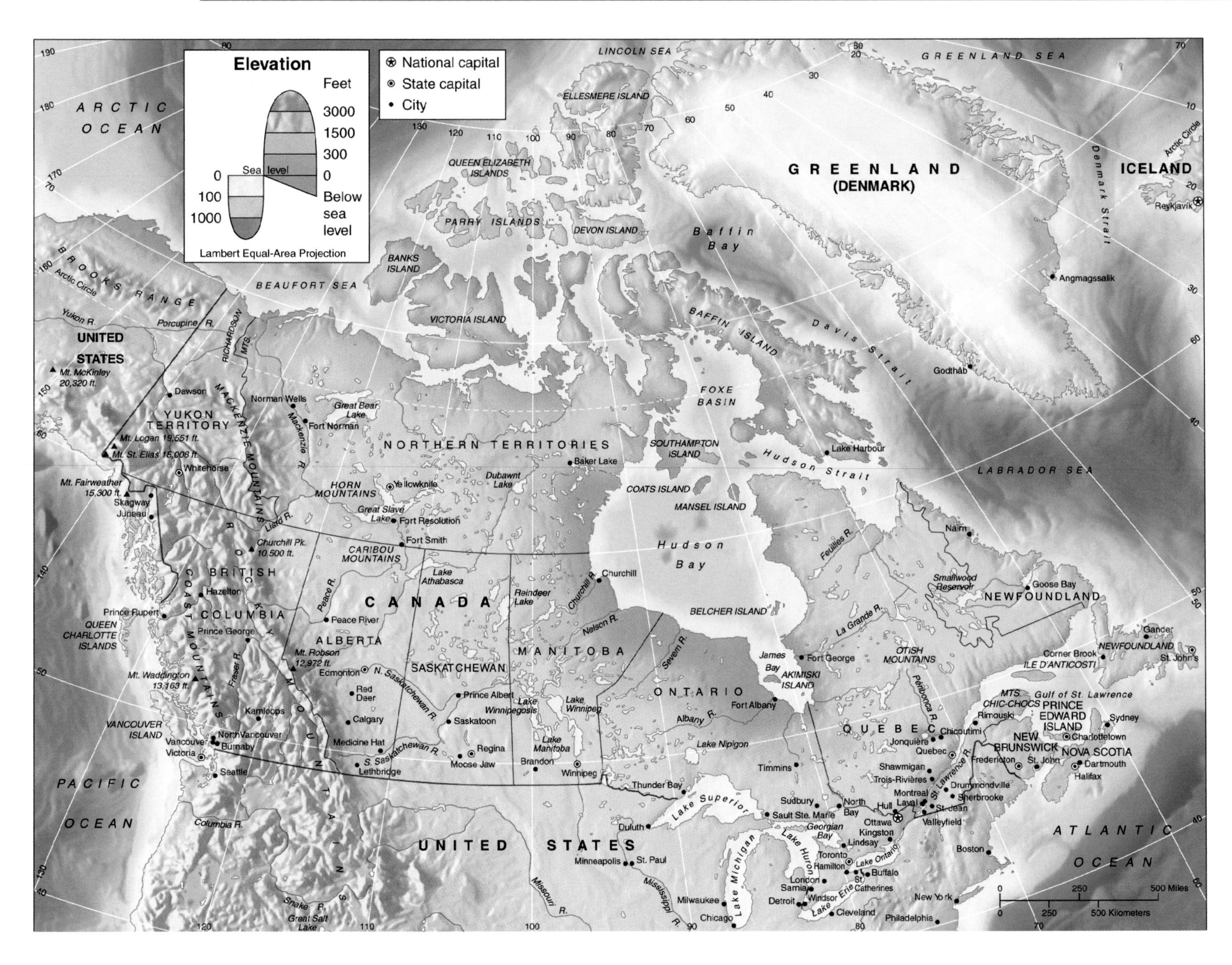

Elevation
Feet
3000
1500
300
0
Sea level
Below sea level
0
100
1000
Lambert Equal-Area Projection
National capital
State capital
City
ARCTIC OCEAN
BEAUFORT SEA
LINCOLN SEA
GREENLAND SEA
GREENLAND (DENMARK)
ICELAND
Reykjavík
Denmark Strait
Angmagssalik
Godthåb
Baffin Bay
Davis Strait
LABRADOR SEA
Hudson Strait
Lake Harbour
BAFFIN ISLAND
FOXE BASIN
SOUTHAMPTON ISLAND
COATS ISLAND
MANSEL ISLAND
Hudson Bay
BELCHER ISLAND
ELLESMERE ISLAND
QUEEN ELIZABETH ISLANDS
PARRY ISLANDS
DEVON ISLAND
BANKS ISLAND
VICTORIA ISLAND
BROOKS RANGE
Arctic Circle
Yukon R.
Porcupine R.
UNITED STATES
Mt. McKinley 20,320 ft.
RICHARDSON MTS.
MACKENZIE MOUNTAINS
Dawson
YUKON TERRITORY
Mt. Logan 19,551 ft.
Mt. St. Elias 18,008 ft.
Whitehorse
Mt. Fairweather 15,300 ft.
Skagway
Juneau
Norman Wells
Mackenzie R.
Fort Norman
Great Bear Lake
NORTHERN TERRITORIES
Baker Lake
HORN MOUNTAINS
Yellowknife
Dubawnt Lake
Great Slave Lake
Fort Resolution
Fort Smith
Liard R.
Churchill Pk. 10,500 ft.
CARIBOU MOUNTAINS
Lake Athabasca
BRITISH COLUMBIA
ROCKY MOUNTAINS
COAST MOUNTAINS
Hazelton
Prince Rupert
QUEEN CHARLOTTE ISLANDS
Prince George
Mt. Waddington 13,163 ft.
Fraser R.
Kamloops
VANCOUVER ISLAND
North Vancouver
Vancouver
Burnaby
Victoria
Seattle
PACIFIC OCEAN
Columbia R.
Peace R.
Peace River
CANADA
ALBERTA
Mt. Robson 12,972 ft.
Edmonton
N. Saskatchewan R.
Red Deer
Calgary
Medicine Hat
S. Saskatchewan R.
Lethbridge
SASKATCHEWAN
Prince Albert
Saskatoon
Regina
Moose Jaw
Reindeer Lake
Churchill R.
Churchill
MANITOBA
Nelson R.
Lake Winnipegosis
Lake Winnipeg
Lake Manitoba
Brandon
Winnipeg
Severn R.
ONTARIO
Albany R.
Fort Albany
Lake Nipigon
Timmins
Thunder Bay
Lake Superior
Duluth
Sudbury
Sault Ste. Marie
North Bay
Georgian Bay
Lake Huron
Lake Michigan
Toronto
Hamilton
Lake Ontario
Buffalo
St. Catherines
London
Sarnia
Detroit
Windsor
Lake Erie
Cleveland
Lindsay
Kingston
Ottawa
Hull
James Bay
AKIMISKI ISLAND
Fort George
La Grande R.
Feuilles R.
OTISH MOUNTAINS
Péribonca R.
QUEBEC
Chicoutimi
Jonquière
Quebec
Shawinigan
Trois-Rivières
Montreal
Laval
St. Jean
Valleyfield
Drummondville
Sherbrooke
St. Lawrence R.
Rimouski
MTS. CHIC-CHOCS
Gulf of St. Lawrence
ILE D'ANTICOSTI
Nain
Smallwood Reservoir
Goose Bay
NEWFOUNDLAND
Gander
Corner Brook
St. John's
PRINCE EDWARD ISLAND
Charlottetown
Sydney
NEW BRUNSWICK
Fredericton
St. John
NOVA SCOTIA
Dartmouth
Halifax
ATLANTIC OCEAN
UNITED STATES
Snake R.
Great Salt Lake
Missouri R.
Mississippi R.
Minneapolis
St. Paul
Milwaukee
Chicago
New York
Philadelphia
Boston
0 250 500 Miles
0 250 500 Kilometers

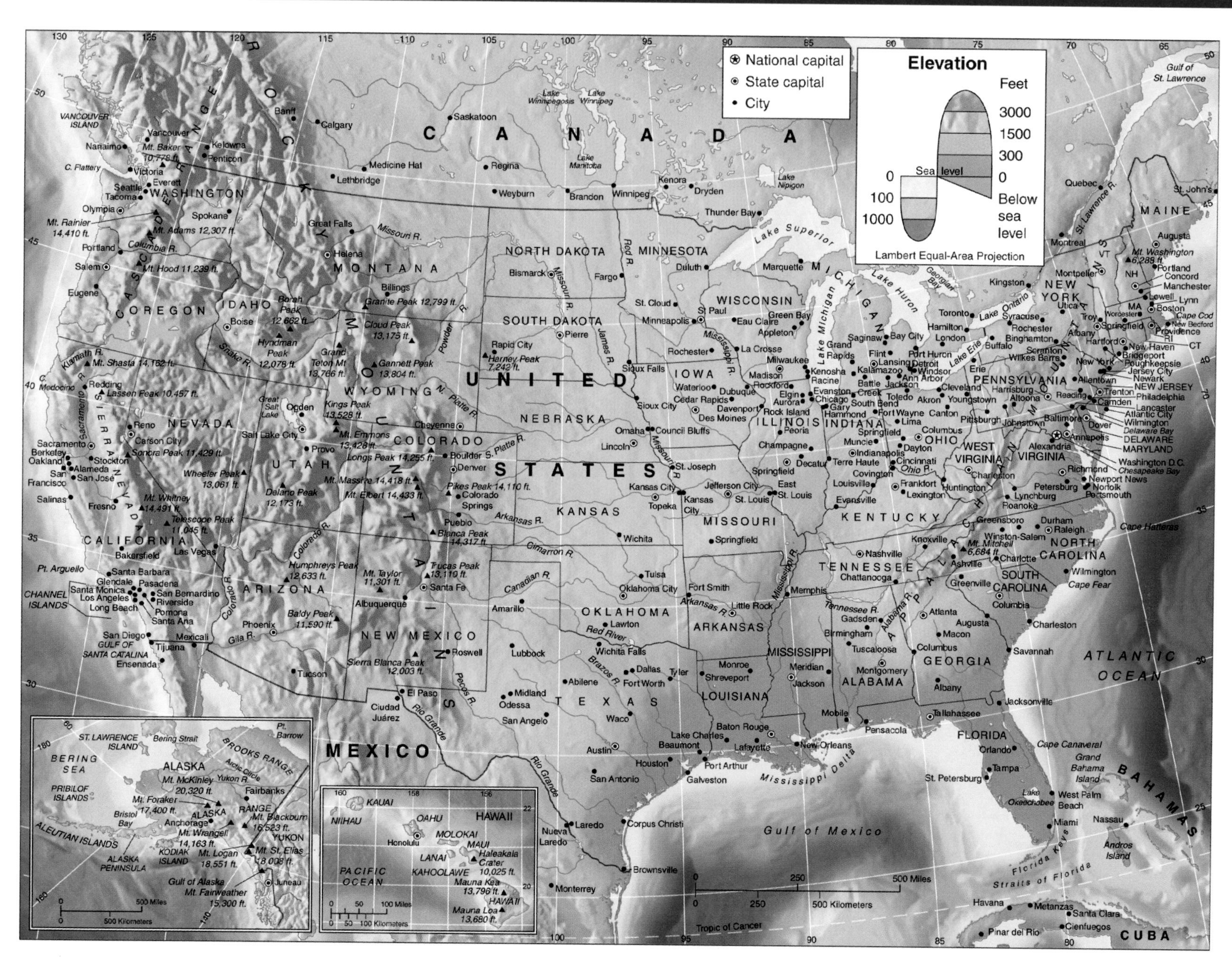
National capital
State capital
City
Elevation
Feet
3000
1500
300
0
Below sea level
Sea level
0
100
1000
Lambert Equal-Area Projection
CANADA
UNITED STATES
MEXICO
CUBA
BAHAMAS
ROCKY MOUNTAINS
CASCADE RANGE
SIERRA NEVADA
APPALACHIAN MTS.
ATLANTIC OCEAN
Gulf of Mexico
Gulf of St. Lawrence
Straits of Florida
Mississippi Delta
Tropic of Cancer
WASHINGTON
OREGON
CALIFORNIA
IDAHO
NEVADA
MONTANA
WYOMING
UTAH
ARIZONA
COLORADO
NEW MEXICO
NORTH DAKOTA
SOUTH DAKOTA
NEBRASKA
KANSAS
OKLAHOMA
TEXAS
MINNESOTA
IOWA
MISSOURI
ARKANSAS
LOUISIANA
WISCONSIN
ILLINOIS
MICHIGAN
INDIANA
OHIO
KENTUCKY
TENNESSEE
MISSISSIPPI
ALABAMA
GEORGIA
FLORIDA
SOUTH CAROLINA
NORTH CAROLINA
VIRGINIA
WEST VIRGINIA
PENNSYLVANIA
NEW YORK
NEW JERSEY
DELAWARE
MARYLAND
MAINE
Washington D.C.
Lake Superior
Lake Michigan
Lake Huron
Lake Erie
Lake Ontario
Mt. Rainier 14,410 ft.
Mt. Whitney 14,491 ft.
Pikes Peak 14,110 ft.
Mt. Washington 6,288 ft.
Mt. Mitchell 6,684 ft.
ALASKA
Mt. McKinley 20,320 ft.
BERING SEA
ALEUTIAN ISLANDS
HAWAII
PACIFIC OCEAN
Mauna Kea 13,796 ft.
Mauna Loa 13,680 ft.
0 250 500 Miles
0 250 500 Kilometers

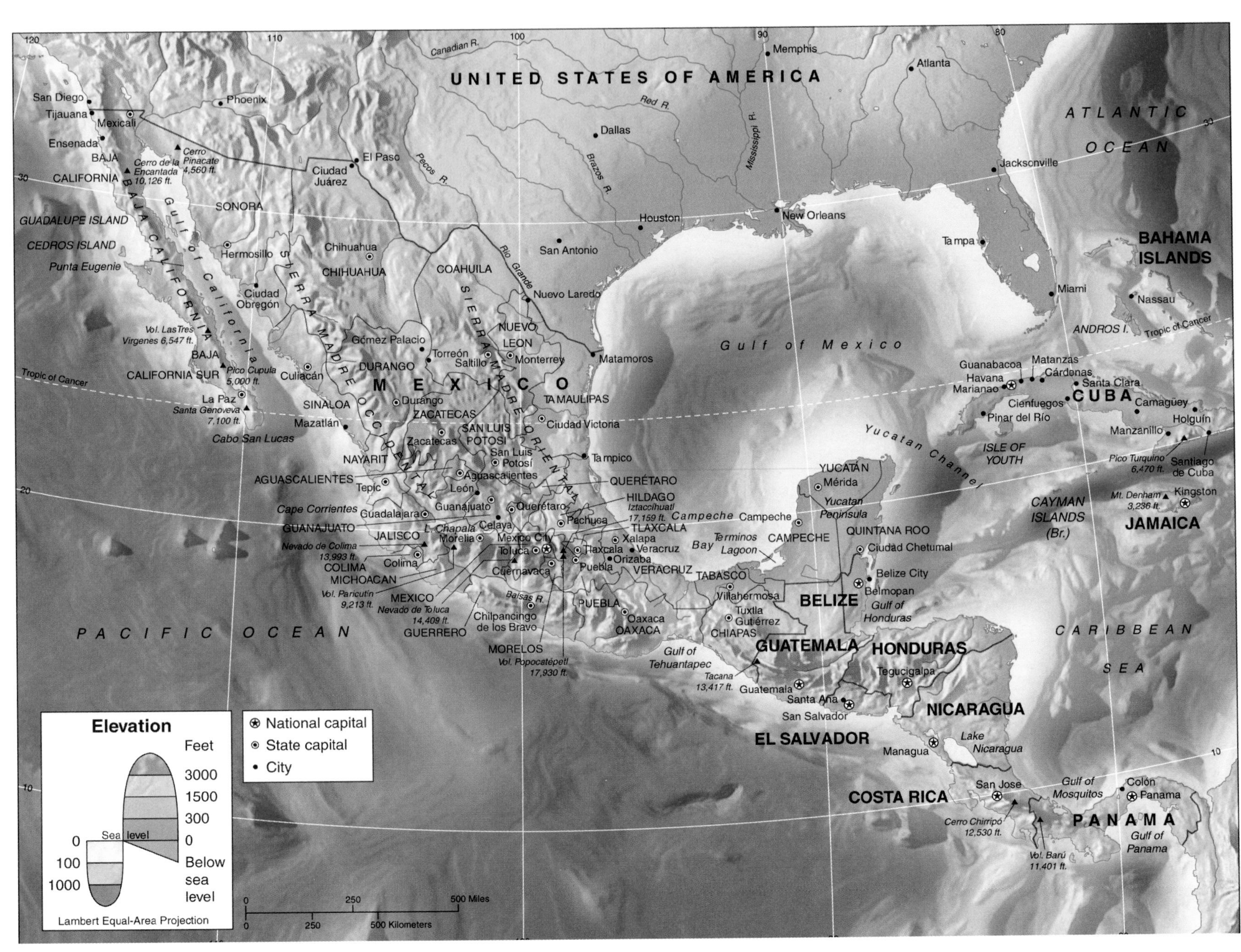
UNITED STATES OF AMERICA
MEXICO
CUBA
BAHAMA ISLANDS
JAMAICA
CAYMAN ISLANDS (Br.)
BELIZE
GUATEMALA
HONDURAS
EL SALVADOR
NICARAGUA
COSTA RICA
PANAMA
ATLANTIC OCEAN
PACIFIC OCEAN
CARIBBEAN SEA
Gulf of Mexico
Gulf of California
Yucatan Channel
Gulf of Tehuantepec
Gulf of Honduras
Gulf of Mosquitos
Gulf of Panama
Campeche Bay
Terminos Lagoon
Lake Nicaragua
L. Chapala
SIERRA MADRE OCCIDENTAL
SIERRA MADRE ORIENTAL
Yucatan Peninsula
BAJA CALIFORNIA
Canadian R.
Red R.
Pecos R.
Brazos R.
Mississippi R.
Rio Grande
Balsas R.
San Diego
Tijauana
Mexicali
Ensenada
Phoenix
El Paso
Ciudad Juárez
Dallas
Houston
San Antonio
Memphis
New Orleans
Atlanta
Jacksonville
Tampa
Miami
Nassau
ANDROS I.
Tropic of Cancer
Cerro Pinacate 4,560 ft.
Cerro de la Encantada 10,126 ft.
BAJA CALIFORNIA
SONORA
GUADALUPE ISLAND
CEDROS ISLAND
Punta Eugenie
Hermosillo
Ciudad Obregón
Chihuahua
CHIHUAHUA
COAHUILA
Nuevo Laredo
NUEVO LEON
Monterrey
Saltillo
Torreón
Gómez Palacio
Matamoros
TAMAULIPAS
Ciudad Victoria
Tampico
Vol. Las Tres Virgenes 6,547 ft.
BAJA CALIFORNIA SUR
Pico Cupula 5,000 ft.
Culiacán
La Paz
Santa Genoveva 7,100 ft.
Cabo San Lucas
Mazatlán
SINALOA
DURANGO
Durango
ZACATECAS
Zacatecas
SAN LUIS POTOSI
San Luis Potosí
NAYARIT
AGUASCALIENTES
Aguascalientes
Tepic
León
QUERÉTARO
Querétaro
HILDAGO
Iztaccihuatl 17,159 ft.
Cape Corrientes
Guadalajara
Guanajuato
GUANAJUATO
Celaya
Pachuca
TLAXCALA
JALISCO
Morelia
Mexico City
Toluca
Tlaxcala
Xalapa
Veracruz
Nevado de Colima 13,993 ft.
COLIMA
Colima
Puebla
Orizaba
VERACRUZ
MICHOACAN
Cuernavaca
Vol. Paricutín 9,213 ft.
MEXICO
Nevado de Toluca 14,409 ft.
GUERRERO
Chilpancingo de los Bravo
PUEBLA
Oaxaca
OAXACA
MORELOS
Vol. Popocatépetl 17,930 ft.
TABASCO
Villahermosa
Tuxtla Gutiérrez
CHIAPAS
Tacana 13,417 ft.
Guatemala
YUCATAN
Mérida
Campeche
CAMPECHE
QUINTANA ROO
Ciudad Chetumal
Belize City
Belmopan
Tegucigalpa
Santa Ana
San Salvador
Managua
San Jose
Cerro Chirripó 12,530 ft.
Vol. Barú 11,401 ft.
Colón
Panama
Guanabacoa
Havana
Marianao
Matanzas
Cárdenas
Santa Clara
Cienfuegos
Pinar del Río
Camagüey
Holguín
Manzanillo
Pico Turquino 6,470 ft.
Santiago de Cuba
ISLE OF YOUTH
Mt. Denham 3,236 ft.
Kingston
Elevation
Feet
3000
1500
300
0
Sea level
Below sea level
0
100
1000
Lambert Equal-Area Projection
National capital
State capital
City
0
250
500 Miles
250
500 Kilometers
120
110
100
90
80
30
20
10

Map 109 The Caribbean

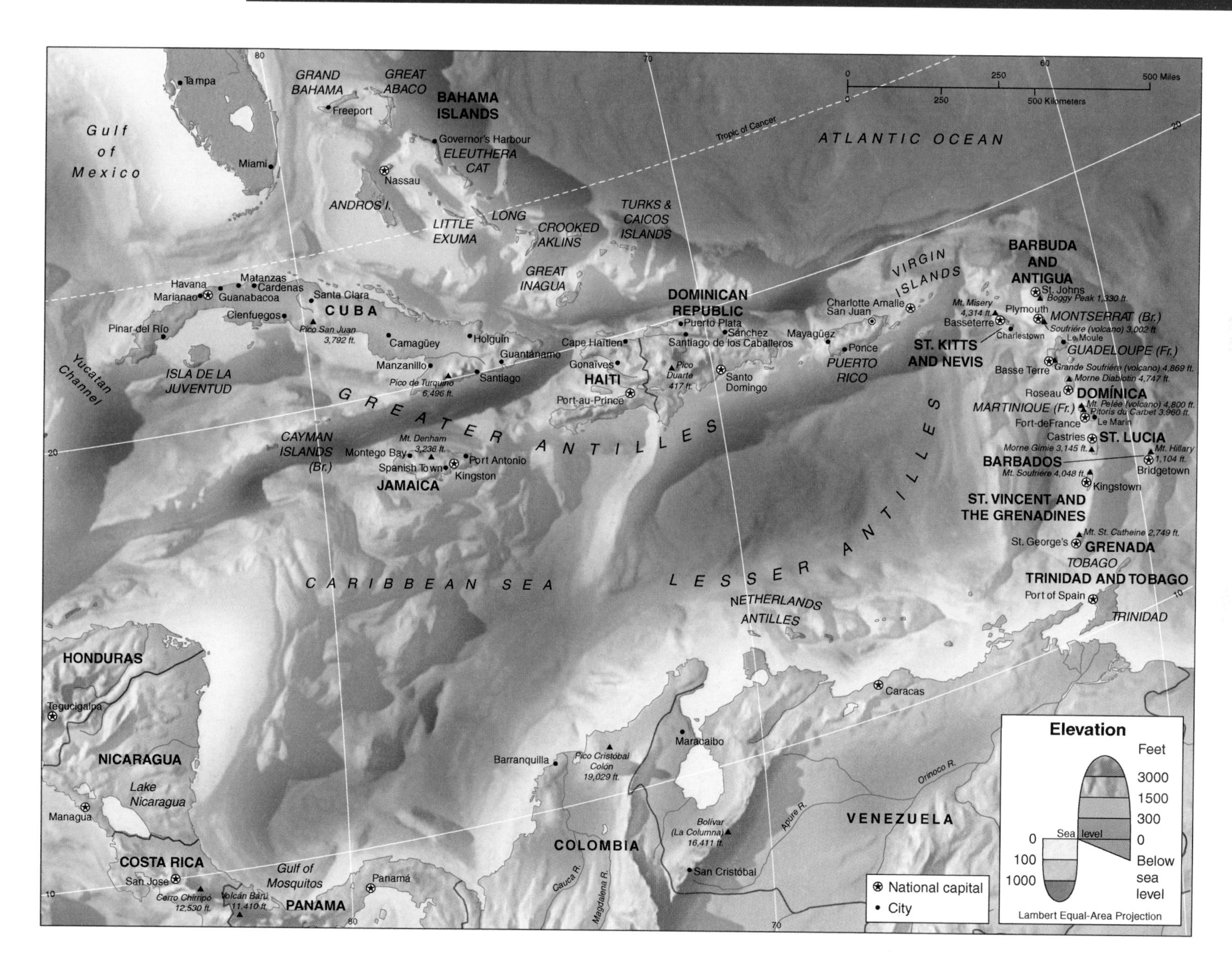

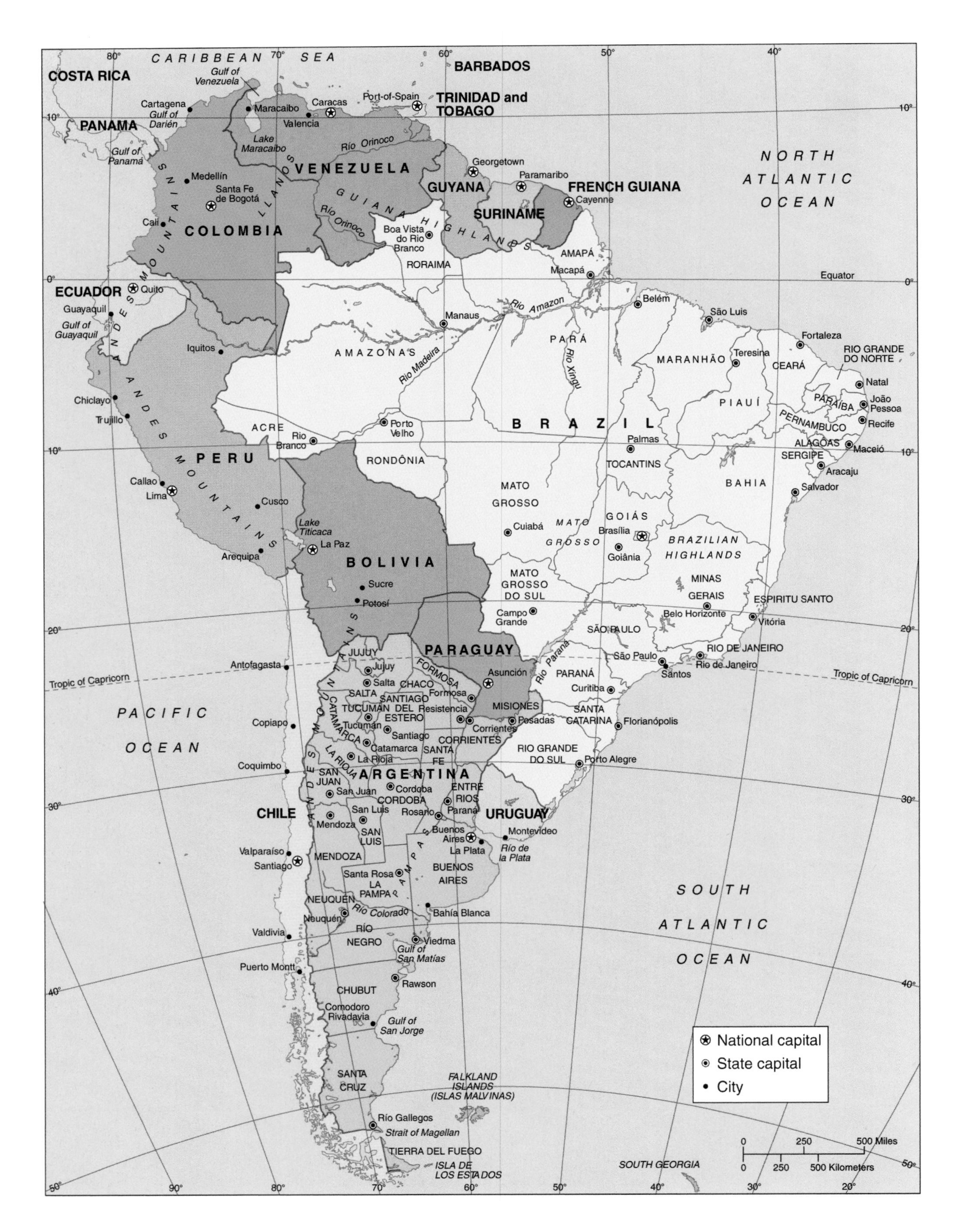
CARIBBEAN SEA
COSTA RICA
PANAMA
BARBADOS
TRINIDAD and TOBAGO
VENEZUELA
COLOMBIA
GUYANA
SURINAME
FRENCH GUIANA
ECUADOR
PERU
BOLIVIA
BRAZIL
PARAGUAY
ARGENTINA
URUGUAY
CHILE
NORTH ATLANTIC OCEAN
PACIFIC OCEAN
SOUTH ATLANTIC OCEAN
Equator
Tropic of Capricorn
ANDES MOUNTAINS
GUIANA HIGHLANDS
BRAZILIAN HIGHLANDS
LLANOS
PAMPAS
Río Orinoco
Rio Amazon
Rio Madeira
Rio Xingu
Rio Paraná
Río Colorado
Lake Maracaibo
Lake Titicaca
Gulf of Venezuela
Gulf of Darién
Gulf of Panamá
Gulf of Guayaquil
Río de la Plata
Gulf of San Matías
Gulf of San Jorge
Strait of Magellan
TIERRA DEL FUEGO
ISLA DE LOS ESTADOS
FALKLAND ISLANDS (ISLAS MALVINAS)
SOUTH GEORGIA
Cartagena
Maracaibo
Caracas
Valencia
Port-of-Spain
Medellín
Santa Fe de Bogotá
Cali
Georgetown
Paramaribo
Cayenne
Boa Vista do Rio Branco
RORAIMA
AMAPÁ
Macapá
Quito
Guayaquil
Iquitos
Chiclayo
Trujillo
Callao
Lima
Cusco
Arequipa
La Paz
Sucre
Potosí
Manaus
AMAZONAS
PARÁ
Belém
São Luis
MARANHÃO
Teresina
Fortaleza
CEARÁ
RIO GRANDE DO NORTE
Natal
PARAÍBA
João Pessoa
PERNAMBUCO
Recife
PIAUÍ
ALAGOAS
Maceió
SERGIPE
Aracaju
BAHIA
Salvador
ACRE
Rio Branco
Porto Velho
RONDÔNIA
Palmas
TOCANTINS
MATO GROSSO
Cuiabá
GOIÁS
Brasília
Goiânia
MATO GROSSO DO SUL
Campo Grande
MINAS GERAIS
Belo Horizonte
ESPIRITU SANTO
Vitória
SÃO PAULO
São Paulo
Santos
RIO DE JANEIRO
Rio de Janeiro
PARANÁ
Curitiba
SANTA CATARINA
Florianópolis
RIO GRANDE DO SUL
Porto Alegre
Antofagasta
Copiapo
Coquimbo
Valparaíso
Santiago
Valdivia
Puerto Montt
JUJUY
Jujuy
SALTA
Salta
FORMOSA
Formosa
CHACO
Resistencia
Asunción
MISIONES
Posadas
SANTIAGO DEL ESTERO
Santiago
TUCUMÁN
Tucumán
CATAMARCA
Catamarca
CORRIENTES
Corrientes
LA RIOJA
La Rioja
SANTA FE
SAN JUAN
San Juan
CORDOBA
Cordoba
ENTRE RIOS
Paraná
Rosario
San Luis
SAN LUIS
Mendoza
MENDOZA
Buenos Aires
La Plata
Montevideo
BUENOS AIRES
Santa Rosa
LA PAMPA
NEUQUÉN
Neuquén
Bahía Blanca
RÍO NEGRO
Viedma
CHUBUT
Rawson
Comodoro Rivadavia
SANTA CRUZ
Río Gallegos
National capital
State capital
City
0 250 500 Miles
0 250 500 Kilometers
80° 70° 60° 50° 40° 90° 30° 20°
10° 0° 10° 20° 30° 40° 50°

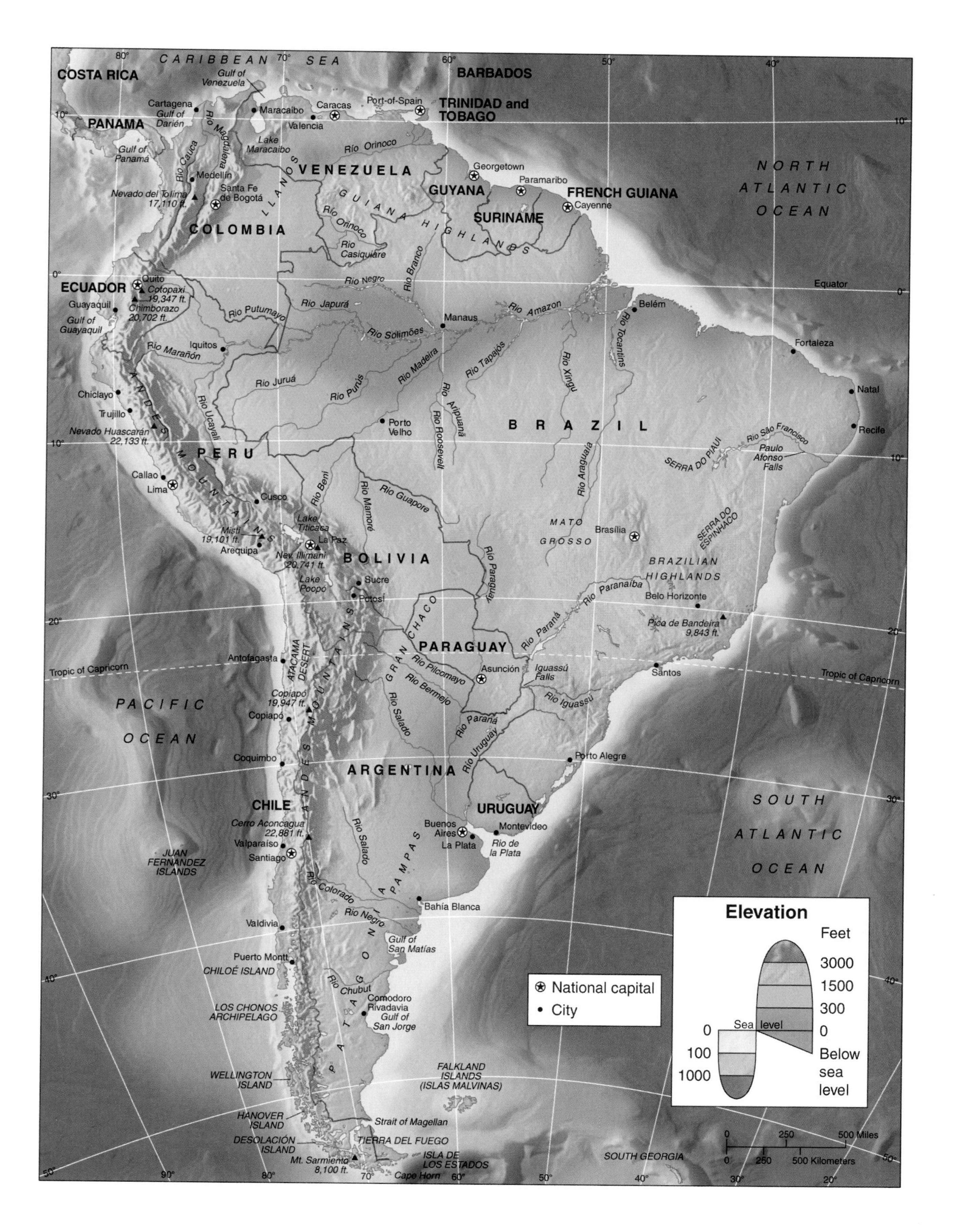
CARIBBEAN SEA
COSTA RICA
Gulf of Venezuela
BARBADOS
Cartagena
Gulf of Darién
PANAMA
Maracaibo
Caracas
Port-of-Spain
TRINIDAD and TOBAGO
Valencia
Lake Maracaibo
Río Orinoco
Gulf of Panamá
Río Cauca
Río Magdalena
Medellín
VENEZUELA
Georgetown
Paramaribo
Nevado del Tolima 17,110 ft.
Santa Fe de Bogotá
LLANOS
GUIANA HIGHLANDS
GUYANA
FRENCH GUIANA
Cayenne
SURINAME
NORTH ATLANTIC OCEAN
COLOMBIA
Río Orinoco
Río Casiquiare
Rio Branco
Rio Negro
Equator
ECUADOR
Quito
Cotopaxi 19,347 ft.
Chimborazo 20,702 ft.
Guayaquil
Gulf of Guayaquil
Río Putumayo
Rio Japurá
Manaus
Rio Amazon
Belém
Rio Tocantins
Rio Solimões
Iquitos
Río Marañón
Fortaleza
Rio Madeira
Rio Tapajós
Rio Xingu
Rio Juruá
Rio Purus
Natal
Chiclayo
Trujillo
Río Ucayali
Rio Aripuanã
Rio Roosevelt
Porto Velho
BRAZIL
Recife
Nevado Huascarán 22,133 ft.
Rio São Francisco
Paulo Afonso Falls
SERRA DO PIAUÍ
PERU
ANDES MOUNTAINS
Callao
Lima
Río Beni
Rio Mamoré
Rio Guaporé
Rio Araguaia
Cusco
Lake Titicaca
MATO GROSSO
Brasília
SERRA DO ESPINHAÇO
Misti 19,101 ft.
Arequipa
La Paz
Nev. Illimani 20,741 ft.
BOLIVIA
Rio Paraguay
BRAZILIAN HIGHLANDS
Sucre
Lake Poopo
Potosí
Rio Paranaíba
Belo Horizonte
GRAN CHACO
Pico de Bandeira 9,843 ft.
Rio Paraná
PARAGUAY
Antofagasta
ATACAMA DESERT
Río Pilcomayo
Asunción
Iguassú Falls
Santos
Tropic of Capricorn
Río Bermejo
Rio Iguassu
PACIFIC OCEAN
Copiapó 19,947 ft.
Copiapó
Río Salado
Río Paraná
Río Uruguay
Coquimbo
Porto Alegre
ARGENTINA
CHILE
URUGUAY
SOUTH ATLANTIC OCEAN
Cerro Aconcagua 22,881 ft.
Valparaíso
Santiago
Río Salado
PAMPAS
Buenos Aires
La Plata
Montevideo
Río de la Plata
JUAN FERNANDEZ ISLANDS
Río Colorado
Bahía Blanca
Valdivia
Río Negro
Gulf of San Matías
PATAGONIA
Puerto Montt
CHILOÉ ISLAND
Río Chubut
Comodoro Rivadavia
Gulf of San Jorge
LOS CHONOS ARCHIPELAGO
National capital
City
Elevation
Feet
3000
1500
300
0
Sea level
0
100
1000
Below sea level
FALKLAND ISLANDS (ISLAS MALVINAS)
WELLINGTON ISLAND
HANOVER ISLAND
Strait of Magellan
DESOLACIÓN ISLAND
TIERRA DEL FUEGO
Mt. Sarmiento 8,100 ft.
ISLA DE LOS ESTADOS
Cape Horn
SOUTH GEORGIA
0 250 500 Miles
0 250 500 Kilometers
80°
70°
60°
50°
40°
90°
30°
20°
10°
0°

Map 112a Environment and Economy

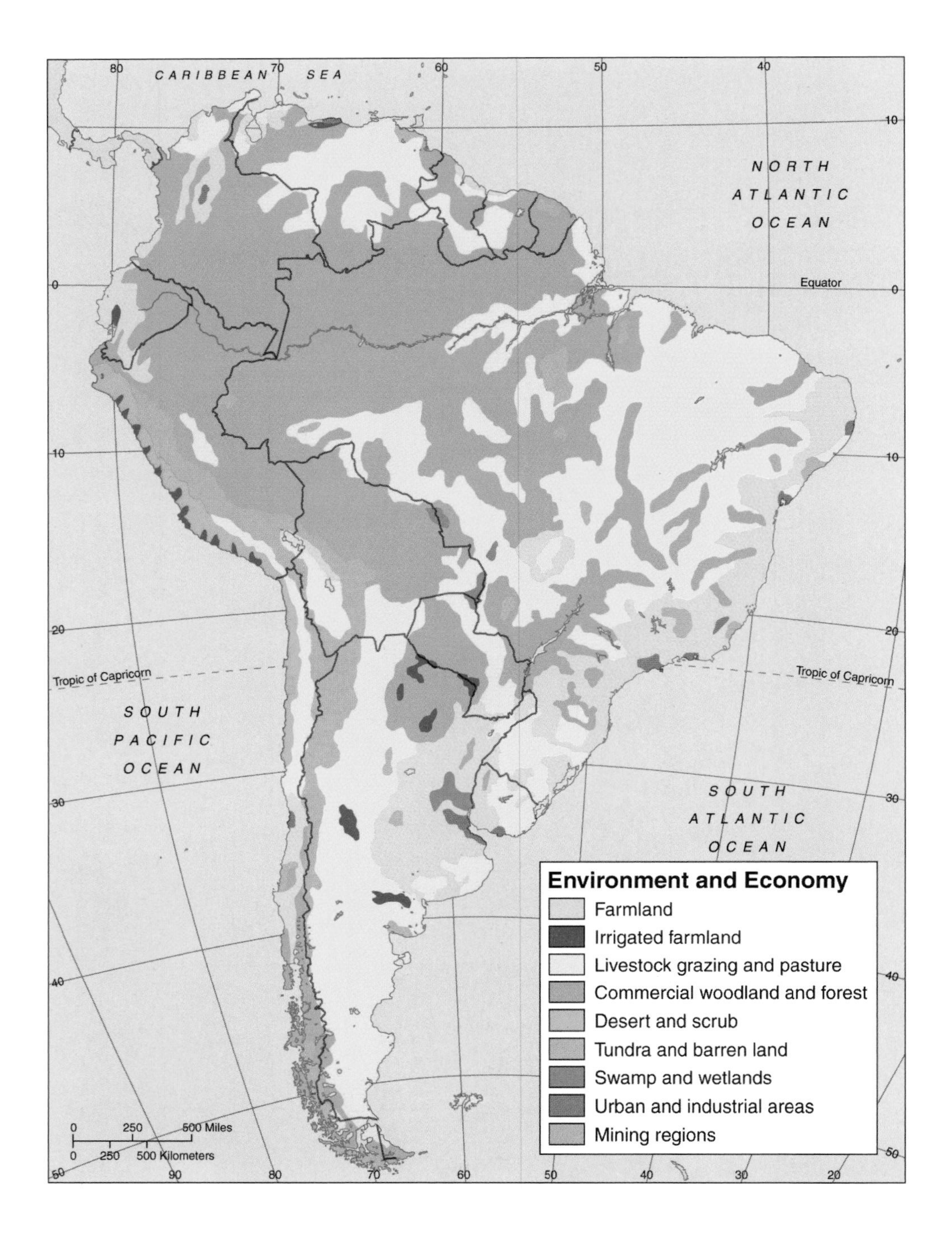

South America is a region just beginning to emerge from a colonial-dependency economy in which raw materials flowed from the continent to more highly developed economic regions. With the exception of Brazil, Argentina, Chile, and Uruguay, most of the continent's countries still operate under the traditional mode of exporting raw materials in exchange for capital that tends to accumulate in the pockets of a small percentage of the population. The land use patterns of the continent are, therefore, still dominated by resource extraction and agriculture. A problem posed by these patterns is that little of the continent's land area is actually suitable for either commercial forestry or commercial crop agriculture without extremely high environmental costs. Much of the agriculture, then, is based on high value tropical crops that can be grown in small areas profitably, or on extensive livestock grazing. Even within the forested areas of the Amazon Basin where forest clearance is taking place at unprecedented rates, much of the land use that replaces forest is grazing.

Map 112b Population Density

Population Density
Persons per square mile (kilometer)
Uninhabited area
Less than 2 (1)
2 – 50 (1 – 20)
50 – 150 (20 – 60)
150 – 300 (60 – 120)
More than 300 (100)
0 250 500 Miles
0 250 500 Kilometers

Since so much of interior South America is uninhabitable (the high Andes) or only sparsely populated (the interior of the Amazon Basin), the continent's population tends to be peripheral—approximately 90 percent of the continent's nearly 375 million people live within 150 miles of the sea. This population also tends to be heavily urbanized. Over 80 percent of South America's population lives in cities and the continent has three of the world's 15 largest cities—Rio de Janeiro, Saõ Paulo, and Buenos Aires. Saõ Paulo is the world's third largest urban agglomeration after Tokyo and New York. As in North America, most of the population of South America can trace at least part of its ancestry to the Old World. Throughout the Spanish-speaking parts of the continent the population is predominantly *mestizo* or mixed European and native South American. In Portuguese-speaking Brazil, in addition to a *mestizo* population, there is a significant admixture of African blood, the result of a large slave labor force imported from Africa to work the sugar, cotton, and other plantations of the colonial period.

Map 112c Environmental Problems

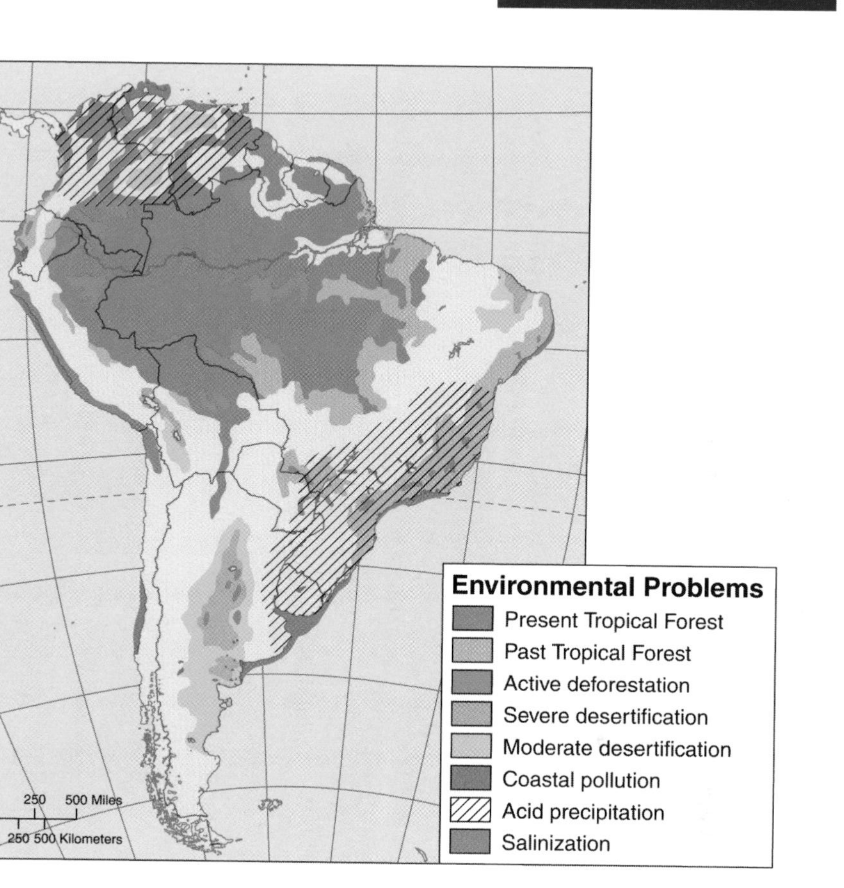

The drainage basin of the Amazon River and its tributaries, along with adjacent regions, is the world's largest remaining area of tropical forest. Much of the periphery of this vast forested region has already been cleared for farming and grazing and the ax and chainsaw and the flames are working their way steadily toward the interior. Tropical deforestation produces a loss of the biological diversity represented by the world's most biologically productive ecosystem, along with changes in soil and soil-water systems. South America has other environmental problems: *desertification* in which grassland and/or scrub vegetation is converted to desert through overgrazing or other unwise agricultural practices; soil *salinization* in which soils become increasingly salty as the consequence of the over application of irrigation water; *coastal and estuarine pollution* resulting from unregulated or unchecked use of coastal waters for industrial, commercial, and transportation purposes; and *acid* ***precipitation*** resulting from the combination of airborne industrial wastes and automobile-truck exhausts with water vapor to produce dry or wet acidic fallout.

Map 112d Protection of Natural Areas

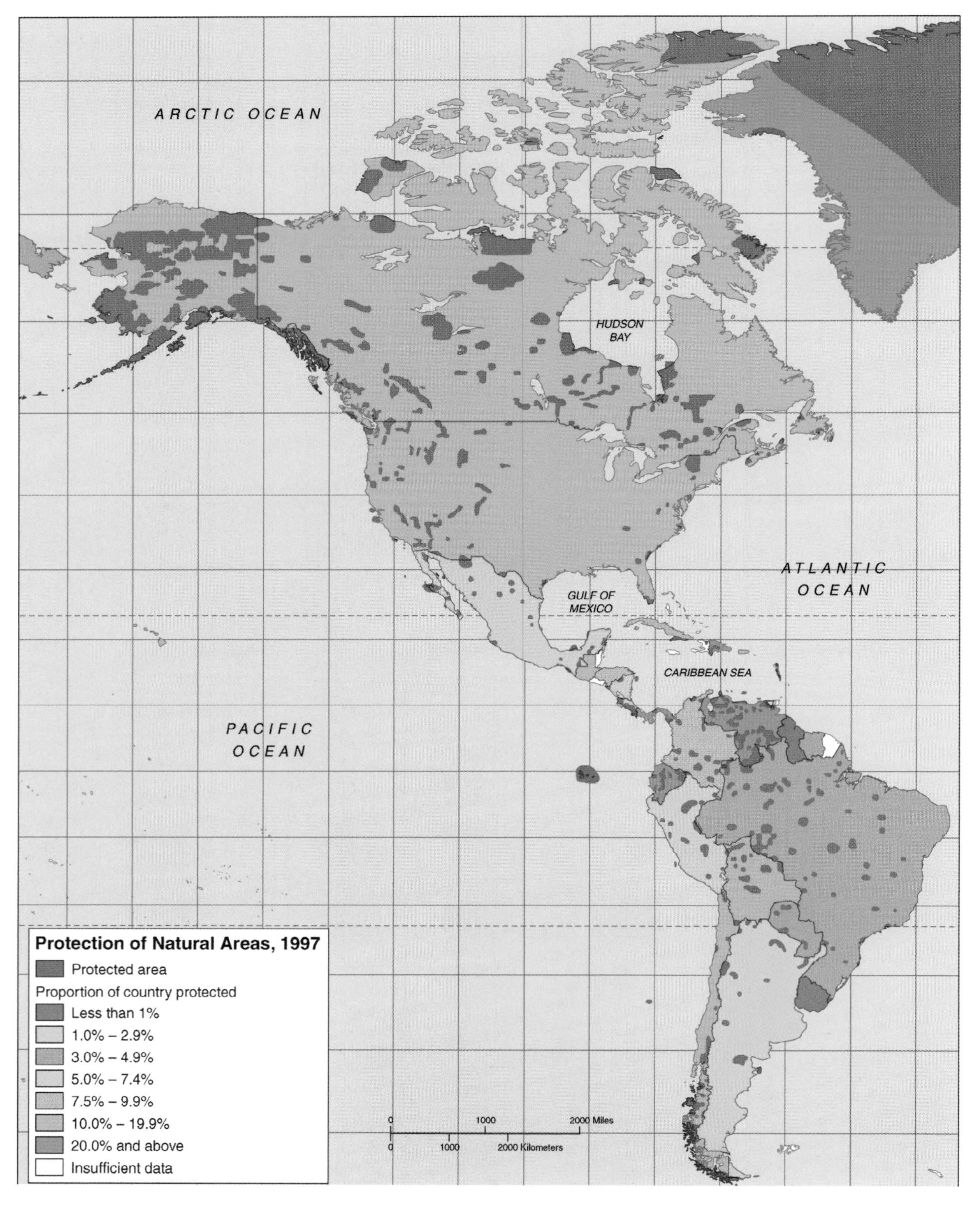

There have been expressed concerns among international conservation organizations that the protection of natural areas is largely an empty measure and that many of the world's national, provincial, regional, or state parks are "paper parks"—more important in conception than in reality. Particularly in Africa, the park systems designated largely for the purposes of wildlife preservation have come under international scrutiny. In the Americas, however, there does seem to be a close correlation between the designation of protected natural areas and the actuality of ecological preservation. With an increasing demand for natural resources and with governmental policies in both the United States and Brazil (the two largest population and economic entities in the Americas) that seem directed against preservation measures, there is fear that American protected areas could become, like their counterparts in many areas of Africa and Asia, simply parks on paper.

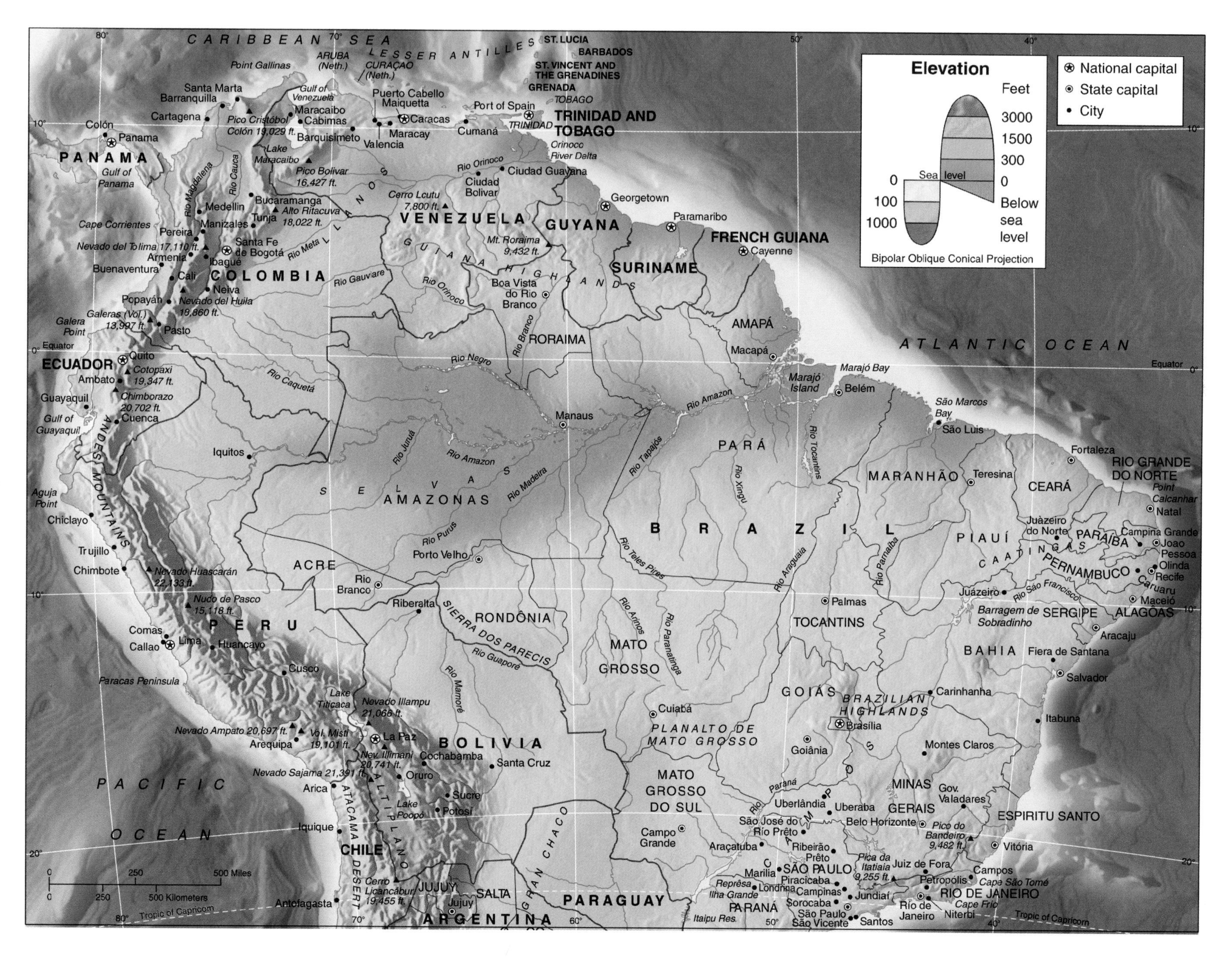
Elevation
Feet
3000
1500
300
0
Below sea level
0
100
1000
Sea level
Bipolar Oblique Conical Projection
National capital
State capital
City
CARIBBEAN SEA
ATLANTIC OCEAN
PACIFIC OCEAN
LESSER ANTILLES
ARUBA (Neth.)
CURAÇAO (Neth.)
ST. LUCIA
BARBADOS
ST. VINCENT AND THE GRENADINES
GRENADA
TOBAGO
TRINIDAD
TRINIDAD AND TOBAGO
PANAMA
COLOMBIA
VENEZUELA
GUYANA
SURINAME
FRENCH GUIANA
ECUADOR
PERU
BRAZIL
BOLIVIA
CHILE
PARAGUAY
ARGENTINA
LLANOS
GUIANA HIGHLANDS
BRAZILIAN HIGHLANDS
CAMPOS
SELVAS
ANDES MOUNTAINS
ALTIPLANO
ATACAMA DESERT
GRAN CHACO
CAATINGAS
PLANALTO DE MATO GROSSO
SIERRA DOS PARECIS
AMAZONAS
RORAIMA
AMAPÁ
PARÁ
MARANHÃO
PIAUÍ
CEARÁ
RIO GRANDE DO NORTE
PARAÍBA
PERNAMBUCO
ALAGOAS
SERGIPE
BAHIA
TOCANTINS
GOIÁS
MINAS GERAIS
ESPIRITU SANTO
RIO DE JANEIRO
SÃO PAULO
PARANÁ
MATO GROSSO
MATO GROSSO DO SUL
RONDÔNIA
ACRE
JUJUY
SALTA
Rio Amazon
Rio Negro
Rio Madeira
Rio Tapajós
Rio Xingu
Rio Tocantins
Rio Araguaia
Rio Orinoco
Rio Branco
Rio Purus
Rio Juruá
Rio Caquetá
Rio Magdalena
Rio Cauca
Rio São Francisco
Rio Paraná
Rio Parnaíba
Rio Mamoré
Rio Guaporé
Rio Teles Pires
Rio Paranatinga
Rio Arinos
Rio Gauviare
Rio Meta
Lake Maracaibo
Lake Titicaca
Lake Poopó
Gulf of Panama
Gulf of Guayaquil
Gulf of Venezuela
Marajó Bay
Marajó Island
São Marcos Bay
Orinoco River Delta
Barragem de Sobradinho
Itaipu Res.
Represa Ilha Grande
Mt. Roraima 9,432 ft.
Cerro Lcutu 7,800 ft.
Pico Bolívar 16,427 ft.
Alto Ritacuva 18,022 ft.
Pico Cristóbol Colón 19,029 ft.
Nevado del Huila 18,860 ft.
Nevado del Tolima 17,110 ft.
Galeras (Vol.) 13,997 ft.
Cotopaxi 19,347 ft.
Chimborazo 20,702 ft.
Nevado Huascarán 22,133 ft.
Nudo de Pasco 15,118 ft.
Nevado Illampu 21,066 ft.
Nev. Illimani 20,741 ft.
Nevado Ampato 20,697 ft.
Vol. Misti 19,101 ft.
Nevado Sajama 21,391 ft.
Cerro Licancabur 19,455 ft.
Pico do Bandeiro 9,482 ft.
Pica da Itatiaia 9,255 ft.
Point Gallinas
Galera Point
Aguja Point
Paracas Peninsula
Cape Corrientes
Point Calcanhar
Cape Frio
Capa São Tomé
Equator
Tropic of Capricorn
Quito
Bogotá
Santa Fe de Bogotá
Caracas
Port of Spain
Georgetown
Paramaribo
Cayenne
Lima
La Paz
Brasília
Panama
Colón
Barranquilla
Santa Marta
Cartagena
Medellín
Manizales
Pereira
Armenia
Ibagué
Tunja
Bucaramanga
Cali
Buenaventura
Neiva
Popayán
Pasto
Maracaibo
Cabimas
Barquisimeto
Puerto Cabello
Maiquetía
Maracay
Valencia
Cumaná
Ciudad Bolívar
Ciudad Guayana
Guayaquil
Ambato
Cuenca
Trujillo
Chiclayo
Chimbote
Comas
Callao
Huancayo
Iquitos
Cusco
Arequipa
Manaus
Belém
Macapá
São Luís
Fortaleza
Natal
Juàzeiro do Norte
Campina Grande
Joao Pessoa
Olinda
Recife
Caruaru
Maceió
Aracaju
Juàzeiro
Fiera de Santana
Salvador
Itabuna
Carinhanha
Montes Claros
Gov. Valadares
Teresina
Palmas
Goiânia
Cuiabá
Campo Grande
Porto Velho
Rio Branco
Riberalta
Boa Vista do Rio Branco
Uberlândia
Uberaba
Belo Horizonte
São José do Rio Prêto
Ribeirão Prêto
Araçatuba
Marília
Londrina
Piracicaba
Campinas
Sorocaba
Jundiaí
São Paulo
São Vicente
Santos
Juiz de Fora
Petrópolis
Campos
Vitória
Rio de Janeiro
Niterói
Santa Cruz
Cochabamba
Sucre
Potosí
Oruro
Arica
Iquique
Antofagasta
Jujuy
0
250
500 Miles
500 Kilometers

Elevation
Feet
3000
1500
300
0
Sea level
Below sea level
0
100
1000
National capital
State capital
City
Bipolar Oblique Conical Projection
500 Miles
500 Kilometers
A T L A N T I C
O C E A N
P A C I F I C
O C E A N
S C O T I A S E A
B R A Z I L
PARAGUAY
URUGUAY
A R G E N T I N A
CHILE
FALKLAND/MALVINAS ISLANDS (Br.)
SOUTH GEORGIA (Br.)
Drake Passage
Tropic of Capricorn
ATACAMA DESERT
GRAN CHACO
PAMPAS
SIERRA DE CÓRDOBA
JUJUY
SALTA
TUCUMÁN
CATAMARCA
LA RIOJA
SAN JUAN
MENDOZA
SAN LUIS
CÓRDOBA
SANTIAGO DEL ESTERO
CHACO
FORMOSA
SANTA FE
CORRIENTES
MISIONES
ENTRE RÍOS
BUENOS AIRES
LA PAMPA
NEUQUÉN
RÍO NEGRO
CHUBUT
SANTA CRUZ
TIERRA DEL FUEGO
PARANÁ
SANTA CATARINA
RIO GRANDE DO SUL
SÃO PAULO
RIO DE JANEIRO
Asunción
Concepción
Montevideo
Buenos Aires
Santiago
Valparaíso
Viña del Mar
Antofagasta
Copiapó
Coquimbo
Rancagua
Talca
Chillán
Concepción
Talcahuano
Temuco
Valdivia
Osorno
Puerto Montt
Punta Arenas
Salta
Jujuy
Tucumán
Catamarca
La Rioja
San Juan
Mendoza
San Luis
San Rafael
Córdoba
Río Cuarto
Rosario
Santa Fe
Paraná
Resistencia
Corrientes
Posadas
Formosa
Concordia
Salto
Paysandú
La Plata
Avellaneda
Mar del Plata
Bahía Blanca
Santa Rosa
Neuquén
Viedma
Rawson
Comodoro Rivadavia
Puerto Deseado
Puerto Santa Cruz
Río Gallegos
Stanley
Pôrto Alegre
Pelotas
Rio Grande
Caxias do Sul
Santa Maria
Florianópolis
Laguna
Curitiba
Ponta Grossa
Londrina
Marília
Piracicaba
Campinas
Jundiaí
Sorocaba
São Paulo
Santos
São Vicente
Rio de Janeiro
Niterói
Petrópolis
Juiz de Fora
Campos
Cape São Tomé
Cape Frio
Cape Santa Marta Grande
Patos Lagoon
Mirim Lagoon
Mangueira Lagoon
Río de La Plata
Norte Point
Cape San Antonio
Cape Corrientes
Blanca Bay
Valdés Peninsula
San Matías Gulf
San Jorge Gulf
C. Tres Puntas
Strait of Magellan
C. San Diego
Cape Horn
Mt. Sarmiento 8,100 ft.
Cerro Chaltel/ Mt. Fitzroy 10,958 ft.
M. San Valentín 13,314 ft.
Lake Buenos Aires
Cerro Tronador 11,660 ft.
Antuco (Vol.) 11,700 ft.
Maipo (Vol.) 17,464 ft.
Cerro Tupungato 21,555 ft.
Cerro Aconcagua 22,831 ft.
Cerro Mercedario 22,211 ft.
Cerro Bonete 22,546 ft.
Nevado Ojos del Salado 22,566 ft.
Nevados de Cachi 22,047 ft.
Volcán Llullaillaco 22,110 ft.
Cerro Licancabur 19,455 ft.
Cerro Azufre (Copiapó) (Vol.) 19,947 ft.
Lengua de Vaca Point
C. Carranza
Lavapié Point
Galera Point
CHILOÉ ISLAND
C. Quilán
CHONOS ARCHIPELAGO
Gulf of Penas
WELLINGTON ISLAND
QUEEN ADELAIDE ARCHIPELAGO
SANTA INÉS ISLAND
West Falkland
East Falkland
Grande Bay
Río Paraná
Río Paraguay
Río Pilcomayo
Río Bermejo
Río Salado
Río Colorado
Río Negro
Río Limay
Río Chubut
Río Deseado
Río Chico
Río Uruguay
Río Iguaçu
Río Paranapanema
Itaipu Res.
Represa Ilha Grande
Pico da Itatiaia 9,255 ft.
Mar Chiquita

National capital
City
0 250 500 Miles
0 250 500 Kilometers
GREENLAND
ARCTIC OCEAN
Arctic Circle
ICELAND
Reykjavik
ATLANTIC
OCEAN
FAEROE ISLANDS
SHETLAND ISLANDS
ORKNEY ISLANDS
HEBRIDES
UNITED KINGDOM
Glasgow
Belfast
Newcastle
Leeds
IRISH SEA
Liverpool
Manchester
Birmingham
London
IRELAND
Dublin
St. George's Channel
Strait of Dover
English Channel
NORTH SEA
NORWAY
Bergen
Oslo
Stavenger
SWEDEN
Uppsala
Stockholm
GOTLAND
Goteborg
OLAND
Gulf of Bothnia
BALTIC SEA
FINLAND
Turku
Helsinki
Tallinn
ESTONIA
Gulf of Riga
LATVIA
Riga
LITHUANIA
Vilnius
RUSSIA
Kaliningrad
Gdansk
DENMARK
Copenhagen
NETHERLANDS
The Hague
Amsterdam
Rotterdam
Antwerp
BELGIUM
Brussels
LUX.
Luxembourg
Essen
Bonn
Frankfurt
Hamburg
Berlin
GERMANY
Rhine R.
Stuttgart
Munich
POLAND
Warsaw
Katowice
BELARUS
Minsk
CZECH REP.
Prague
SLOVAKIA
Bratislava
AUSTRIA
Vienna
HUNGARY
Budapest
SWITZERLAND
Bern
FRANCE
Paris
Lille
Le Havre
Brest
Lyon
Bordeaux
Marseille
Bay of Biscay
MONACO
Monaco
ANDORRA
PORTUGAL
Lisbon
SPAIN
Madrid
Bilbao
Valladolid
Barcelona
Valencia
Seville
BALEARIC IS.
Strait of Gibraltar
CORSICA
SARDINIA
ITALY
Milan
Turin
Venice
Rome
Naples
SICILY
TYRRHENIAN SEA
ADRIATIC SEA
IONIAN SEA
SLOVENIA
Ljubljana
CROATIA
Zagreb
BOSNIA AND HERZEGOVINA
Sarajevo
SERBIA AND MONTENEGRO
Belgrade
Tirane
ALBANIA
MACEDONIA
Skopje
GREECE
Athens
AEGEAN SEA
CRETE
BULGARIA
Sofia
ROMANIA
Bucharest
Danube R.
MOLDOVA
Chisinau
UKRAINE
Kiev
Kharkiv
Dnipropetrovs'k
Donetsk
SEA OF AZOV
BLACK SEA
Bosporus
Istanbul
TURKEY
Ankara
RUSSIA
Murmansk
WHITE SEA
Arkhangelsk
St. Petersburg
Moscow
Nizhniy-Novgorod
Volga R.
Don R.
Volgograd
Astrakhan
Samara
Yekaterinburg
Chelyabinsk
Omsk
URAL MOUNTAINS
KAZAKSTAN
ARAL SEA
UZBEKISTAN
TURKMENISTAN
Ashkhabad
CASPIAN SEA
GEORGIA
Tbilisi
ARMENIA
Yerevan
AZERBAIJAN
Baku
IRAN
Tehran
IRAQ
Baghdad
Tigris R.
Euphrates R.
SYRIA
Damascus
CYPRUS
Nicosia
LEBANON
Beirut
ISRAEL
Tel-Aviv
Jerusalem
JORDAN
Amman
KUWAIT
Kuwait
Persian Gulf
BAHRAIN
QATAR
Abu Dhabi
UNITED ARAB EMIRATES
SAUDI ARABIA
Riyadh
RED SEA
EGYPT
Cairo
Nile R.
LIBYA
Tripoli
TUNISIA
Tunis
MALTA
MEDITERRANEAN SEA
ALGERIA
Algiers
MOROCCO
Rabat
MADEIRA ISLANDS
CANARY ISLANDS
MAURITANIA
MALI

Elevation
Feet
3000
1500
300
0
Below sea level
0
100
1000
Sea level
National capital
City
0 250 500 Miles
0 250 500 Kilometers
GREENLAND
ARCTIC OCEAN
Denmark Strait
Arctic Circle
Reykjavik
ICELAND
ATLANTIC OCEAN
FAEROE ISLANDS
SHETLAND ISLANDS
ORKNEY ISLANDS
HEBRIDES
UNITED KINGDOM
GRAMPIAN MTS.
NORTH SEA
Firth of Forth
North Channel
IRELAND
IRISH SEA
Dublin
Liverpool
Leeds
Newcastle
Manchester
Birmingham
St. George's Channel
Lands End
Thames R.
ISLES OF SCILLY
Strait of Dover
English Channel
London
NORWAY
Oslo
Vestfjord
SWEDEN
Stockholm
Lake Vanern
Lake Vattern
GOTLAND
OLAND
Skagerrak
Kattegat
DENMARK
Copenhagen
FINLAND
Helsinki
Gulf of Bothnia
Gulf of Finland
BALTIC SEA
WHITE SEA
Lake Onega
Lake Ladoga
St. Petersburg
Tallinn
ESTONIA
Gulf of Riga
LATVIA
Riga
LITHUANIA
Vilnius
RUSSIA
BELARUS
Minsk
PRIPYAT MARSHES
NETHERLANDS
The Hague
Amsterdam
Rotterdam
Antwerp
BELGIUM
Brussels
Lille
Weser R.
Hamburg
Elbe
Berlin
Bonn
GERMANY
Frankfurt
ERZGEBIRGE
Luxembourg
LUX.
Rhine R.
Stuttgart
Munich
Paris
Seine R.
FRANCE
Bay of Biscay
MASSIF CENTRAL
Rhone R.
JURA MTS.
VOSGES
SWITZ.
Bern
ALPS
PYRENEES
Marseille
MONACO
Monaco
Gulf of Lyon
Gulf of Genoa
Po
POLAND
Warsaw
Prague
CZECH REP.
Vienna
AUSTRIA
SLOVAKIA
Bratislava
Budapest
HUNGARY
Ljubljana
SLOVENIA
Zagreb
CROATIA
CARPATHIAN MTS.
TRANSYLVANIAN ALPS
ROMANIA
Bucharest
Belgrade
Danube R.
DINARIC ALPS
BOSNIA AND HERZEGOVINA
Sarajevo
SERBIA AND MONTENEGRO
Tirane
ALBANIA
Skopje
MACEDONIA
BULGARIA
Sofia
Bosporus
Istanbul
MOLDOVA
Chisinau
UKRAINE
Kiev
Dnepr R.
Dnipropetrovs'k
SEA OF AZOV
Don R.
Volgograd
Volga R.
RUSSIA
Moscow
Lake Rybinsk
Nizhniy-Novgorod
Dvina R.
URAL MOUNTAINS
Yekaterinburg
Chelyabinsk
KAZAKSTAN
CASPIAN DEPRESSION
ARAL SEA
Syr Darya
UZBEKISTAN
Amu Darya
KARAKUMY DESERT
TURKMENISTAN
Ashkhabad
CASPIAN SEA
CAUCASUS
Tbilisi
GEORGIA
Baku
Yerevan
ARMENIA
AZERBAIJAN
ELBURZ MOUNTAINS
Tehran
IRAN
ZAGROS MOUNTAINS
L.Van
BLACK SEA
Ankara
TURKEY
PORTUGAL
Lisbon
Cape Finisterre
Cabo De Sao Vicente
Strait of Gibraltar
SPAIN
Madrid
ANDORRA
Barcelona
BALEARIC IS.
CORSICA
SARDINIA
ITALY
Rome
APPENNINES
Naples
ADRIATIC SEA
TYRRHENIAN SEA
SICILY
IONIAN SEA
MALTA
MEDITERRANEAN SEA
GREECE
Athens
AEGEAN SEA
CRETE
CYPRUS
Nicosia
Beirut
LEBANON
Damascus
SYRIA
ISRAEL
Tel-Aviv
Jerusalem
Amman
JORDAN
Tigris R.
Baghdad
Euphrates R.
IRAQ
KUWAIT
Kuwait
Persian Gulf
BAHRAIN
QATAR
Abu Dhabi
UNITED ARAB EMIRATES
SAUDI ARABIA
Riyadh
RED SEA
Nile R.
Cairo
EGYPT
LIBYA
Tripoli
TUNISIA
Tunis
Algiers
ALGERIA
ATLAS MOUNTAINS
Rabat
MOROCCO
MADEIRA ISLANDS
CANARY ISLANDS
MAURITANIA

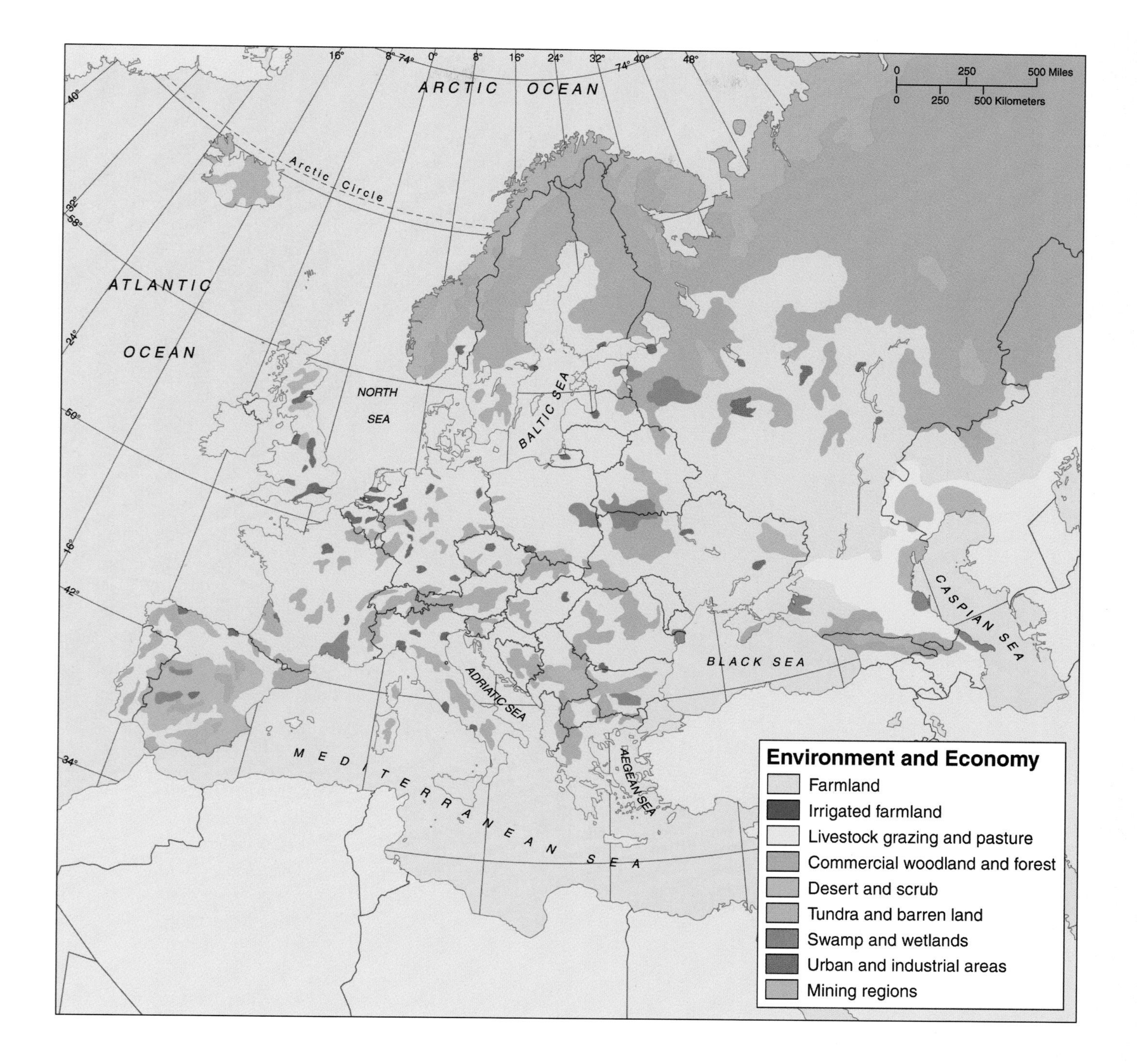

More than any other continent, Europe bears the imprint of human activity—mining, forestry, agriculture, industry, and urbanization. Virtually all of western and central Europe's natural forest vegetation is gone, lost to clearing for agriculture beginning in prehistory, to lumbering that began in earnest during the Middle Ages, or more recently, to disease and destruction brought about by acid precipitation. Only in the far north and the east do some natural stands remain. The region is the world's most heavily industrialized and the industrial areas on the map represent only the largest and most significant. Not shown are the industries that are found in virtually every small town and village and smaller city throughout the industrial countries for Europe. Europe also possesses abundant raw materials and a very productive agricultural base. The mineral resources have long been in a state of active exploitation and the mining regions shown on the map are, for the most part, old regions in upland areas that are somewhat less significant now than they may have been in the past. Agriculturally, the northern European plain is one of the world's great agricultural regions but most of Europe contains decent land for agriculture.

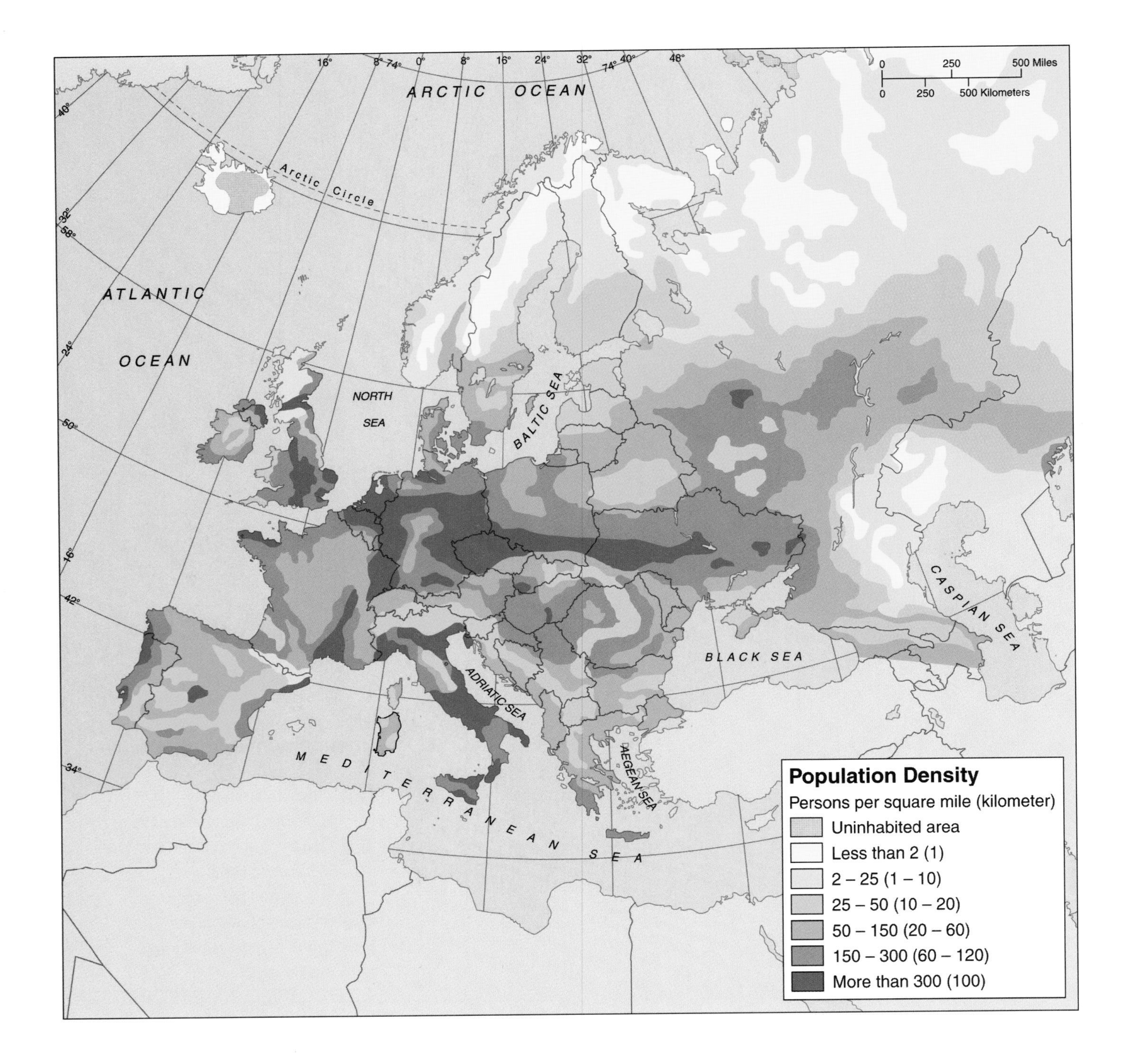

Europe is one of the most densely settled regions of the world with an overall population density nearing 200 persons per square mile (80 per square kilometer), the consequence of a high level of urbanization and an economic system that is heavily industrialized. Even in agricultural regions, the population density is high. Beyond high density, the two chief identifying marks of the European population are remarkable diversity and unusual dynamics. For a small part of the world, Europe has cultural and ethnic diversity that is rarely matched elsewhere; more than 60 languages are spoken in an area not much larger than the United States. The population dynamics of Europe show a mature population that has passed through the "Demographic Transition"—a remarkable increase and then decline in growth rates resulting from the rise of an urban-industrial society. Only in other heavily industrialized regions of the world are found the very small overall growth rates characteristic of Europe.

Map 117c European Political Boundaries, 1914–1948

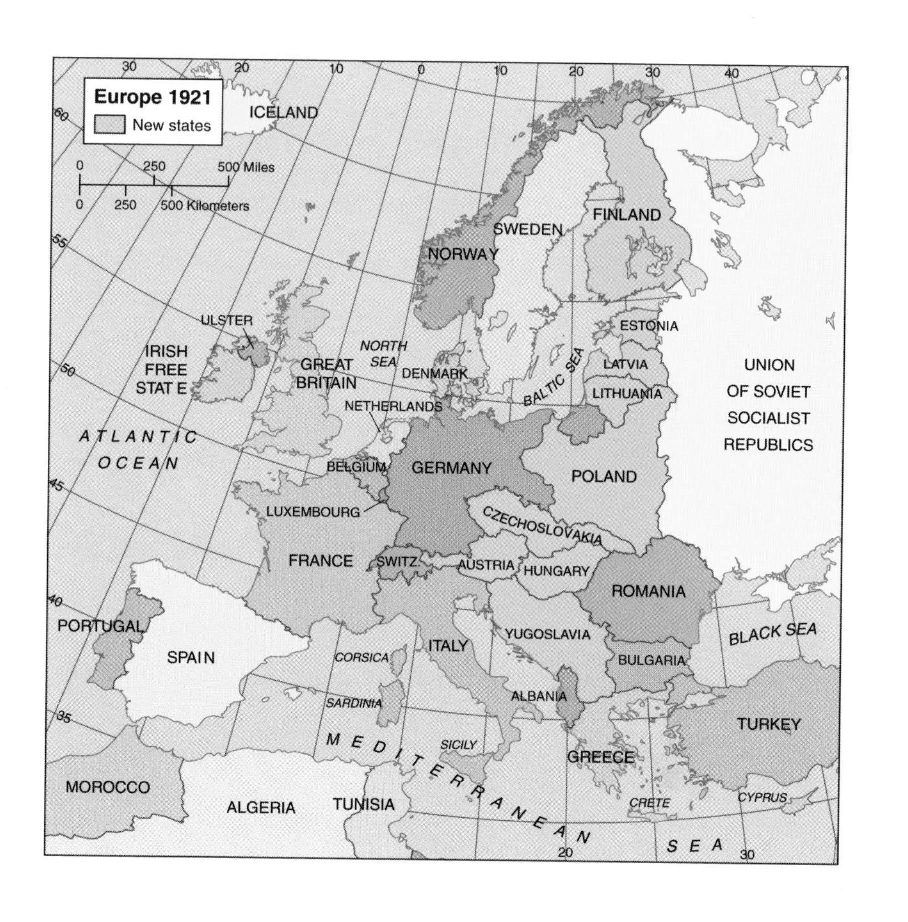

In 1914, on the eve of the First World War, Europe was dominated by the United Kingdom and France in the west, the German Empire and the Austro-Hungarian Empire in central Europe, and the Russian Empire in the east. Battle lines for the conflict that began in 1914 were drawn when the United Kingdom, France, and the Russian Empire joined together as the Triple Entente. In the view of the Germans, this coalition was designed to encircle Germany and its Austrian ally, which, along with Italy, made up the Triple Alliance. The German and Austrian fears were heightened in 1912-14 when a Russian-sponsored "Balkan League" pushed the Ottoman Turkish Empire from Europe, leaving behind the weak and mutually antagonistic Balkan states Serbia and Montenegro. In August 1914, Germany and Austria-Hungary attacked in several directions and World War I began. Four years later, after massive loss of life and destruction, the central European empires were defeated. The victorious French, English, and Americans (who had entered the war in 1917) restructured the map of Europe in 1919, carving nine new states out of the remains of the German and Austro-Hungarian empires and the westernmost portions of the Russian Empire which, by the end of the war, was deep in the Revolution that deposed the czar and brought the Communists to power in a new Union of Soviet Socialist Republics.

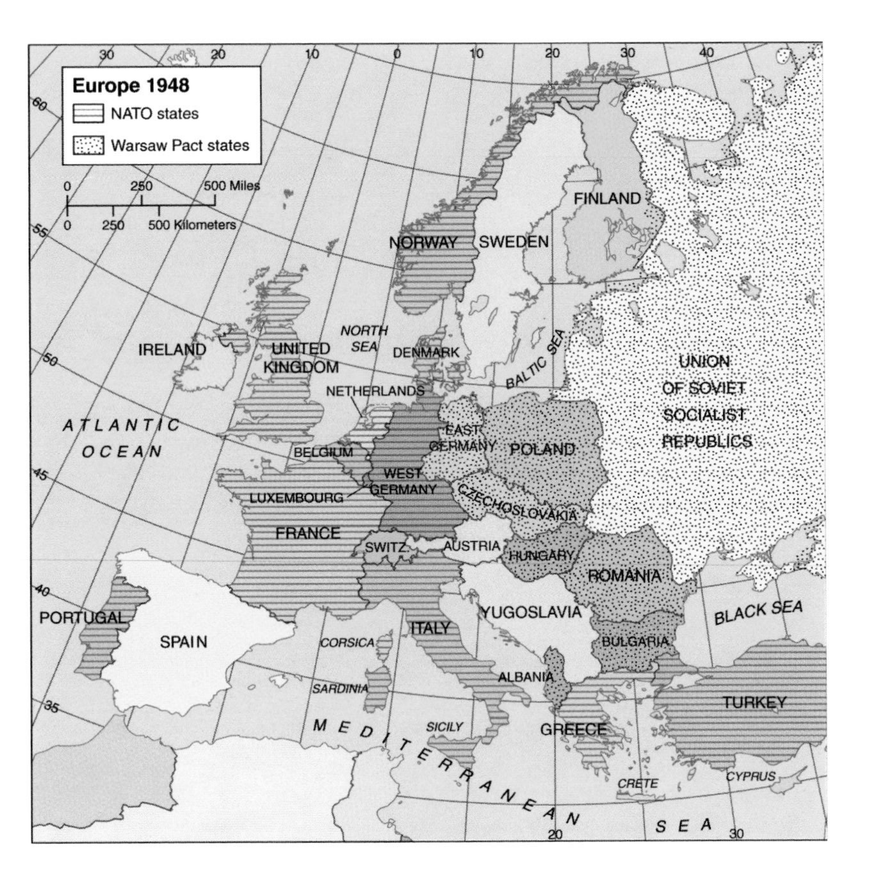

When the victorious Allies redrew the map of central and eastern Europe in 1919, they caused as many problems as they were trying to solve. The interval between the First and Second World Wars was really just a lull in a long war that halted temporarily in 1918 and erupted once again in 1939. Defeated Germany, resentful of the terms of the 1918 armistice and 1919 Treaty of Versailles and beset by massive inflation and unemployment at home, overthrew the Weimar republican government in 1933 and installed the National Socialist (Nazi) party led by Adolf Hitler in Berlin. Hitler quickly began making good on his promises to create a "thousand year realm" of German influence by annexing Austria and the Czech region of Czechoslovakia and allying Germany with a fellow fascist state in Mussolini's Italy. In September 1939 Germany launched the lightning-quick combined infantry, artillery, and armor attack known as **der Blitzkrieg** and took Poland to the east and, in quick succession, the Netherlands, Belgium, and France to the west. By 1943 the greater German Reich extended from the Russian Plain to the Atlantic and from the Black Sea to the Baltic. But the Axis powers of Germany and Italy could not withstand the greater resources and manpower of the combined United Kingdom-United States-USSR–led Allies and, in 1945, Allied armies occupied Germany. Once again, the lines of the central and eastern European map were redrawn. This time, a strengthened Soviet Union took back most of the territory the Russian Empire had lost at the end of the First World War. Germany was partitioned into four occupied sectors (English, French, American, and Russian) and later into two independent countries, the Federal Republic of Germany (West Germany) and the German Democratic Republic (East Germany). Although the Soviet Union's territory stopped at the Polish, Hungarian, Czechoslovakian, and Romanian borders, the eastern European countries (Poland, East Germany, Czechoslovakia, Hungary, Romania, Yugoslavia, Albania, and Bulgaria) became Communist between 1945 and 1948 and were separated from the West by the Iron Curtain.

Map 117d Political Changes 1989–2004

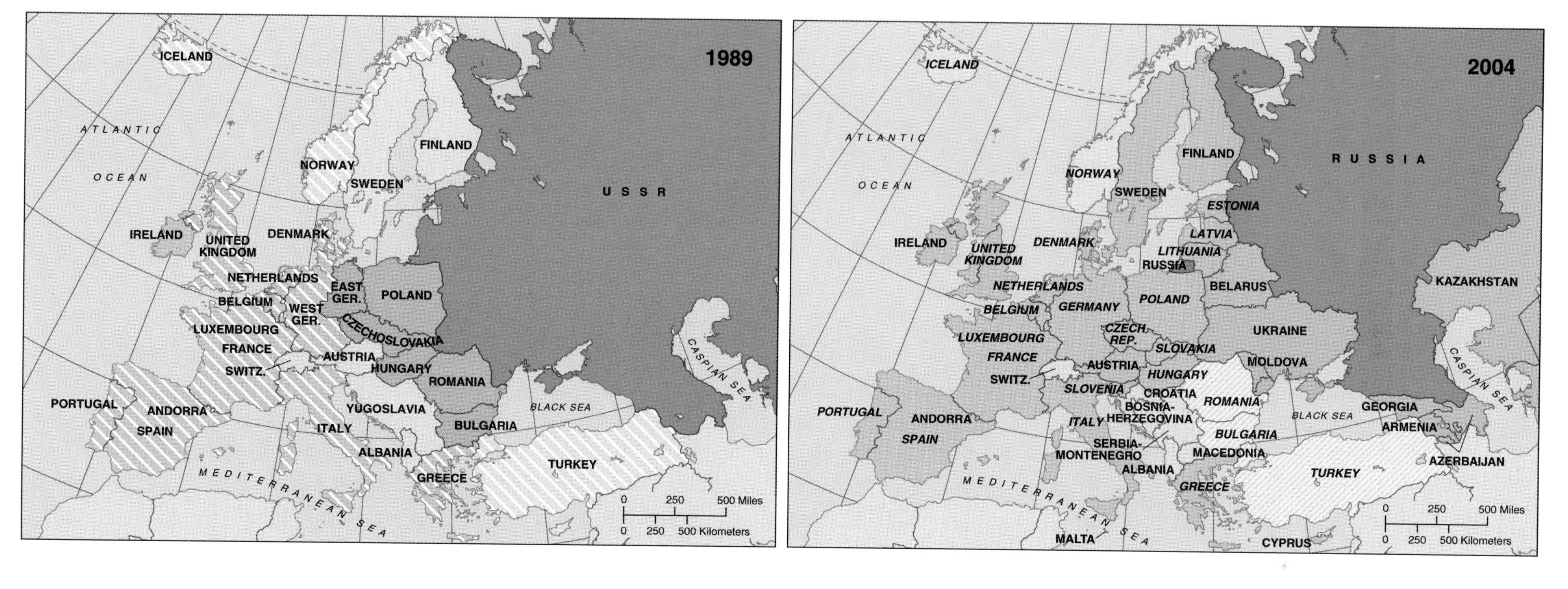

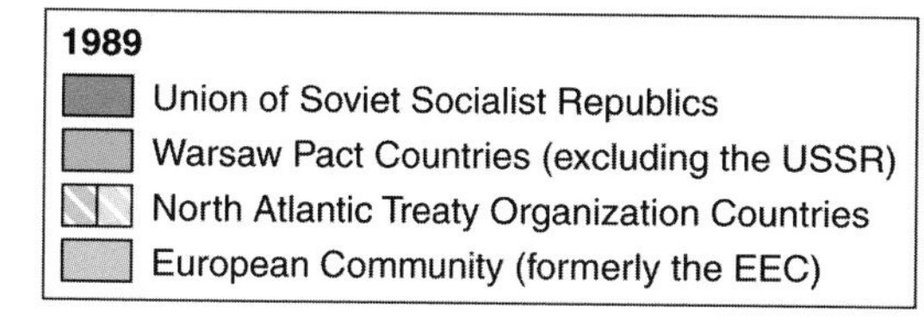

2004

- Former Republics of the USSR, now independent countries
- Russian Federation
- *FRANCE* NATO Countries shown in italics
- European Union (formerly the EC)
- European Union applicant country

During the last decade of the twentieth century, one of the most remarkable series of political geographic changes of the last 500 years took place. The bipolar "East-West" structure that had characterized Europe's political geography since the end of World War II altered in the space of a very few years. In the mid-1980s, as Soviet influence over eastern and central Europe weakened, those countries began to turn to the capitalist West. Between 1989, when the country of Hungary was the first Soviet satellite to open its borders to travel, and 1991, when the Soviet Union dissolved into 15 independent countries, abrupt change in political systems occurred. The result is a new map of Europe that includes a number of countries not present on the map of 1989. These countries have emerged as the result of reunification, separation, or independence from the old Soviet Union.

The new political structure has been accompanied by growing economic cooperation. In May, 2004, ten new applicant countries (the Greek part of Cyprus, Czech Republic, Estonia, Hungary, Latvia, Lithuania, Malta, Poland, Slovakia, and Slovenia) were added to the European Union, increasing the economic strength of the EU, which now has its own currency (the euro) and is presently attempting to develop a constitution acceptable to the 25 member nations. The only remaining formal applicant country to EU membership is Turkey. Concerned about Turkish migration and the potential political problems of adding a country with a population of 60 million Muslims to the European Union, the majority of EU countries have thus far failed to act positively on Turkey's application for membership.

Elevation
Feet
3000
1500
300
0
Sea level
0
100
1000
Below sea level
Azimuthal Equidistant Projection
National capital
City
FAEROES
SHETLAND IS.
HEBRIDES
ORKNEY IS.
UNITED KINGDOM
SCOTLAND
NORTHERN IRELAND
IRELAND
IRISH SEA
ISLE OF MAN
WALES
ENGLAND
PENNINES
NORTH SEA
NORWAY
SWEDEN
DENMARK
Skagerrak
Kattegat
BALTIC SEA
GOTLAND
ÖLAND
BORNHOLM
NETHERLANDS
BELGIUM
LUXEMBOURG
GERMANY
POLAND
CZECH REPUBLIC
SLOVAKIA
AUSTRIA
HUNGARY
SLOVENIA
CROATIA
BOSNIA-HERZEGOVINA
LIECHTENSTEIN
SWITZ.
ALPS
FRANCE
English Channel
Bay of Biscay
ATLANTIC OCEAN
MASSIF CENTRAL
MONACO
ANDORRA
PYRENEES
PORTUGAL
SPAIN
BALEARIC IS.
MAJORCA
MINORCA
IBIZA
CORSICA (France)
SARDINIA (Italy)
ITALY
SAN-MARINO
APENNINES
ADRIATIC SEA
TYRRHENIAN SEA
IONIAN SEA
SICILY
MALTA
MEDITERRANEAN SEA
MOROCCO
ALGERIA
TUNISIA
MIDDLE ATLAS
SAHARAN ATLAS
Strait of Gibraltar
Gulf of Taranto
London
Paris
Madrid
Lisbon
Rome
Berlin
Dublin
Edinburgh
Amsterdam
Brussels
Bern
Vienna
Prague
Copenhagen
Oslo
Stockholm
Rabat
Algiers
Tunis
Valletta
Ljubljana
Zagreb
Sarajevo
Ben Nevis 4,406 ft.
Snowdon 3,560 ft.
Carrauntoohil 3,414 ft.
Scafell Pikes 3,210 ft.
Glittertinden 8,110 ft.
Mt. Perdido 11,000 ft.
Pica de Aneto 11,168 ft.
Mulhacén 11,424 ft.
La Sagra 7,999 ft.
Puy de Sancy 6,185 ft.
Plomb du Cantal 6,096 ft.
Mt. Etna (volcano) 10,902 ft.
Vesuvio 4,190 ft.
Mt. Cinto 8,878 ft.
Punta la Marmora 6,016 ft.
Mt. Corno 9,554 ft.
Mt. Amaro 9,163 ft.
Tagus R.
Duero R.
Ebro R.
Guadiana R.
Guadalquivir R.
Loire R.
Seine R.
Garonne R.
Rhône R.
Rhine R.
Meuse R.
Elbe R.
Oder R.
Danube R.
Thames R.
Severn R.
Po R.
Tiber R.
250
500 Miles
500 Kilometers

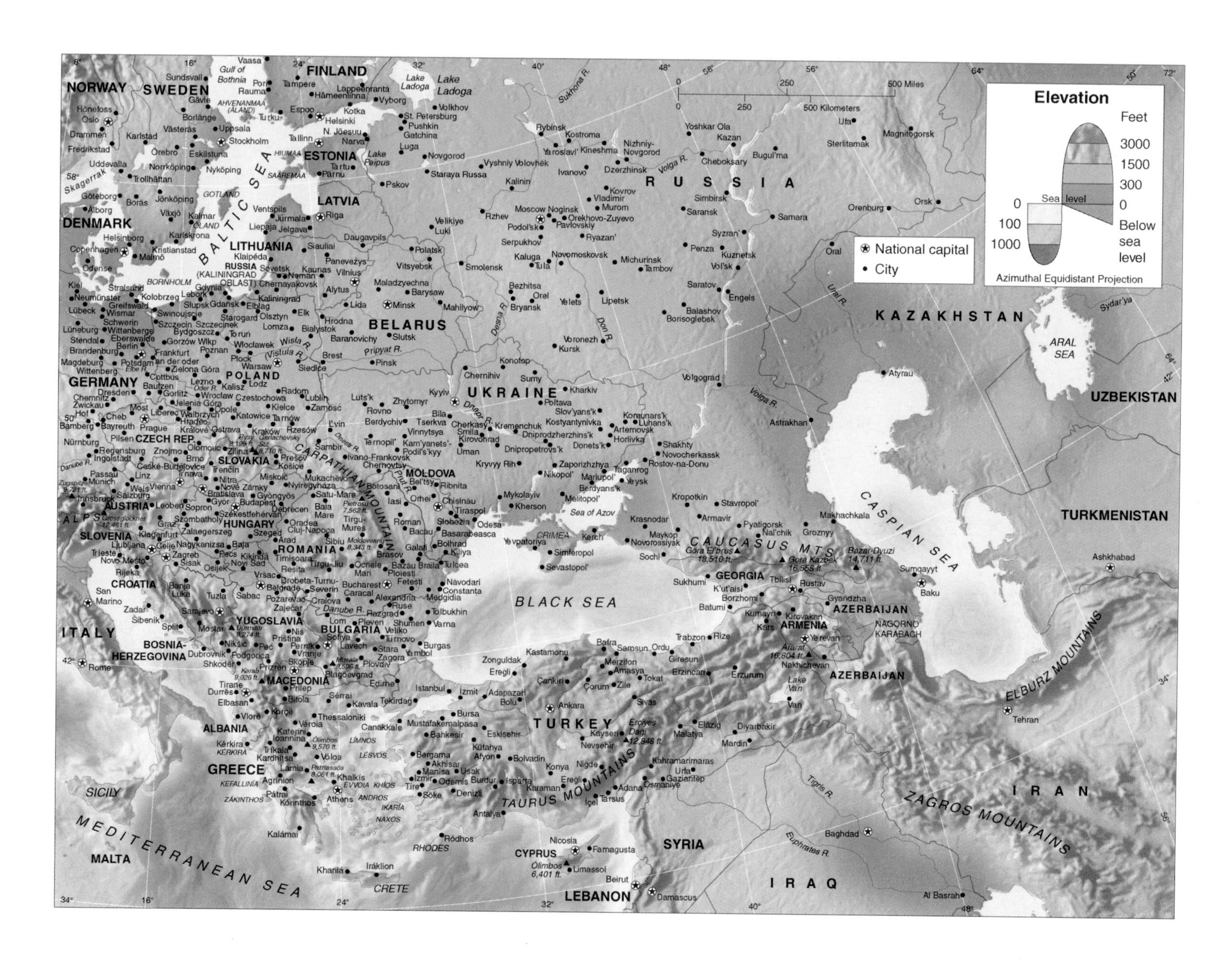

Elevation
Feet
3000
1500
300
0
Below sea level
Sea level
0
100
1000
Azimuthal Equidistant Projection
National capital
City
0
250
500 Miles
0
250
500 Kilometers
NORWAY
SWEDEN
FINLAND
ESTONIA
LATVIA
LITHUANIA
DENMARK
RUSSIA
BELARUS
POLAND
GERMANY
CZECH REP.
SLOVAKIA
AUSTRIA
HUNGARY
SLOVENIA
CROATIA
BOSNIA-HERZEGOVINA
YUGOSLAVIA
ROMANIA
MOLDOVA
UKRAINE
BULGARIA
MACEDONIA
ALBANIA
GREECE
TURKEY
GEORGIA
ARMENIA
AZERBAIJAN
KAZAKHSTAN
UZBEKISTAN
TURKMENISTAN
IRAN
IRAQ
SYRIA
LEBANON
CYPRUS
ITALY
MALTA
BALTIC SEA
BLACK SEA
CASPIAN SEA
ARAL SEA
MEDITERRANEAN SEA
Sea of Azov
Gulf of Bothnia
Lake Ladoga
Lake Peipus
CARPATHIAN MOUNTAINS
CAUCASUS MTS.
TAURUS MOUNTAINS
ZAGROS MOUNTAINS
ELBURZ MOUNTAINS
ALPS
CRIMEA
SICILY
CRETE
RHODES

Elevation
Feet
3000
1500
300
0
Sea level
Below sea level
0
100
1000
Simple Conic Projection
National capital
City
200 Miles
200 Kilometers
Arctic Circle
NORWEGIAN SEA
BARENTS SEA
WHITE SEA
NORTH SEA
BALTIC SEA
Gulf of Bothnia
Gulf of Finland
Gulf of Riga
Skagerrak
Kattegat
KOLA PENINSULA
LOFOTEN
VESTERÅLEN
NORWAY
SWEDEN
FINLAND
RUSSIA
ESTONIA
LATVIA
LITHUANIA
BELARUS
POLAND
GERMANY
DENMARK
RUSSIA (KALININGRAD OBLAST)
HAILUOTO
ÅLAND
HIIUMAA
SAAREMAA
GOTLAND
ÖLAND
BORNHOLM
Lake Ladoga
Lake Onega
Lake Peipus
Vänern
Vättern
Kebnekaise 6,962 ft.
Sulitelma 62,80 ft.
Snöhetta 7,500 ft.
Galdhøpiggen 8,097 ft.
Glittertinden 8,110 ft.
Sylarna 5,761 ft.
Helagsfjället 5,892 ft.
Sånfjället 4,190 ft.
Städjan 3,711 ft.
Nordkapp
Hammerfest
Tromsø
Narvik
Bodø
Mo i Rana
Mosjøen
Trondheim
Kristiansund
Ålesund
Bergen
Stavanger
Oslo
Kristiansand
Mandal
Kiruna
Gällivare
Luleå
Umeå
Sundsvall
Östersund
Gävle
Uppsala
Stockholm
Västerås
Örebro
Norrköping
Linköping
Göteborg
Jönköping
Helsingborg
Malmö
Kalmar
Karlskrona
Visby
Copenhagen
Ålborg
Århus
Odense
Esbjerg
Kolding
Flensburg
Kiel
Lübeck
Hamburg
Rostock
Schwerin
Bremerhaven
Cuxhaven
Emden
Groningen
Szczecin
Kolobrzeg
Koszalin
Slupsk
Gdańsk
Elblag
Grudziadz
Bialystok
Elk
Kaliningrad
Klaipeda
Liepaja
Ventspils
Riga
Jelgava
Daugavpils
Siauliai
Panevezys
Kaunas
Vilnius
Hrodna
Minsk
Navahrudak
Baranavichy
Slutsk
Babruysk
Mahilyow
Homyel'
Novozybkov
Orsha
Smolensk
Vitsyebsk
Polatsk
Tallinn
Tartu
Pärnu
Narva
Helsinki
Turku
Tampere
Pori
Vaasa
Oulu
Kemi
Tornio
Rovaniemi
Kuopio
Joensuu
Jyväskylä
Lahti
Kotka
Hanko
Mariehamn
St. Petersburg
Vyborg
Novgorod
Pskov
Petrozavodsk
Murmansk
Kandalaksha
Apatity
Kirovsk
Olenegorsk
Severodvinsk
Onega
Belomorsk
Kem'
Tikhvin
Volkhov
Vytegra
Lodeynoye Pole
Velikiye-Luki
Ostashkov
Vyshniy-Volochek
Staraya Russa
Gremikha
Ponoy
Nikel'
Kirkenes
Vadsø
Karasjok
Alta
Utsjoki
Lotta R.
Ponoy R.
Kem' R.
Daugava R.
Nemunas R. (Neman R.)
Wisla (Vistula) R.
Elbe R.
Dnepr R.
Volkhov R.
Polist R.
Umeälven R.
Skellefteälv R.
Västerdalälven R.

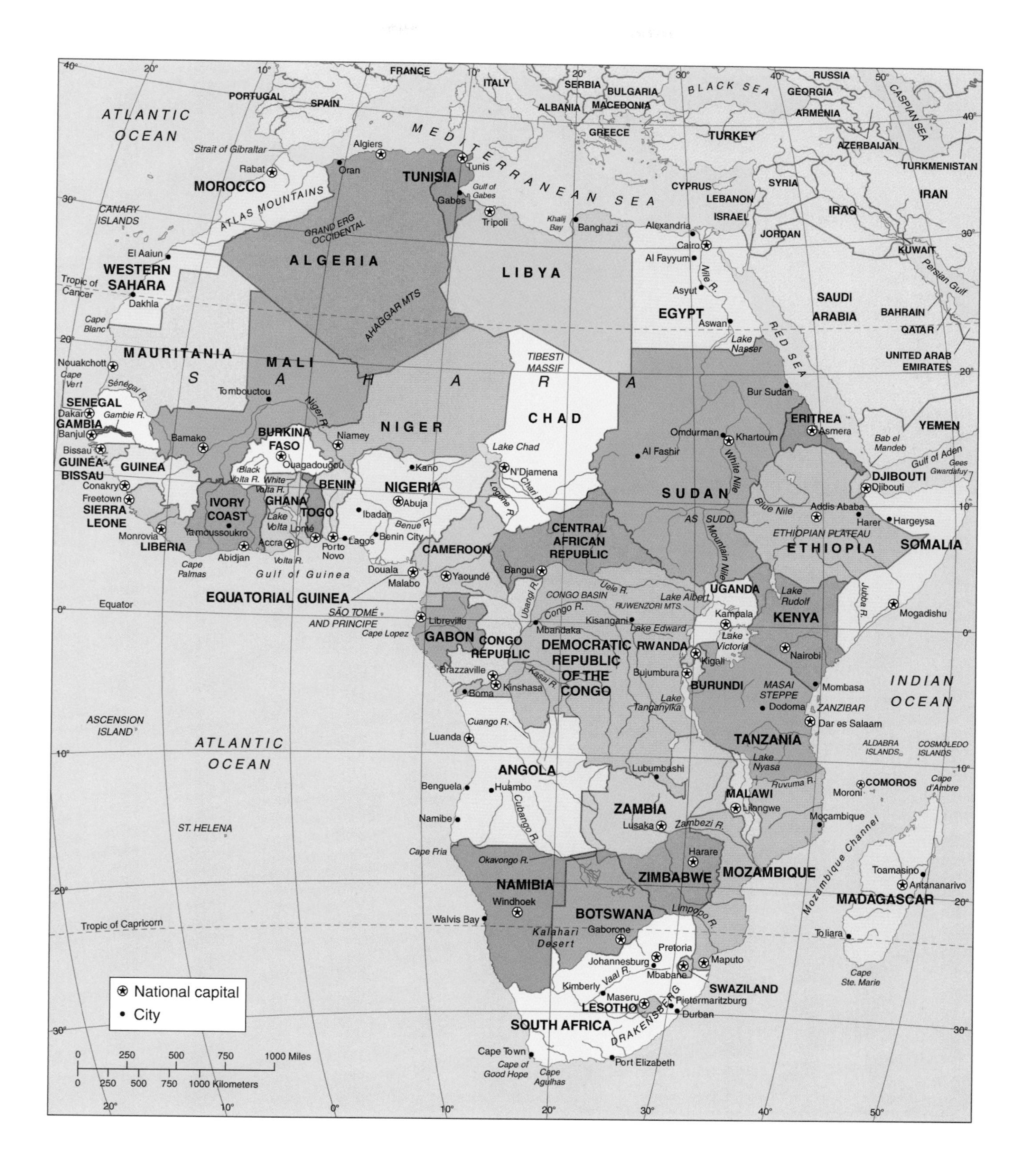
ATLANTIC OCEAN
PORTUGAL
SPAIN
FRANCE
ITALY
SERBIA
BULGARIA
ALBANIA
MACEDONIA
BLACK SEA
RUSSIA
GEORGIA
ARMENIA
CASPIAN SEA
AZERBAIJAN
TURKMENISTAN
GREECE
TURKEY
MEDITERRANEAN SEA
Strait of Gibraltar
Algiers
Tunis
Rabat
Oran
MOROCCO
TUNISIA
Gulf of Gabes
Gabes
ATLAS MOUNTAINS
Tripoli
CYPRUS
SYRIA
LEBANON
ISRAEL
IRAN
IRAQ
JORDAN
Khalij Bay
Banghazi
Alexandria
Cairo
Al Fayyum
KUWAIT
Persian Gulf
CANARY ISLANDS
GRAND ERG OCCIDENTAL
El Aaiun
WESTERN SAHARA
Tropic of Cancer
Dakhla
ALGERIA
LIBYA
Nile R.
Asyut
SAUDI ARABIA
BAHRAIN
QATAR
AHAGGAR MTS
EGYPT
Aswan
Cape Blanc
MAURITANIA
Nouakchott
Cape Vert
Sénégal R.
MALI
SAHARA
TIBESTI MASSIF
Lake Nasser
RED SEA
UNITED ARAB EMIRATES
Tombouctou
Niger R.
Bur Sudan
SENEGAL
Dakar
Gambie R.
GAMBIA
Banjul
Bamako
BURKINA FASO
Niamey
NIGER
CHAD
Omdurman
Khartoum
ERITREA
Asmera
YEMEN
Bab el Mandeb
Gulf of Aden
Gees Gwardafuy
Bissau
GUINEA-BISSAU
GUINEA
Ouagadougou
Lake Chad
Al Fashir
White Nile
N'Djamena
DJIBOUTI
Djibouti
Conakry
Kano
Black Volta R.
White Volta R.
BENIN
NIGERIA
Logone R.
Chari R.
SUDAN
Freetown
SIERRA LEONE
IVORY COAST
GHANA
TOGO
Abuja
Ibadan
Blue Nile
Addis Ababa
Harer
Hargeysa
Lake Volta
Lomé
Benue R.
CENTRAL AFRICAN REPUBLIC
AS SUDD
Mountain Nile
ETHIOPIAN PLATEAU
ETHIOPIA
SOMALIA
Monrovia
Yamoussoukro
Accra
Porto Novo
Lagos
Benin City
LIBERIA
Abidjan
CAMEROON
Cape Palmas
Volta R.
Gulf of Guinea
Douala
Malabo
Yaoundé
Bangui
UGANDA
Lake Rudolf
Uele R.
CONGO BASIN
Lake Albert
EQUATORIAL GUINEA
Equator
Ubangi R.
Congo R.
RUWENZORI MTS.
Jubba R.
Mogadishu
SÃO TOMÉ AND PRINCIPE
Libreville
Kisangani
Kampala
KENYA
Cape Lopez
GABON
CONGO REPUBLIC
Mbandaka
Lake Edward
Lake Victoria
DEMOCRATIC REPUBLIC OF THE CONGO
RWANDA
Kigali
Nairobi
Brazzaville
Kasai R.
Bujumbura
MASAI STEPPE
Mombasa
INDIAN OCEAN
Boma
Kinshasa
BURUNDI
Lake Tanganyika
Dodoma
ZANZIBAR
ASCENSION ISLAND
Cuango R.
Dar es Salaam
Luanda
TANZANIA
ALDABRA ISLANDS
COSMOLEDO ISLANDS
Lake Nyasa
Lubumbashi
ANGOLA
Ruvuma R.
COMOROS
Cape d'Ambre
Benguela
Huambo
MALAWI
Moroni
Lilongwe
ZAMBIA
Cubango R.
Moçambique
Namibe
ST. HELENA
Lusaka
Zambezi R.
Mozambique Channel
Harare
Cape Fria
Okavongo R.
MOZAMBIQUE
Toamasino
NAMIBIA
ZIMBABWE
Antananarivo
MADAGASCAR
Windhoek
BOTSWANA
Limpopo R.
Tropic of Capricorn
Walvis Bay
Kalahari Desert
Gaborone
Toliara
Pretoria
Johannesburg
Maputo
Mbabane
Cape Ste. Marie
Vaal R.
Kimberly
SWAZILAND
Maseru
Pietermaritzburg
LESOTHO
Durban
DRAKENSBERG
SOUTH AFRICA
National capital
City
Cape Town
Cape of Good Hope
Cape Agulhas
Port Elizabeth
0 250 500 750 1000 Miles
0 250 500 750 1000 Kilometers

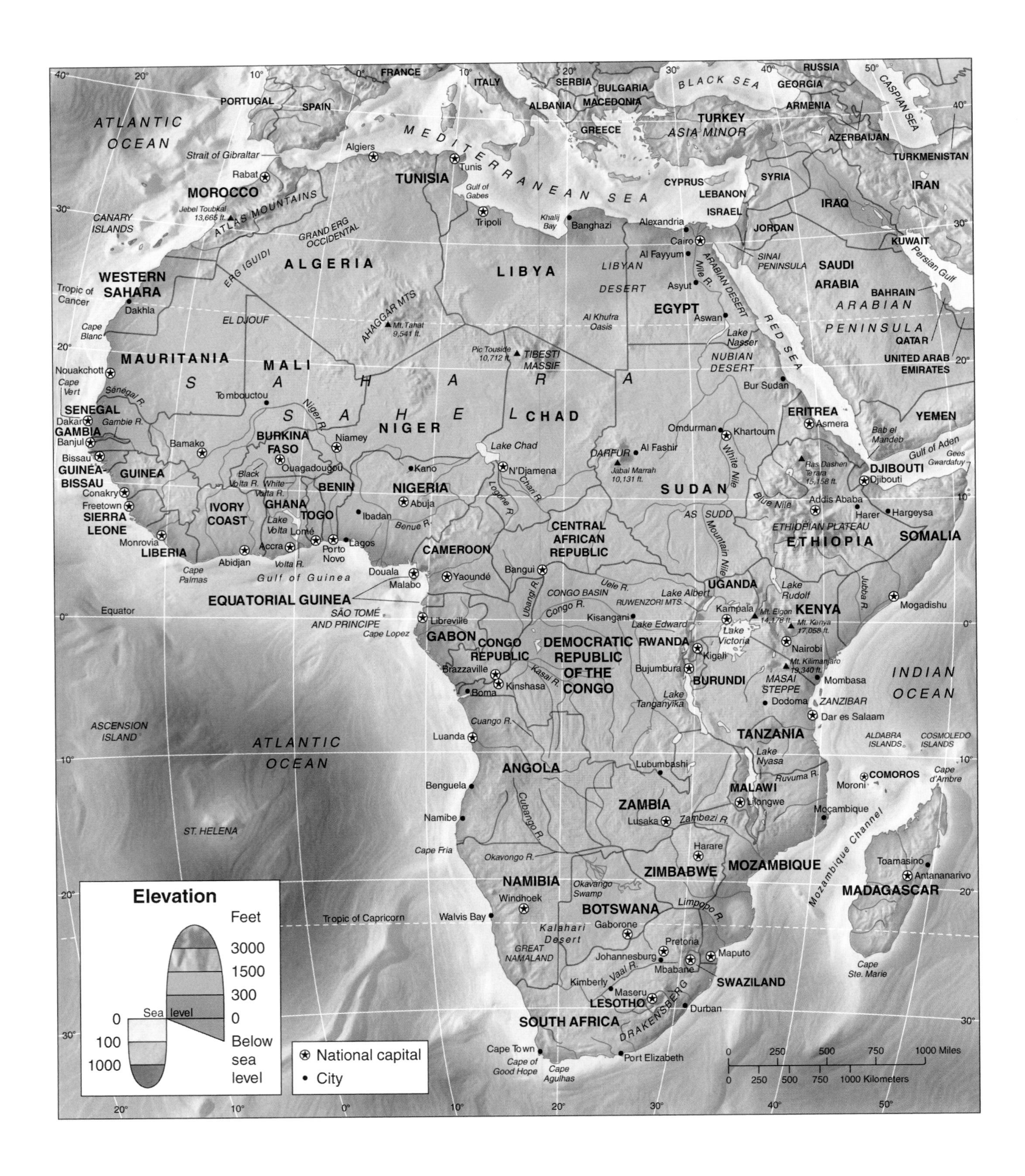
ATLANTIC OCEAN
PORTUGAL
SPAIN
FRANCE
ITALY
SERBIA
BULGARIA
ALBANIA
MACEDONIA
GREECE
BLACK SEA
RUSSIA
GEORGIA
ARMENIA
AZERBAIJAN
CASPIAN SEA
TURKEY
ASIA MINOR
TURKMENISTAN
IRAN
SYRIA
CYPRUS
LEBANON
ISRAEL
JORDAN
IRAQ
KUWAIT
Persian Gulf
SAUDI ARABIA
BAHRAIN
QATAR
UNITED ARAB EMIRATES
ARABIAN PENINSULA
YEMEN
Gulf of Aden
Bab el Mandeb
MEDITERRANEAN SEA
Strait of Gibraltar
Algiers
Tunis
Rabat
MOROCCO
TUNISIA
Gulf of Gabes
Tripoli
Khalij Bay
Banghazi
Alexandria
Cairo
Al Fayyum
Asyut
Aswan
Lake Nasser
SINAI PENINSULA
ARABIAN DESERT
Nile R.
RED SEA
ATLAS MOUNTAINS
Jebel Toubkal 13,665 ft.
CANARY ISLANDS
GRAND ERG OCCIDENTAL
ERG IGUIDI
ALGERIA
LIBYA
LIBYAN DESERT
EGYPT
Al Khufra Oasis
WESTERN SAHARA
Tropic of Cancer
Dakhla
EL DJOUF
AHAGGAR MTS
Mt. Tahat 9,541 ft.
Cape Blanc
Pic Touside 10,712 ft.
TIBESTI MASSIF
NUBIAN DESERT
Bur Sudan
MAURITANIA
MALI
NIGER
CHAD
SAHARA
Nouakchott
Cape Vert
Sénégal R.
Tombouctou
Niger R.
SENEGAL
Dakar
Gambie R.
GAMBIA
Banjul
Bissau
GUINEA-BISSAU
GUINEA
Conakry
Freetown
SIERRA LEONE
Monrovia
LIBERIA
Bamako
BURKINA FASO
Ouagadougou
Niamey
Black Volta R.
White Volta R.
IVORY COAST
GHANA
TOGO
BENIN
Lake Volta
Abidjan
Accra
Lomé
Porto Novo
Lagos
Volta R.
Cape Palmas
Gulf of Guinea
Kano
NIGERIA
Abuja
Ibadan
Benue R.
Lake Chad
N'Djamena
Chari R.
Logone R.
DARFUR
Al Fashir
Jabal Marrah 10,131 ft.
Omdurman
Khartoum
White Nile
Blue Nile
SUDAN
AS SUDD
Mountain Nile
ERITREA
Asmera
Ras Dashen Terara 15,158 ft.
DJIBOUTI
Djibouti
Gees Gwardafuy
Addis Ababa
Harer
Hargeysa
ETHIOPIAN PLATEAU
ETHIOPIA
SOMALIA
Jubba R.
Mogadishu
CENTRAL AFRICAN REPUBLIC
Bangui
CAMEROON
Douala
Malabo
Yaoundé
EQUATORIAL GUINEA
SÃO TOMÉ AND PRINCIPE
Equator
Libreville
Cape Lopez
GABON
CONGO REPUBLIC
Brazzaville
Kinshasa
Boma
Ubangi R.
CONGO BASIN
Congo R.
Uele R.
RUWENZORI MTS.
Lake Albert
UGANDA
Kampala
Lake Rudolf
Mt. Elgon 14,178 ft.
Mt. Kenya 17,058 ft.
KENYA
Nairobi
Kisangani
Lake Edward
Lake Victoria
DEMOCRATIC REPUBLIC OF THE CONGO
RWANDA
Kigali
Bujumbura
BURUNDI
Kasai R.
Lake Tanganyika
Mt. Kilimanjaro 19,340 ft.
MASAI STEPPE
Dodoma
Mombasa
ZANZIBAR
Dar es Salaam
TANZANIA
INDIAN OCEAN
ALDABRA ISLANDS
COSMOLEDO ISLANDS
ASCENSION ISLAND
Cuango R.
Luanda
ATLANTIC OCEAN
ANGOLA
Lubumbashi
Lake Nyasa
Ruvuma R.
MALAWI
Lilongwe
COMOROS
Moroni
Cape d'Ambre
Benguela
Namibe
Cubango R.
ZAMBIA
Lusaka
Zambezi R.
Moçambique
Mozambique Channel
ST. HELENA
Cape Fria
Okavongo R.
Harare
ZIMBABWE
MOZAMBIQUE
Toamasino
Antananarivo
MADAGASCAR
NAMIBIA
Windhoek
Okavango Swamp
BOTSWANA
Limpopo R.
Tropic of Capricorn
Walvis Bay
Kalahari Desert
Gaborone
GREAT NAMALAND
Pretoria
Johannesburg
Maputo
Mbabane
SWAZILAND
Vaal R.
Kimberly
Maseru
LESOTHO
DRAKENSBERG
Durban
SOUTH AFRICA
Cape Ste. Marie
Cape Town
Cape of Good Hope
Cape Agulhas
Port Elizabeth
Elevation
Feet
3000
1500
300
0
Below sea level
Sea level
0
100
1000
National capital
City
0 250 500 750 1000 Miles
0 250 500 750 1000 Kilometers

Map 123a Environment and Economy

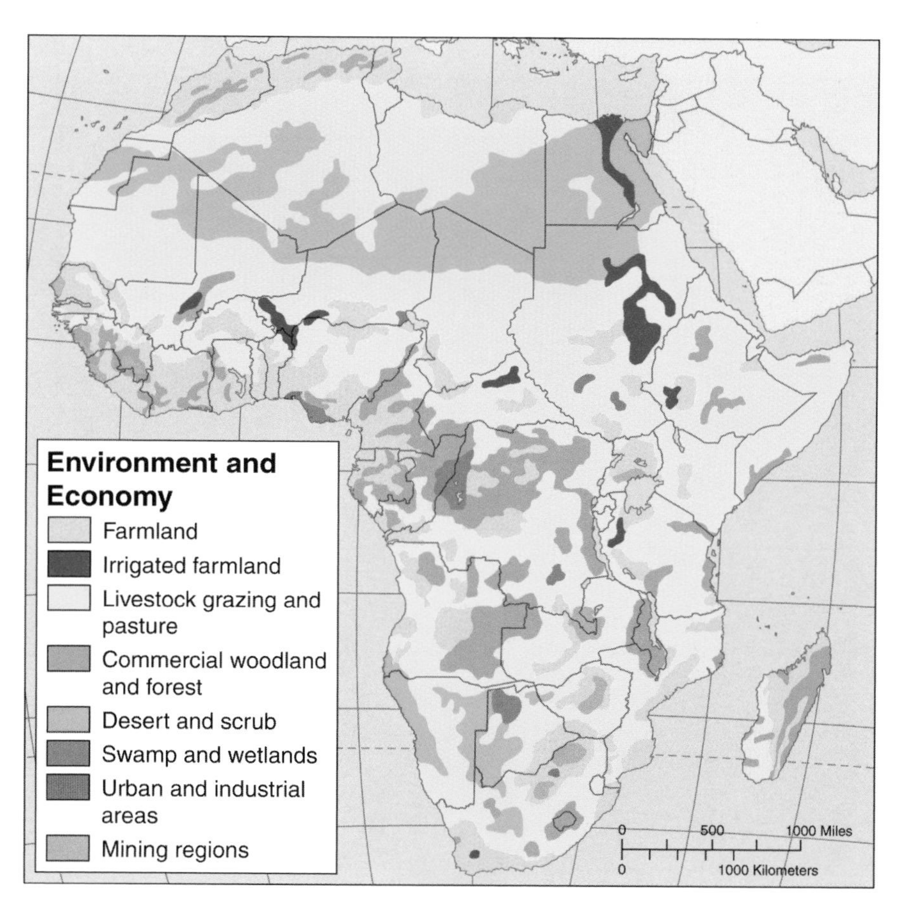

Africa's economic landscape is dominated by subsistence, or marginally-commercial agricultural activities and raw material extraction, engaging three-fourths of Africa's workers. Much of this grazing land is very poor desert scrub and bunch grass that is easily impacted by cattle, sheep, and goats. Growing human and livestock populations place enormous stress on this fragile support capacity and the result is desertification: the conversion of even the most minimal of grazing environments to a small quantity of land suitable for crop farming. Although the continent has approximately 20 percent of the world's total land area, the proportion of Africa's arable land is small. The agricultural environment is also uncertain; unpredictable precipitation and poor soils hamper crop agriculture.

Map 123b Population Density

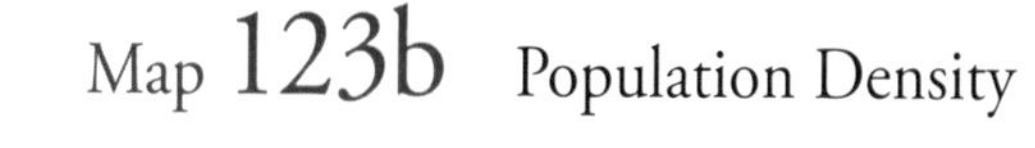

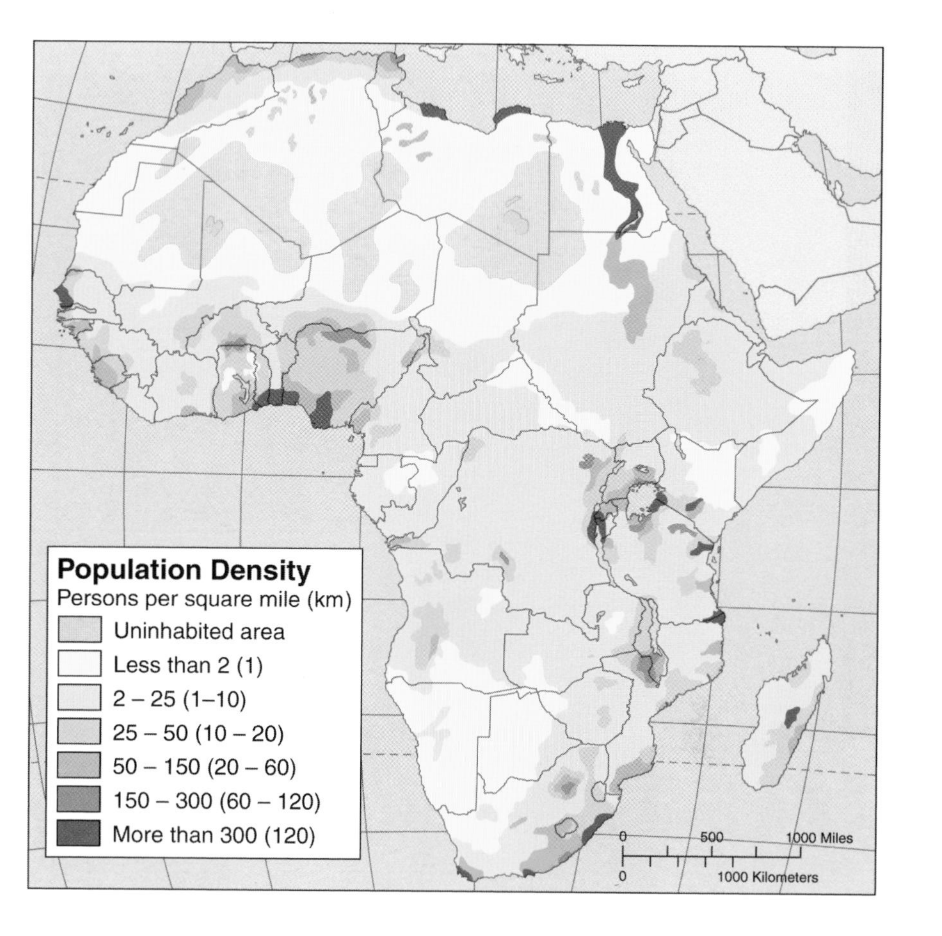

Nearly 800 million people occupy the African continent, approximately one-eighth of the world's population. In general, this population has two chief characteristics: a low level of quality of life and growth rates that are among the world's highest. On a continent beset by poverty, recurrent internal civil and tribal war, and a host of environmental problems, the populations of many African countries are nevertheless increasing at a rate above 3 percent per year. The bulk of the population is concentrated in relatively small areas of the Mediterranean coastal regions of the north, the bulge of West Africa, and the eastern coastal and highland zone stretching from South Africa to Kenya. Where populations in other parts of the world tend to avoid highland locations, in Africa highlands often tend to be the most densely settled regions: moister with better soils, freer of insect pests, and somewhat cooler.

Map 123c Colonial Patterns

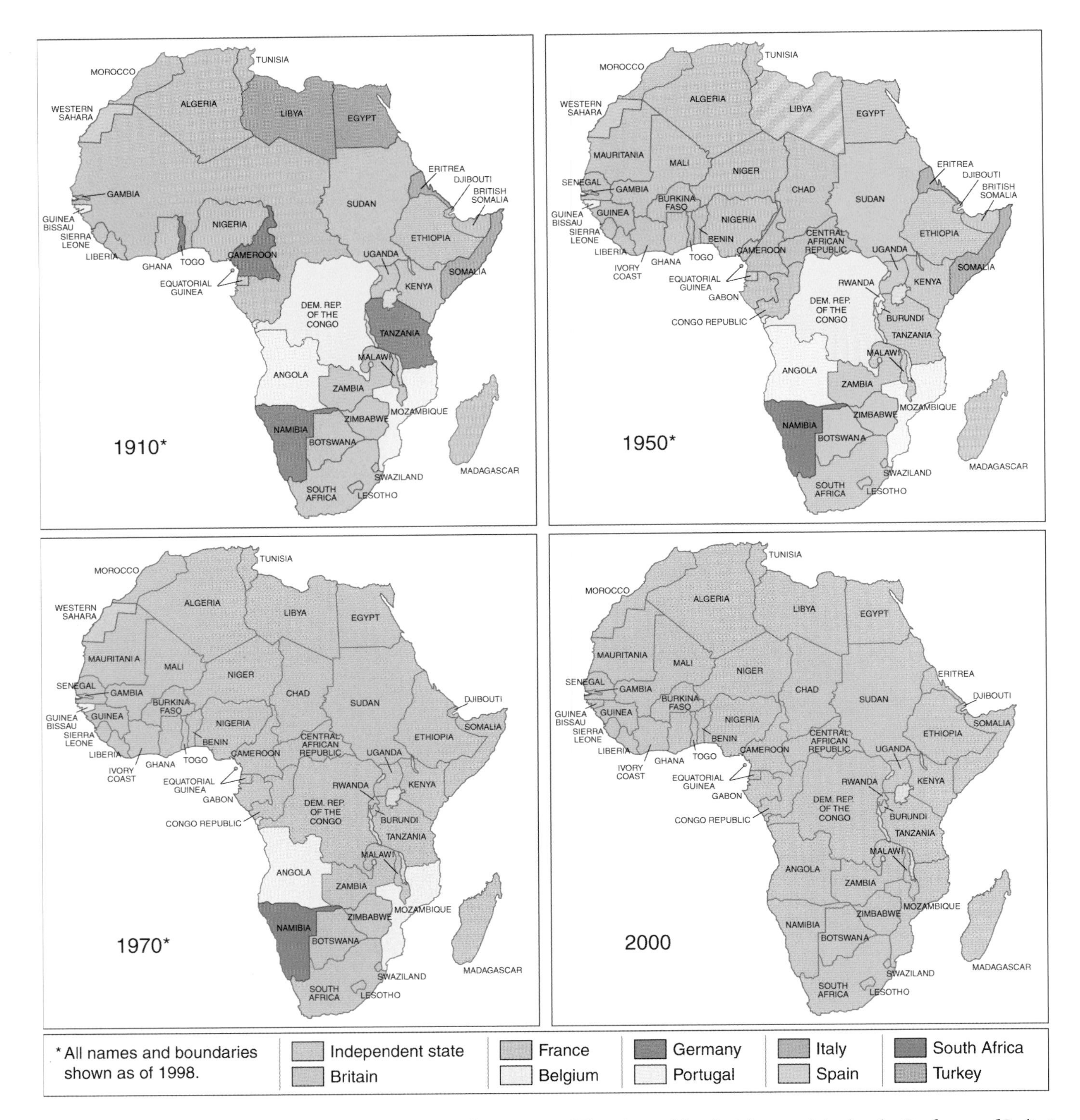

In few parts of the world has the transition from colonialism to independence been as abrupt as on the African continent. Most African states did not become colonies until the nineteenth century and did not become independent until the twentieth, nearly all of them after World War II. Much of the colonial power in Africa is social and economic. The African colony provided the mother country with raw materials in exchange for marginal economic returns, and many African countries still exist in this colonial dependency relationship. An even more important component of the colonial legacy of Europe in Africa is geopolitical. When the world's colonial powers joined at the Conference of Berlin in 1884, they divided up Africa to fit their own needs, drawing boundary lines on maps without regard for terrain or drainage features, or for tribal/ethnic linguistic, cultural, economic, or political borders. Traditional Africa was enormously disrupted by this process. After independence, African countries retained boundaries that are legacies of the colonial past and African countries today are beset by internal problems related to tribal and ethnic conflicts, the disruption of traditional migration patterns, and inefficient spatial structures of market and supply.

Map 123d African Cropland and Dryland Degradation

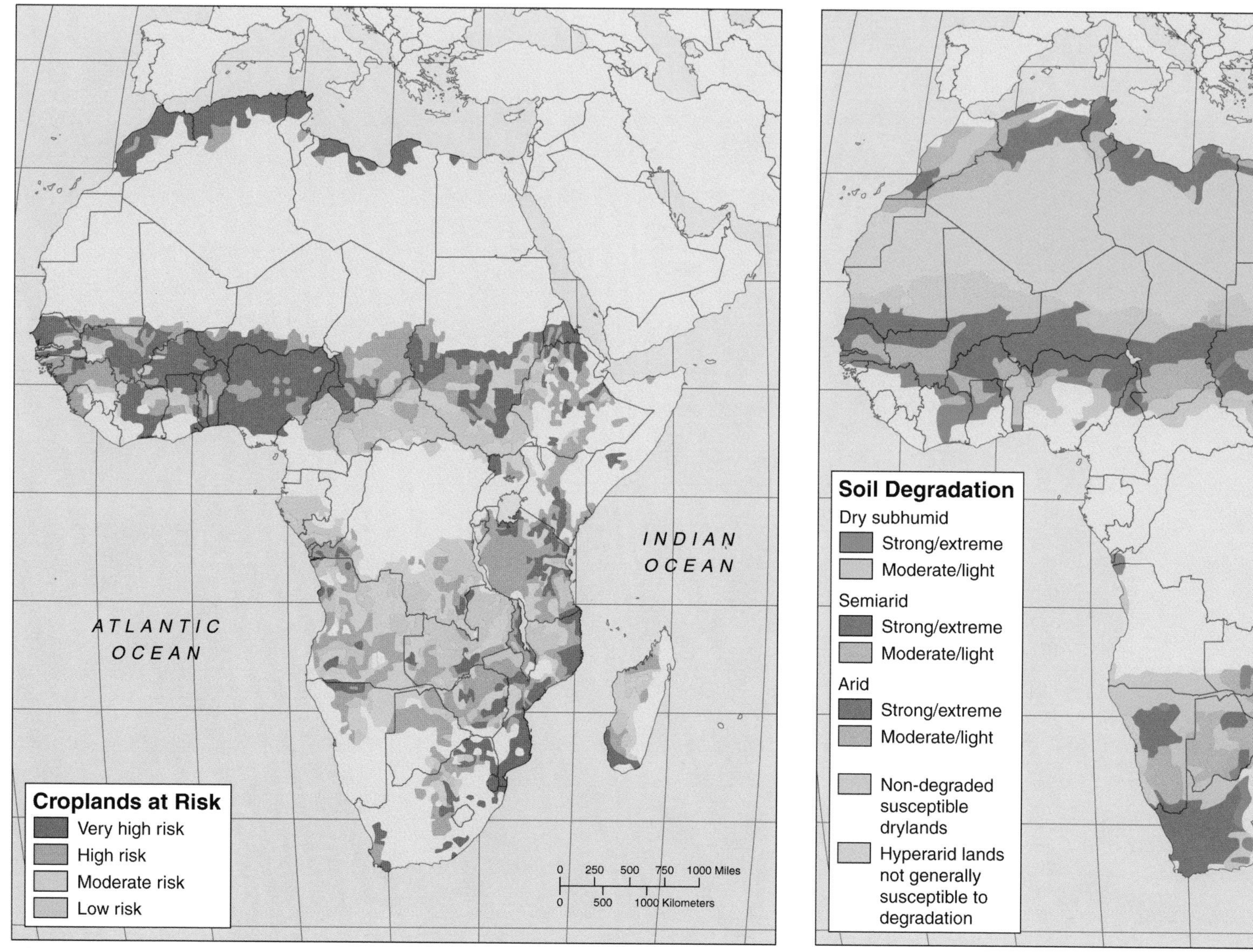

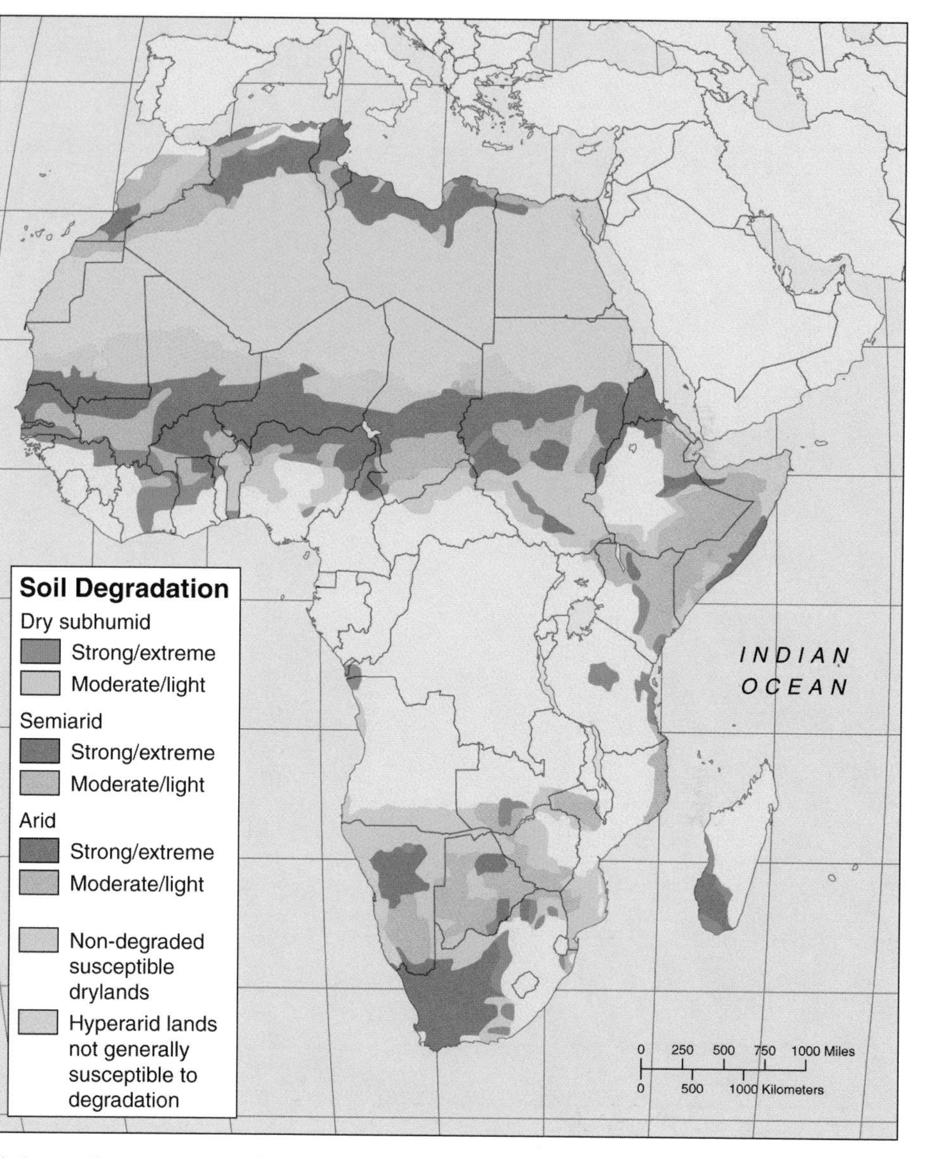

The economy of the African continent is largely agricultural and given an African population in excess of 700 million people, much of the agricultural environment is degraded. Two forms of degradation exist. The first of these is in cropland areas where susceptible tropical soils and high population densities have produced major cropland degradation, largely in the form of loss of fertility. Irrigation and the advent of artificial fertilizer have not helped soils to restore their natural chemical balances after generations of misuse. The second form of degradation is in the dryland areas where livestock grazing rather than cropping is the dominant agricultural form. Populations of domesticated stock that are too large for the carrying capacity of the environment, exacerbated by the view of livestock as wealth and worsened by increasing human populations, have all contributed to the conversion of semi-arid grasslands into desert.

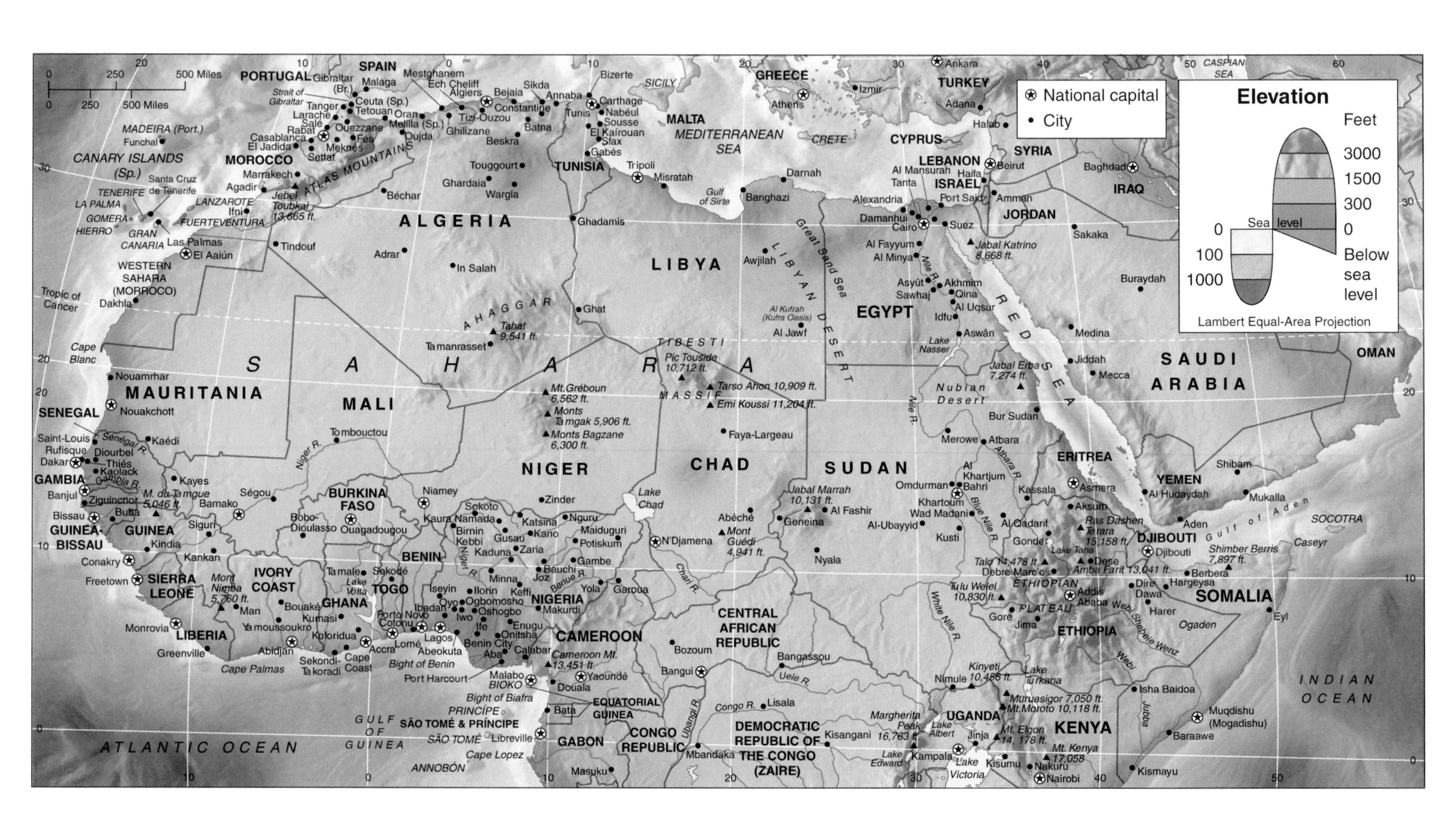
National capital
City
Elevation
Feet
3000
1500
300
0
Sea level
Below sea level
0
100
1000
Lambert Equal-Area Projection
0 250 500 Miles
0 250 500 Miles
PORTUGAL
SPAIN
Gibraltar (Br.)
Strait of Gibraltar
Malaga
Ceuta (Sp.)
Tetouan
Tanger
Larache
Sale
Rabat
Casablanca
El Jadida
Meknes
Fes
Ouezzane
Settat
Melilla (Sp.)
Oran
Oujda
Mestghanem
Ech Cheliff
Algiers
Bejaia
Tizi-Ouzou
Ghilizane
Constantine
Sikda
Annaba
Batna
Beskra
Bizerte
Carthage
Tunis
Nabeul
Sousse
El Kairouan
Sfax
Gabès
SICILY
MALTA
MEDITERRANEAN SEA
CRETE
GREECE
Athens
Izmir
Ankara
TURKEY
Adana
Halab
CYPRUS
LEBANON
Beirut
SYRIA
Baghdad
IRAQ
Al Mansurah
Haifa
Tanta
ISRAEL
Port Said
Amman
JORDAN
Alexandria
Damanhur
Cairo
Suez
Sakaka
CASPIAN SEA
MADEIRA (Port.)
Funchal
CANARY ISLANDS (Sp.)
Santa Cruz de Tenerife
TENERIFE
LA PALMA
GOMERA
HIERRO
GRAN CANARIA
Las Palmas
LANZAROTE
FUERTEVENTURA
Ifni
MOROCCO
Marrakech
Agadir
Jebel Toubkal 13,665 ft.
ATLAS MOUNTAINS
El Aaiún
WESTERN SAHARA (MORROCO)
Dakhla
Tropic of Cancer
Tindouf
Béchar
Touggourt
Ghardaia
Wargla
ALGERIA
Adrar
In Salah
Ghadamis
Tripoli
Misratah
Gulf of Sirte
Banghazi
Darnah
LIBYA
Awjilah
LIBYAN DESERT
Great Sand Sea
Al Kufrah (Kufra Oasis)
Al Jawf
EGYPT
Al Fayyum
Al Minya
Asyût
Sawhaj
Akhmim
Qina
Al Uqsur
Idfu
Aswân
Lake Nasser
Nile R.
Jabal Katrino 8,668 ft.
RED SEA
Buraydah
Medina
Jiddah
Mecca
SAUDI ARABIA
OMAN
AHAGGAR
Tahat 9,541 ft.
Tamanrasset
Ghat
TIBESTI MASSIF
Pic Tousidé 10,712 ft.
Tarso Ahon 10,909 ft.
Emi Koussi 11,204 ft.
SAHARA
Cape Blanc
Nouamrhar
MAURITANIA
Nouakchott
MALI
Mt. Gréboun 6,562 ft.
Monts Tamgak 5,906 ft.
Monts Bagzane 6,300 ft.
Tombouctou
Niger R.
NIGER
Faya-Largeau
CHAD
Nubian Desert
Jabal Erba 7,274 ft.
Bur Sudan
Merowe
Atbara
Atbara R.
SUDAN
ERITREA
Asmera
YEMEN
Al Hudaydah
Shibam
Mukalla
Gulf of Aden
SOCOTRA
SENEGAL
Saint-Louis
Senegal R.
Kaédi
Rufisque
Diourbel
Dakar
Thiés
Kaolack
GAMBIA
Gambia R.
Banjul
Ziguinchor
M. du Tamgue 5,046 ft.
Kayes
Bamako
Ségou
Bissau
Bula
GUINEA-BISSAU
GUINEA
Sigurí
Kindia
Conakry
Kankan
Freetown
SIERRA LEONE
Mont Nimba 5,760 ft.
Man
IVORY COAST
Monrovia
LIBERIA
Greenville
Cape Palmas
BURKINA FASO
Ouagadougou
Bobo-Dioulasso
Niamey
Zinder
Sokoto
Kaura Namada
Birnin Kebbi
Gusau
Katsina
Kano
Nguru
Maiduguri
Potiskum
Kaduna
Zaria
Bauchi
Jos
Gambe
Minna
Benue R.
Yola
Garoua
Lake Chad
N'Djamena
Abéché
Mont Guédi 4,941 ft.
Jabal Marrah 10,131 ft.
Al Fashir
Geneina
Nyala
Omdurman
Al Khartjum Bahri
Khartoum
Wad Madani
Al-Ubayyid
Kusti
Blue Nile R.
Kassala
Aksum
Al Qadarif
Gonder
Ras Dashen Terara 15,158 ft.
Lake Tana
Talo 14,478 ft.
Debre Marc'os
Dese
Amba Farit 13,041 ft.
DJIBOUTI
Djibouti
Aden
Shimber Berris 7,897 ft.
Caseyr
Berbera
Hargeysa
Dire Dawa
Harer
SOMALIA
Eyl
Ogaden
Tamale
Sokodé
Lake Volta
TOGO
BENIN
Iseyin
Ilorin
Keffi
NIGERIA
Makurdi
Oyo
Ogbomosho
Oshogbo
Ibadan
Porto Novo
Cotonou
Iwo
Ife
Enugu
Onitsha
Lagos
Abeokuta
Benin City
Aba
Calabar
Bouaké
GHANA
Kumasi
Yamoussoukro
Koforidua
Abidjan
Accra
Lomé
Sekondi-Takoradi
Cape Coast
Bight of Benin
Port Harcourt
Malabo
BIOKO
CAMEROON
Cameroon Mt. 13,451 ft.
Yaoundé
Douala
Bozoum
CENTRAL AFRICAN REPUBLIC
Bangui
Bangassou
Uele R.
Chari R.
White Nile R.
Tulu Welel 10,830 ft.
ETHIOPIAN PLATEAU
Addis Ababa
Gore
Jima
ETHIOPIA
Webi Shebele Wenz
Webi
Kinyeti 10,456 ft.
Nimule
Lake Turkana
Isha Baidoa
Jubba
Muqdishu (Mogadishu)
Baraawe
Kismayu
INDIAN OCEAN
Bight of Biafra
PRINCIPE
SÃO TOMÉ & PRÍNCIPE
SÃO TOMÉ
Bata
EQUATORIAL GUINEA
Libreville
GABON
Cape Lopez
ANNOBÓN
Masuku
CONGO REPUBLIC
Ubangi R.
Congo R.
Lisala
Mbandaka
DEMOCRATIC REPUBLIC OF THE CONGO (ZAIRE)
Kisangani
Margherita Peak 16,763 ft.
Lake Albert
Lake Edward
UGANDA
Kampala
Jinja
Lake Victoria
Muruasigor 7,050 ft.
Mt. Moroto 10,118 ft.
Mt. Elgon 14,178 ft.
KENYA
Mt. Kenya 17,058
Nakuru
Kisumu
Nairobi
GULF OF GUINEA
ATLANTIC OCEAN

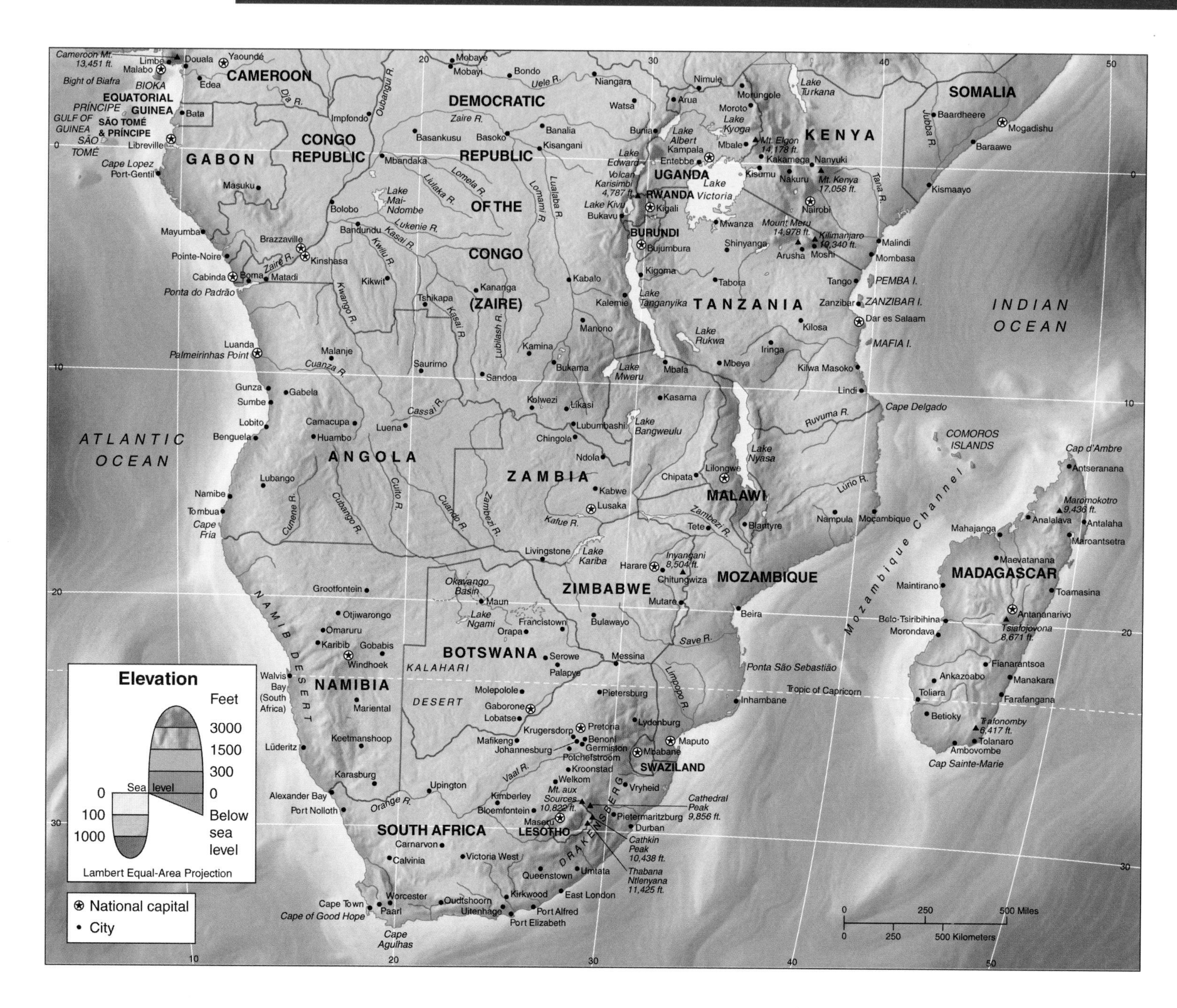
Elevation
Feet
3000
1500
300
0
Below sea level
Sea level
0
100
1000
Lambert Equal-Area Projection
National capital
City
0 250 500 Miles
0 250 500 Kilometers
ATLANTIC OCEAN
INDIAN OCEAN
Mozambique Channel
CAMEROON
EQUATORIAL GUINEA
BIOKA
PRÍNCIPE
GULF OF GUINEA
SÃO TOMÉ & PRÍNCIPE
SÃO TOMÉ
GABON
CONGO REPUBLIC
DEMOCRATIC REPUBLIC OF THE CONGO (ZAIRE)
UGANDA
RWANDA
BURUNDI
KENYA
SOMALIA
TANZANIA
ANGOLA
ZAMBIA
MALAWI
MOZAMBIQUE
ZIMBABWE
BOTSWANA
NAMIBIA
SOUTH AFRICA
LESOTHO
SWAZILAND
MADAGASCAR
COMOROS ISLANDS
KALAHARI DESERT
NAMIB DESERT
DRAKENSBERG
Cameroon Mt. 13,451 ft.
Bight of Biafra
Limbe
Malabo
Douala
Yaoundé
Edea
Bata
Libreville
Cape Lopez
Port-Gentil
Masuku
Mayumba
Pointe-Noire
Brazzaville
Kinshasa
Cabinda
Boma
Matadi
Ponta do Padrão
Luanda
Palmeirinhas Point
Gunza
Sumbe
Gabela
Lobito
Benguela
Namibe
Tombua
Cape Fria
Lubango
Malanje
Camacupa
Huambo
Luena
Saurimo
Kikwit
Bandundu
Bolobo
Impfondo
Mbandaka
Basankusu
Basoko
Banalia
Kisangani
Mobaye
Mobayi
Bondo
Niangara
Watsa
Bunia
Arua
Nimule
Kananga
Tshikapa
Sandoa
Kamina
Bukama
Kolwezi
Likasi
Lubumbashi
Chingola
Ndola
Kabwe
Lusaka
Livingstone
Kabalo
Kalemie
Manono
Kigoma
Bujumbura
Kigali
Bukavu
Volcan Karisimbi 4,787 ft.
Entebbe
Kampala
Mbale
Moroto
Morungole
Mt. Elgon 14,178 ft.
Kakamega
Nanyuki
Kisumu
Nakuru
Mt. Kenya 17,058 ft.
Nairobi
Mwanza
Mount Meru 14,978 ft.
Kilimanjaro 19,340 ft.
Arusha
Moshi
Shinyanga
Tabora
Malindi
Mombasa
Tango
Zanzibar
Dar es Salaam
Kilosa
Iringa
Mbeya
Mbala
Kasama
Kilwa Masoko
Lindi
Cape Delgado
Baardheere
Mogadishu
Baraawe
Kismaayo
PEMBA I.
ZANZIBAR I.
MAFIA I.
Lake Turkana
Lake Kyoga
Lake Albert
Lake Edward
Lake Kivu
Lake Victoria
Lake Tanganyika
Lake Rukwa
Lake Mweru
Lake Bangweulu
Lake Nyasa
Lake Kariba
Lake Mai-Ndombe
Lake Ngami
Okavango Basin
Dja R.
Ubangi R.
Zaire R.
Uele R.
Lomela R.
Liulaka R.
Lukenie R.
Kasai R.
Kwilu R.
Kwango R.
Lomami R.
Lualaba R.
Lubilash R.
Cuanza R.
Cassai R.
Cunene R.
Cubango R.
Cuito R.
Cuando R.
Zambezi R.
Kafue R.
Luangwa R.
Lúrio R.
Ruvuma R.
Tana R.
Jubba R.
Save R.
Limpopo R.
Vaal R.
Orange R.
Chipata
Lilongwe
Blantyre
Tete
Nampula
Moçambique
Harare
Inyangani 8,504 ft.
Chitungwiza
Mutare
Beira
Bulawayo
Maun
Francistown
Orapa
Serowe
Palapye
Messina
Pietersburg
Molepolole
Gaborone
Lobatse
Grootfontein
Otjiwarongo
Omaruru
Karibib
Gobabis
Windhoek
Walvis Bay (South Africa)
Mariental
Keetmanshoop
Lüderitz
Karasburg
Alexander Bay
Port Nolloth
Upington
Kimberley
Bloemfontein
Maseru
Mafikeng
Johannesburg
Krugersdorp
Pretoria
Benoni
Germiston
Potchefstroom
Kroonstad
Welkom
Mt. aux Sources 10,822 ft.
Lydenburg
Maputo
Mbabane
Vryheid
Pietermaritzburg
Durban
Cathedral Peak 9,856 ft.
Cathkin Peak 10,438 ft.
Thabana Ntlenyana 11,425 ft.
Umtata
Queenstown
East London
Kirkwood
Port Alfred
Port Elizabeth
Uitenhage
Oudtshoorn
Victoria West
Carnarvon
Calvinia
Worcester
Paarl
Cape Town
Cape of Good Hope
Cape Agulhas
Inhambane
Ponta São Sebastião
Tropic of Capricorn
Cap d'Ambre
Antseranana
Maromokotro 9,436 ft.
Analalava
Antalaha
Maroantsetra
Mahajanga
Maevatanana
Maintirano
Toamasina
Antananarivo
Tsiafajavona 8,671 ft.
Belo-Tsiribihina
Morondava
Fianarantsoa
Ankazoabo
Manakara
Toliara
Farafangana
Betioky
Trafonomby 6,417 ft.
Tolanaro
Ambovombe
Cap Sainte-Marie

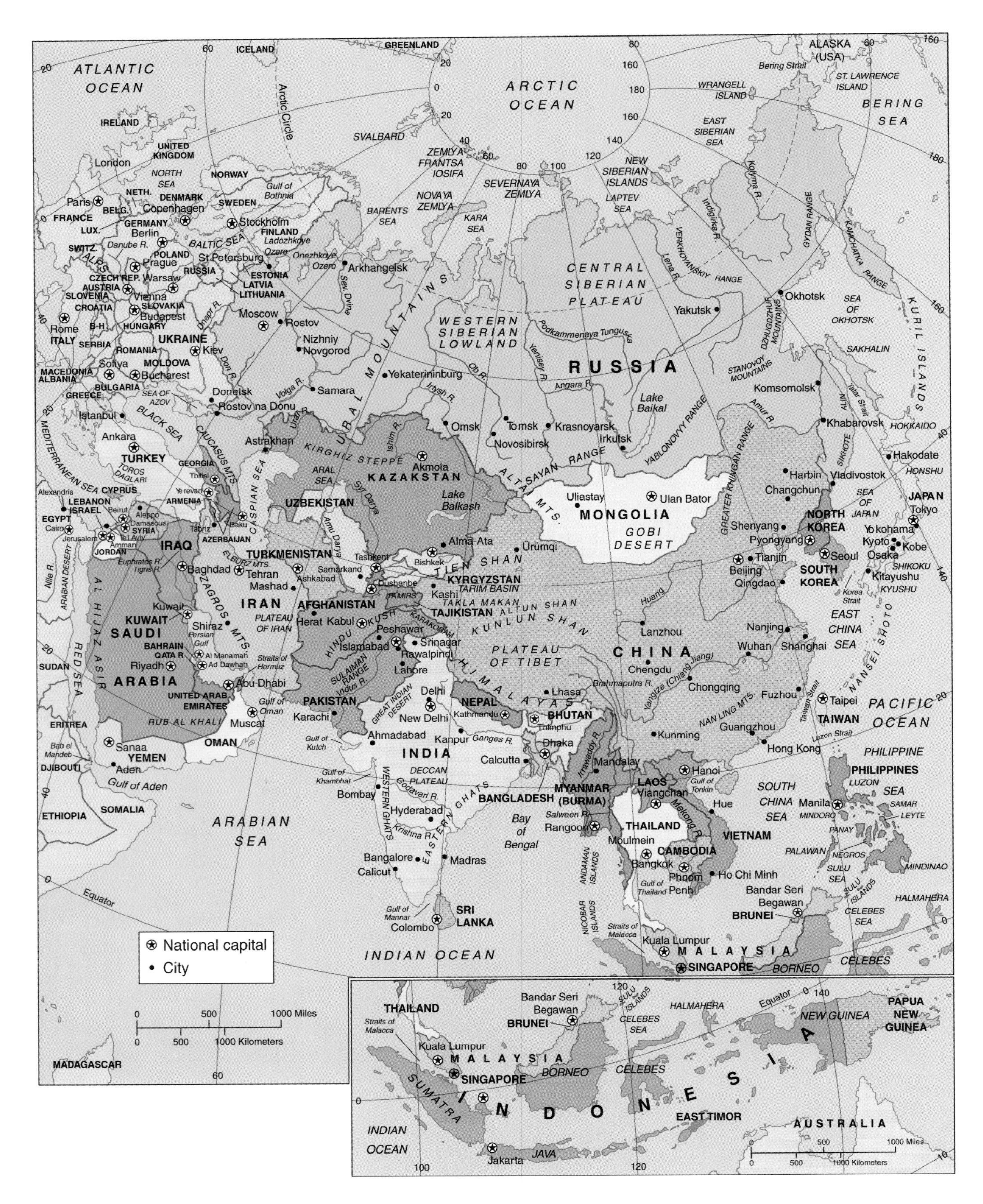
ATLANTIC OCEAN
ICELAND
GREENLAND
ARCTIC OCEAN
ALASKA (USA)
Bering Strait
WRANGELL ISLAND
ST. LAWRENCE ISLAND
BERING SEA
Arctic Circle
IRELAND
SVALBARD
EAST SIBERIAN SEA
UNITED KINGDOM
London
ZEMLYA FRANTSA IOSIFA
NEW SIBERIAN ISLANDS
NORTH SEA
NORWAY
Gulf of Bothnia
NOVAYA ZEMLYA
SEVERNAYA ZEMLYA
LAPTEV SEA
NETH.
DENMARK
SWEDEN
Paris
BELG.
Copenhagen
BARENTS SEA
KARA SEA
FRANCE
GERMANY
Stockholm
LUX.
Berlin
FINLAND
Ladozhkoye Ozero
SWITZ.
Danube R.
BALTIC SEA
ALPS
POLAND
St. Petersburg
Onezhkoye Ozero
Prague
RUSSIA
ESTONIA
Arkhangelsk
CZECH REP.
Warsaw
LATVIA
AUSTRIA
LITHUANIA
Sev. Dvina
SLOVENIA
Vienna
CROATIA
SLOVAKIA
Budapest
Moscow
Rostov
Rome
B-H
HUNGARY
ITALY
SERBIA
UKRAINE
Nizhniy Novgorod
ROMANIA
Kiev
MOLDOVA
MACEDONIA
ALBANIA
Sofiya
Bucharest
BULGARIA
GREECE
SEA OF AZOV
Donetsk
Rostov na Donu
Volga R.
Samara
Istanbul
BLACK SEA
Ankara
TURKEY
TOROS DAGLARI
GEORGIA
Tbilisi
CAUCASUS MTS.
Astrakhan
MEDITERRANEAN SEA
CYPRUS
ARMENIA
Yerevan
Baku
Alexandria
LEBANON
Beirut
ISRAEL
Damascus
EGYPT
Aleppo
SYRIA
Cairo
Jerusalem
Tel Aviv
Amman
JORDAN
Tabriz
AZERBAIJAN
IRAQ
Euphrates R.
Tigris R.
Baghdad
Tehran
ELBURZ MTS.
Nile R.
ARABIAN DESERT
AL HIJAZ
ASIR
RED SEA
Kuwait
KUWAIT
SAUDI ARABIA
Shiraz
ZAGROS MTS.
BAHRAIN
QATAR
Persian Gulf
Riyadh
Al Manamah
Ad Dawhah
Straits of Hormuz
Abu Dhabi
UNITED ARAB EMIRATES
SUDAN
Gulf of Oman
Muscat
RUB AL KHALI
OMAN
ERITREA
Sanaa
YEMEN
Bab el Mandeb
Aden
DJIBOUTI
Gulf of Aden
ETHIOPIA
SOMALIA
ARABIAN SEA
IRAN
Mashad
TURKMENISTAN
Ashkabad
AFGHANISTAN
Herat
Kabul
PLATEAU OF IRAN
CASPIAN SEA
URAL MOUNTAINS
Ural R.
KIRGHIZ STEPPE
ARAL SEA
Syr Darya
Amu Darya
UZBEKISTAN
KAZAKSTAN
Akmola
Lake Balkash
Alma-Ata
Tashkent
Samarkand
Bishkek
KYRGYZSTAN
Dushanbe
PAMIRS
TAJIKISTAN
HINDU KUSH
KARAKORAM
Peshawar
Islamabad
Srinagar
Rawalpindi
Lahore
SULAIMAN RANGE
Indus R.
PAKISTAN
Karachi
GREAT INDIAN DESERT
Delhi
New Delhi
Kathmandu
NEPAL
BHUTAN
Thimphu
Gulf of Kutch
Ahmadabad
Kanpur
Ganges R.
INDIA
Dhaka
Calcutta
Gulf of Khambhat
DECCAN PLATEAU
WESTERN GHATS
EASTERN GHATS
Godavari R.
Bombay
Hyderabad
Krishna R.
BANGLADESH
Bay of Bengal
Bangalore
Madras
Calicut
Gulf of Mannar
Colombo
SRI LANKA
INDIAN OCEAN
Yekaterinburg
WESTERN SIBERIAN LOWLAND
Ob R.
Irtysh R.
Omsk
Tomsk
Novosibirsk
Krasnoyarsk
Yenisey R.
Podkammenaya Tunguska
Angara R.
CENTRAL SIBERIAN PLATEAU
RUSSIA
Lake Baikal
Irkutsk
ALTAI MTS.
SAYAN RANGE
YABLONOVYY RANGE
Ulan Bator
Uliastay
MONGOLIA
GOBI DESERT
Ürümqi
TIEN SHAN
Kashi
TARIM BASIN
TAKLA MAKAN
ALTUN SHAN
KUNLUN SHAN
PLATEAU OF TIBET
HIMALAYAS
Lhasa
Brahmaputra R.
Yakutsk
Lena R.
VERKHOYANSKIY RANGE
Indigirka R.
Kolyma R.
GYDAN RANGE
KAMCHATKA RANGE
Okhotsk
SEA OF OKHOTSK
DZHUGDZHUR MOUNTAINS
STANOVOY MOUNTAINS
KURIL ISLANDS
SAKHALIN
Komsomolsk
Amur R.
Khabarovsk
Tatar Strait
SIKHOTE ALIN
HOKKAIDO
Hakodate
HONSHU
Harbin
Vladivostok
Changchun
GREATER KHINGAN RANGE
SEA OF JAPAN
JAPAN
Tokyo
NORTH KOREA
Shenyang
Pyongyang
Yokohama
Kyoto
Kobe
Osaka
Tianjin
Beijing
Seoul
SHIKOKU
Kitayushu
Qingdao
SOUTH KOREA
KYUSHU
Korea Strait
Huang
Lanzhou
EAST CHINA SEA
Nanjing
CHINA
Wuhan
Shanghai
NANSEI SHOTO
Chengdu
Yangtze (Chiang Jiang)
Chongqing
Fuzhou
Taipei
TAIWAN
Taiwan Strait
PACIFIC OCEAN
NAN LING MTS.
Kunming
Guangzhou
Hong Kong
Luzon Strait
PHILIPPINE SEA
Irrawaddy R.
Mandalay
Hanoi
Gulf of Tonkin
LAOS
Vientiane
PHILIPPINES
LUZON
SOUTH CHINA SEA
MYANMAR (BURMA)
Hue
Manila
SAMAR
Salween R.
Mekong R.
MINDORO
LEYTE
Rangoon
THAILAND
PANAY
Moulmein
VIETNAM
CAMBODIA
PALAWAN
NEGROS
Bangkok
Phnom Penh
Ho Chi Minh
SULU SEA
SULU ISLANDS
MINDINAO
ANDAMAN ISLANDS
Gulf of Thailand
Bandar Seri Begawan
BRUNEI
NICOBAR ISLANDS
HALMAHERA
CELEBES SEA
Straits of Malacca
Kuala Lumpur
MALAYSIA
SINGAPORE
BORNEO
CELEBES
Equator
MADAGASCAR
National capital
City
0
500
1000 Miles
1000 Kilometers
THAILAND
Straits of Malacca
Bandar Seri Begawan
BRUNEI
SULU ISLANDS
CELEBES SEA
HALMAHERA
Equator
NEW GUINEA
PAPUA NEW GUINEA
Kuala Lumpur
MALAYSIA
SINGAPORE
BORNEO
CELEBES
SUMATRA
INDONESIA
EAST TIMOR
AUSTRALIA
INDIAN OCEAN
Jakarta
JAVA

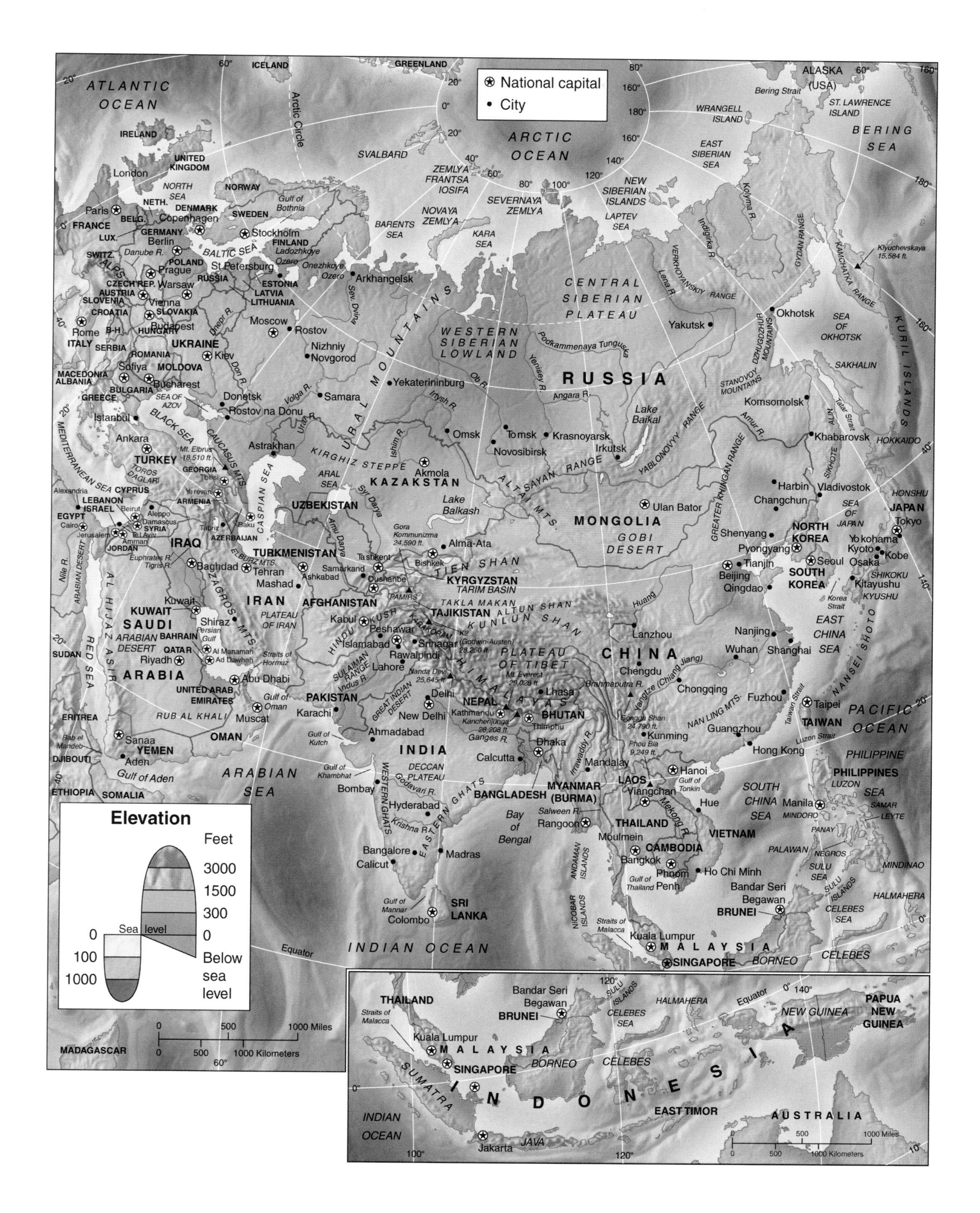
National capital
City
ATLANTIC OCEAN
ICELAND
GREENLAND
Arctic Circle
ARCTIC OCEAN
ALASKA (USA)
Bering Strait
ST. LAWRENCE ISLAND
BERING SEA
WRANGELL ISLAND
EAST SIBERIAN SEA
NEW SIBERIAN ISLANDS
LAPTEV SEA
SEVERNAYA ZEMLYA
ZEMLYA FRANTSA IOSIFA
SVALBARD
NOVAYA ZEMLYA
BARENTS SEA
KARA SEA
IRELAND
UNITED KINGDOM
London
NORTH SEA
NORWAY
Gulf of Bothnia
SWEDEN
NETH.
DENMARK
Copenhagen
Paris
BELG.
FRANCE
GERMANY
LUX.
Berlin
Stockholm
FINLAND
Ladozhkoye Ozero
Onezhkoye Ozero
Arkhangelsk
SWITZ.
Danube R.
ALPS
POLAND
Prague
BALTIC SEA
St Petersburg
CZECH REP.
Warsaw
RUSSIA
ESTONIA
LATVIA
LITHUANIA
AUSTRIA
SLOVENIA
Vienna
CROATIA
SLOVAKIA
Rome
B-H
HUNGARY
Budapest
ITALY
SERBIA
Moscow
Rostov
Dnepr R.
UKRAINE
ROMANIA
Kiev
Nizhniy Novgorod
Sev. Dvina
MACEDONIA
ALBANIA
Sofiya
MOLDOVA
Bucharest
BULGARIA
GREECE
SEA OF AZOV
Don R.
Donetsk
Rostov na Donu
Volga R.
Samara
Yekaterininburg
Istanbul
BLACK SEA
MEDITERRANEAN SEA
Ankara
TURKEY
CAUCASUS MTS.
Mt. Elbrus 18,510 ft.
GEORGIA
Tbilisi
Astrakhan
Ural R.
URAL MOUNTAINS
TOROS DAGLARI
CYPRUS
Alexandria
Yerevan
ARMENIA
LEBANON
ISRAEL
Beirut
Aleppo
Damascus
SYRIA
Tel Aviv
EGYPT
Cairo
Jerusalem
Amman
JORDAN
Tabriz
AZERBAIJAN
Baku
CASPIAN SEA
ARAL SEA
KIRGHIZ STEPPE
Akmola
KAZAKSTAN
UZBEKISTAN
Syr Darya
Amu Darya
IRAQ
Euphrates R.
Tigris R.
Baghdad
ELBURZ MTS.
Tehran
TURKMENISTAN
Ashkabad
Mashad
Tashkent
Samarkand
Dushanbe
Bishkek
Alma-Ata
Gora Kommunizma 24,590 ft.
Lake Balkash
TIEN SHAN
KYRGYZSTAN
TARIM BASIN
Nile R.
ARABIAN DESERT
AL HIJAZ
ASIR
RED SEA
Kuwait
KUWAIT
ZAGROS MTS.
IRAN
PLATEAU OF IRAN
AFGHANISTAN
PAMIRS
TAKLA MAKAN
TAJIKISTAN
ALTUN SHAN
KUNLUN SHAN
SAUDI ARABIA
Shiraz
Persian Gulf
BAHRAIN
QATAR
Al Manamah
Ad Dawhah
Riyadh
Straits of Hormuz
Abu Dhabi
UNITED ARAB EMIRATES
Gulf of Oman
Muscat
OMAN
RUB AL KHALI
SUDAN
ERITREA
Sanaa
YEMEN
Aden
Bab el Mandeb
DJIBOUTI
Gulf of Aden
ETHIOPIA
SOMALIA
Kabul
HINDU KUSH
Peshawar
KARAKORAM
Islamabad
Srinagar
K2 (Godwin-Austen) 28,250 ft.
Rawalpindi
Lahore
SULAIMAN RANGE
Indus R.
PAKISTAN
Karachi
GREAT INDIAN DESERT
Nanda Devi 25,645 ft.
HIMALAYAS
PLATEAU OF TIBET
Mt. Everest 29,028 ft.
Lhasa
Delhi
NEPAL
New Delhi
Kathmandu
Kanchenjunga 28,208 ft.
BHUTAN
Thimphu
Ganges R.
Gulf of Kutch
Ahmadabad
INDIA
Dhaka
Calcutta
ARABIAN SEA
Gulf of Khambhat
DECCAN PLATEAU
Godavari R.
Bombay
WESTERN GHATS
EASTERN GHATS
Hyderabad
Krishna R.
BANGLADESH
Bay of Bengal
Bangalore
Calicut
Madras
Gulf of Mannar
Colombo
SRI LANKA
INDIAN OCEAN
Equator
Brahmaputra R.
Irrawaddy R.
Mandalay
MYANMAR (BURMA)
Salween R.
Rangoon
ANDAMAN ISLANDS
NICOBAR ISLANDS
THAILAND
Moulmein
Bangkok
Gulf of Thailand
CAMBODIA
Phnom Penh
Ho Chi Minh
VIETNAM
LAOS
Vientiane
Mekong R.
Hue
Hanoi
Gulf of Tonkin
Straits of Malacca
Kuala Lumpur
MALAYSIA
SINGAPORE
BORNEO
BRUNEI
Bandar Seri Begawan
CELEBES
CELEBES SEA
SULU SEA
SULU ISLANDS
HALMAHERA
MINDINAO
PALAWAN
NEGROS
PANAY
MINDORO
SAMAR
LEYTE
Manila
LUZON
PHILIPPINES
PHILIPPINE SEA
Luzon Strait
SOUTH CHINA SEA
Hong Kong
Guangzhou
NAN LING MTS.
Kunming
Phou Bia 9,249 ft.
Gongga Shan 24,790 ft.
Chongqing
Yangtze (Chiang jiang)
Chengdu
CHINA
Lanzhou
Huang
Wuhan
Shanghai
Nanjing
Fuzhou
Taiwan Strait
Taipei
TAIWAN
PACIFIC OCEAN
EAST CHINA SEA
NANSEI SHOTO
Qingdao
Tianjin
Beijing
SOUTH KOREA
Seoul
Korea Strait
Pyongyang
NORTH KOREA
Shenyang
Changchun
Harbin
Vladivostok
SEA OF JAPAN
JAPAN
HONSHU
Tokyo
Yokohama
Kyoto
Osaka
Kobe
SHIKOKU
Kitayushu
KYUSHU
HOKKAIDO
GREATER KHINGAN RANGE
MONGOLIA
GOBI DESERT
Ulan Bator
ALTAI MTS.
SAYAN RANGE
Irkutsk
Lake Baikal
YABLONOVYY RANGE
Krasnoyarsk
Tomsk
Novosibirsk
Omsk
Ishim R.
Irtysh R.
Ob R.
Yenisey R.
Angara R.
WESTERN SIBERIAN LOWLAND
Podkammenaya Tunguska
RUSSIA
CENTRAL SIBERIAN PLATEAU
Lena R.
Yakutsk
VERKHOYANSKIY RANGE
Indigirka R.
Kolyma R.
GYDAN RANGE
KAMCHATKA RANGE
Klyuchevskaya 15,584 ft.
Okhotsk
SEA OF OKHOTSK
DZHUGDZHUR MOUNTAINS
STANOVOY MOUNTAINS
Amur R.
Komsomolsk
Khabarovsk
SIKHOTE ALIN
Tatar Strait
SAKHALIN
KURIL ISLANDS
Elevation
Feet
3000
1500
300
0
Sea level
100
1000
Below sea level
MADAGASCAR
0 500 1000 Miles
0 500 1000 Kilometers
INDONESIA
SUMATRA
JAVA
Jakarta
EAST TIMOR
NEW GUINEA
PAPUA NEW GUINEA
AUSTRALIA

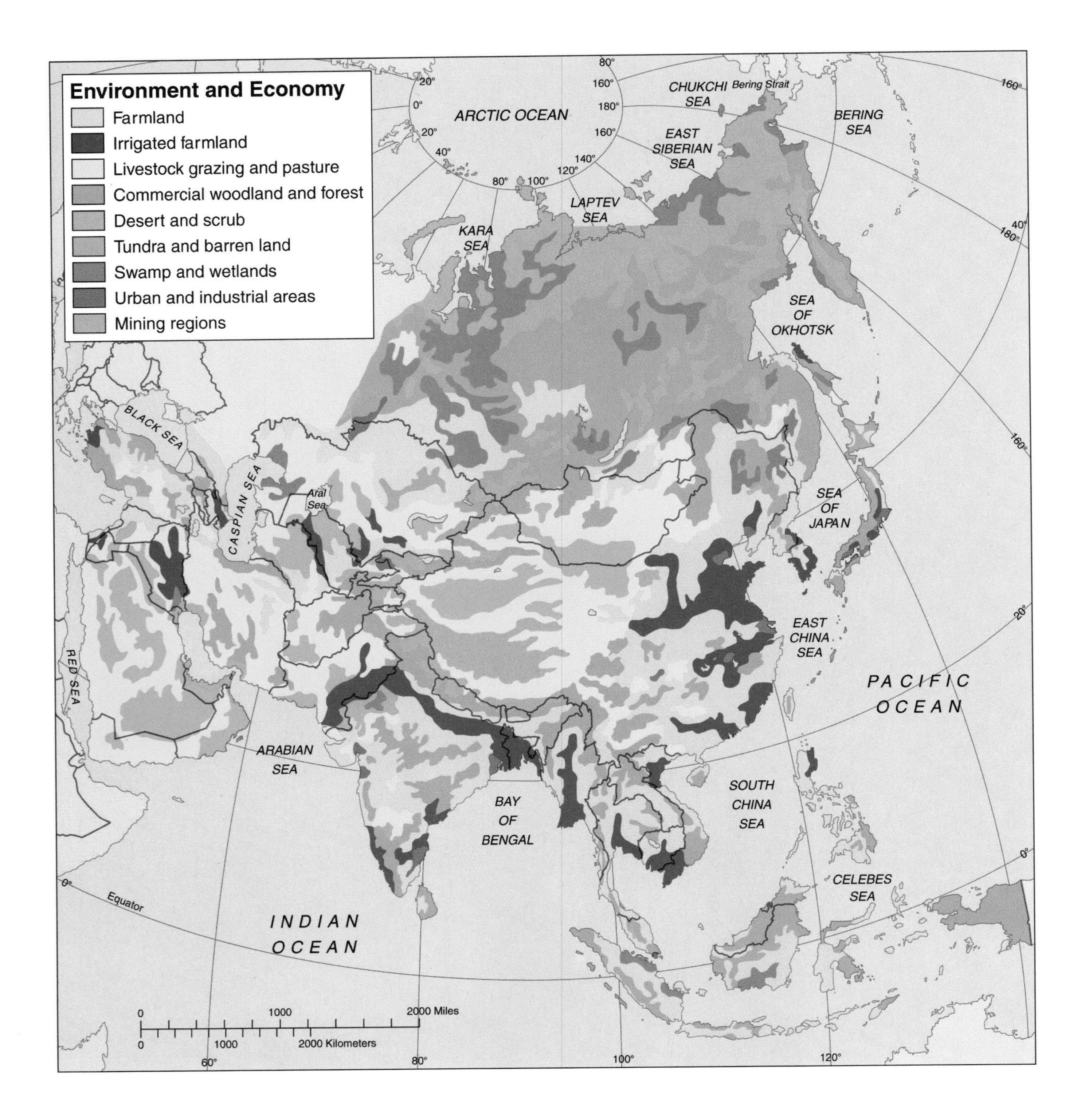

Asia is a land of extremes of land use with some of the world's most heavily industrialized regions, barren and empty areas, and productive and densely populated farm regions. Asia is a region of rapid industrial growth. Yet Asia remains an agricultural region with three out of every four workers engaged in agriculture. Asian commercial agriculture and intensive subsistence agriculture is characterized by irrigation. Some of Asia's irrigated lands are desert requiring additional water. But most of the Asian irrigated regions have sufficient precipitation for crop agriculture and irrigation is a way of coping with seasonal drought—the wet-and-dry cycle of the monsoon—often gaining more than one crop per year on irrigated farms. Agricultural yields per unit area in many areas of Asia are among the world's highest. Because the Asian population is so large and the demands for agricultural land so great, Asia is undergoing rapid deforestation and some areas of the continent have only small remnants of a once-abundant forest reserve.

Map 128b Population Density

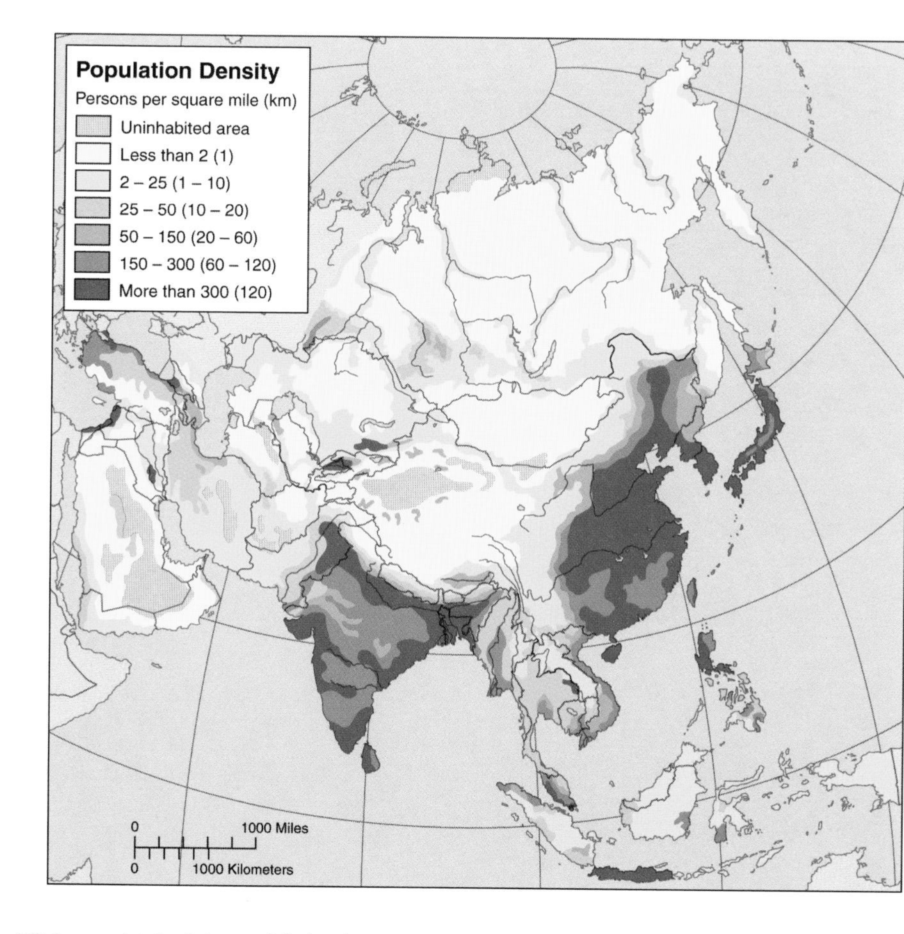

With one-third of the world's land area and nearly two-thirds of the world's population, Asia is more densely settled than any other region of the world. In some of the continent's farming regions, agricultural population density exceeds 2,000 persons per square mile. In some portions of the continent, particularly the Islamic areas of Central and Southwest Asia, this already large population is growing very rapidly with some countries having population growth rates above 3 percent per year and doubling times between 20 and 25 years. The populations of neither China nor India are growing particularly rapidly but since both countries have population bases that are enormous, the absolute number of Indians and Chinese added to the world's population each year is staggering. In spite of these massive populations, Asia also contains areas that are either completely uninhabited or have population densities that are as low as any on earth.

Map 128c Industrialization: Manufacturing and Resources

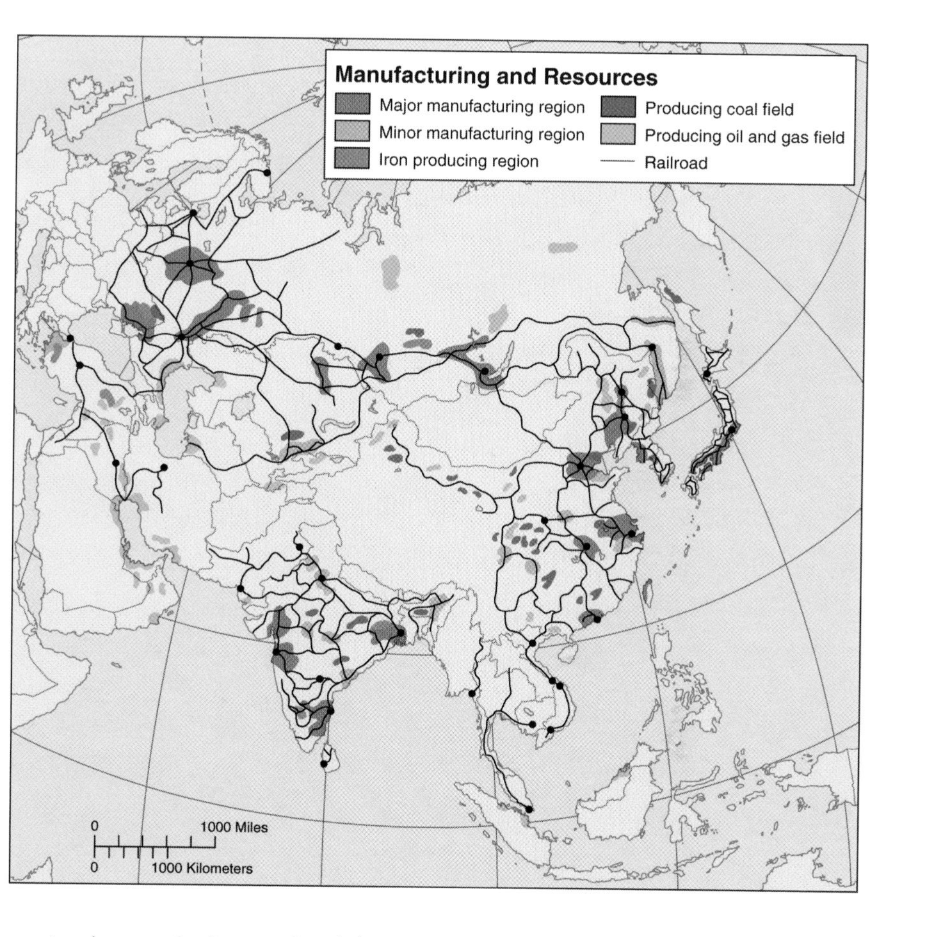

International economists have predicted that the next century will be one of Asian economic dominance, as other Asian nations approach the economic levels of industrial giant Japan. Asian countries have begun to industrialize rapidly and have increased industrial production more than 100 percent in the last decade. But nothing guarantees industrial output; it must master the great distances separating critical raw material and production sites from the locations of the markets for them. Such a mastery is achieved only by the development of efficient transportation systems. Usually that means water travel and here Asia is remarkably deficient when compared with Europe and North America. Nevertheless, the mixed prospects for Asian economic growth into the next century are a great deal better than would have been predicted a decade ago.

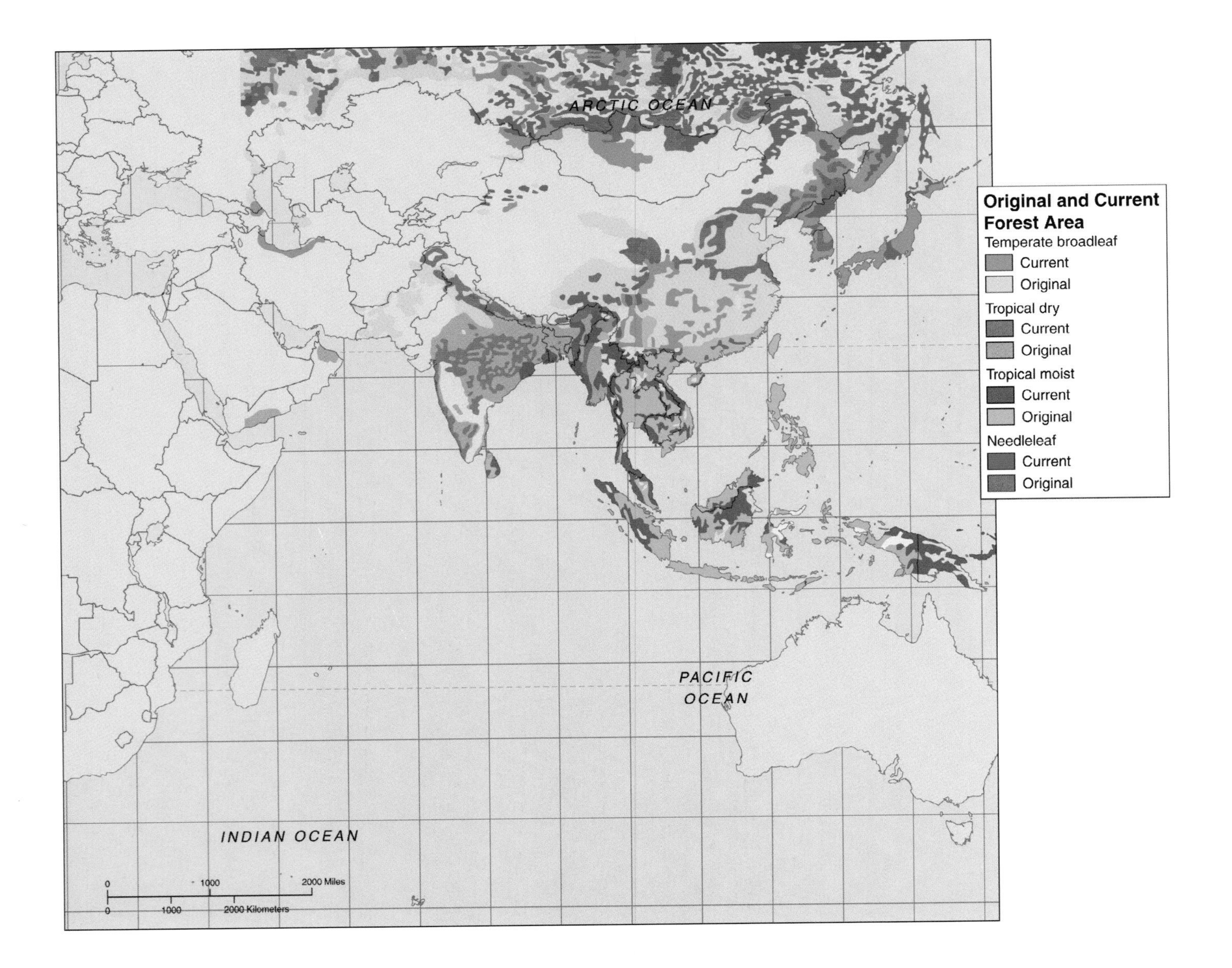

In no part of the world has deforestation produced such dramatic changes in the environment as in Asia. Here, the world's oldest continuous agricultural civilizations have placed enormous demands upon the vegetative and soil environments to the extent that precious little natural forest cover remains in the world's two most populous countries, China and India. While Southeast Asia has been historically less denuded than the major culture centers to its north and west, recent decades have seen significant forest loss to commercial timber operations, stimulated by the high-market value of tropical hardwoods. Also, a fairly recent trend has been the massive deforestation in the northeastern parts of Russia where economic expansion over the last century has stimulated the rapid loss of softwood forests for construction lumber and for paper production.

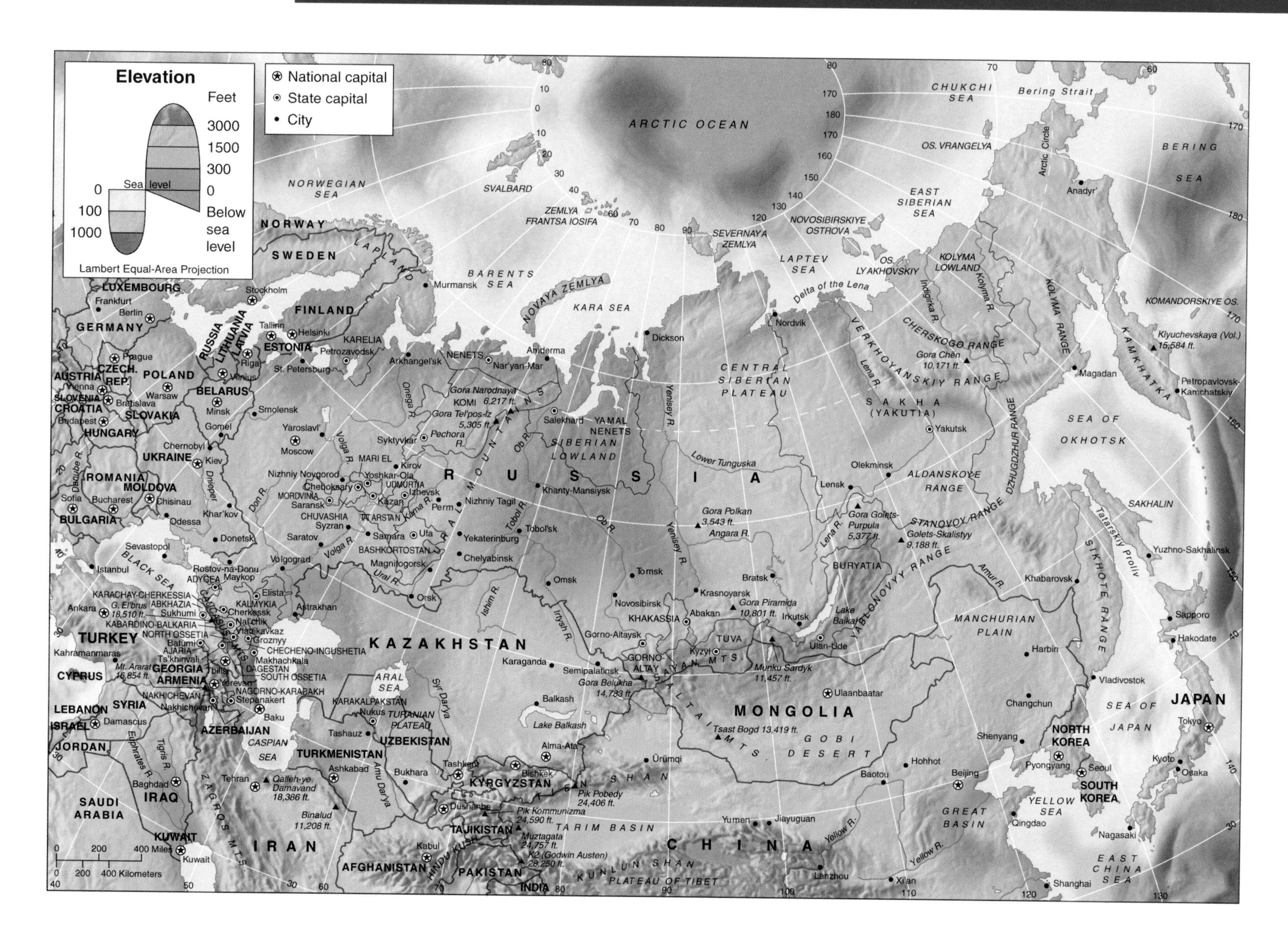
Elevation
Feet
3000
1500
300
0
Below sea level
Sea level
0
100
1000
Lambert Equal-Area Projection
National capital
State capital
City
ARCTIC OCEAN
NORWEGIAN SEA
BARENTS SEA
KARA SEA
LAPTEV SEA
EAST SIBERIAN SEA
CHUKCHI SEA
BERING SEA
SEA OF OKHOTSK
SEA OF JAPAN
YELLOW SEA
EAST CHINA SEA
BLACK SEA
CASPIAN SEA
ARAL SEA
Bering Strait
Arctic Circle
SVALBARD
ZEMLYA FRANTSA IOSIFA
NOVAYA ZEMLYA
SEVERNAYA ZEMLYA
NOVOSIBIRSKIYE OSTROVA
OS. LYAKHOVSKIY
OS. VRANGELYA
KOMANDORSKIYE OS.
SAKHALIN
KAMKHATKA
R U S S I A
KAZAKHSTAN
MONGOLIA
CHINA
JAPAN
NORTH KOREA
SOUTH KOREA
NORWAY
SWEDEN
FINLAND
ESTONIA
LATVIA
LITHUANIA
RUSSIA
BELARUS
POLAND
GERMANY
LUXEMBOURG
CZECH. REP.
AUSTRIA
SLOVAKIA
SLOVENIA
CROATIA
HUNGARY
UKRAINE
MOLDOVA
ROMANIA
BULGARIA
TURKEY
CYPRUS
SYRIA
LEBANON
ISRAEL
JORDAN
IRAQ
KUWAIT
SAUDI ARABIA
IRAN
GEORGIA
ARMENIA
AZERBAIJAN
TURKMENISTAN
UZBEKISTAN
KYRGYZSTAN
TAJIKISTAN
AFGHANISTAN
PAKISTAN
INDIA
CENTRAL SIBERIAN PLATEAU
WEST SIBERIAN LOWLAND
SAKHA (YAKUTIA)
KOLYMA LOWLAND
MANCHURIAN PLAIN
GOBI DESERT
TARIM BASIN
GREAT BASIN
PLATEAU OF TIBET
TURANIAN PLATEAU
URAL MOUNTAINS
VERKHOYANSKIY RANGE
CHERSKOGO RANGE
KOLYMA RANGE
DZHUGDZHUR RANGE
ALDANSKOYE RANGE
STANOVOY RANGE
YABLONOVYY RANGE
SIKHOTE RANGE
SAYAN MTS.
ALTAI MTS.
TIAN SHAN
KUNLUN SHAN
HINDU KUSH
CAUCASUS MTS.
ZAGROS MTS.
Delta of the Lena
Lena R.
Yenisey R.
Ob R.
Irtysh R.
Lower Tunguska
Angara R.
Amur R.
Kolyma R.
Indigirka R.
Volga R.
Don R.
Dnieper
Danube R.
Ural R.
Kama R.
Pechora R.
Onega R.
Tobol R.
Ishim R.
Syr Dar'ya
Amu Dar'ya
Tigris R.
Euphrates R.
Yellow R.
Lake Baikal
Lake Balkhash
Tatarskiy Proliv
Gora Narodnaya 6,217 ft.
Gora Tel'pos-Iz 5,305 ft.
Gora Chen 10,171 ft.
Klyuchevskaya (Vol.) 15,584 ft.
Gora Polkan 3,543 ft.
Gora Golets-Purpula 5,377 ft.
Golets-Skalistyy 9,188 ft.
Gora Piramida 10,801 ft.
Munku Sardyk 11,457 ft.
Gora Belukha 14,783 ft.
Tsast Bogd 13,419 ft.
Pik Pobedy 24,406 ft.
Pik Kommunizma 24,590 ft.
Muztagata 24,757 ft.
K2 (Godwin Austen) 28,250 ft.
G. El'brus 18,510 ft.
Mt. Ararat 16,854 ft.
Qolleh-ye Damavand 18,386 ft.
Binalud 11,208 ft.
Moscow
St. Petersburg
Murmansk
Arkhangel'sk
Nar'yan-Mar
Amderma
Salekhard
Dickson
Nordvik
Yakutsk
Olekminsk
Lensk
Magadan
Anadyr'
Petropavlovsk-Kamchatskiy
Yuzhno-Sakhalinsk
Khabarovsk
Vladivostok
Harbin
Changchun
Shenyang
Beijing
Hohhot
Baotou
Lanzhou
Xi'an
Yumen
Jiayuguan
Ürümqi
Ulaanbaatar
Ulan-Ude
Irkutsk
Bratsk
Krasnoyarsk
Abakan
Kyzyl
Gorno-Altaysk
Novosibirsk
Tomsk
Omsk
Tobol'sk
Khanty-Mansiysk
Yekaterinburg
Chelyabinsk
Nizhniy Tagil
Perm
Izhevsk
Kazan
Ufa
Samara
Saratov
Volgograd
Astrakhan
Orsk
Magnitogorsk
Karaganda
Semipalatinsk
Balkash
Alma-Ata
Bishkek
Tashkent
Dushanbe
Bukhara
Ashkabad
Tashauz
Nukus
Kabul
Tehran
Baku
Yerevan
Tbilisi
Baghdad
Kuwait
Damascus
Ankara
Istanbul
Kahramanmaras
Sevastopol
Odessa
Kiev
Kharkov
Donetsk
Rostov-na-Donu
Chisinau
Bucharest
Sofia
Budapest
Vienna
Bratislava
Prague
Berlin
Frankfurt
Warsaw
Minsk
Vilnius
Riga
Tallinn
Helsinki
Stockholm
Petrozavodsk
Smolensk
Yaroslavl'
Gomel
Chernobyl
Nizhniy Novgorod
Syktyvkar
Kirov
Seoul
Pyongyang
Qingdao
Shanghai
Tokyo
Kyoto
Osaka
Nagasaki
Sapporo
Hakodate
400 Miles
400 Kilometers

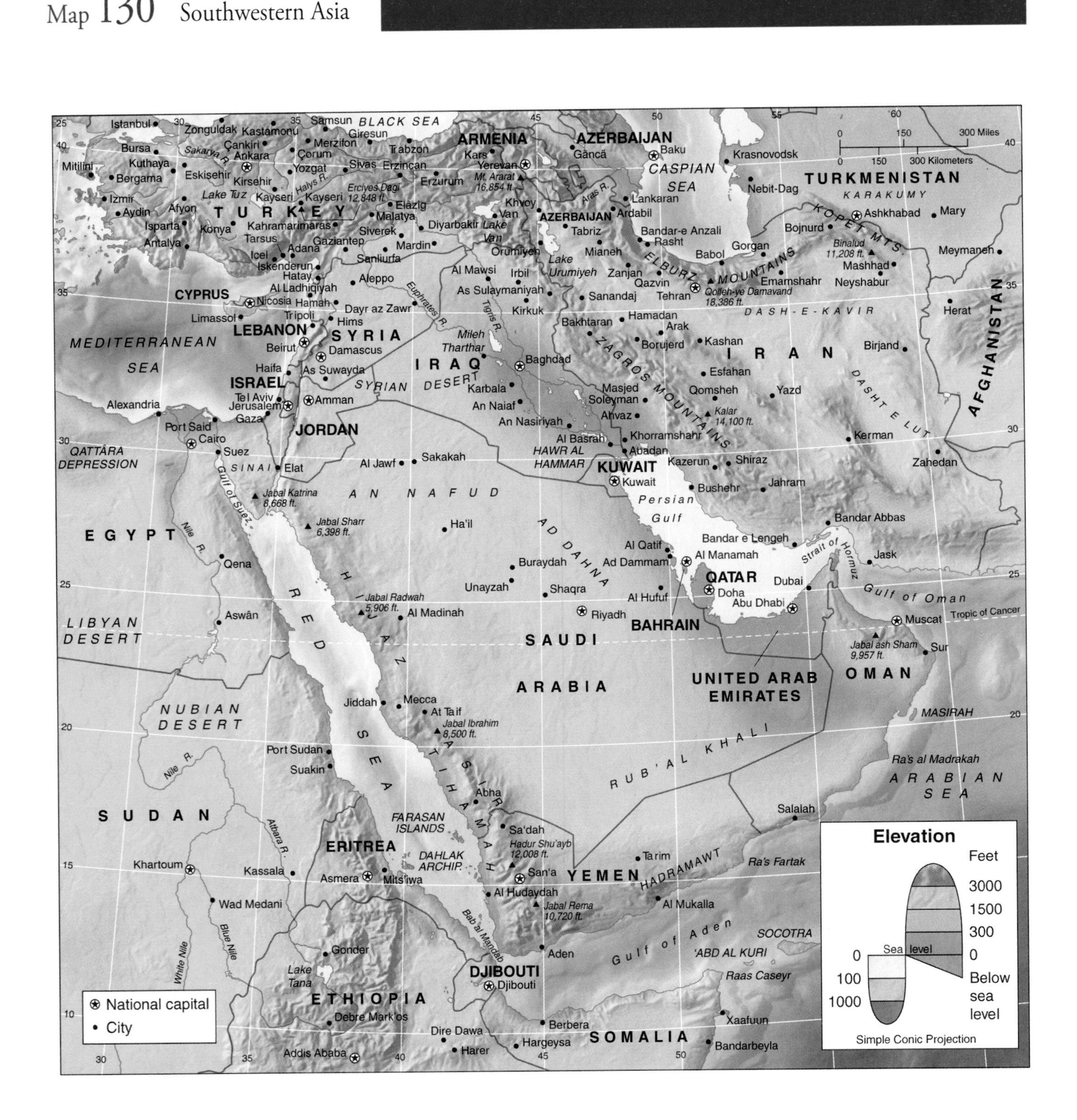

BLACK SEA
MEDITERRANEAN SEA
CASPIAN SEA
TURKEY
ARMENIA
AZERBAIJAN
TURKMENISTAN
CYPRUS
LEBANON
SYRIA
IRAQ
IRAN
AFGHANISTAN
ISRAEL
JORDAN
KUWAIT
EGYPT
SAUDI ARABIA
QATAR
BAHRAIN
UNITED ARAB EMIRATES
OMAN
YEMEN
ERITREA
DJIBOUTI
ETHIOPIA
SOMALIA
SUDAN
RED SEA
Persian Gulf
Gulf of Oman
Gulf of Aden
ARABIAN SEA
AN NAFUD
RUB' AL KHALI
Elevation
Feet
3000
1500
300
0
Sea level
100
1000
Below sea level
Simple Conic Projection
National capital
City
0 150 300 Miles
0 150 300 Kilometers
Tropic of Cancer

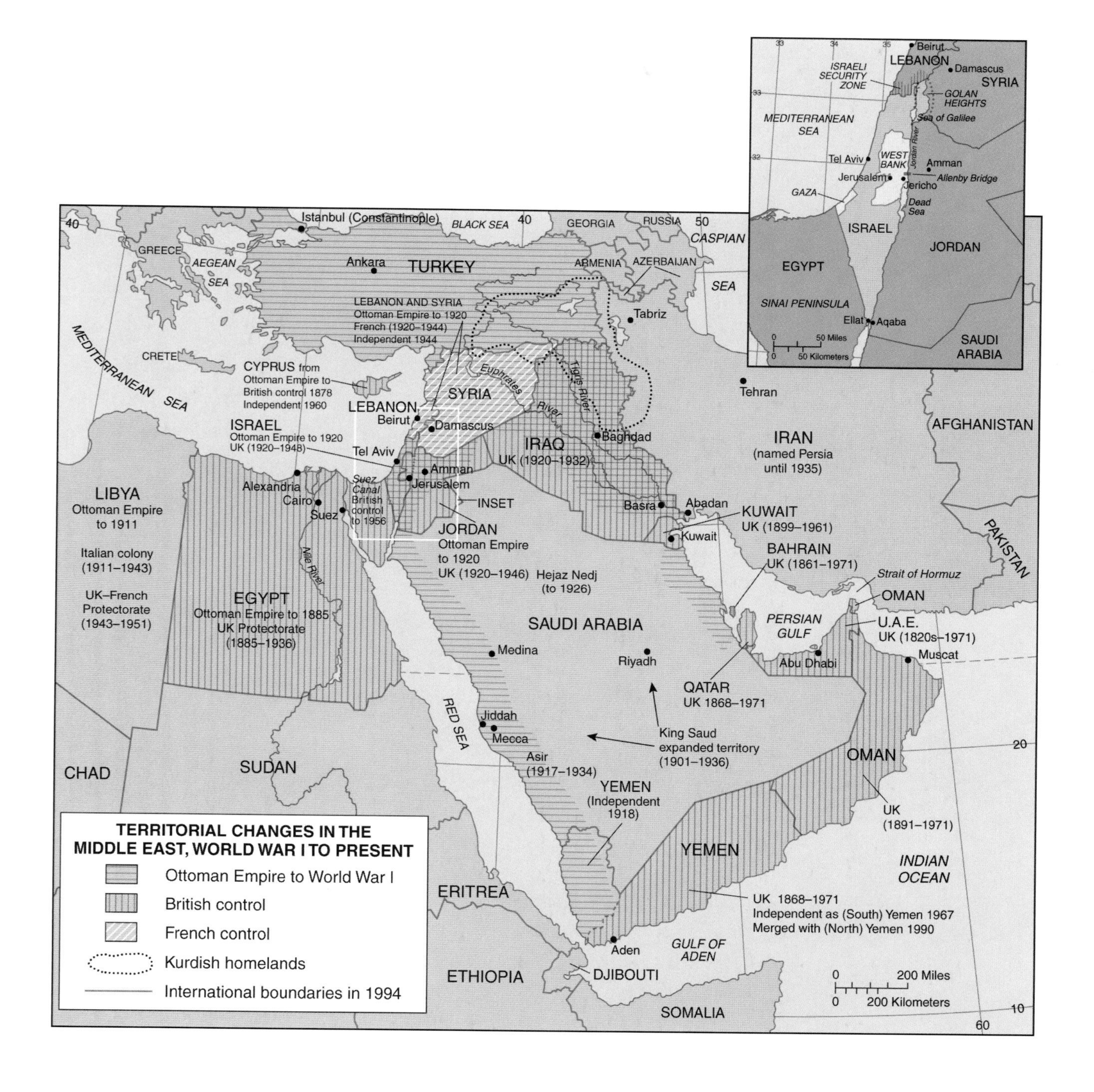

The Middle East, encompassing the northeastern part of Africa and southwestern Asia, has experienced a turbulent history. In the last century alone, many of the region's countries have gone from being ruled by the Turkish Ottoman Empire, to being dependencies of Great Britain or France, to being independent. Having experienced the Crusades and colonial domination by European powers, the region's predominantly Islamic countries are now resentful of interference in the region's affairs by countries with a European and/or Christian heritage. The tension between Israel (settled largely in the late nineteenth and twentieth centuries by Jews of predominantly European background) and its neighbors is a matter of European–Middle Eastern cultural stress as well as a religious conflict between Islamic Arab culture and Judaism. The political boundaries on the map, like those throughout most of Africa, are the invention of European colonial powers and often do not take into account the pre-existing lines of tribal control or authority. In Iraq, for example, three distinct cultural areas exist: Shiite Muslim (Arabic), Sunni Muslim (Arabic), and Kurd (also Islamic but distinctly not Arabic). These three cultures occupying the territory of a state makes stability in that tortured region a distinct problem for the future.

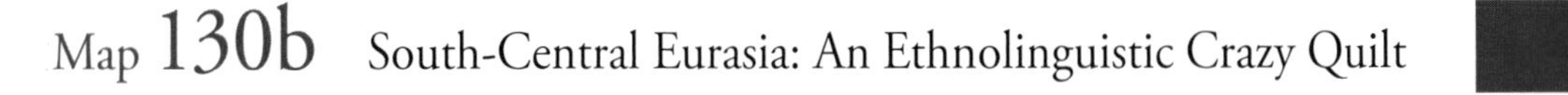

Map 130b South-Central Eurasia: An Ethnolinguistic Crazy Quilt

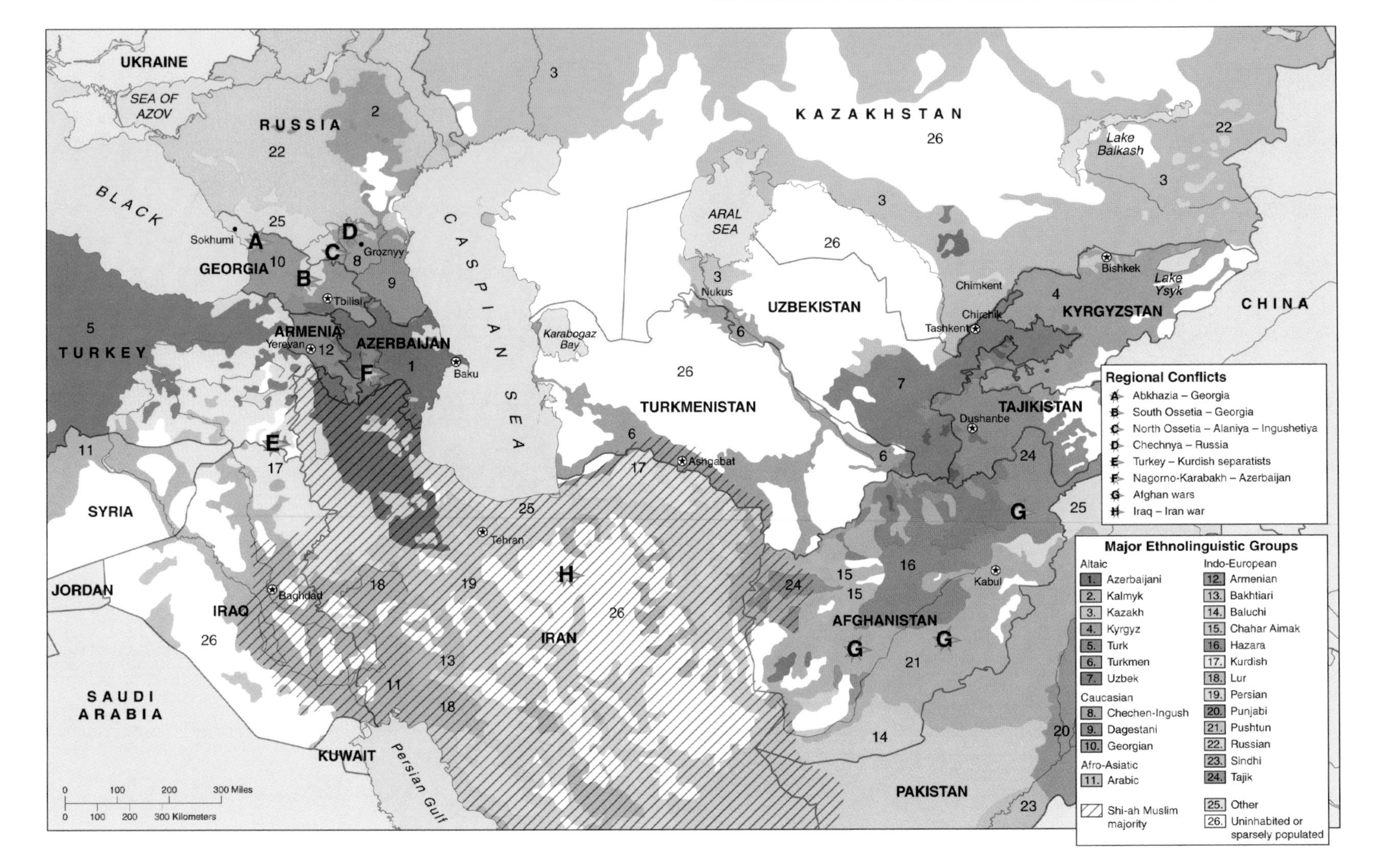

South-Central Eurasia: The area of south-central Eurasia represents what is perhaps the world's most volatile area in terms of potential military conflict. A crazy quilt of ethnic and linguistic groups, this region contains politically-defined "states," but nothing approaching "nation-states" as they are understood elsewhere. Even reasonably well-consolidated states, such as Iran, are hampered by the mixture of languages and ethnic populations within their borders. For some countries, such as Afghanistan, the ethnolinguistic mix is so historically fixed as to render any attempts at modern state building nearly hopeless. This vast area—although nearly universally Muslim—is also split between the two primary Islamic sects, Sunni and Shia, with several smaller sectarian divisions as well. The division of one of the world's global religions into two principal factions occurred in the seventh century, originally over the question of the source of authority in the religious hierarchy. But it has since come to be a theological and metaphysical separation, and Sunni and Shia Muslims now bear somewhat the same relationship to one another as did Catholics and Protestants in seventeenth- and eighteenth-century Europe—a mutual antipathy that periodically flares into civil unrest and conflict that goes far beyond the bounds of religious debate. The area is rich in natural resources, but the mixture of ethnicity, language, and religion has and will continue to produce the human conflicts that inhibit human development.

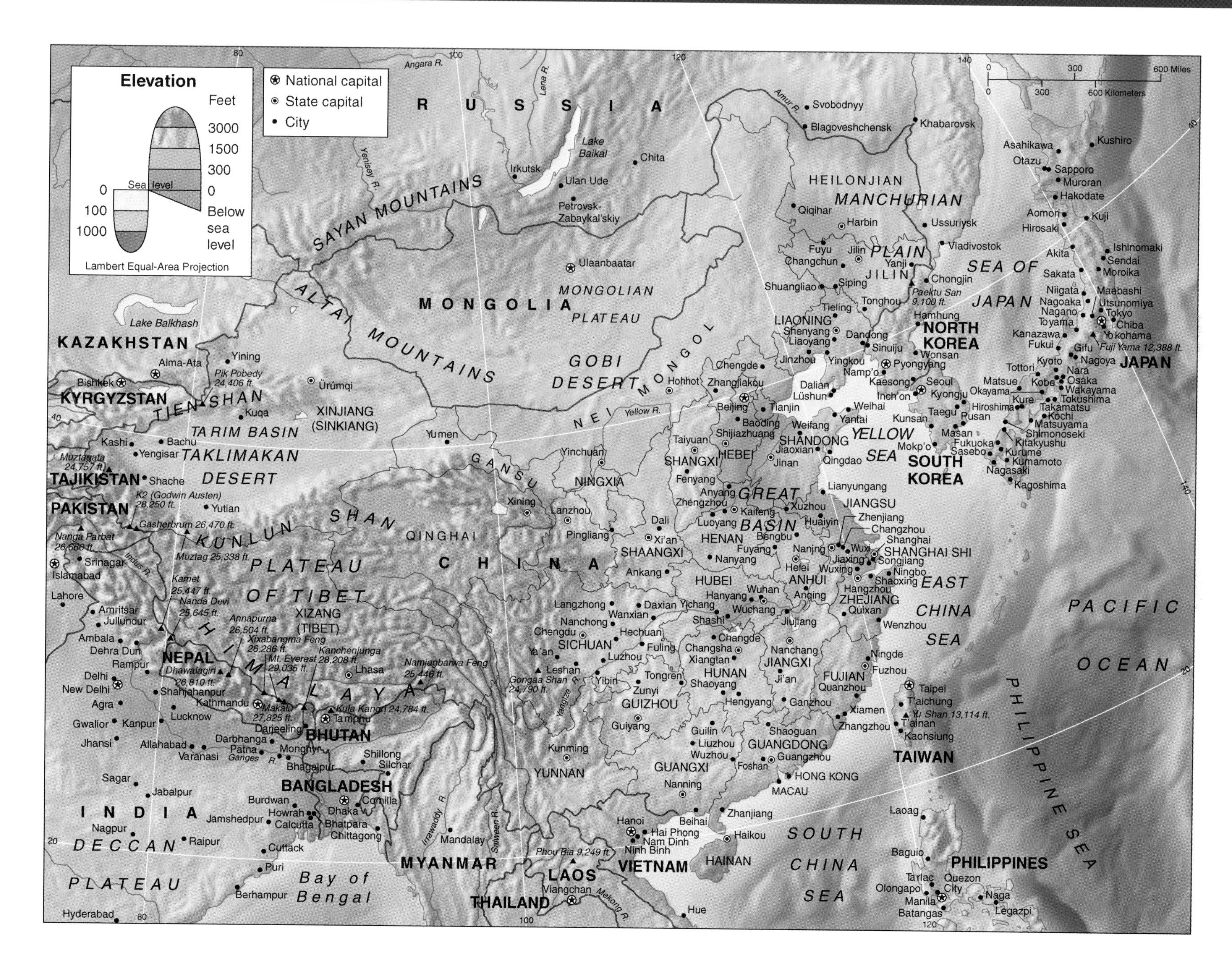
Elevation
Feet
3000
1500
300
0
Sea level
Below sea level
0
100
1000
Lambert Equal-Area Projection
National capital
State capital
City
0
300
600 Miles
0
300
600 Kilometers
R U S S I A
Lake Baikal
Irkutsk
Ulan Ude
Chita
Petrovsk-Zabaykal'skiy
Angara R.
Yenisey R.
Lena R.
Amur R.
Svobodnyy
Blagoveshchensk
Khabarovsk
SAYAN MOUNTAINS
ALTAI MOUNTAINS
M O N G O L I A
Ulaanbaatar
MONGOLIAN PLATEAU
GOBI DESERT
NEI MONGOL
KAZAKHSTAN
Lake Balkhash
Alma-Ata
Yining
Pik Pobedy 24,406 ft.
Ürümqi
XINJIANG (SINKIANG)
Bishkek
KYRGYZSTAN
TIEN SHAN
Kuqa
TARIM BASIN
Kashi
Bachu
Yengisar
TAKLIMAKAN DESERT
Muztagata 24,757 ft.
TAJIKISTAN
Shache
K2 (Godwin Austen) 28,250 ft.
Yutian
PAKISTAN
Gasherbrum 26,470 ft.
Nanga Parbat 26,660 ft.
Srinagar
Islamabad
Lahore
Indus R.
Muztag 25,338 ft.
KUNLUN SHAN
PLATEAU OF TIBET
QINGHAI
C H I N A
XIZANG (TIBET)
Kamet 25,447 ft.
Nanda Devi 25,645 ft.
Annapurna 26,504 ft.
Xixabangma Feng 26,286 ft.
Mt. Everest 29,035 ft.
Kanchenjunga 28,208 ft.
Dhawalagiri 26,810 ft.
Makalu 27,825 ft.
Kula Kangri 24,784 ft.
Namjagbarwa Feng 25,446 ft.
H I M A L A Y A
NEPAL
Kathmandu
Lhasa
Darjeeling
BHUTAN
Tamphu
Amritsar
Jullundur
Ambala
Dehra Dun
Rampur
Delhi
New Delhi
Shahjahanpur
Agra
Lucknow
Gwalior
Kanpur
Jhansi
Allahabad
Varanasi
Darbhanga
Patna
Ganges R.
Monghyr
Bhagalpur
Shillong
Silchar
BANGLADESH
Burdwan
Howrah
Calcutta
Dhaka
Comilla
Bhatpara
Chittagong
Sagar
Jabalpur
I N D I A
Jamshedpur
Nagpur
Raipur
Cuttack
Puri
DECCAN PLATEAU
Berhampur
Bay of Bengal
Hyderabad
MYANMAR
Irrawaddy R.
Mandalay
Salween R.
THAILAND
LAOS
Phou Bia 9,249 ft.
Viangchan
Mekong R.
VIETNAM
Hanoi
Hai Phong
Nam Dinh
Ninh Binh
Hue
Yumen
GANSU
Xining
Lanzhou
Yinchuan
NINGXIA
Pingliang
Dali
Xi'an
SHAANGXI
Ankang
Langzhong
Nanchong
Chengdu
SICHUAN
Ya'an
Leshan
Gongaa Shan 24,790 ft.
Yangtze R.
Yibin
Luzhou
Daxian
Wanxian
Hechuan
Fuling
Zunyi
GUIZHOU
Guiyang
Kunming
YUNNAN
Tongren
Shaoyang
Changsha
Xiangtan
HUNAN
Hengyang
Changde
Yichang
Shashi
Hanyang
Wuhan
Wuchang
HUBEI
Nanyang
HENAN
Luoyang
Zhengzhou
Kaifeng
Anyang
GREAT BASIN
Taiyuan
SHANGXI
Fenyang
Shijiazhuang
HEBEI
Baoding
Beijing
Tianjin
Zhangjiakou
Chengde
Hohhot
Yellow R.
Jinan
SHANDONG
Jiaoxian
Weifang
Qingdao
Yantai
Weihai
YELLOW SEA
Lianyungang
JIANGSU
Xuzhou
Huaiyin
Bengbu
Fuyang
Hefei
ANHUI
Anqing
Nanjing
Zhenjiang
Changzhou
Wuxi
Shanghai
SHANGHAI SHI
Songjiang
Jiaxing
Wuxing
Hangzhou
Shaoxing
Ningbo
ZHEJIANG
Quixan
Wenzhou
Jiujiang
Nanchang
JIANGXI
Ji'an
Ganzhou
FUJIAN
Ningde
Fuzhou
Quanzhou
Xiamen
Zhangzhou
GUANGDONG
Shaoguan
Guangzhou
Foshan
HONG KONG
MACAU
GUANGXI
Guilin
Liuzhou
Wuzhou
Nanning
Beihai
Zhanjiang
Haikou
HAINAN
SOUTH CHINA SEA
EAST CHINA SEA
TAIWAN
Taipei
T'aichung
Yu Shan 13,114 ft.
T'ainan
Kaohsiung
PHILIPPINES
Laoag
Baguio
Tarlac
Quezon City
Olongapo
Manila
Batangas
Naga
Legazpi
PHILIPPINE SEA
PACIFIC OCEAN
HEILONJIAN
MANCHURIAN PLAIN
Qiqihar
Harbin
Fuyu
Jilin
Changchun
JILIN
Shuangliao
Siping
Tieling
Tonghou
Yanji
LIAONING
Shenyang
Liaoyang
Dandong
Jinzhou
Yingkou
Dalian
Lüshun
Ussuriysk
Vladivostok
Chongjin
Paektu San 9,100 ft.
Hamhung
NORTH KOREA
Wonsan
Sinuiju
Pyongyang
Namp'o
Kaesong
Seoul
Inch'on
Kyongju
Kunsan
Taegu
Pusan
Masan
Mokp'o
SOUTH KOREA
SEA OF JAPAN
JAPAN
Asahikawa
Kushiro
Otazu
Sapporo
Muroran
Hakodate
Aomori
Kuji
Hirosaki
Akita
Ishinomaki
Sendai
Moroika
Sakata
Niigata
Maebashi
Nagoaka
Utsunomiya
Nagano
Tokyo
Toyama
Chiba
Kanazawa
Yokohama
Fukui
Gifu
Fuji Yama 12,388 ft.
Tottori
Kyoto
Nagoya
Nara
Matsue
Kobe
Osaka
Okayama
Wakayama
Kure
Tokushima
Hiroshima
Takamatsu
Kochi
Matsuyama
Shimonoseki
Fukuoka
Kitakyushu
Sasebo
Kurume
Kumamoto
Nagasaki
Kagoshima
40
20
80
100
120
140

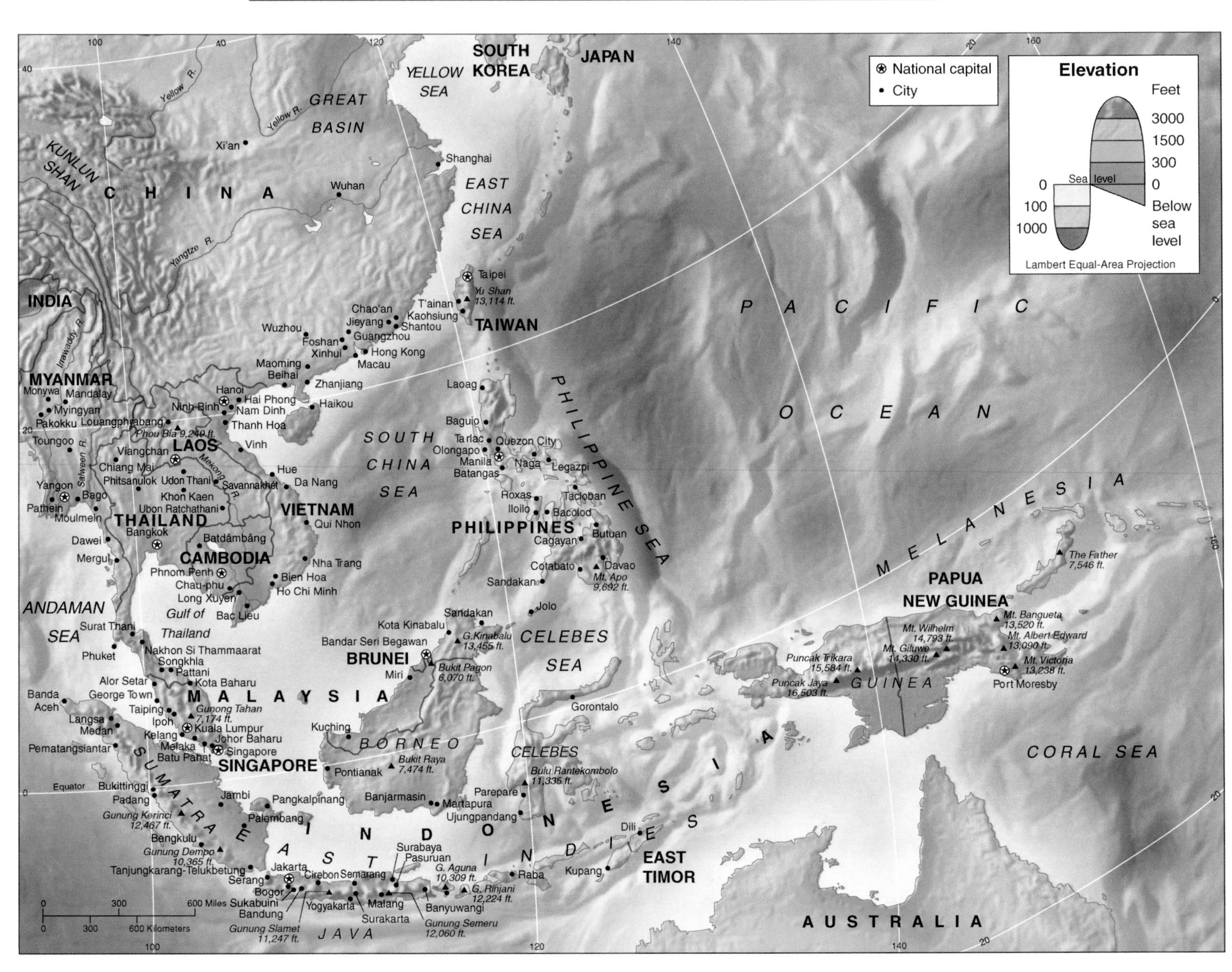
National capital
City
Elevation
Feet
3000
1500
300
0
Below sea level
Sea level
0
100
1000
Lambert Equal-Area Projection
CHINA
INDIA
MYANMAR
THAILAND
LAOS
CAMBODIA
VIETNAM
MALAYSIA
SINGAPORE
BRUNEI
PHILIPPINES
TAIWAN
SOUTH KOREA
JAPAN
INDONESIA
EAST TIMOR
PAPUA NEW GUINEA
AUSTRALIA
KUNLUN SHAN
GREAT BASIN
YELLOW SEA
EAST CHINA SEA
SOUTH CHINA SEA
PHILIPPINE SEA
PACIFIC OCEAN
MELANESIA
CELEBES SEA
CORAL SEA
ANDAMAN SEA
Gulf of Thailand
BORNEO
CELEBES
SUMATRA
JAVA
GUINEA
Yellow R.
Yangtze R.
Irrawaddy R.
Salween R.
Mekong R.
Equator
Xi'an
Wuhan
Shanghai
Taipei
Yu Shan 13,114 ft.
T'ainan
Kaohsiung
Chao'an
Jieyang
Shantou
Guangzhou
Wuzhou
Foshan
Xinhui
Hong Kong
Macau
Maoming
Beihai
Zhanjiang
Haikou
Hanoi
Hai Phong
Nam Dinh
Ninh Binh
Thanh Hoa
Vinh
Hue
Da Nang
Qui Nhon
Nha Trang
Bien Hoa
Ho Chi Minh
Bac Lieu
Long Xuyen
Chau-phu
Phnom Penh
Batdâmbâng
Monywa
Mandalay
Myingyan
Pakokku
Louangphrabang
Phou Bia 9,240 ft.
Toungoo
Viangchan
Chiang Mai
Phitsanulok
Udon Thani
Savannakhet
Khon Kaen
Ubon Ratchathani
Yangon
Bago
Pathein
Moulmein
Bangkok
Dawei
Mergui
Surat Thani
Phuket
Nakhon Si Thammaraat
Songkhla
Pattani
Kota Baharu
Alor Setar
George Town
Taiping
Ipoh
Gunong Tahan 7,174 ft.
Kuala Lumpur
Kelang
Johor Baharu
Melaka
Batu Pahat
Singapore
Banda Aceh
Langsa
Medan
Pematangsiantar
Bukittinggi
Padang
Gunung Kerinci 12,467 ft.
Bengkulu
Gunung Dempo 10,365 ft.
Tanjungkarang-Telukbetung
Jambi
Palembang
Pangkalpinang
Serang
Jakarta
Bogor
Sukabumi
Bandung
Gunung Slamet 11,247 ft.
Cirebon
Semarang
Yogyakarta
Surakarta
Malang
Surabaya
Pasuruan
G. Aguna 10,309 ft.
Banyuwangi
Gunung Semeru 12,060 ft.
G. Rinjani 12,224 ft.
Raba
Kupang
Dili
Kuching
Pontianak
Bukit Raya 7,474 ft.
Banjarmasin
Martapura
Miri
Bandar Seri Begawan
Bukit Pagon 6,070 ft.
Kota Kinabalu
G. Kinabalu 13,455 ft.
Sandakan
Parepare
Ujungpandang
Bulu Rantekombolo 11,335 ft.
Gorontalo
Laoag
Baguio
Tarlac
Olongapo
Quezon City
Manila
Batangas
Naga
Legazpi
Roxas
Iloilo
Bacolod
Tacloban
Cagayan
Butuan
Cotabato
Davao
Mt. Apo 9,692 ft.
Jolo
The Father 7,546 ft.
Mt. Banguela 13,520 ft.
Mt. Albert Edward 13,090 ft.
Mt. Victoria 13,238 ft.
Port Moresby
Mt. Wilhelm 14,793 ft.
Mt. Giluwe 14,330 ft.
Puncak Trikara 15,584 ft.
Puncak Jaya 16,503 ft.
0
300
600 Miles
0
300
600 Kilometers
100
120
140
160
40
20
0

Map 133 Australasia and Oceania: Political Divisions

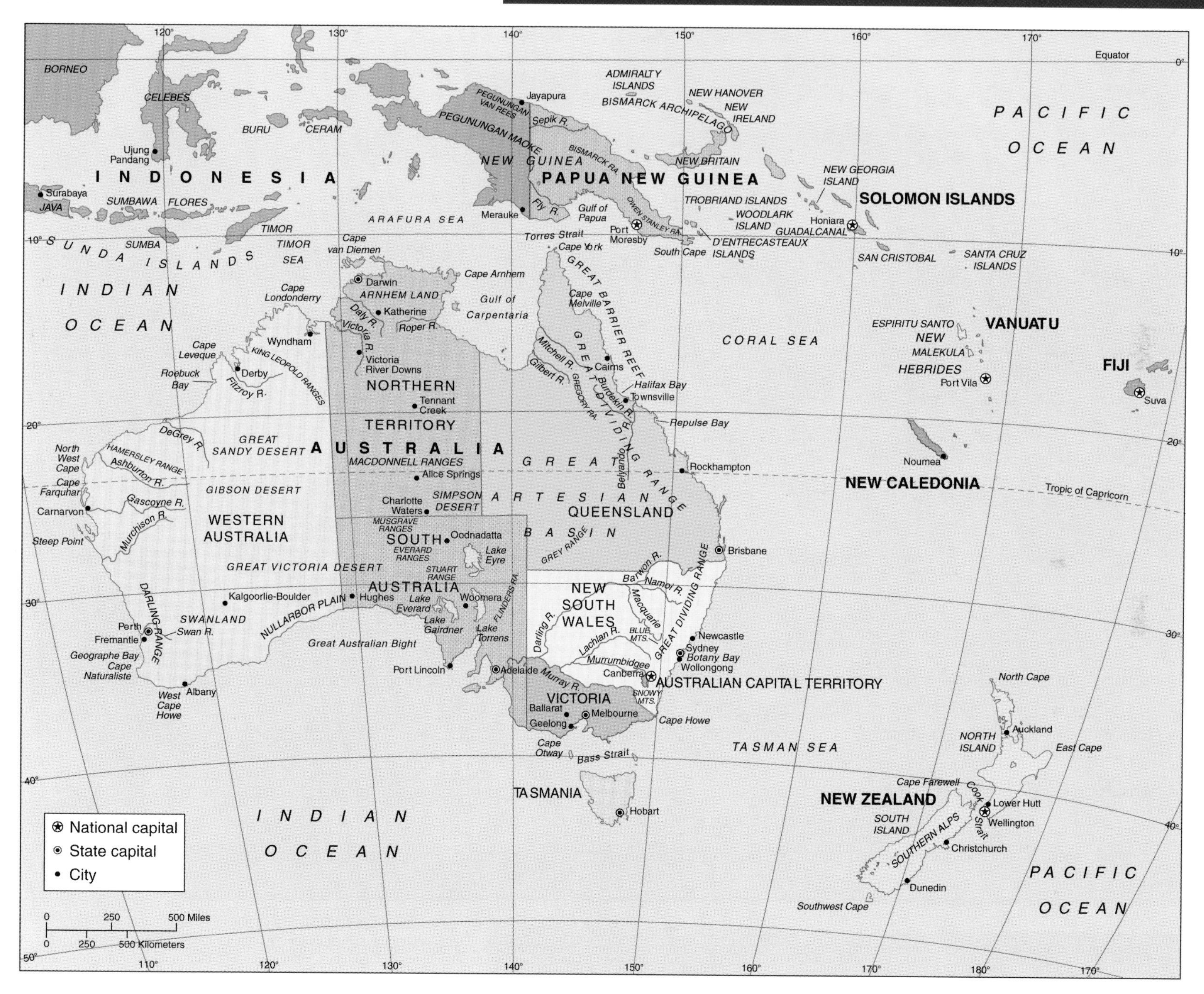

Map 134 Australasia and Oceania: Physical Features

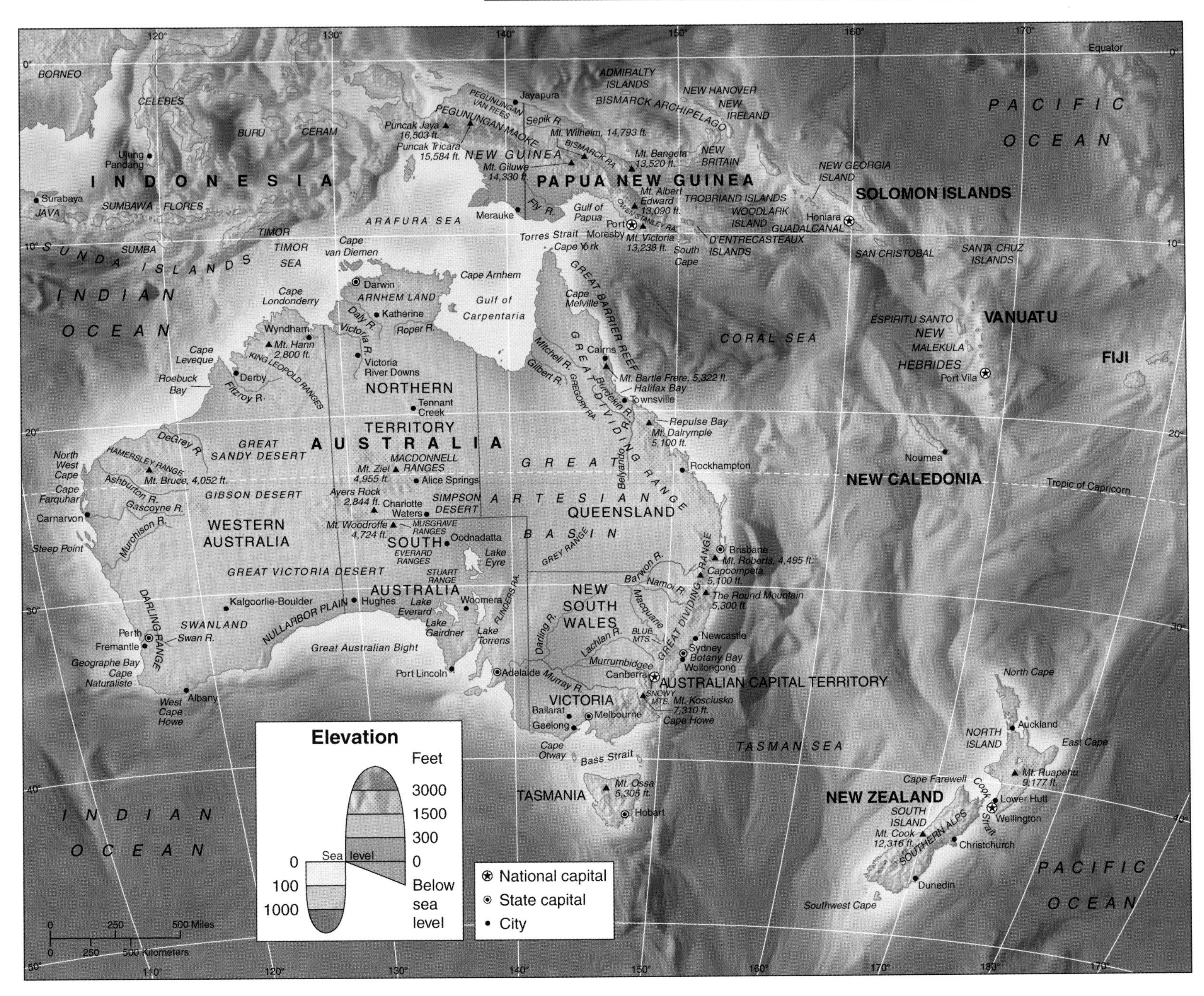

Map 135a Environment and Economy

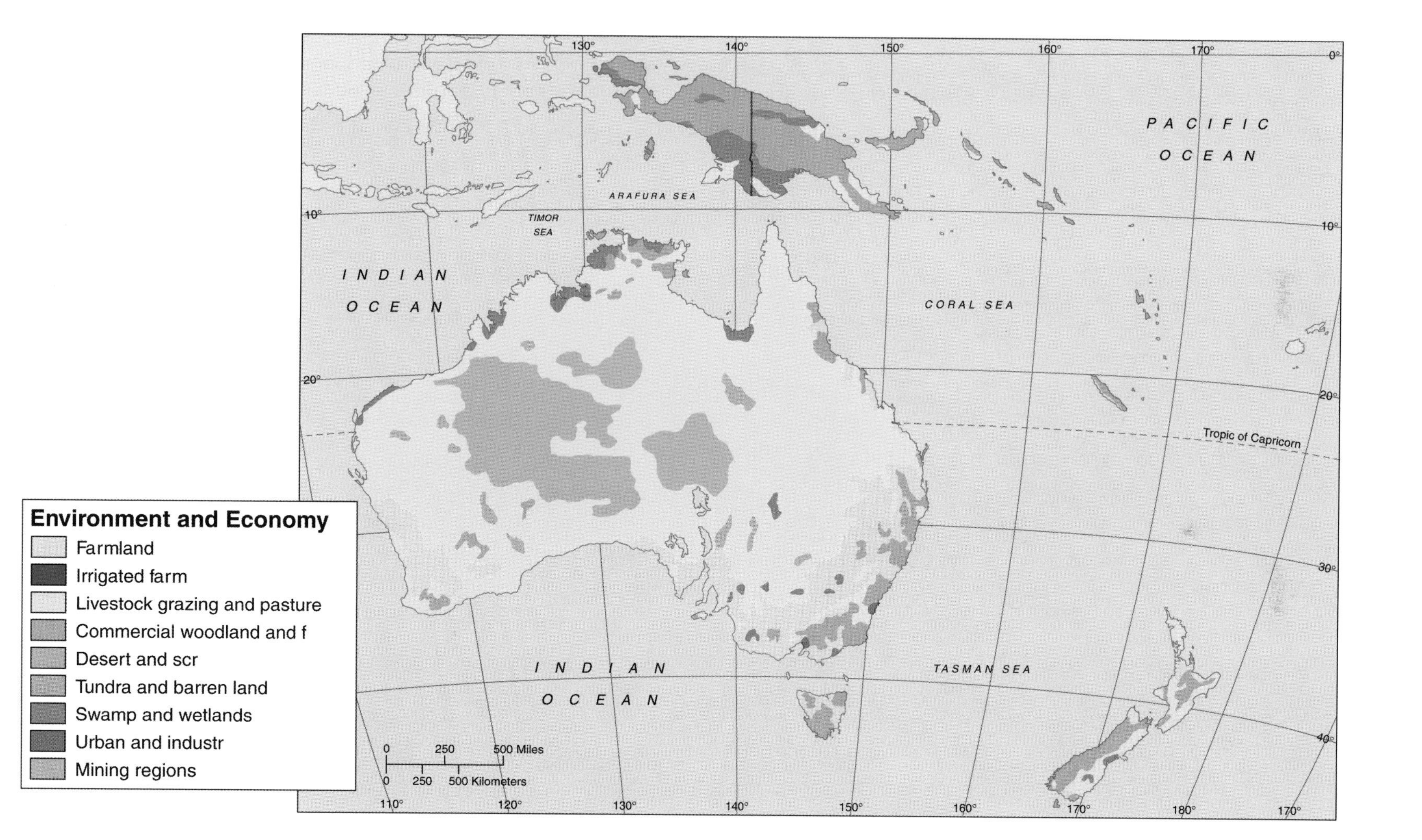

Australasia is dominated by the world's smallest and most uniform continent. Flat, dry and mostly hot, Australia has the simplest of land use patterns: where rainfall exists so does agricultural activity. Two agricultural patterns dominate the map: livestock grazing, primarily sheep, and wheat farming, although some sugar cane production exists in the north and some cotton is grown elsewhere. Only about 6 percent of the continent consists of arable land so the areas of wheat farming, dominant as they may be in the context of Australian agriculture, are small. Australia also supports a healthy mineral resource economy, with iron and copper and precious metals making up the bulk of the extraction. Elsewhere in the region, tropical forests dominate Papua New Guinea, with some subsistence agriculture and livestock. New Zealand's temperate climate with abundant precipitation supports a productive livestock industry and little else besides tourism—which is an important economic element throughout the remainder of the region as well.

Map 135b Population Density

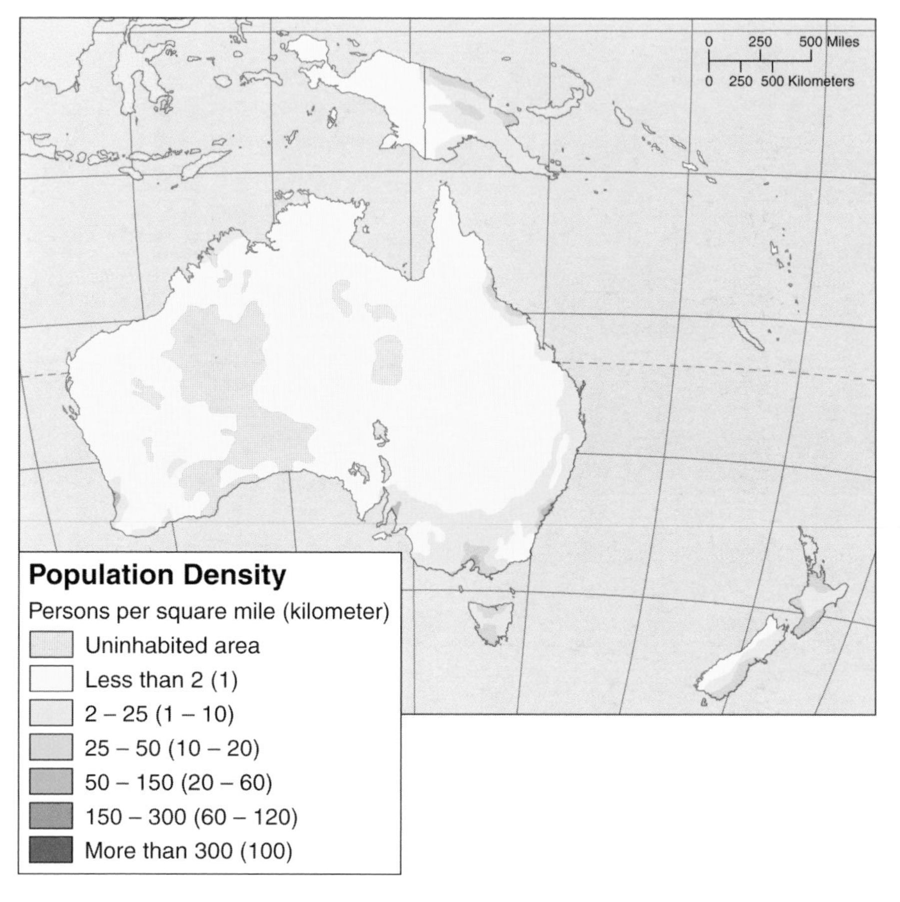

The region's small population is remarkably diverse. To the north are the Melanesian New Guinea peoples, while Europeans dominate the populations of Australia and New Zealand, both of which have significant indigenous populations. Throughout most of the smaller island groups, the bulk of the population is Melanesian but with a scattering of Europeans. The distribution of population is extremely uneven with New Guinea, southeastern Australia, and New Zealand supporting the bulk of the region's people while the remainder of the region—meaning nearly all of Australia—is either sparsely populated or devoid of population altogether. Nowhere do population densities reach the levels they do in other major regions of the world and densities of 50 persons per square mile are the highest to be found. Population location is dependent on precipitation and population growth patterns are culturally variable.

Map 135c Climate Patterns

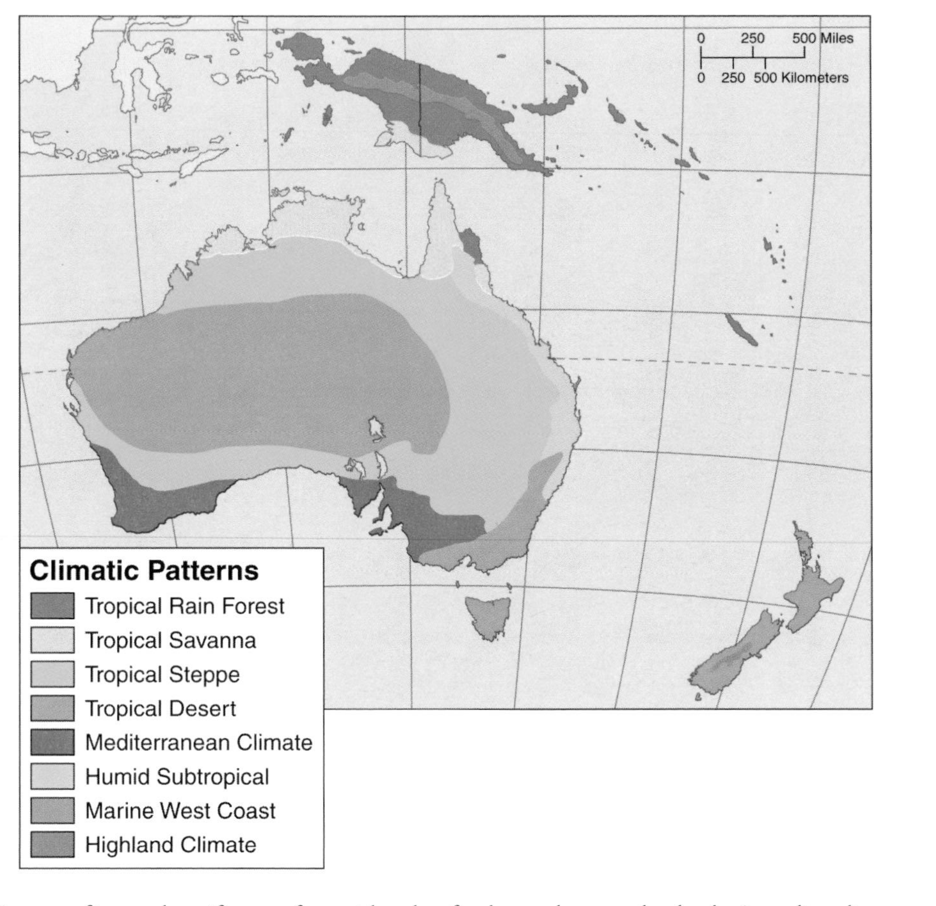

Because of its nearly uniform surface, with only a few low and scattered uplands, Australian climate is a consequence only of the two great climate controls—latitudinal position and location relative to continental margin and interior. The continent bestrides the 30th parallel of latitude and its climatic pattern is dominated by the subtropical high pressure system with dry air masses that are responsible for the existence of great deserts. Toward the equator, the desert grades into steppe, savanna, and tropical forest as the subtropical high gives way to equatorial low pressure and abundant precipitation. Toward the pole, arid land fades into more well-watered steppe grasslands, the Mediterranean type climate of the southern margins of the continent, and the marine west coast climate of the Australian southeast and New Zealand. This latter climate is where most of the region's people live.

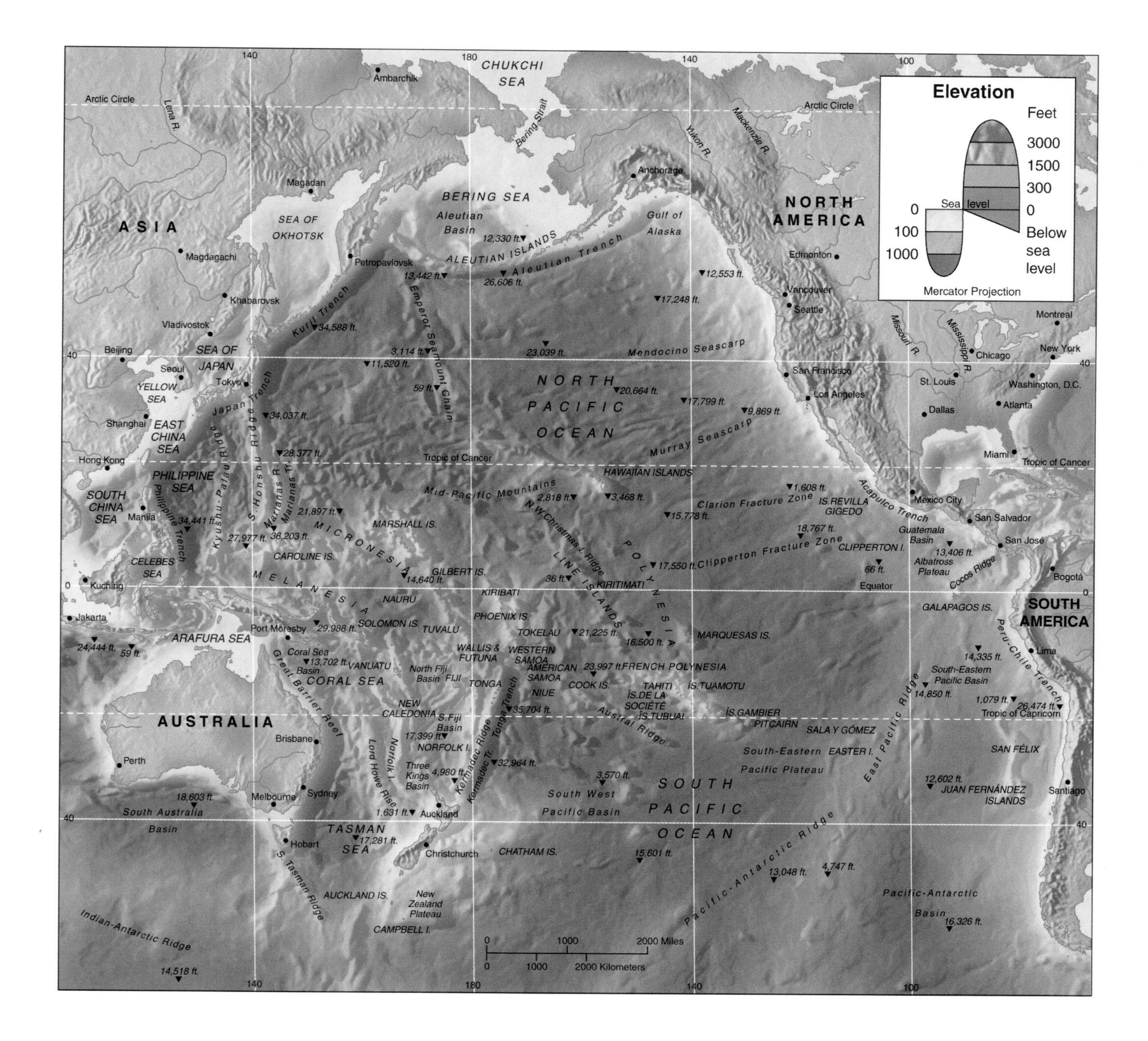
Elevation
Feet
3000
1500
300
0
Sea level
0
100
1000
Below sea level
Mercator Projection
ASIA
NORTH AMERICA
SOUTH AMERICA
AUSTRALIA
NORTH PACIFIC OCEAN
SOUTH PACIFIC OCEAN
CHUKCHI SEA
BERING SEA
SEA OF OKHOTSK
SEA OF JAPAN
YELLOW SEA
EAST CHINA SEA
PHILIPPINE SEA
SOUTH CHINA SEA
CELEBES SEA
ARAFURA SEA
CORAL SEA
TASMAN SEA
Gulf of Alaska
Arctic Circle
Tropic of Cancer
Tropic of Capricorn
Equator
Aleutian Trench
Kuril Trench
Japan Trench
Emperor Seamount Chain
Mid-Pacific Mountains
Mendocino Seascarp
Murray Seascarp
Clarion Fracture Zone
Clipperton Fracture Zone
Acapulco Trench
Peru-Chile Trench
East Pacific Ridge
Pacific-Antarctic Ridge
Indian-Antarctic Ridge
Tonga Trench
Kermadec Tr.
Great Barrier Reef
MICRONESIA
MELANESIA
POLYNESIA
HAWAIIAN ISLANDS
ALEUTIAN ISLANDS
0 1000 2000 Miles
0 1000 2000 Kilometers

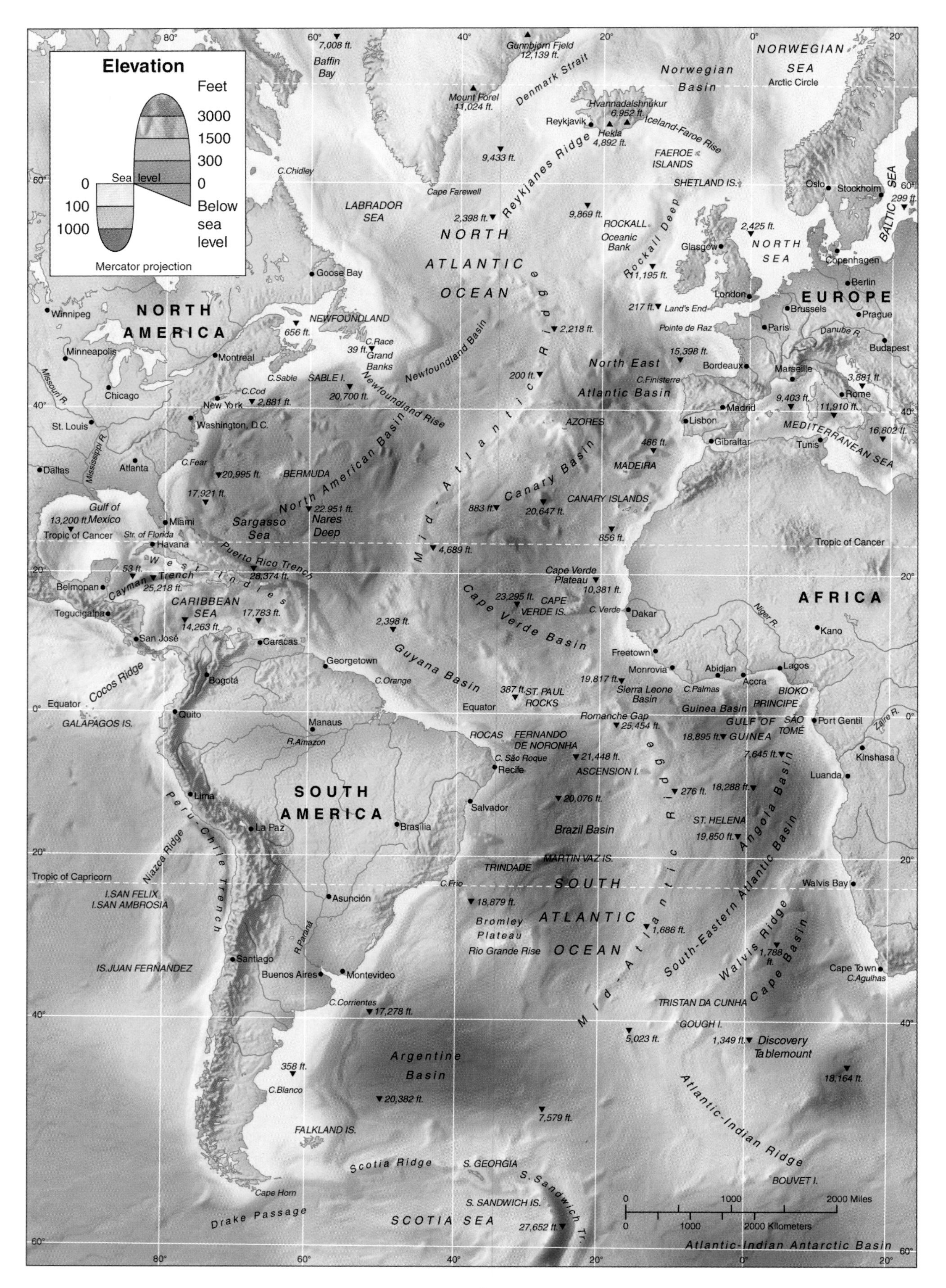
Elevation
Feet
3000
1500
300
0
Sea level
Below sea level
0
100
1000
Mercator projection
80°
60°
40°
20°
0°
7,008 ft.
Baffin Bay
Gunnbjørn Fjeld 12,139 ft.
Denmark Strait
Mount Forel 11,024 ft.
NORWEGIAN SEA
Arctic Circle
Norwegian Basin
Hvannadalshnúkur 6,952 ft.
Reykjavík
Hekla 4,892 ft.
Iceland-Faroe Rise
Reykjanes Ridge
9,433 ft.
FAEROE ISLANDS
SHETLAND IS.
C.Chidley
Cape Farewell
LABRADOR SEA
2,398 ft.
9,869 ft.
ROCKALL
Oceanic Bank
Rockall Deep
Oslo
Stockholm
BALTIC SEA
299 ft.
2,425 ft.
NORTH SEA
Glasgow
Copenhagen
NORTH ATLANTIC OCEAN
Goose Bay
11,195 ft.
Berlin
London
EUROPE
Winnipeg
NORTH AMERICA
NEWFOUNDLAND
656 ft.
217 ft.
Land's End
Brussels
Prague
2,218 ft.
Pointe de Raz
Paris
Danube R.
C.Race
39 ft.
Grand Banks
Newfoundland Basin
Budapest
Minneapolis
Montreal
15,398 ft.
Bordeaux
North East Atlantic Basin
Marseille
Missouri R.
C.Sable
SABLE I.
200 ft.
C.Finisterre
3,881 ft.
Chicago
C.Cod
New York
2,881 ft.
20,700 ft.
Newfoundland Rise
Rome
9,403 ft.
11,910 ft.
Madrid
Mid-Atlantic Ridge
St. Louis
Washington, D.C.
AZORES
Lisbon
MEDITERRANEAN SEA
16,802 ft.
Tunis
Gibraltar
Mississippi R.
486 ft.
C.Fear
Atlanta
Dallas
20,995 ft.
BERMUDA
North American Basin
MADEIRA
Canary Basin
17,921 ft.
CANARY ISLANDS
883 ft.
20,647 ft.
Gulf of Mexico
13,200 ft.
Miami
22,951 ft.
Nares Deep
Sargasso Sea
Tropic of Cancer
Str. of Florida
Havana
856 ft.
Puerto Rico Trench
4,689 ft.
West Indies
Cape Verde Plateau
53 ft.
Cayman Trench
28,374 ft.
10,381 ft.
25,218 ft.
Belmopan
AFRICA
CARIBBEAN SEA
23,295 ft.
CAPE VERDE IS.
C. Verde
Dakar
Tegucigalpa
17,783 ft.
Cape Verde Basin
14,263 ft.
Niger R.
San José
2,398 ft.
Caracas
Kano
Freetown
Georgetown
Guyana Basin
Monrovia
Abidjan
Lagos
19,817 ft.
Cocos Ridge
Bogotá
C.Orange
387 ft.
ST. PAUL ROCKS
Sierra Leone Basin
C.Palmas
Accra
BIOKO
Equator
PRINCIPE
Guinea Basin
Quito
GALAPAGOS IS.
Manaus
Romanche Gap
25,454 ft.
GULF OF GUINEA
SÃO TOMÉ
Port Gentil
Zaire R.
R.Amazon
ROCAS
FERNANDO DE NORONHA
18,895 ft.
C. São Roque
21,448 ft.
7,645 ft.
Kinshasa
Recife
ASCENSION I.
Luanda
SOUTH AMERICA
276 ft.
18,288 ft.
Lima
Salvador
20,076 ft.
Angola Basin
Peru-Chile Trench
ST. HELENA
La Paz
Brasília
Brazil Basin
19,850 ft.
Nazca Ridge
MARTIN VAZ IS.
TRINDADE
Tropic of Capricorn
C.Frio
SOUTH ATLANTIC OCEAN
Walvis Bay
I.SAN FELIX
I.SAN AMBROSIA
Asunción
18,879 ft.
South-Eastern Atlantic Basin
Bromley Plateau
1,686 ft.
R.Paraná
Rio Grande Rise
Walvis Ridge
1,788 ft.
Cape Basin
Santiago
IS.JUAN FERNANDEZ
Buenos Aires
Montevideo
Cape Town
C.Agulhas
C.Corrientes
TRISTAN DA CUNHA
17,278 ft.
GOUGH I.
5,023 ft.
1,349 ft.
Discovery Tablemount
Argentine Basin
358 ft.
18,164 ft.
C.Blanco
20,382 ft.
Atlantic-Indian Ridge
7,579 ft.
FALKLAND IS.
Scotia Ridge
S. GEORGIA
S. Sandwich Tr.
BOUVET I.
Cape Horn
S. SANDWICH IS.
1000
2000 Miles
Drake Passage
SCOTIA SEA
27,652 ft.
2000 Kilometers
Atlantic-Indian Antarctic Basin

Map 138 The Indian Ocean

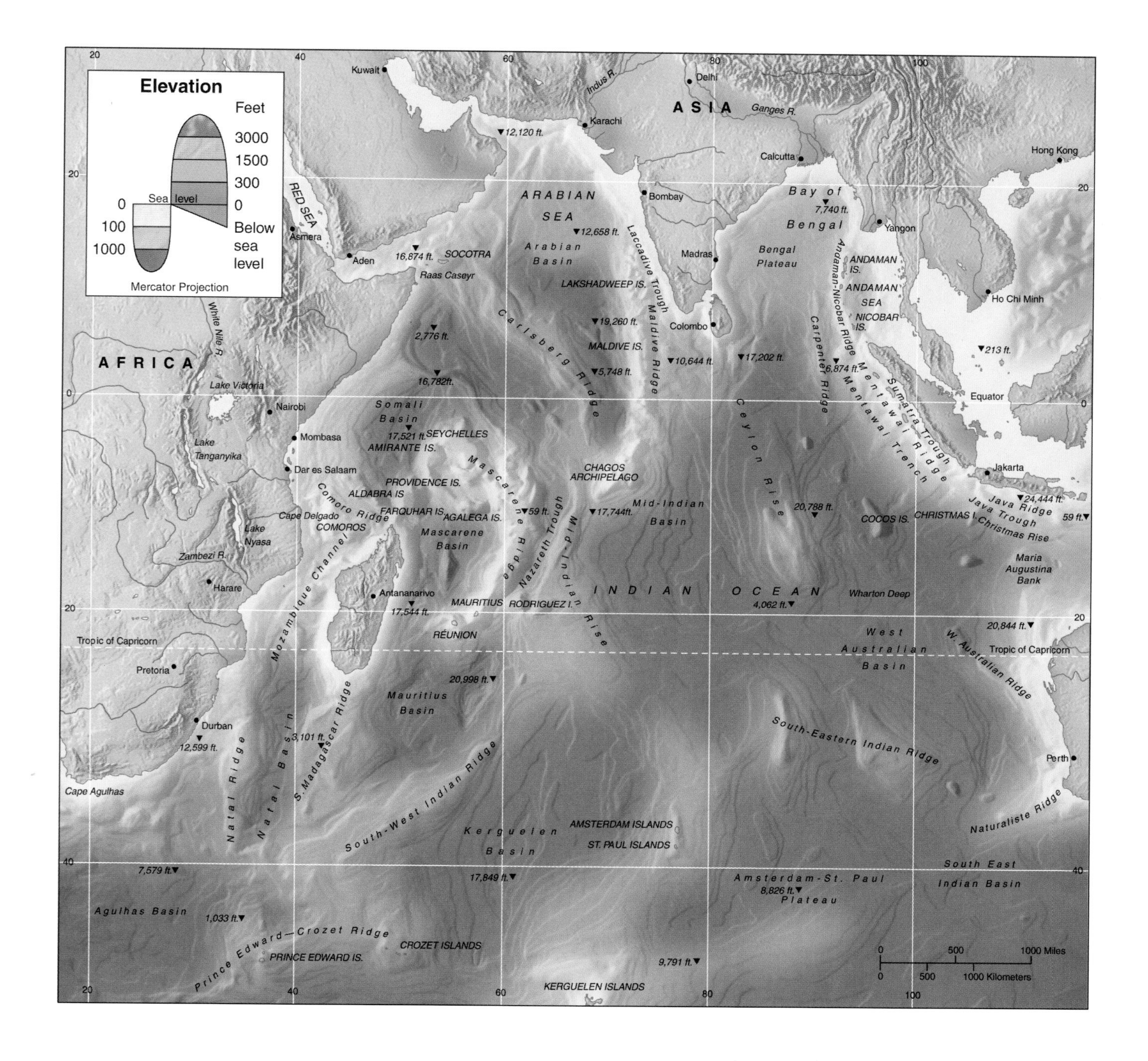

Map 139 The Arctic

Map 140 Antarctica

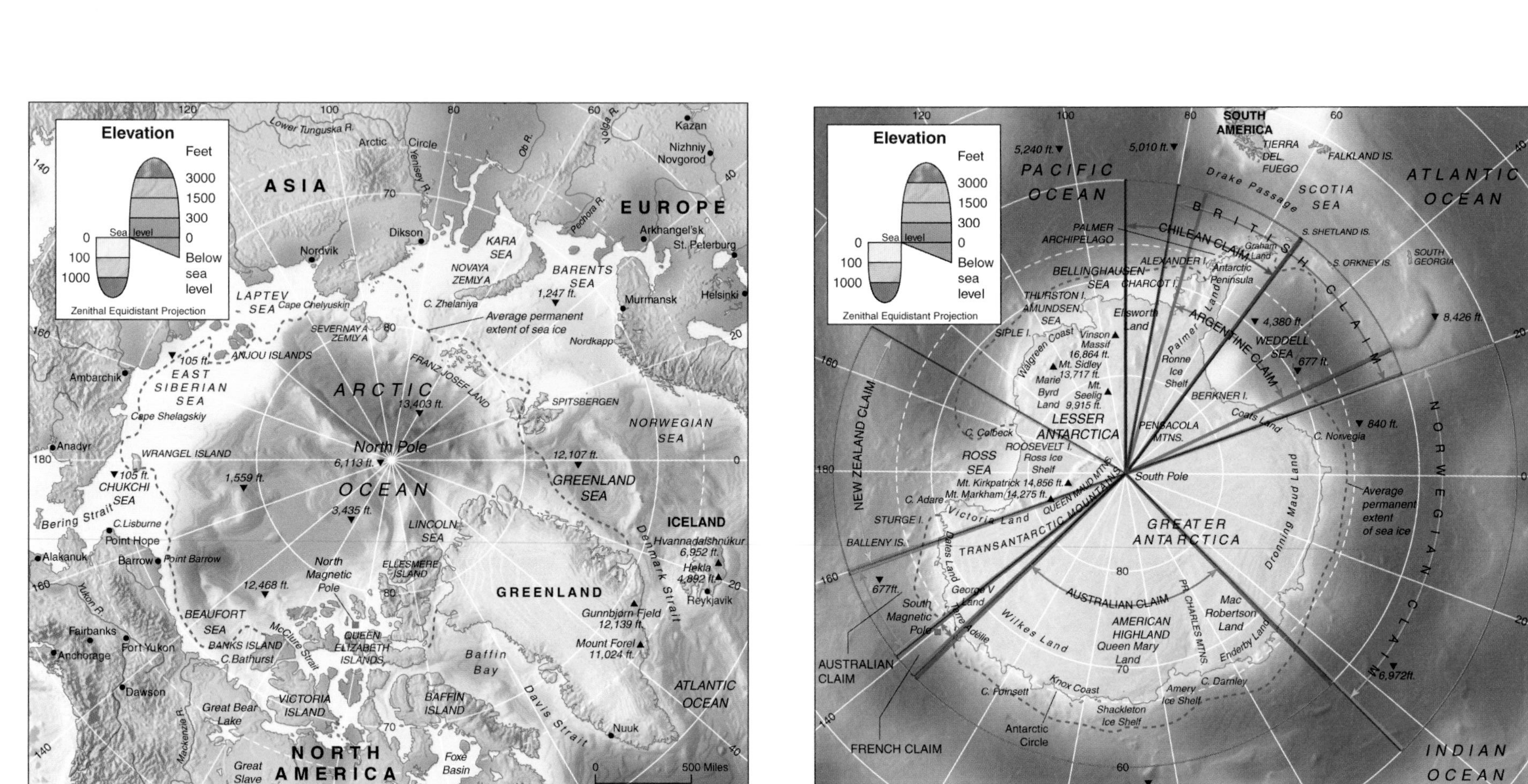

Part VIII

Tables

Table A	World Countries: Area, Population, and Population Density, 2003
Table B	World Countries: Form of Government, Capital City, Major Languages
Table C	World Countries: Basic Economic Indicators
Table D	World Countries: Population Growth, 1950–2025
Table E	World Countries: Basic Demographic Data, 1975–2001
Table F	World Countries: Mortality, Health, and Nutrition
Table G	World Countries: Education and Literacy, 1990–2002
Table H	World Countries: Agricultural Operations, 1996–2002
Table I	World Countries: Land Use and Deforestation, 1980–2000
Table J	World Countries: Energy Production and Use, 1980–2000
Table K	World Countries: Energy Efficiency and Emission, 1980–1999
Table L	World Countries: Power and Transportation
Table M	World Countries: Communications, Information, and Science and Technology
Table N	World Countries: Water Resources
Table O	World Countries: Globally Threatened Plant and Animal Species

Table A
World Countries: Area, Population, and Population Density, 2003

COUNTRY	AREA		POPULATION (in thousands)	DENSITY	
	(Mi²)	(Km²)	(2003[a])	(Pop/Mi²)	(Pop/Km²)
Afghanistan	251,826	652,229	23,897	95	37
Albania	11,100	28,749	3,166	285	110
Algeria	919,595	2,381,750	31,800	35	13
Andorra	175	453	68	389	150
Angola	481,354	1,246,706	13,625	28	11
Antigua and Barbuda	171	443	67	392	151
Argentina	1,073,400	2,780,104	38,428	36	14
Armenia	11,506	29,801	3,061	266	103
Australia	2,966,155	7,682,337	19,731	7	3
Austria	32,377	83,856	8,116	251	97
Azerbaijan	33,436	86,599	8,370	250	97
Bahamas	5,382	13,939	314	58	23
Bahrain	267	692	724	2,712	1,046
Bangladesh	55,598	143,999	146,7316	2,639	1,019
Barbados	166	430	270	1,627	628
Belarus	80,155	207,601	9,895	123	48
Belgium	11,783	30,518	10,318	876	338
Belize	8,866	22,963	256	29	11
Benin	43,475	112,600	6,736	155	60
Bhutan	18,200	47,138	2,257	124	48
Bolivia	424,165	1,098,587	9	0	0
Bosnia and Herzegovina	19,776	51,233	4,161	210	81
Botswana	231,803	600,362	1,785	8	3
Brazil	3,286,488	8,511,999	178,470	54	21
Bruneli Darussalam	2,228	5,770	358	161	62
Bulgaria	42,823	110,912	7,897	184	71
Burkina Faso	105,869	274,201	13,002	123	47
Burundi	10,745	27,830	6,825	635	245
Cambodia	69,898	181,036	14,144	202	78
Cameroon	183,569	475,443	16,018	87	34
Canada	3,849,674	9,970,650	31,510	8	3
Cape Verde	1,557	4,033	463	297	115
Central African Republic	240,535	622,985	3,865	16	6
Chad	495,755	1,284,005	8,598	17	7
Chile	292,259	756,950	15,805	54	21
China	3,705,392	9,596,960	1,304,196	352	136
Hong Kong, China	422	1,092	7,049	16,704	6,455
Colombia	439,734	1,138,910	44,222	101	39
Comoros	838	2,170	768	916	354
Democratic Republic of the Congo (formerly Zaire)	905,564	2,345,410	52,771	58	22
Congo Republic	132,047	342,002	3,724	28	11
Costa Rica	19,730	51,101	4,173	212	82
Côte d'Ivoire	124,502	322,460	16,631	134	52

Table A *(continued)*
World Countries: Area, Population, and Population Density, 2003

COUNTRY	AREA		POPULATION (in thousands)	DENSITY	
	(Mi^2)	(Km^2)	(2003[a])	(Pop/Mi^2)	(Pop/Km^2)
Croatia	21,824	56,538	4,428	203	78
Cuba	42,804	110,862	11,300	264	102
Cyprus	3,571	9,250	802	225	87
Czech Republic	30,387	78,703	10,236	337	130
Denmark	16,629	43,070	5,364	323	125
Djibouti	8,494	22,000	703	83	32
Dominica	290	750	70	241	93
Dominican Republic	18,815	48,730	8,745	465	179
Ecuador	109,484	283,563	13,003	119	46
Egypt	386,662	1,001,454	72	0	0
El Salvador	8,124	21,041	6,515	802	310
Equatorial Guinea	10,831	28,052	494	46	18
Eritrea	46,842	121,320	4,141	88	34
Estonia	17,413	45,100	1,323	76	29
Ethiopia	435,184	1,127,127	70,678	162	63
Fiji	7,054	18,270	839	119	46
Finland	130,127	337,030	5,207	40	15
France	176,460	547,030	60,144	341	110
Gabon	103,347	267,669	1,329	13	5
Gambia	4,363	11,300	1,426	327	126
Georgia	26,911	69,699	5,126	190	74
Germany	137,803	356,910	82,476	599	231
Ghana	92,098	238,534	20,922	227	88
Greece	50,942	131,940	10,976	215	83
Grenada	131	340	89	679	262
Guatemala	42,042	108,889	12,347	294	113
Guinea	94,926	245,858	8,480	89	34
Guinea-Bissau	13,948	36,125	1,493	107	41
Guyana	83,000	214,970	765	9	4
Haiti	10,714	27,749	8,326	777	300
Honduras	43,277	112,087	7	0	0
Hungary	35,920	93,033	9,877	275	106
Iceland	39,768	103,000	290	7	3
India	1,269,340	3,287,590	1,065,462	839	324
Indonesia	741,097	1,919,440	219,883	297	115
Iran	636,294	1,648,000	68,920	108	42
Iraq	168,754	437,072	25,175	149	58
Ireland	27,137	70,285	3,956	146	56
Israel[b]	8,019	20,769	6,433	802	310
Italy	116,305	301,230	57,423	494	191
Jamaica	4,244	10,992	2,651	625	241
Japan	145,882	377,835	127,654	875	338
Jordan	35,445	89,213	5,473	154	61
Kazakhstan	1,049,156	2,717,313	15,433	15	6

Table A *(continued)*
World Countries: Area, Population, and Population Density, 2003

COUNTRY	AREA		POPULATION (in thousands)	DENSITY	
	(Mi²)	(Km²)	(2003[a])	(Pop/Mi²)	(Pop/Km²)
Kenya	224,961	582,649	31,987	142	55
Kiribati	277	717	96	347	134
Korea, North	46,540	120,539	22,664	487	188
Korea, South	38,023	98,480	47,700	1,255	484
Kuwait	6,880	17,819	2,521	366	141
Kyrgyzstan	76,641	198,500	5,138	67	26
Laos	91,429	236,801	5,657	62	24
Latvia	24,749	64,100	2,307	93	36
Lebanon	4,015	10,399	3,653	910	351
Lesotho	11,720	30,355	1,802	154	59
Liberia	43,000	111,370	3,367	78	30
Libya	679,362	1,759,547	5,551	8	3
Liechtenstein	62	161	33	532	205
Lithuania	25,174	65,201	3,444	137	53
Luxembourg	998	2,585	453	454	175
Macedonia	9,781	25,333	2,056	210	81
Madagascar	226,658	587,044	17,404	77	30
Malawi	45,747	118,485	12,105	265	102
Malaysia	127,317	329,750	24,425	192	74
Maldives	115	298	318	2,765	1,067
Mali	478,767	1,240,006	13,007	27	10
Malta	124	320	318	2,565	994
Marshall Islands	70	181	74	1,057	409
Mauritania	397,954	1,030,700	2,893	7	3
Mauritius	718	1,860	1,221	1,701	656
Mexico	761,603	1,972,550	103,457	136	52
Micronesia	271	702	109	402	155
Moldova	13,012	33,701	4,435	341	132
Monaco	1.21	1.95	32	26,446	16,410
Mongolia	604,427	1,565,000	2,594	4	2
Morocco	172,413	446,550	30,566	177	68
Mozambique	309,494	801,590	18,863	61	24
Myanmar (Burma)	261,969	678,500	49,485	189	73
Namibia	318,259	824,290	1,987	6	2
Nauru	8	21	12	1,500	571
Nepal	54,363	140,800	25,164	463	179
Netherlands	14,413	37,330	16,149	1,120	433
New Zealand	103,738	268,680	3,875	37	14
Nicaragua	49,998	129,494	5,466	109	42
Niger	489,191	1,267,004	11,972	24	9
Nigeria	356,669	923,772	124,009	348	134
Norway	125,182	324,220	4,533	36	14
Oman	82,030	212,458	2,851	35	13
Pakistan	310,402	803,940	153,578	495	191

Table A (*continued*)
World Countries: Area, Population, and Population Density, 2003

COUNTRY	AREA		POPULATION (in thousands)	DENSITY	
	(Mi²)	(Km²)	(2003[a])	(Pop/Mi²)	(Pop/Km²)
Palau	177	458	19	107	41
Panama	30,193	78,200	3,120	103	40
Papua New Guinea	178,259	461,690	5,711	32	12
Paraguay	157,048	406,754	5,878	37	14
Peru	496,225	1,285,222	27,167	55	21
Philippines	115,831	300,002	79,999	691	267
Poland	120,728	312,685	38,537	319	123
Portugal	35,552	92,080	10,062	283	109
Qatar	4,247	11,000	610	144	55
Romania	91,699	237,500	22,334	244	94
Russian Federation	6,592,745	17,075,200	143,246	22	8
Rwanda	10,169	26,338	8,387	825	318
St. Kitts and Nevis	104	269	39	375	145
St. Lucia	238	616	149	626	242
St. Vincent/Grenadines	131	340	120	916	353
Samoa	1,104	2,860	178	161	62
San Marino	23	60	28	1,217	467
São Tomé and Principe	372	963	161	433	167
Saudi Arabia	756,982	1,960,582	24,217	32	12
Senegal	75,749	196,190	10,095	133	51
Serbia-Montenegro	39,517	102,350	10,527	266	103
Seychelles	175	453	80	457	177
Sierra Leone	27,699	71,740	4,971	179	69
Singapore	244	633	4,253	17,430	6,719
Slovakia	18,859	48,845	5,042	267	103
Slovenia	7,836	20,296	1,984	253	98
Solomon Islands	10,985	28,450	477	43	17
Somalia	246,201	637,660	9,890	40	16
South Africa	471,444	1,221,040	45,026	96	37
Spain	194,885	504,752	41,060	211	81
Sri Lanka	25,332	65,610	19,065	753	291
Sudan	967,500	2,505,824	33,610	35	13
Suriname	63,039	163,270	436	7	3
Swaziland	6,704	17,363	1,077	161	62
Sweden	173,732	449,966	8,876	51	20
Switzerland	15,943	41,292	7,169	450	174
Syria	71,498	185,180	17,800	249	96
Taiwan	13,892	35,980	22,548	1,623	627
Tajikistan	55,251	143,100	6,245	113	44
Tanzania	364,900	945,090	36,977	101	39
Thailand	198,456	514,000	62,833	317	122
Togo	21,925	56,786	4,909	224	86
Tonga	290	751	104	359	138
Trinidad and Tobago	1,980	5,128	1,303	658	254

Table A *(continued)*
World Countries: Area, Population, and Population Density, 2003

COUNTRY	AREA		POPULATION (in thousands)	DENSITY	
	(Mi^2)	(Km^2)	(2003[a])	(Pop/Mi^2)	(Pop/Km^2)
Tunisia	63,170	163,610	9,832	156	60
Turkey	301,382	780,580	71,325	237	91
Turkmenistan	188,456	488,101	4,867	26	10
Tuvalu	10	26	11	1,100	423
Uganda	93,135	236,040	25,827	277	109
Ukraine	233,090	603,703	48,523	208	80
United Arab Emirates	31,969	82,880	2,995	94	36
United Kingdom	94,525	244,820	59,251	627	242
United States	3,717,797	9,629,091	294,043	79	31
Uruguay	68,039	176,220	3,415	50	19
Uzbekistan	172,742	447,402	26,093	151	58
Vanuatu	5,699	14,760	212	37	14
Venezuela	352,145	912,055	25,699	73	28
Vietnam	127,243	329,560	81,377	640	247
Yemen	203,850	527,970	20,010	98	38
Zambia	290,586	752,617	10,812	37	14
Zimbabwe	150,803	390,580	12,891	85	33

[a]Primary source for population figures: United Nations Population Division.

[b]The figures for Israel do not include the West Bank and Gaza. These territories combined have a population of 3,557,000 (2003), and an area of approximately 2,400 square miles. The West Bank has an estimated population density of 956 persons per square mile and Gaza an estimated population density of 8,820 persons per square mile.

Sources: World Development Indicators 2003 (The World Bank); World Almanac and Book of Facts; World Population Prospects: The 2000 Revision (United Nations Population Information Network, 2001).

Table B
World Countries: Form of Government, Capital City, Major Languages

Notes: Unless indicated otherwise, republics are multi-party. Theocratic normally refers to fundamentalist Islamic rule. Transitional governments are those still in the process of change from a previous form (eg. Single-party communist state to multi-party republic).

COUNTRY	GOVERNMENT	CAPITAL	MAJOR LANGUAGES
Afghanistan	Transitional	Kabul	Dari, Pashtu, Uzbek, Turkmen
Albania	Multi-party democracy	Tiranë	Albanian, Greek
Algeria	Republic	Algiers	Arabic, Berber, dialects, French
Andorra	Parliamentary democracy	Andorra	Cataln, French, Spanish, Portuguese
Angola	Multi-party republic	Luanda	Portugese; Bantu and other African
Antigua and Barbuda	Parliamentary democracy	St. John's	English, local dialects
Argentina	Federal republic	Buenos Aires	Spanish, English, Italian, German, other
Armenia	Republic	Yerevan	Armenian, Russian, other
Australia	Federal parliamentary democracy	Canberra	English, indigenous
Austria	Federal republic	Vienna	German
Azerbaijan	Republic	Baku	Azerbaijani, Russian, Armenian, other
Bahamas	Parliamentary democracy; independent Commonwealth	Nassau	English, Creole
Bahrain	Constitutional hereditary monarchy	Al Manamah	Arabic, English, Farsi, Urdu
Bangladesh	Parliamentary democracy	Dhaka	Bangla, English
Barbados	Parliamentary democracy	Bridgetown	English
Belarus	Republic	Minsk	Byelorussian, Russian, other
Belgium	Constitutional monarchy	Brussels	Dutch (Flemish), French, German
Belize	Parliamentary democracy	Belmopan	English, Spanish, Garifuna, Mayan
Benin	Multi-party republic	Porto-Novo	French, Fon, Yoruba
Bhutan	Monarchy; special treaty relationship with India	Timphu	Dzongkha, Tibetan, Nepalese
Bolivia	Republic	La paz, Sucre	Spanish, Quechua, Aymara
Bosnia-Herzegovina	Emerging federal democratic republic	Sarajevo	Croatian, Serbian, Bosnian
Botswana	Parliamentary republic	Gaborone	English, Setswana
Brazil	Federal republic	Brasilia	Portugese, Spanish, English, French
Brunel	Constitutional monarchy	Bandar Seri Begawan	Malay, English, Chinese
Bulgaria	Parliamentary democracy	Sofia	Bulgarian
Burkina Faso	Parliamentary republic	Ouagadougou	French, indigenous
Burundi	Republic	Bujumbura	French, Kirundi, Swahili
Cambodia	Multi-party Democracy (under UN supervision)	Phnom Penh	Khmer, French, indigenous
Cameroon	Multi-party republic	Yaoundé	English, French, English
Canada	Federal parliamentary	Ottawa	English, French, other
Cape Verde	Republic	Cidade de Praia	Portuguese, Crioulu
Central African Republic	Republic	Bangui	French, Sangho
Chad	Republic	N'Djamena	French, Arabic, Sara, other indigenous
Chile	Republic	Santiago	Spanish
China	Single-party communist state	Beijing	Various Chinese dialects
Colombia	Republic	Bogota	Spanish
Comoros	Republic	Moroni	Arabic, French, Shikomoro

Table B *(continued)*
World Countries: Form of Government, Capital City, Major Languages

Notes: Unless indicated otherwise, republics are multi-party. Theocratic normally refers to fundamentalist Islamic rule. Transitional governments are those still in the process of change from a previous form (eg. Single-party communist state to multi-party republic).

COUNTRY	GOVERNMENT	CAPITAL	MAJOR LANGUAGES
Democratic Republic of the Congo (formerly Zaire)	Republic/transitional from military dictatorship	Kinshasa	French, Lingala, kingwana, Kikongo, Tshilluba
Congo Republic	Multi-party republic	Brazzaville	French, Lingala, Monokutuba, Kikongo
Costa Rica	Democratic republic	San José	Spanish
Côte d'Ivoire	Multi-party republic	Abidjan, Yamoussoukro	French, indigenous
Croatia	Parliamentary democracy	Zagreb	Croatian
Cuba	Single-party communist state	Havana	Spanish
Cyprus	Republic	Nicosia	Greek, Turkish, English
Czech Republic	Parliamentary democracy	Prague	Czech
Denmark	Constitutional monarchy	Copenhagen	Danish, Faroese, Greenlandic, German
Djibouti	Republic	Djibouti	French, Somali, Afar, Arabic
Dominica	Parliamentary democracy, republic within Commonwealth	Roseau	English, French
Dominican Republic	Republic	Santo Domingo	Spanish
East Timor	Republic	Dili	Tetum, Portuguese, Indonesian, English
Ecuador	Republic	Quito	Spanish, Quechua, indigenous
Egypt	Republic	Cairo	Arabic
El Salvador	Republic	San Salvador	Spanish, Nahua
Equatorial Guinea	Republic	Malabo	Spanish, French, indigenous, English
Eritrea	Transitional government	Asmara	Afar, Amharic, Arabic, Tigre, other indigenous
Estonia	Parliamentary republic	Tallinn	Estonian, Russian, Ukranian, Finnish
Ethiopia	Federal republic	Addis Ababa	Amharic, Tigrinya, Orominga, Somali, Arabic, English
Fiji	Republic	Suva	English, Fijian, Hindustani
Finland	Republic	Helsinki	Finnish, Swedish
France	Republic	Paris	French
Gabon	Multi-party republic	Libreville	French, Fang, indigenous
The Gambia	Multi-party democratic republic	Banjul	English, Mandinka, Wolof, Fula
Georgia	Republic	Tbilisi	Georgian, Russian, Armenian
Germany	Federal republic	Berlin	German
Ghana	Parliamentary democracy	Accra	English, indigenous
Greece	Parliamentary republic	Athens	Greek
Grenada	Parliamentary democracy	St.George's	English, French
Guatemala	Republic	Guatemala City	Spanish, Quiche, Cakchiquel, other indigenous
Guinea	Republic	Conakry	French, indigenous
Guinea-Bissau	Multi-party republic	Bissau	Portuguese, Crioulo, indigenous
Guyana	Republic within Commonwealth	Georgetown	English, Creole, Hindi, Urdu, indigenous
Haiti	Republic	Port-au-Prince	Creole, French
Honduras	Republic	Tegucigalpa	Spanish, indigenous
Hungary	Parliamentary democracy	Budapest	Hungarian

Table B *(continued)*
World Countries: Form of Government, Capital City, Major Languages

Notes: Unless indicated otherwise, republics are multi-party. Theocratic normally refers to fundamentalist Islamic rule. Transitional governments are those still in the process of change from a previous form (eg. Single-party communist state to multi-party republic).

COUNTRY	GOVERNMENT	CAPITAL	MAJOR LANGUAGES
Iceland	Republic	Reykjavk	Icelandic
India	Federal republic	New Delhi	English, Hindi, 14 other official
Indonesia	Republic	Jakarta	Bahasa Indonesian, English, Dutch, Javanese
Iran	Theocratic republic	Tehran	Farsi, Turkish, Kurdish
Iraq	In transition following US-led invasion	Baghdad	Arabic, Kurdish, Assyrian, Armenian
Ireland	Republic	Dublin	English, Irish Gaelic
Israel	Parliamentary democracy	Jerusalem	Hebrew, Arabic, English
Italy	Republic	Rome	Italian
Jamaica	Parliamentary democracy	Kingston	English, Creole
Japan	Constitutional monarchy	Tokyo	Japanese
Jordan	Constitutional monarchy	Amman	Arabic
Kazakhstan	Republic	Astana	Kazakh, Russian
Kenya	Republic	Nairobi	English, Swahili, indigenous
Kiribati	Republic	Tarawa	English, I-Kiribati
Korea, North	Single-party communist state	Pyongyang	Korean
Korea, South	Republic	Seoul	Korean
Kuwait	Constitutional monarchy	Kuwait	Arabic, English
Kyrgyzstan	Republic	Bishkek	Kirghiz, Russian
Laos	Single-party communist state	Vientiane	Lao, French, English, indigenous
Lativia	Parliamentary democracy	Riga	Latvian, Lithuanian, Russian
Lebanon	Republic	Beirut	Arabic, French, Armenian, English
Lesotho	Constitutional monarchy	Maseru	English, Sesotho, Zulu, Xhosa
Liberia	Republic	Monrovia	English, indigenous
Libya	Single party/military dictatorship	Tripoli	Arabic
Liechtenstein	Constitutional monarchy	Vaduz	German
Lithuania	Parliamentary democracy	Vilnius	Lithuanian, Russian, Polish
Luxembourg	Constitutional monarchy	Luxembourg	French, Luxembourgian, German
Macedonia	Parliamentary democracy	Skopje	Macedonian, Albanian, Turkish, Serbo-Croatian
Madagascar	Republic	Antananarivo	Malagasy, French
Malawi	Multi-party democracy	Lilongwe	Chichewa, English, Tombuka
Malaysia	Constitutional monarchy	Kuala Lumpur	Malay, Chinese, English, indigenous
Maldives	Republic	Male	Dhivehi
Mali	Republic	Bamako	French, Bambara, indigenous
Malta	Parliamentary democracy	Valletta	English, Maltese
Marshall Islands	Constitutional government (free association with U.S.)	Majuro	English, Polynesian dialects, Japanese
Mauritania	Republic	Nouakchott	Arabic, Wolof, Pular, French, Solinke
Mauritius	Parliamentary democracy	Port Louis	English, Creole, French, Hindi, Urdu, Bojpoori, Hakka
Mexico	Federal republic	Mexico City	Spanish, indigenous
Micronesia	Constitutional government (free association with U.S.)	Palikir	English, Trukese, Pohnpeian, Yapese, others
Moldova	Republic	Chisinau	Moldavian, Russian, Gagauz

Table B *(continued)*
World Countries: Form of Government, Capital City, Major Languages

Notes: Unless indicated otherwise, republics are multi-party. Theocratic normally refers to fundamentalist Islamic rule. Transitional governments are those still in the process of change from a previous form (eg. Single-party communist state to multi-party republic).

COUNTRY	GOVERNMENT	CAPITAL	MAJOR LANGUAGES
Monaco	Constitutional monarchy	Monaco	French, English, Italian, Monegasque
Mongolia	Republic	Ulaanbaatar	Khalkha Mongol, Turkic, Russian
Morocco	Constitutional monarchy	Rabat	Arabic, Berber dialects, French
Mozambique	Republic	Maputo	Portuguese, indigenous
Myanmar (Burma)	Military regime	Rangoon	Burmese, indigenous
Namibia	Republic	Windhoek	Afrikaans, English, German, indigenous
Nauru	Republic	Yaren district	Nauruan, English
Nepal	Constitutional monarchy	Kathmandu	Nepali, indigenous
Netherlands	Constitutional monarchy	Amsterdam	Dutch
New Zealand	Parliamentary democracy	Wellington	English, Maori
Nicaragua	Republic	Managua	Spanish, English, indigenous
Niger	Provisional military	Niamey	French, Hausa, Djerma
Nigeria	Military/transitional	Abuja	English, Hausa, Fulani, Yorbua, Ibo
Norway	Constitutional monarchy	Oslo	Norwegian, Sami
Oman	Monarchy	Muscat	Arabic, English, Baluchi, Urdu
Pakistan	Federal republic	Islamabad	Punjabi, Sindhi, Siraiki, Pashtu, Urbu, English, others
Palau	Constitutional government (free association with U.S.)	Koror	English, Palauan, Sonsolorese, Tobi, Angaur, Japanese
Panama	Constitutional democracy	Panama	Spanish, English
Papua New Guinea	Parliamentary democracy	Port Moresby	Various indigenous, English, Motu
Paraguay	Constitutional republic	Asunción	Spanish, Guarani
Peru	Republic	Lima	Quechua, Spanish, Aymara
Philippines	Republic	Manila	English, Filipino, 8 major dialects
Poland	Republic	Warsaw	Polish
Portugal	Parliamentary democracy	Lisbon	Portuguese
Qatar	Traditional monarchy	Doha	Arabic, English
Romania	Republic	Bucharest	Romanian, Hungarian, German
Russia	Federation	Moscow	Russian, numerous other
Rwanda	Republic	Kigali	French, Kinyarwanda, English, Kiswahili
St. Kitts and Nevis	Constitutional monarchy	Basseterre	English
St. Lucia	Parliamentary democracy	Castries	English, French
St. Vincent/Grenadines	Parliamentary monarchy independent within Commonwealth	Kingstown	English, French
Samoa	Constitutional monarchy	Apia	Samoan, English
San Marino	Republic	San Marino	Italian
São Tomé and Principe	Republic	São Tome	Portuguese
Saudi Arabia	Monarchy	Riyadh	Arabic
Senegal	Republic	Dakar	French, Wolof, indigenous
Serbia and Montenegro	Republic	Belgrade	Serbian, Albanian
Seychelles	Republic	Victoria	English, French, Creole
Sierra Leone	Constitutional democracy	Freetown	English, Krio, Mende,Temne
Singapore	Parliamentary republic	Singapore	Chinese, English, Malay, Tamil
Slovakia	Parliamentary democracy	Bratislava	Slovak, Hungarian

Table B *(continued)*
World Countries: Form of Government, Capital City, Major Languages

Notes: Unless indicated otherwise, republics are multi-party. Theocratic normally refers to fundamentalist Islamic rule. Transitional governments are those still in the process of change from a previous form (eg. Single-party communist state to multi-party republic).

COUNTRY	GOVERNMENT	CAPITAL	MAJOR LANGUAGES
Slovenia	Republic	Ljubjana	Slovenian, Serbo-Croatian, other
Solomon Islands	Parliamentary democracy	Honiara	English, indigenous
Somalia	No permanent national government	Mogadishu	Arabic, Somali, English, Italian
South Africa	Republic	Pretoria	Afrikaans, English, Zulu, Xhosa, other
Spain	Parliamentary monarchy	Madrid	Castilian Spanish, Catalan, Galician, Basque
Sri Lanka	Republic	Colombo	English, Sinhala, Tamil
Sudan	Provisional military	Khartoum	Arabic, Nubian, others
Suriname	Constitutional democracy	Paramaribo	Dutch, Sranang Tongo, English, Hindustani, Javanese
Swaziland	Monarchy within Commonwealth	Mbabane	English, siSwati
Sweden	Constitutional monarchy	Stockholm	Swedish
Switzerland	Federal republic	Bern	German, French, Italian, Romansch
Syria	Republic (under military regime)	Damascus	Arabic, Kurdish, Armenian, Aramaic
Taiwan	Multi-party democracy	Taipei	Mandarin Chinese, Taiwanese, Hakka
Tajikistan	Republic	Dushanbe	Tajik, Russian
Tanzania	Republic	Dar es Salaam	Kiswahili, English, Arabic, indigenous
Thailand	Constitutional monarchy	Bangkok	Thai, English
Togo	Republic/transitional	Lomé	French, indigenous
Tonga	Constitutional monarchy	Nuku'alofa	Tongan, English
Trinidad and Tobago	Parliamentary democracy	Port-of-Spain	English, Hindi, French, Spanish, Chinese
Tunisia	Republic	Tunis	Arabic, French
Turkey	Parliamentary republic	Ankara	Turkish, Kurdish, Arabic, Armenian, Greek
Turkmenistan	Republic	Ashkhabad	Turkmen, Russian, Uzbek, other
Tuvalu	Constitutional monarchy	Funafuti	Tuvaluan, English, Samoan, Kiribati
Uganda	Republic	Kampala	English, Luganda, Swahili, Arabic, indigenous
Ukraine	Republic	Kiev	Ukranian, Russian, Romanian, Polish, Hungarian
United Arab Emirates	Federated monarchy	Abu Dhabi	Arabic, English, Farsi, Hindi, Urdu
United Kingdom	Constitutional monarchy	London	English, Welsh, Scottish Gaelic
United States	Federal republic	Washington	English, Spanish
Uruguay	Republic	Montevideo	Spanish, Portunol/Brazilero
Uzbekistan	Republic	Tashkent	Uzbek, Russian, Kazakh, Tajik, other
Vanuatu	Republic	Port-Vila	English, French, Bislama (pidgin)
Venezuela	Federal republic	Caracas	Spanish, indigenous
Vietnam	Single-party communist state	Hanoi	Vietnamese, French, Chinese, English, Khmer
Yemen	Republic	San'aa	Arabic
Zambia	Republic	Lusaka	English, Tonga, Lozi, other indigenous
Zimbabwe	Parliamentary democracy	Harare	English, Shona, Sindebele, other

Source: The World Factbook 2002 (C/A, Washington, DC).

Table C
World Countries: Basic Economic Indicators

COUNTRY	GROSS NATIONAL INCOME (GNI) 2001[a]		PURCHASING POWER PARITY GNI 2001[b]			AVERAGE ANNUAL % GROWTH IN GDP			STRUCTURE OF ECONOMIC OUTPUT (GDP) 2001 (value added in % of GDP)			
	Total ($U.S. billions)	Per Capita ($U.S.)	Total ($U.S. billions)	Per Capita ($U.S.)	Rank	1980–1990	1990–1999	2000–01	Agriculture	Industry	Manufacturing	Services
Afghanistan	—	—	—	—	—	—	—	—	—	—	—	—
Albania	4.2	1,340	12	3,810	130	1.5	3.2	6.5	50	23	13	26
Algeria	51.0	1,650	182[c]	5,910[c]	99	2.7	1.6	2.1	10	55	8	36
Angola	6.7	500	23[c]	1,690[c]	171	3.7	0.4	3.2	8	67	4	25
Argentina	260.3	7	412	10,980	63	–0.4	4.9	–4.5	5	27	17	69
Armenia	2.2	570	10	2,730	145	—	–3.2	9.6	28	34	22	38
Australia	385.9	19,900	478	24,630	24	3.4	4.1	3.9	4	26	13	70
Austria	194.7	23,940	215	26,380	17	2.2	1.9	1.0	2	33	22	65
Azerbaijan	5.3	650	23	2,890	141	—	–9.6	9.9	17	46	6	36
Bangladesh	48.6	360	213	1,600	173	4.3	4.7	5.3	23	25	15	52
Belarus	12.9	1,290	76	7,630	83	—	–3.0	4.1	11	39	33	50
Belgium	245.3	29,850	269	26,150	18	2.0	1.7	1.0	2	27	20	71
Benin	2.4	380	6	970	190	2.9	4.7	5.0	36	14	9	50
Bolivia	8.1	950	19	2,240	155	–0.2	4.2	1.2	16	29	15	56
Bosnia-Herzegovina	5.0	1,240	25	6,250	92	—	35.2	6.0	15	31	16	55
Botswana	5.3	3,100	13	7,410	84	10.3	4.3	6.3	2	47	4	51
Brazil	528.9	3,070	1,219	7,070	86	2.7	3.0	1.5	9	34	21	57
Bulgaria	13.2	1,650	54	6,740	89	3.4	–2.7	4.0	14	29	18	57
Burkina Faso	2.5	220	13[c]	1,120[c]	185	3.6	3.8	5.6	38	21	15	41
Burundi	0.7	100	5[c]	680[c]	203	4.4	–2.9	3.2	50	19	9	31
Cambodia	3.3	270	22	1,790	168	—	4.8	6.3	37	22	—	41
Cameroon	8.7	580	24	1,580	174	3.4	1.3	5.3	43	20	11	38
Canada	681.6	21,930	825[c]	26,530[c]	15	3.3	2.7	1.5	—	—	—	—
Central African Republic	1.0	260	5[c]	1,300[c]	181	1.4	1.8	1.5	55	21	9	24
Chad	1.6	200	8	1,060	187	3.7	2.1	8.5	39	14	10	48
Chile	70.6	4,590	136	8,840	76	4.2	7.2	2.8	9	34	16	57
China	1,131.2	890	5,027	3,950	127	10.2	10.7	7.3	15	51	35	34
Hong Kong China	170.3	25,30	172	25,560	19	6.9	3.9	0.1	0	14	6	86
Colombia	81.6	1,890	292	6,790	88	3.6	3.3	1.4	13	30	16	57
Democratic Republic of the Congo (formerly Zaire)	4.2	80	33	630	205	1.6	–5.1	–4.5	56	19	4	25
Congo Republic	2.0	640	2	680	203	3.3	–0.5	2.9	6	66	4	28
Costa Rica	15.7	4,060	36	9,260	74	3.0	5.1	0.9	9	29	21	62
Côte d'Ivoire	10.3	630	23	1,400	179	0.7	3.7	–0.9	24	22	19	54
Croatia	19.9	4,550	39	8,930	75	—	0.2	4.1	9	33	23	58
Cuba	—	–9	—	—	—	—	—	—	7	46	37	47
Czech Republic	54.3	5,310	146	14,320	55	1.7	0.8	3.3	4	41	—	55
Denmark	164.0	30,600	153	23,490	9	2.3	2.4	1.0	33	26	17	71

Table C *(continued)*
World Countries: Basic Economic Indicators

COUNTRY	GROSS NATIONAL INCOME (GNI) 2001[a]		PURCHASING POWER PARITY GNI 2001[b]			AVERAGE ANNUAL % GROWTH IN GDP			STRUCTURE OF ECONOMIC OUTPUT (GDP) 2001 (value added in % of GDP)			
	Total ($U.S. billions)	Per Capita ($U.S.)	Total ($U.S. billions)	Per Capita ($U.S.)	Rank	1980–1990	1990–1999	2000–01	Agriculture	Industry	Manufacturing	Services
Dominican Republic	19.0	2,230	57	6,650	90	3.1	5.8	2.7	11	33	16	55
Ecuador	14.0	1,080	38	2,960	140	2.0	2.2	5.6	11	33	18	56
Egypt	99.6	1,530	232	3,560	131	5.4	4.4	2.9	17	33	19	50
El Salvador	13.0	2,040	33	5,160	107	0.2	5.0	1.8	9	30	23	61
Eritrea	0.7	160	4	1,030	189	—	5.0	9.7	19	22	11	59
Estonia	5.3	3,870	13	9,650	71	2.2	–1.3	5.0	6	29	19	65
Ethiopia	6.7	100	53	800	198	2.3	4.6	7.7	52	11	7	37
Finland	123.4	23,780	125	24,030	28	3.3	2.4	0.4	3	33	26	63
France	1,380.7[h]	22,730[h]	1,425	24,080	27	2.3	1.5	1.8	3	26	18	72
Gabon	4.0	3,160	7	5,190	105	0.9	3.2	2.5	8	51	5	42
The Gambia	0.4	320	3[c]	2,010[c]	160	3.6	2.8	6.0	40	14	5	46
Georgia	3.1	590	14	2,580	148	0.4	—	4.5	21	23	—	57
Germany	1,939.6	24	2,078	25,240	21	2.2	1.3	0.6	1	31	24	68
Ghana	5.7	290	43[c]	2,170[c]	157	3.0	4.3	4.0	36	25	9	39
Greece	121.0	11,430	186	17,520	47	1.8	2.2	4.1	8	21	12	71
Guatemala	19.6	1,680	51	4,380	120	0.8	4.2	2.1	23	19	13	58
Guinea	3.1	410	14	2	164	—	4.2	3.6	24	38	4	38
Guinea-Bissau	0.2	160	1	890	193	4.0	0.3	0.2	56	13	10	37
Haiti	3.9	480	15[c]	1,870[c]	166	–0.2	–1.3	–1.7	—	—	—	—
Honduras	5.9	900	18	2,760	144	2.7	3.3	2.6	14	32	20	55
Hungary	49.2	4,830	122	11,990	59	1.3	1.0	3.8	—	—	—	—
India	477.4	460	2,913	2,820	143	5.8	6.0	5.4	25	26	16	48
Indonesia	144.7	690	591	2,830	142	6.1	4.7	3.3	16	47	26	37
Iran	108.7	1,680	383	5,940	98	1.7	3.6	4.8	19	33	16	48
Iraq	—	—[c]	—	—	—	–6.8	—		—	—	—	—
Ireland	87.7	22,850	104	27,170	14	3.2	6.9	5.8	4	42	33	55
Israel	106.6	17	125	19,630	40	3.5	5.2	–0.9	—	—	—	—
Italy	1,128.8	19,390	1,422	24,530	25	2.4	1.4	1.8	3	29	21	68
Jamaica	7.3	2,800	9	3	133	2.0	0.3	1.7	6	31	13	63
Japan	4,523.3	36	3,246	25,550	20	4.0	1.3	–0.6	1	32	22	67
Jordan	8.8	1,750	20	3,880	128	2.5	5.3	4.2	2	25	15	73
Kazakhstan	20.1	1,350	92	6,150	94	—	–5.9	13.2	9	39	16	52
Kenya	10.7	350	30	970	190	4.2	2.2	1.1	19	18	13	63
Korea, North	—	—[d]	—	—	—	—	—		—	—	—	—
Korea, South	447.6	9,460	713	15	54	9.4	5.7	3.0	4	41	30	54
Kuwait	37.4	18,270	44	21,530	35	1.3	—	–1.0	—	—	—	—
Kyrgyzstan	1.4	280	13	2,630	147	—	–5.4	5.3	38	27	8	35
Laos	1.6	300	8[c]	1,540[c]	175	—	6.6	5.7	51	23	18	26
Latvia	7.6	3,230	18	7,760	82	3.5	–4.8	7.6	5	26	15	69
Lebanon	17.6	4,010	19	4,400	119	—	7.7	1.3	12	22	10	66
Lesotho	1.1	530	6[c]	2,980[c]	139	4.4	4.4	4.0	16	42	14	42
Liberia	0.5	140	—	—	196	—	—	5.3	—	—	—	—
Libya	—	—[g]	—	—	—	–5.7	—		—	—	—	—
Lithuania	11.7	3,350	29	8,350	78	—	–4.0	5.9	7	35	23	58
Luxembourg	18.8	41,550						1	1	20		79

Table C *(continued)*
World Countries: Basic Economic Indicators

COUNTRY	GROSS NATIONAL INCOME (GNI) 2001[a]		PURCHASING POWER PARITY GNI 2001[b]			AVERAGE ANNUAL % GROWTH IN GDP			STRUCTURE OF ECONOMIC OUTPUT (GDP) 2001 (value added in % of GDP)			
	Total ($U.S. billions)	Per Capita ($U.S.)	Total ($U.S. billions)	Per Capita ($U.S.)	Rank	1980–1990	1990–1999	2000–01	Agriculture	Industry	Manufacturing	Services
Macedonia	3.5	1,690	12	6,040	97	—	−0.8	−4.1	11	31	20	58
Madagascar	4.2	260	13	820	197	1.1	1.7	6.0	30	14	12	56
Malawi	1.7	151	6	560	206	2.5	3.6	−1.5	34	18	13	48
Malaysia	79.3	3,330	188	7,910	81	5.3	7.3	0.4	9	49	31	42
Mali	2.5	230	9	770	200	2.8	3.6	1.4	38	26	4	36
Mauritania	1.0	360	5	1,940	162	1.8	4.2	4.6	21	29	8	50
Mauritius	4.6	3,830	12	9,860	70	6.2	5.1	7.2	6	31	23	62
Mexico	550.2	5,530	820	8	80	0.7	2.7	−0.3	4	27	19	69
Moldova	1.5	400	10	2,300	154	3.0	−11.0	6.1	26	24	18	50
Mongolia	1.0	400	4	1,710	170	5.4	0.7	1.4	30	17	5	53
Morocco	34.7	1,190	102	3,500	132	4.2	2.3	6.5	16	18	50	53
Mozambique	3.8	210	19[c]	1,050[c]	188	−0.1	6.2	13.9	22	26	12	52
Myanmar (Burma)	—	—[d]	—	—	—	0.6	6.3	—	57	10	7	33
Namibia	3.5	1,960	13[c]	7,410[c]	85	0.9	3.4	2.7	11	33	11	56
Nepal	5.8	250	32	1,360	180	4.6	4.9	4.8	39	22	9	39
Netherlands	390.3	24,330	439	27,390	13	2.3	2.7	1.1	3	27	17	70
New Zealand	51.0	13,250	70	18	43	1.8	3.1	3.2	—	—	—	—
Nicaragua	—	—	—	—	158	−2.0	3.2	—	—	—	—	—
Niger	2.0	180	10[c]	880[c]	194	−0.1	2.4	7.6	40	17	7	43
Nigeria	37.1	290	102	790	199	1.6	2.4	3.9	30	46	4	25
Norway	160.8	36	132	29,340	7	2.8	3.8	1.4	2	43	—	55
Oman	—	—[g]		—	—	8.4	5.9	—	—	—	—	—
Pakistan	60.0	420	263	1,860	167	6.3	3.8	2.7	25	23	16	52
Panama	9.5	3,260	16[c]	5,440[c]	104	0.5	4.2	0.3	7	9	7	77
Papua New Guinea	3.0	580	13[c]	2,450[c]	149	1.9	4.7	−3.5	26	42	8	32
Paraguay	7.6	1,350	29[c]	5,180[c]	106	2.5	2.4	2.7	20	26	13	54
Peru	52.2	1,980	118	4,470	117	−0.3	5.0	0.2	9	30	15	62
Philippines	80.8	1,030	319	4	125	1.0	3.2	3.4	15	31	22	54
Poland	163.6	4,230	362	9,370	73	1.8	4.5	1.0	4	37	20	59
Portugal	109.3	10,900	178	17,710	46	3.1	2.5	1.7	4	30	19	66
Puerto Rico	42.1	10,950	69	18,090	44	4.0	3.1	5.6	1	43	40	56
Romania	38.6	1,720	130	5,780	101	0.5	−0.8	5.3	15	35	—	50
Russian Federation	253.4	1,750	995	6,880	87	—	−6.1	5.0	7	37	—	56
Rwanda	1.9	220	11	1,240	183	2.2	−1.5	6.7	40	22	10	38
Saudi Arabia	181.1	8	284	13,290	54	0.0	1.6	1.2	—	—	—	—
Senegal	4.7	490	14	1,480	176	3.1	3.3	5.7	18	27	18	55
Serbia-Montenegro	9.9	930	—	—	—	—		—	15	32	—	53
Sierra Leone	0.7	140	2	460	208	0.3	−4.7	5.4	50	30	5	20
Singapore	88.8	21,500	94	22,850	32	6.6	8.0	−2.0	0	34	23	68
Slovak Republic	20.3	4	64	11,780	60	2.0	1.8	3.3	4	29	21	67
Slovenia	19.4	9,760	34	17,060	49	—	2.4	3.0	3	38	28	58
Somalia	—	[d]	—	—	—	—	—	—	—	—	—	—

Table C (continued)
World Countries: Basic Economic Indicators

COUNTRY	GROSS NATIONAL INCOME (GNI) 2001[a]		PURCHASING POWER PARITY GNI 2001[b]			AVERAGE ANNUAL % GROWTH IN GDP			STRUCTURE OF ECONOMIC OUTPUT (GDP) 2001 (value added in % of GDP)			
	Total ($U.S. billions)	Per Capita ($U.S.)	Total ($U.S. billions)	Per Capita ($U.S.)	Rank	1980–1990	1990–1999	2000–01	Agriculture	Industry	Manufacturing	Services
South Africa	121.9	2,820	472[c]	10,910[c]	64	1.2	1.9	2.2	3	31	19	66
Spain	588.0	14,800	816	19,860	39	3.0	2.2	2.8	4	30	19	66
Sri Lanka	16.4	880	61	3,260	134	4.0	5.3	−1.4	19	27	16	54
Sudan	10.7	340	56	1,750	169	0.4	8.2	6.9	39	19	10	42
Swaziland	1.4	1,300	5	4,430	118	—	—	1.6	17	44	36	39
Sweden	225.9	25,400	212	23,800	29	2.3	1.6	1.2	2	32	—	71
Switzerland	277.2	38,330	224	30,970	5	2.0	0.6	1.3	—	—	—	—
Syria	17.3	1,040	52	3,160	136	1.5	5.7	2.8	22	28	20	50
Tajikistan	1.1	180	7	1,140	184	—	—	10.2	29	29	25	41
Tanzania	9.4[i]	270[i]	18	520	207	—[i]	2.8	5.7	45	16	7	39
Thailand	118.5	1,940	381	6,230	93	7.6	4.7	1.8	10	40	32	49
Togo	1.3	270	8	1,620	172	1.7	2.4	2.7	39	21	10	39
Trinidad and Tobago	7.8	5,960	11	8,620	77	−0.8	2.7	5.0	2	44	8	55
Tunisia	20.0	2,070	59	6,090	96	3.3	4.6	4.9	12	29	18	60
Turkey	167.3	2,530	386	5,830	100	5,4	3.8	−7.4	14	26	15	61
Turkmenistan	5.1	950	23	4,240	124	—	−6.8	20.5	29	51	—	20
Uganda	5.9	260	33[c]	1,460[c]	177	2.9	7.2	4.6	36	21	10	43
Ukraine	35.2	720	210	4,270	131	—	−10.7	9.1	17	39	23	44
United Arab Emirates	—	—[f]	—	—	—	−3.5	2.9	—	—	—	—	—
United Kingdom	1,476.8	25,120	1,431	24,340	26	3.2	2.5	2.2	1	27	19	72
United States	9,780.8	34,280	10	34,280	3	3.0	3.3	0.3	2	25	17	73
Uruguay	19.2	5,710	28	8,250	79	0.4	3.8	−3.1	6	28	56	67
Uzbekistan	13.8	550	60	2,410	152	—	−1.2	4.5	34	23	9	43
Venezuela	117.2	4,760	138	5,590	102	1.1	1.7	2.7	5	50	20	45
Vietnam	32.8	410	164	2,070	159	4.6	8.1	6.8	24	38	20	39
West Bank and Gaza	4.2	1,350	—	—	—	—	3.7	−11.9	8	27	15	66
Yemen, Rep.	8.2	450	13	730	202	—	3.2	3.1	16	50	7	35
Zambia	3.3	320	8	750	201	1.0	0.2	4.9	22	26	11	52
Zimbabwe	6.2	480	28	2,220	156	3.6	2.8	−8.4	18	24	14	58

[a]Gross National Income (GNI) has replaced GNP in the World Bank Atlas Method's estimate of national income.
[b]Calculated using the World Bank Atlas method.
[c]The estimate is based on regression; others are extrapolated from the latest International Comparison Programme benchmark estimates.
[d]Estimated to be low income ($745 or less).
[e]Estimated to be lower middle income ($746 to $2,975).
[f]Estimated to be high income ($9,206 or more).
[g]Estimated to be upper middle income ($2,976 to $9,205).
[h]GNI and GNI per capita estimates include the French overseas departments of French Guiana, Guadeloupe, Martinique, and Reunion.
[i]Data refer to mainland Tanzania only.

Sources: World Development Indicators, 2003 (World Bank).

Table D
World Countries: Population Growth, 1950–2025

COUNTRY	POPULATION (thousands)			AVERAGE ANNUAL POPULATION CHANGE (percent)		AVERAGE ANNUAL INCREMENT TO THE POPULATION (mid-year population, in thousands)		
	1950	2000[a]	2025[a]	1975–1980	2001–2015[a]	1985–1990	1995–2000	2005–2010
WORLD	2,518,629.0	6,070,581.0	7,851,455.0	1.7	2.5	85,831.0		
AFRICA								
Algeria	8,753.0	30,245.0	42,429.0	3.1	1.5	631.8	566.0	539.3
Angola	4,131.0	12,386.0	19,268.0	2.7	2.6	130.2	187.2	267.5
Benin	2,046.0	6,222.0	11,120.0	2.5	2.4	135.6	184.8	204.6
Botswana	419.0	1,725.0	1,614.0	3.5	0.5	43.3	21.5	–15.4
Burkina Faso	3,960.0	11,905.0	24,527.0	2.5	2.1	233.8	307.0	360.3
Burundi	2,456.0	6,267.0	12,328.0	2.3	1.7	95.2	123.0	166.5
Cameroon	4,466.0	15,127.0	20,831.0	2.8	1.7	326.5	370.9	374.8
Central African Republic	1,314.0	3,715.0	5,193.0	2.3	1.5	57.5	62.0	60.0
Chad	2,658.0	7,861.0	15,770.0	2.1	2.9	171.2	259.8	340.4
Congo Democratic Republic	12,184.0	48,571.0	95,448.0	3.0	2.6	1,146.2	1,234.4	1,907.8
Congo Republic	808.0	3,447.0	6,750.0	2.9	2.7	56.3	62.4	67.2
Côte d'Ivoire	2,775.0	15,827.0	22,140.0	3.9	1.6	411.1	350.2	394.2
Egypt	21,834.0	67,784.0	103,165.0	2.4	1.5	1,318.5	1,199.9	1,125.0
Equatorial Guinea	226.0	456.0	812.0	–0.7	—	8.7	11.1	13.6
Eritrea	1,140.0	3,712.0	7,261.0	2.6	2.3	35.4	134.9	153.2
Ethiopia	18,434.0	65,590.0	116,006.0	2.4	2.1	1,530.7	1,670.9	1,848.4
Gabon	469.0	1,258.0	1,915.0	3.1	2.2	11.6	13.9	8.6
Gambia	294.0	1,312.0	2,177.0	3.1	2.0	33.0	42.3	48.0
Ghana	4,900.0	19,593.0	30,618.0	1.9	1.6	435.4	380.2	285.2
Guinea	2,550.0	8,117.0	13,704.0	1.5	1.9	173.7	63.1	193.9
Guinea-Bissau	505.0	1,367.0	2,774.0	4.7	2.2	22.1	28.4	35.1
Kenya	6,265.0	30,549.0	39,917.0	3.8	1.4	723.5	604.9	206.8
Lesotho	734.0	1,785.0	1,608.0	2.5	0.8	41.5	39.6	11.5
Liberia	824.0	2,943.0	6,081.0	3.1	2.3	–2.9	236.3	107.3
Libya	1,029.0	5,237.0	7,785.0	4.4	1.9	92.8	92.2	136.3
Madagascar	4,230.0	15,970.0	30,249.0	2.5	2.5	308.1	433.2	590.5
Malawi	2,881.0	11,370.0	18,245.0	3.3	1.8	416.6	169.9	103.8
Mali	3,520.0	11,904.0	25,679.0	2.1	2.1	164.0	305.1	390.7
Mauritania	825.0	2,645.0	4,973.0	2.5	2.3	47.5	65.2	94.9
Morocco	8,953.0	29,108.0	40,721.0	2.3	1.4	565.7	535.1	515.0
Mozambique	6,442.0	17,861.0	25,350.0	2.8	1.6	78.7	359.2	75.1
Namibia	511.0	1,894.0	2,350.0	2.7	1.2	58.6	33.5	7.5
Niger	2,500.0	10,742.0	25,722.0	3.2	2.8	207.6	259.7	322.0
Nigeria	29,790.0	114,746.0	192,115.0	2.8	1.9	2,530.7	3,225.5	3,161.6
Rwanda	2,162.0	7,724.0	12,509.0	3.3	1.6	187.9	311.1	49.1
Senegal	2,500.0	9,393.0	15,663.0	2.8	2.0	191.6	278.9	336.2
Sierra Leone	1,944.0	4,415.0	7,593.0	2.0	1.9	106.3	143.8	162.5
Somalia	2,264.0	8,720.0	20,978.0	7.0	3.1	45.8	192.4	266.1
South Africa	13,683.0	44,000.0	42,962.0	2.2	0.4	943.4	383.4	–419.6

Table D *(continued)*
World Countries: Population Growth, 1950–2025

COUNTRY	POPULATION (thousands)			AVERAGE ANNUAL POPULATION CHANGE (percent)		AVERAGE ANNUAL INCREMENT TO THE POPULATION (mid-year population, in thousands)		
	1950	2000[a]	2025[a]	1975–1980	2001–2015[a]	1985–1990	1995–2000	2005–2010
Sudan	9,190.0	31,437.0	47,536.0	3.1	2.0	634.6	902.5	1,059.5
Tanzania	7,886.0	34,837.0	53,435.0	3.1	1.7	799.7	832.1	970.9
Togo	1,329.0	4,562.0	7,551.0	2.7	1.9	122.0	160.9	115.3
Tunisia	3,530	9,586.0	12,843.0	2.6	1.3	168.9	124.3	104.8
Uganda	5,310.0	23,487.0	54,883.0	3.2	2.4	590.9	98.7	877.5
Zambia	2,440.0	10,419.0	14,401.0	3.4	1.2	210.9	174.1	190.9
Zimbabwe	2,744.0	12,650.0	12,857.0	3.0	0.6	308.9	78.1	–58.6
NORTH AND CENTRAL AMERICA						353.9		
Belize	69.0	240.0	356.0	1.7	—	5.0	6.3	7.2
Canada	13,737.0	30,769.0	36,128.0	1.2	0.6	369.8	331.8	289.5
Costa Rica	966.0	3,929.0	5,621.0	3.0	1.4	76.6	65.4	58.0
Cuba	5,850.0	11,202.0	11,479.0	0.9	0.3	93.2	48.4	37.3
Dominican Republic	2,353.0	8,353.0	10,955.0	2.4	1.3	141.1	136.4	147.1
El Salvador	1,951.0	6.209.0	8,418.0	2.1	1.6	87.1	110.8	117.7
Guatemala	2,969.0	11,428.0	19,456.0	2.5	2.4	255.9	317.6	366.3
Haiti	3,261.0	8,005.0	10,670.0	2.1	1.7	111.8	89.1	114.6
Honduras	1,380.0	6,457.0	10,115.0	3.4	2.1	117.2	151.1	134.9
Jamaica	1,403.0	2,580.0	3,263.0	1.2	1.1	18.3	16.8	23.9
Mexico	27,737.0	98,933.0	129,866.0	2.7	1.4	1,594.3	1,572.4	1,425.0
Nicaragua	1,134.0	5,073.0	8,318.0	3.1	2.1	91.0	107.6	100.9
Panama	860.0	2,950.0	4,290.0	2.5	1.3	44.8	39.8	32.5
Trinidad and Tobago	636.0	1,289.0	1,340.0	1.3	0.8	6.5	–4.9	–6.1
United States	157,813.0	285,003.0	358,030.0	0.9	0.8	2,296.3	2,503.8	2,429.2
SOUTH AMERICA								
Argentina	17,150.0	37,074.0	47,043.0	1.5	1.0	445.4	427.4	401.1
Bolivia	2,714.0	8,317.0	12,495.0	2.4	1.8	127.7	155.2	128.3
Brazil	53,975.0	171,796.0	216,372.0	2.4	1.1	2,756.3	1,975.6	1,285.4
Chile	6,082.0	15,224.0	19,651.0	1.5	1.0	212.2	189.7	148.3
Colombia	12,568.0	4,120.0	58,157.0	2.3	1.3	636.0	681.0	630.9
Ecuador	3,387.0	12,420.0	16,704.0	2.8	1.5	262.0	264.1	257.2
Guyana	423.0	759.0	724.0	0.7	—	–3.2	–3.5	4.1
Paraguay	1,488.0	5,470.0	9,173.0	3.2	2.1	113.5	141.6	162.8
Peru	7,632.0	25,952.0	35,622.0	2.7	1.3	472.9	491.4	432.4
Suriname	215.0	425.0	486.0	–0.5	—	3.9	3.3	1.4
Uruguay	2,239.0	3,342.0	3,875.0	0.6	0.6	19.4	23.7	26.7
Venezuela	5,094.0	24,277.0	31,189.0	3.4	1.5	465.5	397.3	351.7
ASIA								
Afghanistan	8,151.0	21,391.0	44,940.0	0.9	2.5	170.3	879.9	724.7
Armenia	1,354.0	3,112.0	2,866.0	1.8	0.3	–0.7	–13.8	7.6
Azerbaijan	2,896.0	8,157.0	10,222.0	1.6	0.7	103.6	23.6	61.8

Table D *(continued)*
World Countries: Population Growth, 1950–2025

COUNTRY	POPULATION (thousands)			AVERAGE ANNUAL POPULATION CHANGE (percent)		AVERAGE ANNUAL INCREMENT TO THE POPULATION (mid-year population, in thousands)		
	1950	2000[a]	2025[a]	1975–1980	2001–2015[a]	1985–1990	1995–2000	2005–2010
Bangladesh	41,783.0	137,952.0	208,268.0	2.8	1.6	2,028.8	2,001.0	2,119.6
Bhutan	734.0	2,063.0	3,701.0	2.3	—	34.7	42.4	48.8
Cambodia	4,346.0	13,147.0	21,899.0	–1.8	1.5	313.1	271.6	314.6
China[b]	556,924.0	1,282,472.0	1,454,141.0	1.5	0.6	16,833.4	11,408.3	8,726.8
Georgia	3,527.0	5,262.0	4,429.0	0.7	–0.7	49.9	–53.5	–14.4
India	357,561.0	1,016,938.0	1,369,284.0	2.1	1.2	16,448.0	16,317.4	15,140.5
Indonesia	79,538.0	211,559.0	270,113.0	2.1	1.1	3,283.4	3,702.8	3,388.7
Iran	16,913.0	664,423.0	90,927.0	3.3	1.6	1,632.8	818.3	1,000.3
Iraq	5,158.0	23,224.0	41,707.0	3.3	1.9	488.2	623.7	719.5
Israel	1,258.0	6,042.0	8,598.0	2.3	1.5	87.4	107.5	73.6
Japan	83,625.0	127,034.0	123,444.0	0.9	–0.2	556.6	252.5	–30.4
Jordan	472.0	5,035.0	8,116.0	2.3	2.2	126.9	159.4	145.2
Kazakhstan	6,703.0	15,640.0	15,388.0	1.1	0.1	148.5	–42.0	85.9
Korea, North	10,815.0	22,268.0	24,665.0	1.6	0.6	307.4	27.2	4,294.4
Korea, South	18,859.0	46,835.0	50,165.0	1.6	0.4	943.5	459.2	322.0
Kuwait	152.0	2,247.0	3,930.0	6.2	2.1	81.8	70.6	90.4
Kyrgyzstan	1,740.0	4,921.0	6,484.0	1.9	1.1	76.8	30.0	80.9
Laos	1,755.0	5,279.0	8,635.0	1.2	2.2	110.7	130.3	155.3
Lebanon	1,443.0	3,478.0	4,554.0	–0.7	1.2	11.8	48.7	46.0
Malaysia	6,110.0	23,001.0	33,479.0	2.3	1.5	391.7	436.4	438.2
Mongolia	761.0	2,500.0	3,368.0	2.8	1.3	62.1	37.4	44.4
Myanmar (Burma)	17,832.0	47,544.0	59,760.0	2.1	1.0	452.8	317.4	162.3
Nepal	8,643.0	23,518.0	37,831.0	2.5	2.0	457.5	559.0	616.3
Oman	456.0	2,609.0	4,785.0	5.0	2.2	58.3	80.5	104.3
Pakistan	39,659.0	142,654.0	249,766.0	2.6	2.2	2,984.4	2,984.8	2,936.8
Philippines	1,996.0	75,711.0	108,589.0	2.3	1.6	1,450.5	1,658.1	1,666.1
Saudi Arabia	3,201.0	22,147.0	39,751.0	5.6	2.9	527.8	678.3	922.2
Singapore	1,022.0	4,016.0	4,905.0	1.3	1.1	56.1	134.2	169.1
Sri Lanka	7,483.0	18,595.0	21,464.0	1.7	1.1	234.4	186.9	153.5
Syria	3,495.0	16,560.0	26,979.0	3.1	2.1	391.1	399.2	431.5
Tajikistan	1,532.0	6,089.0	8,193.0	2.8	1.5	149.0	115.3	168.7
Thailand	19,626.0	60,925.0	73,869.0	2.4	0.6	755.5	598.8	468.5
Turkey	21,484.0	68,281.0	88,995.0	2.1	1.1	1,083.1	895.5	732.4
Turkmenistan	1,211.0	4,643.0	6,549.0	2.5	1.1	85.4	83.3	95.8
United Arab Emirates	70.0	2,820.0	3,944.0	14.0	1.8	76.1	38.6	40.0
Uzbekistan	6,314.0	24,913.0	33,774.0	2.6	1.3	473.1	381.7	485.8
Vietnam	27,369.0	78,137.0	104,649.0	2.2	1.2	1,321.7	1,178.3	1,123.4
Yemen	4,316.0	18,017.0	43,204.0	3.2	3.0	436.3	524.0	782.1
EUROPE								
Albania	1,215.0	3,113.0	3,629.0	1.9	1.0	60.3	50.7	34.4
Austria	6,935.0	8,102.0	7,979.0	–0.1	–0.1	32.1	17.8	11.3
Belarus	7,745.0	10,034.0	8,950.0	0.6	–0.5	46.7	–7.5	–1.4

Table D (continued)
World Countries: Population Growth, 1950–2025

COUNTRY	POPULATION (thousands)			AVERAGE ANNUAL POPULATION CHANGE (percent)		AVERAGE ANNUAL INCREMENT TO THE POPULATION (mid-year population, in thousands)		
	1950	2000[a]	2025[a]	1975–1980	2001–2015[a]	1985–1990	1995–2000	2005–2010
Belgium	8,639.0	10,251.0	10,516.0	0.1	0.0	22.2	20.9	5.3
Bosnia-Herzegovina	2,661.0	3,977.0	4,183.0	0.9	0.5	29.7	96.0	15.5
Bulgaria	7,251.0	8,099.0	6,609.0	0.3	−0.7	−9.9	−95.1	−74.3
Croatia	3,850.0	4,446.0	4,088.0	0.5	−0.3	10.1	−34.6	11.2
Czech Republic	8,925.0	10,269.0	9,806.0	0.6	−0.2	−0.1	−10.6	−14.7
Denmark	4,271.0	5,322.0	5,469.0	0.2	0.1	5.5	20.8	12.0
Estonia	1,101.0	1,369.0	1,017.0	0.6	−0.5	7.0	−10.5	−4.9
Finland	4,009.0	5,177.0	5,289.0	0.3	0.1	16.9	12.3	4.8
France	41,829.0	59,296.0	64,165.0	0.4	0.3	312.8	236.0	142.8
Germany	68,376.0	82,282.0	81,959.0	−0.1	−0.2	339.1	229.8	152.4
Greece	7,566.0	10,903.0	10,707.0	1.3	0.1	44.5	22.4	11.0
Hungary	9,338.0	10,012.0	8,865.0	0.3	−0.6	−55.4	−31.4	−31.1
Iceland	143.0	282.0	325.0	0.9	—	2.7	1.8	1.1
Ireland	2,969.0	3,819.0	4,668.0	1.4	0.8	−6.4	37.2	31.9
Italy	47,104.0	57,536.0	52,939.0	0.4	−0.4	4.0	74.2	−67.5
Latvia	1,949.0	2,373.0	1,857.0	0.4	−0.7	12.3	−23.5	−12.9
Lithuania	2,567.0	3,501.0	3,035.0	0.7	−0.2	22.2	−10.4	−3.6
Macedonia	1,230.0	2,024.0	2,199.0	1.4	0.4	6.9	11.0	7.2
Moldova	2,341.0	4,283.0	4,096.0	0.9	−0.2	49.9	−5.8	16.0
Netherlands	10,114.0	15,898.0	17,123.0	0.7	0.4	92.0	86.6	62.5
Norway	3,265.0	4,473.0	4,859.0	0.4	0.4	17.9	24.4	18.2
Poland	24,824.0	38,671.0	37,337.0	0.9	0.0	178.7	8.5	11.2
Portugal	8,405.0	10,016.0	9,834.0	1.4	−0.1	5.1	15.9	9.6
Romania	16,311.0	22,480.0	20,806.0	0.9	−0.3	69.0	−56.3	−49.8
Russian Federation	102,192.0	145,612.0	124,428.0	0.6	−0.5	820.8	−422.7	−281.7
Serbia-Montenegro	7,131.0	10,555.0	10,230.0	0.9	0.1	21.0	−5.2	−4.6
Slovak Republic	3,463.0	5,391.0	5,397.0	1.0	0.0	23.6	9.3	6.1
Slovenia	1,473.0	1,990.0	1,859.0	1.0	−0.2	4.6	3.6	1.2
Spain	28,009.0	40,752.0	40,369.0	1.1	0.0	163.2	49.1	−2.8
Sweden	7,014.0	8,856.0	9,055.0	0.3	0.0	40.5	9.5	0.5
Switzerland	4,694.0	7,173.0	6,801.0	−0.1	−0.1	54.8	19.2	7.9
Ukraine	37,298.0	49,688.0	40,775.0	0.4	−0.7	142.7	−432.6	−264.4
United Kingdom	49,816.0	58,689.0	63,275.0	0.0	0.0	189.0	178.9	94.6
OCEANIA								
Australia	8,219.0	19,153.0	23,205.0	0.9	0.7	246.8	209.7	167.0
Fiji	289.0	814.0	965.0	1.9	—	7.8	11.3	12.8
New Zealand	1,908.0	3,784.0	4,379.0	0.2	0.5	12.3	50.8	38.5
Papua New Guinea	1,798.0	5,334.0	8,443.0	2.5	1.9	89.3	116.4	125.1
Solomon Islands	90.0	437.0	783.0	3.5	—	11.0	13.8	14.3

[a]Data include projections based on 1990 base year population data.
[b]Includes Hong Kong and Macao.

Source: United Nations Population Division and International Labour Organisation. World Resources 2000–2001 (World Resources Institute), U.S. Bureau of the Census International Data Base (2000).

Table E
World Countries: Basic Demographic Data, 1975–2001

COUNTRY	CRUDE BIRTH RATE (births per 1,000 population)		LIFE EXPECTANCY AT BIRTH (years)		LIFE EXPECTANCY OF FEMALES AS A PERCENTAGE OF MALES (years)		TOTAL FERTILITY RATE		PERCENTAGE OF POPULATION IN SPECIFIC AGE GROUPS					
									1980			2001		
	1975–1980	2001	1975–80	2001	1975–80	1995–00	1975–80	2001	<15	15–65	>65	<15	15–65	>65
WORLD	28.3	21	59.7	67	106.0	107.9	3.9	2.6	35.2	58.9	5.9	29.6	63.4	7.0
AFRICA	46.0	39	47.9	46	106.7	105.6	6.5	—	44.7	52.2	3.1			
Algeria	45.0	23	57.5	71	103.5	102.9	7.2	2.9	46.5	49.6	3.9	35.4	60.7	3.9
Angola	50.2	47	40.0	47	108.1	106.6	6.8	6.6	44.7	52.4	2.9	47.4	49.7	2.9
Benin	51.4	39	47.0	53	108.2	105.7	7.1	5.4	45.1	50.8	4.1	45.9	51.4	2.7
Botswana	46.6	31	56.5	39	106.6	104.3	6.4	3.9	48.7	49.3	2.0	42.0	56.7	2.2
Burkina Faso	50.8	44	42.9	44	106.0	102.2	7.8	6.4	47.4	49.8	2.8	47.1	50.2	2.7
Burundi	44.7	39	46.0	42	107.2	107.3	6.8	5.9	44.7	51.8	3.4	46.0	51.4	2.6
Cameroon	45.5	36	48.5	49	106.4	105.6	6.5	4.7	44.4	52.0	3.6	41.6	54.8	3.7
Central African Republic	44.1	36	44.5	44	111.9	109.3	5.9	4.7	41.7	54.4	4.0	42.3	54.2	3.5
Chad	44.1	45	41.0	48	108.1	106.5	5.9	6.3	41.9	54.5	3.6	49.6	47.4	3.0
Dem. Republic of the Congo (formerly Zaire)	47.8	45	48.0	45	107.1	106.1	6.5	6.1	46.0	51.1	2.8	47.6	49.8	2.6
Congo Republic	45.8	42	48.7	51	111.3	110.8	6.3	5.9	45.1	51.5	3.4	46.3	50.5	3.2
Côte d'Ivoire	50.7	37	47.9	46	107.1	102.0	7.4	4.7	46.6	51.0	2.5	42.1	55.3	2.6
Egypt	38.9	25	54.1	68	104.5	104.6	5.3	3.2	39.5	56.5	4.0	34.7	61.1	4.2
Equatorial Guinea	42.7	—	42.0	—	107.9	108.3	5.7	—	41.0	54.8	4.1	—	—	—
Eritrea	45.1	38	45.3	51	106.8	106.1	6.1	5.4	44.2	53.1	2.6	45.2	52.2	2.6
Ethiopia	49.0	43	42.0	42	107.9	104.7	6.8	5.6	46.1	51.2	2.7	46.0	51.2	2.8
Gabon	32.9	35	47.0	53	107.3	105.8	4.1	34.4	59.5	6.1	40.3	54.1	5.7	
The Gambia	48.8	38	39.0	53	108.3	108.8	6.5	4.9	42.6	54.4	2.8	40.3	56.5	3.2
Ghana	45.1	29	51.0	56	106.9	106.9	6.5	4.1	44.9	52.3	2.8	43.3	52.2	4.6
Guinea	51.6	38	38.8	46	102.6	102.1	7.0	5.1	45.8	51.6	2.6	44.3	53.1	2.6
Guinea-Bissau	42.4	40	37.5	45	108.6	106.9	5.6	5.7	39.0	57.0	4.0	43.6	52.9	3.6
Kenya	53.6	35	53.4	46	107.8	103.9	8.1	4.3	50.1	46.5	3.4	43.0	54.3	2.7
Lesotho	41.9	32	51.8	43	107.0	103.6	5.7	4.3	41.9	53.9	4.2	39.7	56.1	4.3
Liberia	47.4	44	49.5	47	106.3	106.5	6.8	5.9	44.3	52.0	3.7	44.5	52.8	2.8
Libya	47.3	27	55.8	72	106.3	105.8	7.4	3.4	46.7	51.1	2.2	33.6	62.9	3.5
Madagascar	46.9	39	49.5	55	106.3	105.3	6.6	5.3	45.9	51.4	2.7	44.8	52.2	3.0
Malawi	57.2	45	43.1	38	103.3	102.5	7.6	6.2	47.5	50.3	2.3	44.4	52.0	3.6
Mali	50.7	46	40.0	41	108.1	105.8	7.1	6.2	46.8	50.7	2.5	47.1	49.9	3.0
Mauritania	44.7	41	45.5	51	107.3	105.8	6.5	4.5	43.7	53.3	3.0	43.9	52.9	3.2
Mauritius	26.7	—	64.9	—	108.3	—	3.1	—	35.6	60.8	3.6	25.5	68.3	6.2
Morocco	39.4	22	55.8	68	106.3	106.1	5.9	2.8	43.2	52.7	4.1	34.1	61.7	4.2
Mozambique	45.4	40	43.5	42	107.6	106.8	6.5	5.1	43.4	53.4	3.2	42.8	53.5	3.7
Namibia	41.9	35	51.3	44	105.0	101.9	6.0	4.9	43.1	53.4	3.5	41.7	54.6	3.8
Niger	59.7	50	40.5	46	108.2	106.4	8.1	7.2	46.8	50.8	2.5	49.0	48.7	2.4

Table E (continued)
World Countries: Basic Demographic Data, 1975–2001

COUNTRY	CRUDE BIRTH RATE (births per 1,000 population)		LIFE EXPECTANCY AT BIRTH (years)		LIFE EXPECTANCY OF FEMALES AS A PERCENTAGE OF MALES (years)		TOTAL FERTILITY RATE		PERCENTAGE OF POPULATION IN SPECIFIC AGE GROUPS					
									1980			2001		
	1975–1980	2001	1975–80	2001	1975–80	1995–00	1975–80	2001	<15	15-65	>65	<15	15-65	>65
Nigeria	46.6	39	45.0	46	107.4	106.1	6.5	5.2	44.3	3.1	2.6	43.9	53.5	2.6
Rwanda	52.8	44	45.0	40	107.4	107.7	8.5	5.8	48.8	48.8	2.4	47.6	49.3	3.1
Senegal	49.3	36	42.8	52	104.8	105.8	7.0	5.0	45.3	51.8	2.8	44.4	52.9	2.7
Sierra Leone	48.8	44	35.2	37	108.9	108.3	6.5	5.7	43.0	53.9	3.1	44.6	52.9	2.6
Somalia	50.4	50	42.0	47	107.9	108.8	7.0	7.0	46.0	51.0	3.0	47.9	49.7	2.4
South Africa	37.3	25	55.9	47	111.3	111.5	5.1	2.8	40.3	55.9	3.9	32.3	63.1	4.6
Sudan	47.1	34	46.7	58	106.2	103.7	6.7	4.5	44.9	52.4	2.7	39.9	56.6	3.5
Swaziland	46.2	35	49.9	45	109.9	—	6.5	4.3	45.9	51.3	2.9	42.2	55.0	2.8
Tanzania	47.5	39	49.0	44	107.2	104.2	6.8	5.2	47.6	50.1	2.3	45.2	52.4	2.4
Togo	45.2	35	48.0	49	107.1	104.1	6.6	5.0	44.6	52.3	3.2	43.9	53.0	3.2
Tunisia	36.4	17	60.0	72	101.7	104.4	5.7	2.1	41.6	54.6	3.8	28.9	65.1	5.9
Uganda	50.3	45	47.0	43	107.0	102.5	6.9	6.1	47.8	49.7	2.5	49.0	49.1	1.9
Zambia	51.6	39	49.3	37	106.9	102.5	7.2	5.2	49.4	48.2	2.4	45.1	52.7	2.2
Zimbabwe	44.2	29	53.8	39	106.9	102.3	6.6	3.7	47.9	49.5	2.6	44.6	52.2	3.2
NORTH AMERICA			73.3		111.2	108.7	1.8		22.5	66.4	11.0			
Canada	15.4	11	74.2	79	110.8	107.8	1.8	1.5	22.7	67.9	9.4	18.8	68.6	12.6
United States	15.1	15	73.2	78	111.2	109.6	1.8	2.1	22.5	66.3	11.2	21.2	66.2	12.6
CENTRAL AMERICA			63.7		110.2	106.7	5.4		45.1	51.2	3.6			
Belize	40.9	—	69.7	—	102.5	104.1	6.2	—	47.2	48.6	4.8	—	—	—
Costa Rica	31.7	20	71.0	78	106.4	106.7	3.9	2.4	38.9	57.5	3.6	31.2	63.1	5.7
Cuba	17.2	12	73.0	77	104.8	105.4	2.1	1.6	32.9	60.5	7.6	21.1	68.8	10.1
Dominican Republic	34.9	23	62.0	67	106.1	105.8	4.7	2.7	42.3	54.6	3.1	33.0	62.7	4.4
EL Salvador	41.5	26	57.2	70	119.5	108.9	5.7	3.0	46.1	50.8	3.1	35.3	59.7	5.0
Guatemala	44.3	33	56.4	65	107.2	109.8	6.4	4.4	45.9	51.3	2.8	43.2	53.3	3.6
Haiti	36.8	32	50.7	52	106.3	109.8	5.4	4.3	40.7	54.8	4.4	40.2	56.3	3.5
Honduras	44.9	30	57.7	66	107.7	105.9	6.6	4.1	47.2	50.1	2.7	41.1	55.2	3.4
Jamaica	28.8	21	70.1	76	106.3	105.5	4.0	2.4	40.2	53.0	6.8	30.6	62.4	7.0
Mexico	37.1	24	65.3	73	110.3	107.1	5.3	2.5	45.1	51.1	3.8	33.6	61.4	5.0
Nicaragua	45.7	30	57.6	69	108.5	107.6	6.4	3.5	47.7	49.8	2.5	42.1	54.9	3.1
Panama	31.0	21	69.0	75	105.8	105.5	4.1	2.5	40.5	55.0	4.5	30.9	63.5	5.6
Trinidad and Tobago	29.3	15	68.3	72	107.6	105.5	3.4	1.8	34.2	60.2	5.5	24.9	68.8	6.3
SOUTH AMERICA			62.9		108.8	109.0	4.3		37.8	57.6	4.6			
Argentina	25.7	19	68.7	74	110.4	110.0	3.4	2.5	30.5	61.4	8.1	27.5	62.8	9.7
Bolivia	41.0	30	50.1	63	108.8	105.0	5.8	3.8	42.6	53.9	3.5	39.1	56.5	4.4
Brazil	32.6	19	61.8	68	108.1	112.7	4.3	2.2	38.1	57.8	4.2	28.4	66.4	5.2
Chile	24.0	17	67.2	76	110.5	108.3	3.0	2.1	33.5	60.9	5.6	27.8	65.0	7.2
Colombia	31.7	22	64.0	72	107.3	110.4	4.1	2.5	40.0	56.2	3.7	32.4	62.9	4.7

Table E *(continued)*
World Countries: Basic Demographic Data, 1975–2001

COUNTRY	CRUDE BIRTH RATE (births per 1,000 population)		LIFE EXPECTANCY AT BIRTH (years)		LIFE EXPECTANCY OF FEMALES AS A PERCENTAGE OF MALES (years)		TOTAL FERTILITY RATE		PERCENTAGE OF POPULATION IN SPECIFIC AGE GROUPS					
									1980			2001		
	1975–1980	2001	1975–80	2001	1975–80	1995–00	1975–80	2001	<15	15-65	>65	<15	15-65	>65
Ecuador	38.2	24	61.4	70	105.9	107.5	5.4	2.9	42.8	53.2	4.0	33.4	61.9	4.7
Guyana	31.5	—	60.7	—	108.4	111.4	3.9	—	40.8	55.2	4.0	—	—	—
Paraguay	35.9	30	66.6	71	106.7	107.4	5.2	3.9	42.2	53.3	4.5	39.1	57.3	3.5
Peru	38.0	24	58.5	70	106.7	107.6	5.4	2.7	41.9	54.5	3.6	32.9	62.2	4.9
Suriname	29.5	—	65.1	—	107.8	107.3	4.2	—	39.7	55.8	4.5	—	—	—
Uruguay	20.3	16	69.6	74	110.2	111.4	2.9	2.2	26.9	62.5	10.5	24.7	62.7	12.6
Venezuela	34.2	23	67.6	74	109.1	108.6	4.5	2.8	40.7	66.1	3.2	33.5	62.1	4.4
ASIA			58.5		102.6	106.2	4.2		37.6	58.0	4.4			
Afghanistan	50.8	48	40.0	43	100.0	102.2	7.2	6.8	43.0	54.5	2.5	43.7	53.5	2.8
Armenia	20.9	11	72.3	74	109.0	110.0	2.5	1.4	30.4	63.6	6.0	23.0	67.7	9.3
Azerbaijan	24.7	16	68.5	65	112.1	110.4	3.6	2.1	34.5	60.0	5.4	28.4	64.4	7.2
Bangladesh	47.2	28	46.6	62	97.9	100.0	6.7	3.0	46.0	50.5	3.4	37.0	59.7	3.3
Bhutan	42.8	—	42.6	—	104.8	103.3	5.9	—	41.3	55.6	3.1	—	—	—
Cambodia	30.0	29	31.2	54	108.3	107.8	4.1	3.9	39.2	57.9	2.9	43.0	54.2	2.8
China	21.5	15	65.3	70	103.0	105.8	3.3	1.9	35.5	59.7	4.7	24.8	68.1	7.1
Georgia	18.1	8	70.7	73	111.9	111.6	2.4	1.1	25.7	65.1	9.1	20.1	66.8	13.1
India	34.7	25	52.9	63	96.3	101.6	4.8	3.0	38.5	57.4	4.0	33.1	61.9	5.0
Indonesia	35.4	21	52.8	66	104.9	106.3	4.7	2.4	41.0	55.6	3.3	30.2	65.1	4.7
Iran	44.7	22	58.6	69	101.4	101.4	6.5	2.6	44.9	51.8	3.3	32.6	62.7	4.7
Iraq	41.9	30	61.4	62	103.0	104.9	6.6	4.2	46.0	51.3	2.7	40.9	56.2	2.9
Israel	26.0	21	73.1	79	104.9	105.2	3.4	2.8	33.2	58.2	8.6	27.6	62.6	9.7
Japan	15.2	9	75.5	81	107.4	107.8	1.8	1.4	23.6	67.4	9.0	14.5	67.9	17.6
Jordan	45.0	29	61.2	72	106.1	104.3	7.4	3.6	49.4	47.5	3.1	38.2	58.7	3.1
Kazakhstan	24.9	15	65.4	63	117.1	114.2	3.1	1.8	32.4	61.5	6.1	26.1	66.4	7.5
Korea, North	“22.2	18	65.8	61	110.3	108.7	3.3	2.1	39.5	57.0	3.5	27.1	67.1	5.8
Korea, South	“23.9	13	64.8	74	111.6	110.1	2.9	1.4	34.0	62.2	3.8	21.3	71.8	7.0
Kuwait	40.1	20	69.6	77	106.2	105.4	5.9	2.6	40.2	58.3	1.4	32.2	65.5	2.3
Kyrgyzstan	29.9	20	64.2	66	114.2	114.2	4.1	2.5	37.1	57.1	5.8	33.4	60.6	6.0
Laos	45.1	36	43.5	54	106.9	105.7	6.7	4.9	42.0	55.2	2.8	42.4	54.1	3.5
Lebanon	30.1	20	65.0	71	106.2	105.8	4.3	2.3	40.1	54.5	5.4	31.5	62.6	5.9
Malaysia	30.4	22	65.3	55	105.7	105.7	4.2	2.9	39.3	57.0	3.7	33.7	62.1	4.2
Mongolia	39.2	22	56.3	65	104.5	104.6	6.7	2.5	43.2	53.9	2.9	33.2	62.7	4.0
Myanmar (Burma)	37.5	24	51.2	57	106.4	105.0	5.3	2.9	39.6	56.4	4.0	32.7	62.7	4.6
Nepal	43.4	32	46.2	59	96.6	98.2	6.2	4.2	42.9	54.1	3.0	40.7	55.5	3.8
Oman	46.1	27	54.9	74	104.3	105.8	7.2	4.1	44.6	52.8	2.6	43.2	54.2	2.5
Pakistan	47.3	33	53.4	63	101.3	103.1	7.0	4.6	44.4	52.7	2.9	41.2	55.5	3.3
Philippines	35.9	26	59.9	70	105.5	104.4	5.0	3.3	41.9	55.3	2.8	36.9	59.2	3.9
Saudi Arabia	45.9	33	58.8	73	104.0	104.2	7.3	5.9	44.3	52.9	2.8	40.9	56.2	2.9
Singapore	17.2	—	70.8	78	106.6	105.3	1.9	1.4	27.0	68.2	4.7	21.5	71.2	7.3
Sri Lanka	28.5	18	66.8	73	105.4	105.6	3.8	2.1	35.3	60.4	4.3	26.0	67.6	6.4

Table E *(continued)*
World Countries: Basic Demographic Data, 1975–2001

COUNTRY	CRUDE BIRTH RATE (births per 1,000 population)		LIFE EXPECTANCY AT BIRTH (years)		LIFE EXPECTANCY OF FEMALES AS A PERCENTAGE OF MALES (years)		TOTAL FERTILITY RATE		PERCENTAGE OF POPULATION IN SPECIFIC AGE GROUPS					
									1980			2001		
	1975–1980	2001	1975–80	2001	1975–80	1995–00	1975–80	2001	<15	15-65	>65	<15	15-65	>65
Syria	46.0	29	60.1	70	106.2	105.9	7.4	3.6	48.5	48.3	3.2	40.0	56.9	3.1
Tajikistan	37.2	21	64.5	67	108.3	109.3	5.9	3.0	42.9	52.5	4.6	38.6	57.0	4.4
Thailand	31.6	15	61.2	69	106.6	109.0	4.3	1.8	40.0	56.5	3.5	23.6	70.1	6.3
Turkey	32.0	20	60.3	70	107.8	107.4	4.5	2.3	39.2	56.0	4.7	28.3	65.9	5.8
Turkmenistan	35.3	20	61.6	65	11.2	111.2	5.3	2.3	41.3	54.4	4.3	36.6	59.1	4.3
United Arab Emirates	30.5	17	66.8	75	106.5	102.7	5.7	3.1	28.6	70.1	1.3	26.1	71.2	2.7
Uzbekistan	34.4	21	65.1	67	111.0	110.9	5.1	2.5	40.9	54.0	5.1	36.5	59.0	4.5
Vietnam	38.3	19	55.8	69	108.2	107.7	5.6	2.2	42.5	52.7	4.8	32.4	62.3	5.3
Yemen	53.7	41	44.1	57	101.1	101.7	7.6	6.1	50.2	47.2	2.6	46.2	51.0	2.8
EUROPE			71.3		111.4	110.6	2.0		22.2	65.5	12.3			
Albania	30.3	17	68.9	74	106.6	108.6	4.2	2.2	35.9	58.9	5.2	28.6	64.5	6.9
Austria	11.5	9	72.0	78	110.4	108.1	1.7	1.3	20.4	64.2	15.4	16.5	67.8	15.7
Belarus	16.1	9	71.1	68	114.7	119.3	2.1	1.3	22.9	66.4	10.7	18.0	68.3	13.6
Belgium	12.4	11	72.3	78	109.6	109.4	1.7	1.6	20.2	65.5	14.3	17.2	66.1	16.6
Bosnia-Herzegovina	19.6	12	69.9	74	107.0	107.0	2.2	1.6	27.3	66.7	6.1	18.6	71.3	10.2
Bulgaria	15.8	9	71.3	72	107.9	110.2	2.2	1.3	22.1	66.0	11.9	15.2	68.6	16.2
Croatia	15.6	9	70.6	74	110.4	111.6	2.0	1.4	21.2	67.2	11.7	16.7	68.1	15.2
Czech Republic	17.4	9	70.6	75	110.4	111.0	2.3	1.6	23.5	63.2	13.4	16.1	70.1	13.8
Denmark	12.3	12	74.2	77	108.4	106.8	1.7	1.8	20.8	64.7	14.4	18.4	66.7	14.9
Estonia	15.0	9	69.7	71	115.3	119.0	2.1	1.2	21.7	65.8	12.5	17.1	68.0	14.9
Finland	13.6	11	72.7	78	112.6	110.9	1.6	1.7	20.3	67.7	12.0	17.9	67.0	15.0
France	14.0	13	73.7	79	111.6	110.8	1.9	1.9	22.3	63.8	14.0	18.7	65.1	16.1
Germany	10.4	9	72.5	78	109.4	108.1	1.5	1.4	18.5	65.9	15.6	15.3	68.3	16.4
Greece	15.6	11	73.7	78	105.7	106.5	2.3	1.3	22.8	64.0	13.1	14.9	67.0	18.1
Hungary	16.2	10	69.4	72	109.8	111.9	2.1	1.3	21.9	64.6	13.4	16.8	68.7	14.5
Iceland	19.2	—	76.3	—	108.0	105.1	2.3	—	27.6	62.7	10.1	—	—	—
Ireland	21.2	15	72.0	77	107.2	106.7	3.5	1.9	30.6	58.7	10.7	21.5	67.2	11.2
Italy	13.0	9	73.6	79	109.2	108.0	1.9	1.2	22.3	64.6	13.1	14.2	67.4	18.4
Latvia	14.0	8	69.2	70	115.6	119.3	2.0	1.2	20.4	66.5	13.0	16.5	68.6	14.9
Lithuania	15.4	9	70.8	73	114.2	118.7	2.1	1.3	23.6	65.1	11.3	18.9	67.4	13.7
Macedonia	22.6	13	69.6	73	105.2	105.6	2.7	1.8	28.6	64.5	6.9	22.2	67.5	10.2
Moldova	19.6	9	64.8	67	111.2	112.5	2.4	1.4	26.7	65.5	7.8	21.8	67.1	11.0
Netherlands	12.7	13	75.3	78	109.0	108.0	1.6	1.7	22.3	66.2	11.5	18.5	67.9	13.7
Norway	12.9	13	75.3	79	108.9	108.0	1.8	1.8	22.2	63.1	14.8	20.0	65.0	15.1
Poland	19.2	10	70.9	74	111.9	113.2	2.3	1.3	24.3	65.6	10.1	18.8	69.0	12.2
Portugal	18.2	11	70.2	76	110.6	109.7	2.4	1.5	25.9	63.6	10.5	17.1	67.6	15.2
Romania	19.1	10	69.6	70	107.0	112.1	2.6	1.3	26.7	63.1	10.3	17.7	68.8	13.5
Russian Federation	15.8	9	67.4	66	118.1	119.6	1.9	1.2	21.6	68.1	10.2	17.5	69.9	12.5

Table E *(continued)*
World Countries: Basic Demographic Data, 1975–2001

COUNTRY	CRUDE BIRTH RATE (births per 1,000 population)		LIFE EXPECTANCY AT BIRTH (years)		LIFE EXPECTANCY OF FEMALES AS A PERCENTAGE OF MALES (years)		TOTAL FERTILITY RATE		PERCENTAGE OF POPULATION IN SPECIFIC AGE GROUPS					
									1980			2001		
	1975–1980	2001	1975–80	2001	1975–80	1995–00	1975–80	2001	<15	15-65	>65	<15	15-65	>65
Serbia-Montenegro	18.4	12	70.3	73	106.5	107.1	2.4	1.7	24.1	66.1	9.8	20.1	66.2	13.7
Slovak Republic	20.6	10	70.4	73	110.9	111.5	2.5	1.3	26.1	63.5	10.4	19.3	69.3	11.3
Slovenia	16.6	9	71.0	76	111.6	109.8	2.2	1.2	23.4	65.3	11.4	15.7	70.2	14.2
Spain	17.4	10	74.3	78	108.4	109.3	2.6	1.2	26.6	62.7	10.7	15.0	68.1	16.9
Sweden	11.7	10	75.2	80	109.3	106.6	1.7	1.6	19.6	64.1	16.3	17.9	64.6	17.5
Switzerland	11.6	10	75.2	80	109.2	109.3	1.5	1.4	19.7	66.4	13.8	16.8	67.8	15.4
Ukraine	14.9	8	69.3	68	114.8	115.6	2.0	1.2	21.4	66.7	11.9	17.1	68.6	14.3
United Kingdom	12.4	11	72.8	77	109.0	106.6	1.7	1.7	20.9	64.0	15.1	18.6	65.4	16.1
OCEANIA			68.2		108.7	106.7	2.8		29.3	62.7	8.0			
Australia	16.0	13	73.4	79	109.8	108.0	2.1	1.8	25.3	65.1	9.6	20.5	67.5	12.0
Fiji	33.2	—	67.1	—	105.3	105.6	4.0	—	39.1	58.0	2.8	—	—	—
New Zealand	17.2	15	72.4	78	109.2	108.1	2.2	2.0	26.7	63.3	10.0	22.3	65.9	11.8
Papua New Guinea	39.6	32	49.7	57	101.0	103.5	5.9	4.4	43.0	55.5	1.6	39.9	57.6	2.5
Solomon Islands	45.0	—	65.3	—	106.2	105.7	7.1	—	47.6	49.3	3.1	—	—	—
DEVELOPING COUNTRIES			56.7		103.8	104.8	4.7		39.3	56.6	4.1			
DEVELOPING COUNTRIES			72.2		110.8	111.2	1.9		22.5	65.9	11.6			

Sources: United Nations Population Division; World Resources 2000–2001 (World Resources Institute); World Development Indicators, 2003 (World Bank).

Table F
World Countries: Mortality, Health, and Nutrition

COUNTRY	MORTALITY						HEALTH			NUTRITION	
	Infant Mortality		Adult Mortality								
	Infant Mortality (per 1,000 live births)	Under-five Mortality (per 1,000)	Male (per 1,000)		Female (per 1,000)						
	2001	2001	1980	2000–2001	1980	2000–2001	Health Expenditure per Capita ($US) 1997–2000[a]	Physicians per 1,000 people 1995–2000[a]	Hospital Beds per 1,000 people 1995–2000[a]	Average daily per capita supply of calories (kilocalories)[c] 1987	Average daily per capita supply of calories (kilocalories)[c] 1997
Afghanistan	—	257	—	437	—	376	8	0.1	—	—	—
Albania	23	25	140	209	82	95	41	1.3	3.2	2,556	2,961
Algeria	39	49	226	155	197	119	64	1.0	2.1	2,757	2,853
Angola	154	260	569	492	458	386	24	0.1	—	2,063	2,183
Argentina	16	19	205	184	102	92	658	2.7	3.3	3,096	3,093
Armenia	31	35	158	223	85	106	38	3.2	0.7	—	2,371
Australia	6	6	178	100	53	55	1,698	2.5	7.9	3,159	3,224
Austria	5	5	197	124	92	59	1,872	3.1	8.6	3,419	3,536
Azerbaijan	77	96	262	261	127	150	8	3.6	9.7	—	2,236
Bangladesh	51	77	383	262	388	252	14	0.2	—	2,062	2,086
Belarus	17	20	255	381	95	133	57	4.4	12.2	—	3,226
Belgium	5	6	173	127	90	66	1,936	3.9	7.3	3,454	3,619
Benin	94	158	486	384	397	328	11	0.1	—	1,961	2,487
Bolivia	60	77	357	264	273	219	67	1.3	1.7	2,153	2,174
Bosnia-Herzegovina	15	18	181	200	108	93	50	1.4	1.8	—	2,266
Botswana	80	110	341	703	278	669	191	—	—	2,337	2,183
Brazil	31	36	221	259	161	136	267	1.3	3.1	2,745	2,974
Bulgaria	14	16	190	239	106	109	59	3.4	7.4	3,699	2,686
Burkina Faso	104	197	467	559	362	507	8	0.0[b]	1.4	2,181	2,121
Burundi	114	190	489	648	400	603	3	—	—	2,023	1,685
Cambodia	97	138	473	373	355	264	19	0.3	—	1,868	2,048
Cameroon	96	155	489	488	415	440	24	0.1	—	2,178	2,111
Canada	5	7	161	101	85	57	2,058	6.8	3.9	3,105	3,119
Central African Republic	115	80	540	620	424	573	8	0.0[b]	0.9	1,914	2,016
Chad	117	200	556	449	449	361	6	—	—	1,596	2,032
Chile	10	12	218	151	120	67	336	1.1	2.7	2,518	2,796
China	31	39	185	161	148	110	45	1.7	2.4	2,608[d]	2,897[d]
Hong Kong, China	3	—	150	99	87	51	—	1.3	—	—	—
Colombia	19	23	237	238	162	115	186	1.2	1.5	2,341	2,597
Democratic Republic of the Congo (formerly Zaire)	129	205	—	571	—	493	9	0.1	—	2,132	1,755
Congo, Rep.	81	108	408	475	298	406	22	0.3	—	2,326	2,144
Costa Rica	9	11	159	131	100	78	273	0.9	1.7	2,717	2,649
Côte d'Ivoire	102	175	421	553	346	494	16	0.1	—	2,677	2,610
Croatia	7	8	233	152	106	114	434	2.3	—	—	2,445

Table F (*continued*)
World Countries: Mortality, Health, and Nutrition

COUNTRY	MORTALITY						HEALTH			NUTRITION	
	Infant Mortality		Adult Mortality								
	Infant Mortality (per 1,000 live births)	Under-five Mortality (per 1,000)	Male (per 1,000)		Female (per 1,000)						
	2001	2001	1980	2000–2001	1980	2000–2001	Health Expenditure per Capita ($US) 1997–2000[a]	Physicians per 1,000 people 1995–2000[a]	Hospital Beds per 1,000 people 1995–2000[a]	Average daily per capita supply of calories (kilocalories)[c] 1987	Average daily per capita supply of calories (kilocalories)[c] 1997
Cuba	7	9	135	143	94	94	169	5.3	5.1	3,125	2,480
Czech Republic	4	5	225	167	102	75	358	3.1	8.8	—	3,244
Denmark	8	4	163	129	102	81	2,512	3.4	4.5	3,211	3,407
Dominican Republic	41	47	183	234	138	146	151	2.2	1.5	2,330	2,288
Ecuador	24	30	229	199	176	120	26	1.7	1.6	2,430	2,679
Egypt	35	41	257	210	204	147	51	1.6	2.1	3,120	3,287
El Salvador	33	39	410	250	178	148	184	1.1	1.6	2,312	2,562
Eritrea	72	111	—	493	—	441	9	0.0[b]	—	—	1,622
Estonia	11	12	291	316	110	114	218	3.0	7.4	—	2,849
Ethiopia	116	172	491	594	401	535	5	—	—	1,677	1,858
Finland	4	5	206	144	74	61	1,559	3.1	7.5	2,941	3,100
France	4	6	190	137	85	59	2,057	3.0	8.2	3,543	3,518
Gabon	60	90	474	380	387	330	120	—	—	2,460	2,556
The Gambia	91	126	584	373	466	320	10	0.0[b]	—	2,498	2,350
Georgia	24	29	210	250	94	133	41	4.4	4.8	—	2,614
Germany	4	5	177	126	90	60	2,422	3.6	9.1	3,478	3,382
Ghana	57	100	400	379	334	326	11	0.1	—	1,979	2,611
Greece	5	5	134	114	86	47	884	4.4	4.9	3,481	3,649
Guatemala	43	58	336	286	266	182	79	0.9	1.0	2,384	2,339
Guinea	109	169	589	432	507	366	13	0.1	—	2,060	2,232
Guinea-Bissau	130	211	535	495	517	427	9	0.2	—	2,378	2,430
Haiti	79	123	348	524	275	373	21	0.2	0.7	1,848	1,869
Honduras	31	38	306	221	237	157	62	0.8	1.1	2,206	2,403
Hungary	8	9	270	295	130	123	315	3.2	8.2	3,768	3,313
India	67	93	261	250	279	191	23	—	—	2,228	2,496
Indonesia	33	45	368	230	308	178	19	—	—	2,458	2,886
Iran	35	42	221	170	190	139	258	0.9	1.6	2,659	2,836
Iraq	107	133	207	258	191	208	375	0.6	1.5	3,418	2,619
Ireland	6	6	175	108	103	62	1,692	2.3	9.7	3,623	3,565
Israel	6	6	138	99	85	56	2,021	3.8	6.0	3,075	3,278
Italy	4	6	163	110	80	53	1,498	6.0	4.9	3,512	3,507
Jamaica	17	20	186	169	121	127	165	1.4	2.1	2,630	2,553
Japan	3	5	129	98	70	44	2,908	1.9	16.5	2,870	2,932
Jordan	27	33	—	199	—	144	137	1.7	1.8	2,780	3,014
Kazakhstan	81	99	312	366	140	201	44	3.5	8.5	—	3,085
Kenya	78	122	417	578	339	529	28	0.1	—	2,025	1,977
Korea, North	42	55	270	238	156	192	18	3.0	—	2,509	1,837

Table F *(continued)*
World Countries: Mortality, Health, and Nutrition

COUNTRY	MORTALITY						HEALTH			NUTRITION	
	Infant Mortality		Adult Mortality								
	Infant Mortality (per 1,000 live births)	Under-five Mortality (per 1,000)	Male (per 1,000)		Female (per 1,000)		Health Expenditure per Capita ($US) 1997–2000[a]	Physicians per 1,000 people 1995–2000[a]	Hospital Beds per 1,000 people 1995–2000[a]	Average daily per capita supply of calories (kilocalories)[c] 1987	Average daily per capita supply of calories (kilocalories)[c] 1997
	2001	2001	1980	2000–2001	1980	2000–2001					
Korea, South	5	5	270	186	156	71	584	1.3	6.1	3,110	3,155
Kuwait	9	10	172	100	116	68	586	1.9	2.8	3,021	3,096
Kyrgyzstan	52	61	296	335	131	299	12	3.0	9.5	—	2,447
Laos	87	100	531	355	439	299	11	0.2	—	2,102	2,108
Latvia	17	21	281	328	106	122	174	2.8	10.3	—	2,864
Lebanon	28	32	241	192	181	136	499	2.1	2.7	3,040	3,277
Lesotho	91	132	371	667	279	630	28	0.1	—	2,216	2,244
Liberia	157	235	—	448	—	385	2	0.0	—	—	—
Libya	16	19	276	210	218	157	246	1.3	4.3	3,308	3,289
Lithuania	8	9	243	286	92	106	185	4.0	9.2	—	3,261
Macedonia	22	26	—	160		89	106	2.2	4.9	—	2,664
Madagascar	84	136	353	385	278	322	9	0.1	—	2,292	2,022
Malawi	114	183	429	701	349	653	11	—	1.3	2,027	2,043
Malaysia	8	8	230	202	149	113	101	0.7	2.0	2,616	2,977
Mali	141	231	454	518	362	446	10	0.1	0.2	1,967	2,030
Mauritania	120	183	505	357	416	302	14	0.1	—	2,509	2,622
Mauritius	17	19	277	228	181	109	134	0.9	—	—	—
Mexico	24	29	205	180	121	101	311	1.8	1.1	3,022	3,097
Moldova	27	32	289	325	173	165	11	3.5	12.1	—	2,567
Mongolia	61	76	320	280	273	199	23	2.4	—	2,034	1,917
Morocco	39	44	264	174	207	113	50	0.5	1.0	3,047	3,078
Mozambique	125	197	468	674	361	612	9	—	—	1,785	1,832
Myanmar (Burma)	77	109	284	343	313	245	153	0.3	—	2,697	2,862
Namibia	55	67	427	695	366	661	136	0.3	—	2,199	2,183
Nepal	66	91	376	314	395	314	—	0.0[b]	0.2	2,144	2,366
Netherlands	5	6	133	95	74	64	1,900	3.2	10.8	3,076	3,284
New Zealand	6	6	177	108	91	69	1,062	2.2	6.2	3,201	3,395
Nicaragua	36	43	277	225	189	161	43	0.9	1.5	2,330	2,186
Niger	156	265	562	473	453	308	5	0.0[b]	0.1	2,033	2,097
Nigeria	110	183	535	443	453	393	8	—	—	2,103	2,735
Norway	4	4	144	106	71	60	2,832	2.9	14.6	3,304	3,357
Oman	12	13	389	187	326	135	295	1.3	2.2	—	—
Pakistan	84	109	283	221	291	198	18	0.6	—	2,224	2,476
Panama	19	25		145		93	268	1.7	2.2	2,302	2,430
Papua New Guinea	70	94	514	359	478	329	31	0.1	—	2,137	2,224
Paraguay	26	30	198	173	144	129	112	1.1	1.3	2,564	2,566

Table F *(continued)*
World Countries: Mortality, Health, and Nutrition

COUNTRY	MORTALITY						HEALTH			NUTRITION	
	Infant Mortality		Adult Mortality								
	Infant Mortality (per 1,000 live births)	Under-five Mortality (per 1,000)	Male (per 1,000)		Female (per 1,000)						
	2001	2001	1980	2000–2001	1980	2000–2001	Health Expenditure per Capita ($US) 1997–2000[a]	Physicians per 1,000 people 1995–2000[a]	Hospital Beds per 1,000 people 1995–2000[a]	Average daily per capita supply of calories (kilocalories)[c] 1987	Average daily per capita supply of calories (kilocalories)[c] 1997
Peru	30	39	287	190	229	139	100	0.9	1.5	2,276	2,302
Philippines	29	38	323	249	259	142	33	1.2	—	2,244	2,366
Poland	8	9	254	226	105	88	246	2.2	4.9	3,441	3,366
Portugal	5	6	199	164	95	66	862	3.2	4.0	3,400	3,667
Puerto Rico	10	—	159	149	78	56	—	1.8	3.3	—	—
Romania	19	21	216	260	119	117	48	1.8	7.6	2,944	3,253
Russian Federation	18	21	341	424	120	153	92	4.2	12.1	—	2,904
Rwanda	96	183	503	667	409	599	12	—	—	2,042	2,057
Saudi Arabia	23	28	283	181	241	116	448	1.7	2.3	2,488	2,783
Senegal	79	138	586	355	516	303	22	0.1	0.4	2,104	2,418
Serbia-Montenegro	17	19	164	180	106	100	50	2.0	5.3	—	3,031
Sierra Leone	182	316	540	587	527	531	6	0.1	—	2,126	2,035
Singapore	3	4	199	114	115	61	814	1.6	—	—	—
Slovak Republic	8	9	226	210	105	83	210	3.5	7.1	—	2,984
Slovenia	4	5	250	170	105	76	788	2.3	5.7	—	3,101
Somalia	47	225	—	516	—	452	19	0.0	—	—	—
South Africa	56	71	—	594	—	543	255	0.6	—	2,976	2,990
Spain	4	6	144	122	69	49	1,073	3.3	4.1	3,150	3,310
Sri Lanka	17	19	210	244	152	124	31	0.4	—	2,253	2,302
Sudan	65	107	537	341	462	291	13	0.1	—	2,208	2,395
Swaziland	106	149	—	635	—	595	56	0.2	—	—	—
Syria	23	28	—	170	—	132	30	1.3	1.4	3,166	3,352
Tajikistan	91	116	190	293	129	204	6	2.0	—	—	2,001
Tanzania	104	165	451	569	370	520	12	0.0[b]	—	2,228	1,995
Thailand	24	28	280	245	210	150	71	0.4	2.0	2,133	2,360
Togo	79	141	457	460	375	406	8	0.1	—	1,946	2,469
Trinidad and Tobago	17	20	234	209	166	133	268	0.8	5.1	2,975	2,661
Tunisia	21	27	227	169	224	99	110	0.7	1.7	3,067	3,283
Turkey	36	43	—	218	—	120	150	1.2	2.6	3,496	3,525
Turkmenistan	69	87	263	280	154	157	52	3.0	—	—	2,306
Uganda	79	124	463	617	395	567	10	—	—	2,113	2,085
Ukraine	17	20	282	365	112	135	26	3.0	11.8	—	2,795
United Arab Emirates	8	9	153	143	106	93	767	1.8	2.6	3,038	3,390
United Kingdom	6	7	160	109	96	66	1,747	1.8	4.1	3,215	3,276
United States	7	8	194	141	102	82	4,499	2.8	3.6	3,430	3,699

Table F *(continued)*
World Countries: Mortality, Health, and Nutrition

COUNTRY	MORTALITY						HEALTH			NUTRITION	
	Infant Mortality		Adult Mortality								
	Infant Mortality (per 1,000 live births)	Under-five Mortality (per 1,000)	Male (per 1,000)		Female (per 1,000)						
	2001	2001	1980	2000–2001	1980	2000–2001	Health Expenditure per Capita ($US) 1997–2000[a]	Physicians per 1,000 people 1995–2000[a]	Hospital Beds per 1,000 people 1995–2000[a]	Average daily per capita supply of calories (kilocalories)[c] 1987	Average daily per capita supply of calories (kilocalories)[c] 1997
Uruguay	14	16	176	185	91	89	653	3.7	4.4	2,613	2,816
Uzbekistan	52	68	219	282	116	176	29	3.3	8.3	—	2,433
Venezuela	19	22	219	178	123	99	233	2.4	1.5	2,602	2,321
Vietnam	30	38	262	203	204	139	21	0.5	1.7	2,193	2,484
West Bank and Gaza	21	25	—	157	—	100	—	0.5	1.2	—	—
Yemen	79	107	382	278	304	226	20	0.2	0.6	2,126	2,051
Zambia	112	202	482	725	413	687	18	0.1	—	2,017	1,970
Zimbabwe	76	123	389	650	321	612	43	0.1	—	2,112	2,145

[a]Data are for the most recent year available.
[b]Less than 0.05.
[c]Has replaced "calories available as a percentage of need" as a benchmark for nutrition.
[d]Data for China include Taiwan.
Sources: World Development Indicators, 2003 (World Bank); World Resources 2000–2001 (World Resources Institute).

Table G
World Countries: Education and Literacy, 1990–2002

COUNTRY	EDUCATIONAL INPUTS					OUTCOMES							
	Public Expenditure on Education 2000[b] (% of government expenditure)	Primary Pupil-Teacher Ratio 2000 (pupils per teacher)	Primary School Enrollment 2000 (% of age group)	Secondary School Enrollment 2000 (% of age group)	Tertiary School Enrollment 2000 (% of age group)	ADULT ILLITERACY (% above age 15)				YOUTH ILLITERACY RATE (% aged 15–24)			
						Male		Female		Male		Female	
						1990	2001	1990	2001	1990	2001	1990	2001
Afghanistan	—	43	15	—	—	—	—	—	—	—	—	—	—
Albania	—	22	107	78	15	14	8	32	22	3	1	7	3
Algeria	—	28	112	71	15	32	23	59	42	13	6	32	15
Angola	—	35	74	15	1	—	—	—	—	—	—	—	—
Argentina	11.8	22	120	97	48	4	3	4	3	2	2	2	1
Armenia	—	—	78	73	20	1	1	4	2	0[a]	0[a]	1	0[a]
Australia	—	—	102	161[d]	63	—	—	—	—	—	—	—	—
Austria	12.4	13	104	99	58	—	—	—	—	—	—	—	—
Azerbaijan	24.4	19	98	80	22	—	—	—	—	—	—	—	—
Bangladesh	15.7	57	100	46	7	54	50	77	69	45	43	68	60
Belarus	—	17	109	84	56	0[a]	0[a]	1	0[a]	0[a]	0[a]	0[a]	0[a]
Belgium	11.6	12	105	—	57	—	—	—	—	—	—	—	—
Benin	—	54	95	22	4	59	47	84	75	37	28	74	63
Bolivia	23.1	24	116	80	36	13	8	30	20	4	2	11	6
Bosnia-Herzegovina	—	—	—	—	—	—	—	—	—	—	—	—	—
Botswana	—	27	108	93	5	34	25	30	19	21	15	13	8
Brazil	12.9	26	162	108	17	18	13	20	13	12	6	9	3
Bulgaria	—	18	103	94	41	2	1	4	2	1	0[a]	1	0[a]
Burkina Faso	—	47	44	10	—	75	65	92	85	64	53	86	75
Burundi	—	50	65	10	1	50	43	73	58	42	33	55	36
Cambodia	10.1	53	110	19	3	49	20	86	42	34	16	73	25
Cameroon	12.5	63	108	20	5	28	20	46	35	10	8	15	11
Canada	—	15	99	103	60	—	—	—	—	—	—	—	—
Central African Republic	—	74	75	—	2	53	39	79	63	34	23	61	39
Chad	—	71	73	11	1	63	47	81	64	42	25	62	38
Chile	17.5	25	103	75	38	6	4	6	4	2	1	2	1
China	—	22	106	63	7	14	7	33	21	3	1	8	3
Hong Kong, China	—	—	—	—	—	5	3	16	10	2	1	1	0[a]
Colombia	—	26	112	70	23	11	8	12	8	6	4	4	2
Democratic Republic of the Congo (formerly Zaire)	—	26	47	18	1	38	26	66	48	19	11	42	24
Congo Republic	12.6	51	97	42	5	23	12	42	24	5	2	10	3
Costa Rica	—	25	107	60	16	6	4	6	4	3	2	2	1
Côte d'Ivoire	21.5	48	81	23	7	56	40	77	62	40	29	59	46
Croatia	10.4	18	—	—	—	1	1	5	3	0[a]	0[a]	0[a]	0[a]
Cuba	15.1	11	102	85	24	5	3	5	3	1	0[a]	1	0[a]

Table G *(continued)*
World Countries: Education and Literacy, 1990–2002

COUNTRY	EDUCATIONAL INPUTS					OUTCOMES							
	Public Expenditure on Education 2000[b] (% of government expenditure)	Primary Pupil-Teacher Ratio 2000 (pupils per teacher)	Primary School Enrollment 2000 (% of age group)	Secondary School Enrollment 2000 (% of age group)	Tertiary School Enrollment 2000 (% of age group)	ADULT ILLITERACY (% above age 15)				YOUTH ILLITERACY RATE (% aged 15–24)			
						Male		Female		Male		Female	
						1990	2001	1990	2001	1990	2001	1990	2001
Czech Republic	9.5	18	104	95	30	—	—	—	—	—	—	—	—
Denmark	15.3	10	102	128	59	—	—	—	—	—	—	—	—
Dominican Republc	15.7	40	12	59	—	20	16	21	16	13	9	12	8
Ecuador	8.0	23	115	57	—	10	7	15	10	4	2	5	3
Egypt	—	22	100	86	39	40	33	66	55	29	23	49	36
El Salvador	13.4	26	109	54	18	24	18	31	23	15	11	17	12
Eritrea	—	45	59	28	2	42	32	72	54	27	19	54	39
Estonia	—	14	103	92	58	0[a]	0[a]	0[a]	0[a]	—	0[a]	—	0[a]
Ethiopia	13.8	55	64	18	2	64	52	80	68	52	38	64	50
Finland	12.5	16	102	126	—	—	—	—	—	—	—	—	—
France	11.5	19	105	108	54	—	—	—	—	—	—	—	—
Gabon	—	49	144	60	8	—	—	—	—	—	—	—	—
The Gambia	14.2	37	82	36	—	68	55	80	69	49	33	66	49
Georgia	—	16	95	73	35	—	—	—	—	—	—	—	—
Germany	9.7	15	104	99	46	—	—	—	—	—	—	—	—
Ghana	—	33	80	36	3	30	19	53	35	12	6	25	11
Greece	7.0	13	99	98	—	2	1	8	4	1	0[a]	0[a]	0[a]
Guatemala	11.4	33	102	37	—	31	23	47	38	20	14	34	27
Guinea	25.6	44	67	14	—	—	—	—	—	—	—	—	—
Guinea-Bissau	4.8	4	83	20	0	54	45	89	75	30	26	79	54
Haiti	10.9	—	—	—	—	57	47	63	51	44	35	46	34
Honduras	—	34	106	—	15	31	25	32	24	23	16	20	13
Hungary	14.1	11	02	99	40	1	1	1	1	0[a]	0[a]		0[a]
India	12.7	40	102	49	10	38	31	64	54	27	20	46	34
Indonesia	—	22	110	57	15	13	8	27	17	3	2	7	3
Iran	20.4	25	86	78	10	27	16	45	30	8	4	18	8
Iraq	—	21	102	38	14	43	45	67	76	29	40[a]	48	70
Ireland	13.2	20	119	—	48	—	—	—	—	—	—	—	—
Israel	—	12	114	93	53	3	3	9	7	1	0[a]	2	1
Italy	9.5	11	101	96	50	2	1	3	2	0[a]	0[a]	0[a]	0[a]
Jamaica	11.1	36	100	83	16	22	17	14	9	13	9	5	2
Japan	9.3	20	101	102	48	—	—	—	—	—	—	—	—
Jordan	5.0	—	101	88	29	10	5	28	15	2	1	4	1
Kazakhstan	—	19	99	88	31	1	0[a]	2	1	—	0[a]	—	0[a]
Kenya	22.5	30	94	31	3	19	11	39	23	7	4	13	5
Korea, North	—	—	—	—	—	—	—	—	—	—	—	—	—
Korea, South	17.4	32	101	94	78	2	1	7	3	0[a]	0[a]	0[a]	0[a]
Kuwait	—	14	85	56	21	20	16	27	20	12	8	13	6
Kyrgyzstan	—	24	101	86	41	—	—	—	—	—	—	—	—

Table G (continued)
World Countries: Education and Literacy, 1990–2002

COUNTRY	EDUCATIONAL INPUTS					OUTCOMES							
	Public Expenditure on Education 2000[b] (% of government expenditure)	Primary Pupil-Teacher Ratio 2000 (pupils per teacher)	Primary School Enrollment 2000 (% of age group)	Secondary School Enrollment 2000 (% of age group)	Tertiary School Enrollment 2000 (% of age group)	ADULT ILLITERACY (% above age 15)				YOUTH ILLITERACY RATE (% aged 15–24)			
						Male		Female		Male		Female	
						1990	2001	1990	2001	1990	2001	1990	2001
Laos	8.8	30	113	38	3	47	23	80	46	28	15	62	28
Latvia	—	15	100	91	63	0[a]	0[a]	0[a]	0[a]	0[a]	0[a]	0[a]	0[a]
Lebanon	11.1	17	99	76	42	12	8	27	19	5	3	11	7
Lesotho	18.5	48	115	33	3	35	27	11	6	23	17	3	1
Liberia	—	36	118	38	—	45	29	77	62	25	14	61	46
Libya	—	8	116	90	—	17	9	49	31	1	0[a]	17	6
Lithuania	—	16	101	95	52	1	0[a]	1	0[a]	0[a]	0[a]	0[a]	0[a]
Macedonia	—	22	99	84	24	—	—	—	—	—	—	—	—
Madagascar	10.2	50	103	14	2	34	26	50	39	22	16	33	23
Malawi	24.6	56	137	36	0	31	25	64	52	24	19	49	38
Malaysia	26.7	18	99	70	28	13	8	25	16	5	2	6	2
Mali	—	63	61	15	2	67	63	81	83	62	52	83	74
Mauritania	18.9	42	83	21	4	53	49	74	69	44	43	65	59
Mauritius	12.1	26	109	77	11	15	12	25	18	9	6	9	5
Mexico	22.6	27	113	75	21	10	7	15	11	4	2	6	3
Moldova	15.0	20	84	71	28	1	0[a]	4	2	0[a]	0[a]	0[a]	0[a]
Mongolia	2.2	32	99	61	33	2	1	3	2	1	1	1	1
Morocco	26.1	28	94	39	10	47	37	75	63	32	23	58	40
Mozambique	12.3	64	92	12	1	51	39	82	70	34	24	68	52
Myanmar (Burma)	9.0	32	89	39	12	13	11	26	19	10	9	14	9
Namibia	—	32	112	62	6	23	17	28	18	14	10	11	6
Nepal	14.1	37	118	51	5	53	39	86	75	34	23	73	56
Netherlands	10.7	10	108	124[d]	55	—	—	—	—	—	—	—	—
New Zealand	—	16	100	112	69	—	—	—	—	—	—	—	—
Nicaragua	13.8	36	104	54	—	36	33	34	33	32	29	28	27
Niger	—	42	35	6	1	82	76	95	91	75	67	91	86
Nigeria	—	—	—	—	—	41	27	62	42	19	10	34	15
Norway	16.2	—	101	115	70	—	—	—	—	—	—	—	—
Oman	—	24	72	68	8	33	19	62	36	5	0[a]	25	3
Pakistan	7.8	44	75	—	—	50	42	79	71	36	28	67	57
Panama	—	25	112	69	35	10	7	12	9	4	3	5	4
Papua New Guinea	17.5	36	84	21	2	34	29	52	42	25	20	38	28
Paraguay	11.2	20	111	60	1	8	5	12	8	4	3	5	3
Peru	21.1	25	128	81	29	8	5	21	14	3	2	8	5
Philippines	20.6	35	113	77	31	7	5	8	5	3	1	3	1
Poland	11.4	11	100	101	56	0[a]	0[a]	1	0[a]	0[a]	0[a]	0[a]	0[a]
Portugal	13.1	13	121	114[d]	50	9	5	16	10	1	0[a]	0[a]	0[a]
Puerto Rico	—	—	—	—	—	9	6	9	6	5	3	3	2

Table G *(continued)*
World Countries: Education and Literacy, 1990–2002

COUNTRY	EDUCATIONAL INPUTS					OUTCOMES							
	Public Expenditure on Education 2000[b] (% of government expenditure)	Primary Pupil-Teacher Ratio 2000 (pupils per teacher)	Primary School Enrollment 2000 (% of age group)	Secondary School Enrollment 2000 (% of age group)	Tertiary School Enrollment 2000 (% of age group)	ADULT ILLITERACY (% above age 15)				YOUTH ILLITERACY RATE (% aged 15–24)			
						Male		Female		Male		Female	
						1990	2001	1990	2001	1990	2001	1990	2001
Romania	—	20	99	82	27	1	1	5	3	1	0[a]	1	0[a]
Russian Federation	—	17	—	83	64	0[a]	0[a]	1	1	0[a]	0[a]	0[a]	0[a]
Rwanda	—	51	119	12	2	37	26	56	38	22	14	33	17
Saudi Arabia	—	12	68	68	22	22	16	49	32	9	5	21	9
Senegal	—	51	75	18	4	62	52	81	71	50	40	70	57
Serbia and Montenegro	—	20	—	—	—	—	—	—	—	—	—	—	—
Sierra Leone	—	44	93	26	2	—	—	—	—	—	—	—	—
Singapore	23.6	—	—	—	—	6	4	17	11	1	0[a]	1	0[a]
Slovak Republic	13.8	19	103	87	30	—	—	—	—	—	—	—	—
Slovenia	—	14	100	—	61	0[a]	0[a]	0[a]	0[a]	0[a]	0[a]	0[a]	0[a]
Somalia	—	—	—	—	—	—	—	—	—	—	—	—	—
South Africa	25.8	—	111	87	15	18	14	20	15	11	8	12	9
Spain	11.3	33	105	116	59	2	1	5	3	0[a]	0[a]	0[a]	0[a]
Sri Lanka	—	14	106	72	—	7	5	15	11	4	3	6	3
Sudan	—	—	55	29	7	39	30	68	52	24	17	46	27
Swaziland	—	33	125	60	5	26	19	30	21	15	10	15	8
Sweden	13.4	11	110	149[d]	70	—	—	—	—	—	—	—	—
Switzerland	15.2	14	107	100	42	—	—	—	—	—	—	—	—
Syria	11.1	24	109	43	6	18	11	53	38	8	4	33	20
Tajikistan	11.8	22	104	79	14	1	0[a]	3	1	0[a]	0[a]	0[a]	0[a]
Tanzania	—	40	63	6	1	23	15	49	32	10	1	22	11
Thailand	31.0	21	95	82	35	5	3	11	6	1	1	2	2
Togo	23.2	34	124	39	4	36	27	71	56	19	12	55	35
Trinidad and Tobago	16.7	20	100	81	6	6	1	11	2	0[a]	0[a]	0[a]	0[a]
Tunisia	17.4	23	117	78	22	28	18	54	38	7	2	25	10
Turkey	—	—	101	58	15	11	6	33	23	3	1	12	6
Turkmenistan	—	—	—	—	—	—	—	—	—	—	—	—	—
Uganda	—	59	136	19	3	31	22	57	42	20	14	39	27
Ukraine	15.7	20	78	105	43	0[a]	0[a]	1	0[a]	0[a]	0[a]	0[a]	0[a]
United Arab Emirates	—	16	99	75	12	29	25	30	20	19	12	11	5
United Kingdom	11.4	18	99	156[d]	60	—	—	—	—	—	—	—	—
United States	—	15	101	95	73	—	—	—	—	—	—	—	—
Uruguay	—	21	109	98	36	4	3	3	2	1	1	1	1
Uzbekistan	—	—	—	—	—	1	0[a]	2	1	0[a]	0[a]	0[a]	0[a]
Venezuela	—	—	102	59	28	10	7	12	8	5	3	3	1
Vietnam	—	28	106	67	10	6	5	13	9	5	5	5	4

Table G *(continued)*
World Countries: Education and Literacy, 1990–2002

COUNTRY	EDUCATIONAL INPUTS					OUTCOMES							
	Public Expenditure on Education 2000[b] (% of government expenditure)	Primary Pupil-Teacher Ratio 2000 (pupils per teacher)	Primary School Enrollment 2000 (% of age group)	Secondary School Enrollment 2000 (% of age group)	Tertiary School Enrollment 2000 (% of age group)	ADULT ILLITERACY (% above age 15)				YOUTH ILLITERACY RATE (% aged 15–24)			
						Male		Female		Male		Female	
						1990	2001	1990	2001	1990	2001	1990	2001
West Bank and Gaza	—	—	—	—	—	—	—	—	—	—	—	—	—
Yemen	32.8	30	79	48	11	45	32	87	73	26	16	75	51
Zambia	17.6	45	78	24	2	22	14	41	27	14	9	24	14
Zimbabwe	—	37	95	44	4	13	7	25	15	3	1	9	4

[a]Less than 0.5.
[b]Data are for the most recent year available.
[c]Gross National Income (GNI) has replaced GNP in the World bank Atlas Method's estimate of national income.
[d]Includes training for the unemployed.
Source: World Development Indicators 2003 (World Bank).

Table H
World Countries: Agricultural Operations, 1996–2002

COUNTRY			AGRICULTURAL INPUTS				AGRICULTURAL OUTPUT AND PRODUCTIVITY			
			Agricultural Machinery							
	Arable Land (hectares per capita)	Irrigated Land (% of cropland)	Land under Cereal Production (thousand hectares)	Fertilizer Consumption (hundreds of grams per hectare of arable land)	Tractors per Thousand Agricultural Workers	Tractors per Hundred Hectares of Agricultural Land	Crop Production Index (1989–91 = 100)	Food Production index (1989–91 = 100)	Livestock Production Index (1989–91 = 100)	Agricultural Productivity (agricultural value added per worker in 1995$)
	1998–2000	1998–2000	1999–2001	1998–2000	1998–2000	1998–2000	1990–2001	1990–2001	1999–2001	1999–2001
Afghanistan	0.31	29.6	2,400	7	0	1	—	—	—	—
Albania	0.19	48.6	213	261	11	142	—	—	—	2,101
Algeria	0.26	6.8	1,965	121	38	121	127.9	133.1	128.3	1,939
Angola	0.24	2.2	904	9	2	34	151.2	146.3	136.8	131
Argentina	0.68	5.7	11,004	323	191	112	163.9	142.9	112.1	10,351
Armenia	0.13	51.3	178	153	69	354	98.5	73.9	60.2	5,435
Australia	2.67	4.6	17,514	455	700	62	171.2	145.7	114.8	35,225
Austria	0.17	0.3	820	1,665	1,728	2,512	102.6	107.6	106.3	31,091
Azerbaijan	0.21	74.3	635	62	34	192	57.4	78.0	79.6	768
Bangladesh	0.06	47.6	11,736	1,593	0	7	132.4	135.6	140.9	311
Belarus	0.61	1.8	2,462	1,403	107	132	86.7	59.5	56.8	2,180
Belgium	0.08	4.2	324	3,687	1,234	1,298	146.1	113.6	109.7	29,098
Benin	0.31	0.6	843	228	0	1	181.9	157.6	110.8	608
Bolivia	0.24	5.9	775	25	4	29	163.0	144.8	126.8	748
Bosnia-Herzegovina	0.13	0.4	367	872	284	580	—	—	—	7,811
Botswana	0.21	0.3	159	127	20	175	80.4	95.5	97.4	575
Brazil	0.32	4.4	17,629	1,200	59	152	129.1	145.9	162.2	4,798
Bulgaria	0.53	17.6	1,867	333	77	58	59.8	66.0	60.7	8,277
Burkina Faso	0.34	0.7	2,948	114	0	5	137.8	135.4	141.4	183
Burundi	0.13	5.9	202	41	0	2	92.3	91.8	77.3	150
Cambodia	0.31	7.1	2,050	7	0	5	149.0	153.3	166.1	363
Cameroon	0.41	0.5	734	76	0	1	132.9	130.8	122.0	1,189
Canada	1.49	1.6	18,016	572	1,761	156	123.7	126.7	129.6	43,428
Central African Republic	0.53	—	151	3	0	0	133.3	139.4	137.3	490
Chad	0.47	0.6	2,156	49	0	0	144.5	138.0	119.0	213
Chile	0.13	78.4	586	2,416	55	273	131.2	137.8	147.4	6,040
China	0.10	39.6	86,688	2,871	2	62	146.1	175.9	217.9	334
Hong Kong, China	—	—	—	—	—	—	59.3	58.0	57.0	—
Colombia	0.06	19.6	1,119	2,342	6	83	104.1	120.2	122.9	3,590
Democratic Republic of Congo (formerly Zaire)	0.14	0.1	2,066	2	0	4	81.9	85.3	101.5	218
Congo Republic	0.06	0.5	10	287	1	40	126.0	128.3	133.4	471
Costa Rica	0.06	21.2	79	8,572	21	311	149.9	148.0	132.7	5,272
Côte d'Ivoire	0.19	1.0	1,439	264	1	13	135.8	138.0	122.9	1,057
Croatia	0.33	0.2	665	1,354	12	17	86.6	67.3	50.0	9,449
Cuba	0.33	19.5	208	423	98	215	58.4	62.2	68.3	—
Czech Republic	0.30	0.7	1,624	936	173	273	92.0	77.1	66.9	6,235

Table H *(continued)*
World Countries: Agricultural Operations, 1996–2002

COUNTRY			AGRICULTURAL INPUTS				AGRICULTURAL OUTPUT AND PRODUCTIVITY			
			Agricultural Machinery							
	Arable Land (hectares per capita)	Irrigated Land (% of cropland)	Land under Cereal Production (thousand hectares)	Fertilizer Consumption (hundreds of grams per hectare of arable land)	Tractors per Thousand Agricultural Workers	Tractors per Hundred Hectares of Agricultural Land	Crop Production Index (1989–91 = 100)	Food Production index (1989–91 = 100)	Livestock Production Index (1989–91 = 100)	Agricultural Productivity (agricultural value added per worker in 1995$)
	1998–2000	1998–2000	1999–2001	1998–2000	1998–2000	1998–2000	1990–2001	1990–2001	1999–2001	1999–2001
Denmark	0.43	19.4	1,523	1,667	1,104	557	91.9	104.2	116.8	57,896
Dominican Republic	0.13	17.2	160	860	4	22	87.3	110.1	141.0	3,179
Ecuador	0.13	28.8	897	1,062	7	57	157.6	156.1	153.1	1,716
Egypt	0.05	100.0	2,716	4,284	10	303	151.2	155.9	161.4	1,324
El Salvador	0.09	4.9	380	1,497	4	61	104.0	117.0	120.2	1,710
Eritrea	0.12	4.3	333	189	0	12	148.9	126.8	111.0	85
Estonia	0.81	0.4	326	268	556	454	69.5	43.4	37.2	4,265
Ethiopia	0.16	1.8	7,562	163	0	3	153.2	141.1	118.4	141
Finland	0.42	2.9	1,154	1,419	1,268	893	94.7	89.8	91.7	40,463
France	0.31	10.8	8,963	2,491	1,326	692	107.9	105.3	104.6	58,177
Gabon	0.27	3.0	16	11	7	46	119.9	115.9	119.0	2,047
The Gambia	0.17	0.9	133	70	0	2	145.0	139.8	106.1	298
Georgia	0.15	44.2	352	480	19	131	52.5	78.7	94.9	—
Germany	0.14	4.0	6,911	2,460	955	887	119.4	98.2	87.8	32,814
Ghana	0.19	0.2	1,304	39	1	10	178.6	169.6	117.7	569
Greece	0.26	37.2	1,254	1,688	302	877	111.6	103.4	96.1	14,079
Guatemala	0.12	6.8	655	1,511	2	32	131.8	134.7	134.1	2,115
Guinea	0.12	6.4	803	36	0	6	153.8	155.3	174.4	274
Guinea-Bissau	0.26	4.9	123	40	0	1	144.7	140.0	126.3	322
Haiti	0.07	8.2	460	203	0	3	87.6	99.8	145.6	—
Honduras	0.22	4.6	446	1,242	7	35	105.1	112.1	149.5	990
Hungary	0.47	4.2	2,758	897	171	192	80.1	76.3	70.1	5,159
India	0.16	32.2	100,967	1,063	6	93	124.6	129.9	144.1	402
Indonesia	0.10	14.7	15,270	1,326	1	36	119.4	118.3	116.4	744
Iran	0.25	42.5	7,234	833	37	139	135.9	139.0	145.1	3,698
Iraq	0.23	63.6	2,646	732	76	95	73.6	73.3	68.1	—
Ireland	0.29	—	282	6,252	1,035	1,590	111.0	113.5	110.6	—
Israel	0.05	46.4	80	3,406	343	734	100.3	115.0	112.5	—
Italy	0.14	24.3	4,185	2,116	1,154	1,979	104.7	104.9	106.6	26,690
Jamaica	0.07	9.1	2	1,320	12	177	118.1	118.8	122.3	1,540
Japan	0.04	54.6	2,026	3,192	707	4,691	88.0	92.5	94.1	30,828
Jordan	0.05	19.1	34	882	28	197	114.9	134.1	176.6	825
Kazakhstan	1.47	10.4	12,121	12	44	28	84.1	70.3	45.3	1,649
Kenya	0.14	1.5	1,945	349	1	36	110.4	107.7	106.5	216
Korea, North	0.08	73	1,287	1,530	21	441	—	—	—	—
Korea, South	0.04	60.4	1,163	4,794	66	981	113.0	131.0	161.6	13,782
Kuwait	0.00	81.0	1	1,341	10	132	166.2	208.1	210.0	—
Kyrgyzstan	0.28	74.8	607	210	47	190	143.5	121.4	78.1	1,636
Laos	0.17	18.1	765	88	1	12	153.7	163.4	185.7	614
Latvia	0.77	1.1	422	278	324	293	74.3	41.9	31.7	2,671
Lebanon	0.04	32.3	39	3,416	113	298	138.3	143.7	164.7	28,322
Lesotho	0.16	—	233	170	6	62	164.6	116.9	87.5	553

Table H (continued)
World Countries: Agricultural Operations, 1996–2002

COUNTRY			AGRICULTURAL INPUTS				AGRICULTURAL OUTPUT AND PRODUCTIVITY			
			Agricultural Machinery							
	Arable Land (hectares per capita)	Irrigated Land (% of cropland)	Land under Cereal Production (thousand hectares)	Fertilizer Consumption (hundreds of grams per hectare of arable land)	Tractors per Thousand Agricultural Workers	Tractors per Hundred Hectares of Agricultural Land	Crop Production Index (1989–91 = 100)	Food Production index (1989–91 = 100)	Livestock Production Index (1989–91 = 100)	Agricultural Productivity (agricultural value added per worker in 1995$)
	1998–2000	1998–2000	1999–2001	1998–2000	1998–2000	1998–2000	1990–2001	1990–2001	1999–2001	1999–2001
Liberia	0.12	0.5	147	0	0	9	—	—	—	525
Libya	0.35	21.9	337	356	306	187	133.8	157.5	162.1	—
Lithuania	0.83	0.2	947	509	394	337	74.6	59.8	49.4	3,131
Macedonia	0.29	8.8	220	744	420	919	105.2	94.6	86.1	4,155
Madagascar	0.19	31.3	1,409	30	1	12	103.3	109.5	110.0	155
Malawi	0.20	1.3	1,622	217	0	7	149.4	160.8	113.2	123
Malaysia	0.08	4.8	719	7,843	24	238	119.7	143.2	155.1	6,843
Mali	0.44	3.0	2,411	107	1	6	135.0	119.5	111.8	290
Mauritania	0.19	9.8	246	29	1	8	133.9	109.4	106.0	500
Mauritius	0.09	18.9	0	3,463	6	37	90.4	101.7	141.3	5,580
Mexico	0.26	23.8	10,269	726	20	69	121.9	133.6	145.1	1,801
Moldova	0.42	14.1	854	28	82	239	56.8	45.9	34.5	1,661
Mongolia	0.52	6.8	222	28	22	56	30.9	103.4	109.2	1,428
Morocco	0.31	13.2	5,258	393	10	48	93.1	103.8	121.9	1,512
Mozambique	0.23	2.6	1,793	26	1	14	140.8	128.5	103.2	138
Myanmar (Burma)	0.21	17.8	6,919	186	1	10	167.9	163.6	157.9	—
Namibia	0.47	0.9	308	3	11	39	121.5	118.0	117.7	1,618
Nepal	0.13	38.2	3,289	325	0	16	127.8	127.6	125.5	200
Netherlands	0.06	59.8	209	5,082	590	1,673	115.4	102.9	100.5	58,280
New Zealand	0.41	8.7	138	4,743	447	489	142.7	128.0	117.9	28,791
Nicaragua	0.50	3.2	398	141	7	11	138.4	144.1	139.5	—
Niger	0.43	1.5	7,569	7	0	0	144.5	136.0	125.2	201
Nigeria	0.23	0.8	18,995	63	2	11	159.7	156.2	127.4	714
Norway	0.20	—	328	2,230	1,266	1,549	79.2	91.9	99.1	34,535
Oman	0.01	77.2	2	3,586	1	77	166.1	162.5	133.6	—
Pakistan	0.16	82.1	12,443	1,312	13	151	126.0	145.2	155.1	712
Panama	0.18	5.3	129	716	20	100	94.0	106.7	126.1	2,738
Papua New Guinea	0.04	—	3	480	1	58	122.2	122.8	140.9	815
Paraguay	0.42	2.9	552	299	24	74	114.6	137.5	132.6	3,389
Peru	0.14	28.5	1,210	619	5	36	173.0	171.8	160.0	1,834
Philippines	0.07	15.4	6,581	1,269	1	21	119.9	131.1	162.7	1,428
Poland	0.36	0.7	8,778	1,092	293	930	85.3	85.9	83.4	1,601
Portugal	0.20	24.3	566	1,224	242	844	93.8	102.9	119.9	7,552
Puerto Rico	0.01	49.4	4	—	—	—	67.9	83.8	89.2	—
Romania	0.42	27.8	5,732	343	96	178	92.4	96.3	88.2	3,193
Russian Federation	0.86	3.6	37,953	108	96	65	78.7	63.7	51.5	2,648
Rwanda	0.10	0.4	266	3	0	1	100.4	103.8	116.8	251
Saudi Arabia	0.18	42.8	629	1,002	15	26	88.1	83.3	144.6	—
Senegal	0.25	3.0	1,247	162	0	2	129.5	136.6	147.4	334
Serbia and Montenegro	—	—	—	—	—	—	—	—	—	—

Table H *(continued)*
World Countries: Agricultural Operations, 1996–2002

COUNTRY			AGRICULTURAL INPUTS				AGRICULTURAL OUTPUT AND PRODUCTIVITY			
			Agricultural Machinery							
	Arable Land (hectares per capita)	Irrigated Land (% of cropland)	Land under Cereal Production (thousand hectares)	Fertilizer Consumption (hundreds of grams per hectare of arable land)	Tractors per Thousand Agricultural Workers	Tractors per Hundred Hectares of Agricultural Land	Crop Production Index (1989–91 = 100)	Food Production index (1989–91 = 100)	Livestock Production Index (1989–91 = 100)	Agricultural Productivity (agricultural value added per worker in 1995$)
	1998–2000	1998–2000	1999–2001	1998–2000	1998–2000	1998–2000	1990–2001	1990–2001	1999–2001	1999–2001
Sierra Leone	0.10	5.4	220	4	0	2	72.9	81.9	126.0	359
Singapore	0.00	—	—	36,137	19	650	48.2	39.5	39.5	44,907
Slovak Republic	0.27	11.0	—	711	—	167	—	—	—	—
Slovenia	0.09	1.1	100	4,468	4,476	6,108	94.4	108.9	104.7	4,912
Somalia	0.12	18.7	507	5	1	18	—	—	—	—
South Africa	0.35	8.9	4,741	523	46	55	106.9	107.7	103.0	3,837
Spain	0.34	19.9	6,637	1,688	636	642	112.3	114.8	128.4	22,088
Sri Lanka	0.05	34.7	888	2,791	2	88	120.3	122.2	135.0	734
Sudan	0.54	11.8	6,764	23	1	6	159.6	161.7	158.3	—
Swaziland	0.17	36.7	58	320	26	164	87.9	89.0	82.5	1,922
Sweden	0.31	—	1,188	1,049	1,083	623	89.2	97.1	102.5	36,365
Switzerland	0.06	5.7	181	2,607	673	2,699	90.1	94.1	93.5	—
Syria	0.29	22.1	3,060	766	68	201	156.3	147.0	134.6	2,618
Tajikistan	0.12	83.5	373	257	37	410	51.6	51.5	39.5	1,322
Tanzania	0.12	3.3	3,005	64	1	20	97.2	103.8	121.8	185
Thailand	0.25	27.1	11,154	1,113	10	148	118.5	118.3	130.1	904
Togo	0.56	0.3	680	73	0	0	140.7	132.9	107.5	531
Trinidad and Tobago	0.06	2.5	4	836	53	360	108.2	114.7	100.8	3,036
Tunisia	0.31	7.6	1,411	388	38	121	122.1	132.4	162.8	3,168
Turkey	0.38	16.7	13,174	890	63	372	113.8	111.5	106.8	1,852
Turkmenistan	0.32	—	779	558	75	307	91.8	134.1	136.8	1,518
Uganda	0.23	0.1	1,371	8	1	9	135.5	131.8	127.3	342
Ukraine	0.66	7.2	12,801	140	93	110	63.9	49.5	46.4	1,521
United Arab Emirates	0.02	47.2	1	8,034	4	70	284.6	270.7	203.4	—
United Kingdom	0.10	1.8	3,168	3,195	908	822	97.9	95.5	95.7	33,520
United States	0.64	12.5	57,456	1,090	1,542	271	119.8	122.4	122.1	50,777
Uruguay	0.39	13.5	542	974	174	257	150.4	136.6	199.9	8,010
Uzbekistan	0.18	88.3	1,520	1,733	57	380	88.8	118.0	116.3	1,088
Venezuela	0.10	16.9	736	920	60	201	118.2	123.2	118.6	5,304
Vietnam	0.07	40.9	8,322	3,420	5	224	166.8	153.9	154.9	253
West Bank and Gaza	—	—	—	—	—	—	—	—	—	—
Yemen	0.09	29.5	621	111	2	37	129.6	134.1	145.9	406
Zambia	0.53	0.9	735	64	2	11	100.7	106.0	117.5	195
Zimbabwe	0.26	3.5	1,787	544	7	73	121.3	110.0	114.5	361

Source: World Development Indicators 2003 (World Bank).

Table I
World Countries: Land Use and Deforestation, 1980–2000

COUNTRY	LAND AREA	RURAL POPULATION DENSITY	LAND USE						FOREST AREA	AVERAGE ANNUAL DEFORESTATION
	(thousands of sq. km.)	(people per sq. km. of arable land)	Arable Land (% of land area)		Permanent Cropland (% of land area)		Other (% of land area)		(thousands of sq. km)	Decline in forest area %
	2000	2000	1980	2000	1980	2000	1980	2000	2000	1990–2000
Afghanistan	652	262	12.1	12.1	0.2	0.2	87.7	87.6	14	0.8
Albania	27	313	21.4	21.1	4.3	4.4	74.4	74.5	10	0.8
Algeria	2,382	170	2.9	3.2	0.3	0.2	96.8	96.6	21	–1.3
Angola	1,247	288	2.3	2.4	0.4	0.2	97.3	97.4	698	0.2
Argentina	2,737	17	9.1	9.1	0.8	0.8	90.1	90.1	346	0.8
Armenia	28	252	—	17.6	—	2.3	—	80.1	4	–1.3
Australia	7,682	4	5.7	6.5	0.0	0.0	94.2	93.4	1,581	0.0
Austria	83	190	18.6	16.9	1.2	0.9	80.2	82.2	39	–0.2
Azerbaijan	87	236	—	19.0	—	3.0	—	78.0	11	–1.3
Bangladesh	130	1,208	68.3	62.5	2.0	2.7	29.6	34.8	13	–1.3
Belarus	207	50	—	29.6	—	0.6	—	69.8	94	–3.2
Belgium[a]	33	34	23.2	24.8	0.4	0.7	76.4	74.5	7	0.2
Benin	111	186	12.2	17.6	4.0	2.4	83.8	80.0	27	2.3
Bolivia	1,084	161	1.7	1.8	0.2	0.2	98.1	98.0	531	0.3
Bosnia-Herzegovina	51	454	—	9.8	—	2.9	—	87.3	23	0.0
Botswana	567	231	0.7	0.7	0.0	0.0	99.3	99.4	124	0.9
Brazil	8,457	60	4.6	6.3	1.2	1.4	94.2	92.3	5,325	0.4
Bulgaria	111	60	34.6	40.0	3.2	19	62.2	58.1	37	–0.6
Burkina Faso	274	248	10.0	13.9	0.1	0.2	89.8	85.9	71	0.2
Burundi	26	689	35.8	35.0	10.1	14.0	54.0	50.9	1	9.0
Cambodia	177	270	11.3	21.0	0.4	0.6	88.3	78.4	93	0.6
Cameroon	465	128	12.7	12.8	2.2	2.6	85.1	84.6	239	0.9
Canada	9,221	14	4.9	4.9	0.0	0.0	95.0	95.0	2,446	0.0
Central African Republic	623	113	3.0	3.1	0.1	0.1	96.9	96.8	229	0.1
Chad	1,259	167	2.5	2.8	0.0	0.0	97.5	97.2	127	0.6
Chile	749	109	5.4	2.6	0.3	0.4	94.3	96.9	155	0.1
China[b]	9,327	653	10.4	13.3	0.4	1.2	89.3	85.5	1,635	–1.2
Hong Kong, China	1	—	7.0	—	1.0	—	92.0	—	—	—
Colombia	1,039	376	3.6	2.7	1.4	1.7	95.0	95.6	496	0.4
Congo, Democratic Republic	2,267	—	3.0	3.0	0.3	0.5	96.6	96.5	1,352	0.4
Congo, Rep.	342	597	0.4	0.5	0.1	0.1	99.6	99.4	221	0.1
Costa Rica	51	694	5.5	4.4	4.4	5.5	90.1	90.1	20	0.8
Côte d'Ivoire	318	306	6.1	9.3	7.2	13.8	86.6	76.9	71	3.1
Croatia	56	127	—	26.1	—	2.3	—	71.6	18	–0.1
Cuba	110	76	23.9	33.1	6.4	7.6	69.7	59.3	23	–1.3
Czech Republic	77	85	—	39.9	—	3.1	—	57.1	26	0.0
Denmark	42	35	62.3	53.8	0.3	0.2	37.4	46.1	5	–0.2
Dominican Republic	48	264	22.1	22.7	7.2	10.3	70.6	67.0	14	0.0
Ecuador	277	277	5.6	5.7	3.3	5.2	91.1	89.2	106	1.2

Table I *(continued)*
World Countries: Land Use and Deforestation, 1980–2000

COUNTRY	LAND AREA	RURAL POPULATION DENSITY	LAND USE						FOREST AREA	AVERAGE ANNUAL DEFORESTATION
	(thousands of sq. km.)	(people per sq. km. of arable land)	Arable Land (% of land area)		Permanent Cropland (% of land area)		Other (% of land area)		(thousands of sq. km)	Decline in forest area %
	2000	2000	1980	2000	1980	2000	1980	2000	2000	1990–2000
Egypt	995	1,298	2.3	2.8	0.2	0.5	97.5	96.7	1	–3.4
El Salvador	21	445	27.0	27.0	8.0	12.1	65.0	60.9	1	4.6
Eritrea	101	669	—	4.9	—	0.0	—	95.0	16	0.3
Estonia	42	37	—	26.5	—	0.3	—	73.2	21	–0.6
Ethiopia	1,000	543	—	10.0	—	0.7	—	89.3	46	0.8
Finland	305	97	8.4	7.2	—	0.0	—	92.8	219	0.0
France	550	79	31.8	33.5	2.5	2.1	65.7	64.4	153	–0.4
Gabon	258	70	1.1	1.3	0.6	0.7	98.2	98.1	218	0.0
Gambia	10	393	15.9	23.0	—	0.5	—	76.5	5	–1.0
Georgia	70	303	—	11.4	—	3.9	—	84.7	30	0.0
Germany	349	87	34.4	33.1	1.4	0.6	64.1	66.3	107	0.0
Ghana	228	342	8.4	15.9	7.5	9.7	84.2	74.5	63	1.7
Greece	129	154	22.5	21.3	7.9	8.6	69.6	70.1	36	–0.9
Guatemala	108	505	11.7	12.5	4.4	5.0	83.9	82.4	29	1.7
Guinea	246	607	2.0	3.6	0.9	2.4	97.1	94.0	69	0.5
Guinea-Bissau	28	274	9.1	10.7	1.1	1.8	89.9	87.6	22	0.9
Haiti	28	914	19.8	20.3	12.5	12.7	67.7	67.0	1	5.7
Honduras	112	284	13.9	9.5	1.8	3.2	84.3	87.2	54	1.0
Hungary	92	78	54.4	49.8	3.3	2.2	42.2	48.0	18	–0.4
India	2,973	454	54.8	54.4	1.8	2.7	43.4	42.9	641	–0.1
Indonesia	1,812	594	9.9	11.3	4.4	7.2	85.6	81.5	1,050	1.2
Iran	1,622	160	8.0	8.8	0.5	1.2	91.5	90.0	73	0.0
Iraq	437	145	12.0	11.9	0.4	0.8	87.6	87.3	8	0.0
Ireland	69	148	16.1	15.2	0.0	0.0	83.9	84.7	7	–3.0
Israel	21	156	15.8	16.1	4.3	4.1	80.0	79.7	1	–4.9
Italy	294	239	32.2	27.1	10.0	9.7	57.7	63.2	100	–0.3
Jamaica	11	649	16.6	16.1	5.5	9.2	77.8	74.7	3	1.5
Japan	377	603	11.4	12.3	1.6	1.0	87.0	86.7	241	0.0
Jordan	89	427	3.4	2.7	0.4	1.8	96.2	95.5	1	0.0
Kazakhstan	2,671	31	—	8.0	—	0.1	—	92.0	121	–2.2
Kenya	569	501	6.7	7.0	0.8	0.9	92.5	92.1	171	0.5
Korean, North	120	521	13.4	14.1	2.4	2.5	84.2	83.4	82	0.0
Korea, South	99	496	20.9	17.4	1.4	2.0	77.8	80.6	63	0.1
Kuwait	18	989	0.1	0.4	—	0.1	—	99.4	0	–5.2
Kyrgyzstan	192	236	—	7.1	—	0.3	—	92.5	10	–2.6
Laos	231	486	2.9	3.8	0.1	0.4	97.0	95.8	126	0.4
Latvia	62	51	—	29.7	—	0.5	—	69.8	29	–0.4
Lebanon	10	234	20.5	18.6	8.9	13.9	70.6	67.5	0	0.3
Lesotho	30	451	9.6	10.7	—	—	—	—	0	0.0
Liberia	96	454	3.9	3.9	2.1	2.2	94.0	93.8	35	2.0

Table I *(continued)*
World Countries: Land Use and Deforestation, 1980–2000

COUNTRY	LAND AREA	RURAL POPULATION DENSITY	LAND USE						FOREST AREA	AVERAGE ANNUAL DEFORESTATION
	(thousands of sq. km.)	(people per sq. km. of arable land)	Arable Land (% of land area)		Permanent Cropland (% of land area)		Other (% of land area)		(thousands of sq. km)	Decline in forest area %
	2000	2000	1980	2000	1980	2000	1980	2000	2000	1990–2000
Libya	1,760	36	1.0	1.0	0.2	0.2	98.8	98.8	4	-1.4
Lithuania	65	38	—	45.3	—	0.9	—	53.8	20	-0.2
Macedonia	25	149	—	21.8	—	1.7	—	76.4	9	0.0
Madagascar	582	377	4.3	5.0	0.9	1.0	94.8	94.0	117	0.9
Malawi	94	419	13.3	22.3	0.9	1.5	85.8	76.2	26	2.4
Malaysia	329	544	3.0	5.5	11.6	17.6	85.4	76.9	193	1.2
Mali	1,220	163	1.6	3.8	0.0	0.0	98.3	96.2	132	0.7
Mauritania	1,025	231	0.2	0.5	0.0	0.0	99.8	99.5	3	2.7
Mauritius	2	697	49.3	49.3	3.4	3.0	47.3	47.8	0	0.6
Mexico	1,909	101	12.1	13.0	0.8	1.3	87.1	85.7	552	1.1
Moldova	33	137	—	55.3	—	11.2	—	33.5	3	-0.2
Mongolia	1,567	89	0.8	0.8	0.0	0.0	99.2	99.2	106	0.5
Morocco	446	146	16.6	19.6	1.1	2.2	82.3	78.2	30	0.0
Mozambique	784	308	3.6	5.0	0.3	0.3	96.1	94.7	306	0.2
Myanmar (Burma)	658	349	14.6	15.1	0.7	0.9	84.8	84.0	344	1.4
Namibia	823	149	0.8	1.0	0.0	0.0	99.2	99.0	80	0.9
Nepal	143	701	16.0	20.3	0.2	0.5	83.8	79.2	39	1.8
Netherlands	34	184	23.3	26.8	0.9	1.0	75.8	72.1	4	-0.3
New Zealand	268	35	9.3	5.8	3.7	6.4	86.9	87.8	79	-0.5
Nicaragua	121	91	9.5	20.2	1.5	2.4	89.1	77.4	33	3.0
Niger	1,267	192	2.8	3.5	0.0	0.0	97.2	96.4	13	3.7
Nigeria	911	252	30.6	31.0	2.8	2.9	66.6	66.1	135	2.6
Norway	307	129	2.7	2.9	—	—	—	—	89	-0.4
Oman	212	3,049	0.1	0.1	0.1	0.2	99.8	99.7	0	0.0
Pakistan	771	434	25.9	27.6	0.4	0.9	73.7	71.5	25	1.1
Panama	74	250	5.8	6.7	1.6	2.1	92.5	91.2	29	1.6
Papua New Guinea	453	2,067	0.4	0.5	1.1	1.4	98.5	98.1	306	0.4
Paraguay	397	106	4.1	5.8	0.3	0.2	95.6	94.0	234	0.5
Peru	1,280	191	2.5	2.9	0.3	0.4	97.2	96.7	652	0.4
Philippines	298	572	14.5	18.6	14.8	15.1	70.8	66.3	58	1.4
Poland	304	104	48.0	46.0	1.1	1.1	50.9	52.9	93	-0.1
Portugal	92	179	26.5	21.7	7.8	7.8	65.7	70.4	37	-1.7
Puerto Rico	9	2,701	8.3	3.9	5.6	5.2	88.7	90.9	2	0.2
Romania	230	108	42.7	40.7	2.9	2.2	54.4	57.2	64	-0.2
Russian Federation	16,889	32	—	7.4	—	0.1	—	92.5	8,514	0.0
Rwanda	25	887	30.8	36.5	10.3	10.1	58.9	53.4	3	3.9
Saudi Arabia	2,150	80	0.9	1.7	0.0	0.1	99.1	98.2	15	0.0
Senegal	193	212	12.2	12.3	0.0	0.2	87.8	87.5	62	0.7
Serbia-Montenegro	—	—	—	—	—	—	—	—	29	0.0
Sierra Leone	72	651	6.3	6.8	0.7	0.8	93.0	92.3	11	2.9

Table I *(continued)*
World Countries: Land Use and Deforestation, 1980–2000

COUNTRY	LAND AREA	RURAL POPULATION DENSITY	LAND USE						FOREST AREA	AVERAGE ANNUAL DEFORESTATION
	(thousands of sq. km.)	(people per sq. km. of arable land)	Arable Land (% of land area)		Permanent Cropland (% of land area)		Other (% of land area)		(thousands of sq. km)	Decline in forest area %
	2000	2000	1980	2000	1980	2000	1980	2000	2000	1990–2000
Singapore	1	0	3.3	1.6	9.8	0.0	86.9	98.4	0	0.0
Slovak Republic	48	158	—	30.4	—	2.8	—	66.8	20	-0.3
Slovenia	20	584	—	8.6	—	1.5	—	89.9	11	-0.2
Somalia	627	610	1.6	1.7	0.0	0.0	98.4	98.3	75	1.0
South Africa	1,221	125	10.2	12.1	0.7	0.8	89.1	87.1	89	0.1
Spain	499	68	31.1	26.7	9.9	9.8	59.0	63.5	144	-0.6
Sri Lanka	65	1,602	13.2	13.8	15.9	15.8	70.8	70.4	19	1.6
Sudan	2,376	122	5.2	6.8	0.0	0.1	94.8	93.1	616	1.4
Swaziland	17	432	10.8	10.3	0.2	0.7	89.0	89.0	5	-1.2
Sweden	412	55	7.2	6.6	—	—	—	—	271	0.0
Switzerland	40	567	9.9	10.4	0.5	0.6	89.6	89.0	12	-0.4
Syria	184	173	28.5	24.7	2.5	4.4	69.1	70.9	5	0.0
Tajikistan	141	614	—	5.2	—	0.9	—	93.9	4	-0.5
Tanzania	884	571	2.5	4.5	1.0	1.1	96.5	94.4	388	0.2
Thailand	511	331	32.3	28.8	3.5	6.5	64.2	64.8	148	0.7
Togo	54	120	36.8	46.1	6.6	2.2	56.6	51.6	5	3.4
Trinidad Tobago	5	450	13.6	14.6	9.0	9.2	77.4	76.2	3	0.8
Tunisia	155	113	20.5	18.7	9.7	13.5	69.7	67.7	5	-0.2
Turkey	770	93	32.9	31.4	4.1	3.3	63.0	65.3	102	-0.2
Turkmenistan	470	179	—	3.5	—	0.1	—	96.4	38	0.0
Uganda	200	377	20.4	25.7	8.0	9.6	71.6	64.7	42	2.0
Ukraine	579	49	—	56.2	—	1.6	—	42.2	96	-0.3
United Arab Emirates	84	644	0.2	0.7	0.1	22	99.7	97.0	3	-2.8
United Kingdom	241	105	28.7	24.4	0.3	0.2	71.1	75.4	26	-0.8
United States	9,159	36	20.6	19.3	0.2	0.2	79.2	80.5	2,260	-0.2
Uruguay	175	21	8.0	7.4	0.3	0.2	91.7	92.3	13	-5.0
Uzbekistan	414	350	—	10.8	—	0.9	—	88.3	20	-0.2
Venezuela	882	130	3.2	2.8	0.9	1.1	95.9	96.1	495	0.4
Vietnam	325	1,037	18.2	17.7	1.9	4.9	79.8	77.4	98	-0.5
West Bank and Gaza	—	—	—	—	—	—	—	—	—	—
Yemen	528	853	2.6	2.9	0.2	0.2	97.2	96.8	4	1.8
Zambia	743	116	6.9	7.1	0.0	0.0	93.1	92.9	312	2.4
Zimbabwe	387	254	6.4	8.3	0.3	0.3	93.4	91.3	190	1.5

[a]Includes Luxembourg.
[b]Includes Taiwan.

Source: World Development Indicators 2003 (World Bank).

Table J
World Countries: Energy Production and Use, 1980–2000

COUNTRY	COMMERCIAL ENERGY PRODUCTION		COMMERCIAL ENERGY USE			COMMERCIAL ENERGY USE PER CAPITA			NET ENERGY IMPORTS[a]	
	Thousand Metric Tons of Oil Equivalent		Thousand Metric Tons of Oil Equivalent		Average Annual % Growth	Kg. of Oil Equivalent		Average Annual % Growth	% of Commercial Energy Use	
	1980	2000	1980	2000	1980–2000	1980	2000	1980–2000	1980	2000
Afghanistan	—	—	—	—	—	—	—	—	—	—
Albania	3,428	814	3,049	1,634	-5.6	1,142	521	-6.3	-12	50
Algeria	66,741	149,629	12,089	29,060	3.5	648	956	1.0	-452	-415
Angola	11,301	43,669	4,437	7,667	2.8	632	584	-0.3	-155	-470
Argentina	38,813	81,221	41,868	61,469	2.2	1,490	1,660	0.8	7	-32
Armenia	1,263	632	1,070	2,061	—	346	542	—	—	69
Australia	86,096	232,552	70,372	110,174	2.4	4,790	5,744	1.0	-22	-111
Austria	7,655	9,686	23,450	28,582	1.6	3,105	3,524	1.1	67	66
Azerbaijan	14,821	18,951	15,001	11,703	—	2,433	1,454	—	1	-62
Bangladesh	9,234	15,053	10,930	18,666	4.1	99	142	1.9	20	19
Belarus	2,566	3,466	2,385	24,330	—	247	2,432	—	-8	86
Belgium	7,986	13,233	46,100	59,217	1.8	4,682	5,776	1.6	84	78
Benin	1,212	1,821	1,363	2,362	2.3	394	377	-0.8	11	23
Bolivia	4,372	5,901	2,287	7,929	3.8	455	592	1.5	-79	-20
Bosnia and Herzegovina	—	3,277	—	4,359	—	—	1096	—	—	25
Botswana	—	—	—	—	—	—	—	—	—	—
Brazil	63,372	142,078	111,471	183,165	2.7	917	1,077	1.0	44	22
Bulgaria	7,737	10,005	28,673	18,784	-2.6	3,235	2,299	-2.2	73	47
Burkina Faso	—	—	—	—	—	—	—	—	—	—
Burundi	—	—	—	—	—	—	—	—	—	—
Cambodia	—	—	—	—	—	—	—	—	—	—
Cameroon	6,707	12,729	3,676	6,355	2.5	421	427	-0.2	-82	-100
Canada	207,417	374,864	193,000	250,967	1.6	7,848	8,156	0.4	-7	-49
Central African Republic	—	—	—	—	—	—	—	—	—	—
Chad	—	—	—	—	—	—	—	—	—	—
Chile	5,801	8,299	9,662	24,403	5.7	867	1,604	4.0	40	66
China	615,475	1,107,636	598,498	1,142,439	3.7	610	905	2.4	-3	3
Hong Kong, China	39	48	5,439	15,453	5.6	1,079	2,319	4.2	99	100
Colombia	18,040	74,584	19,349	28,786	2.5	680	681	0.5	7	-159
Democratic Republic of the Congo (formerly Zaire)	8,697	13,546	8,706	14,888	2.7	324	292	-0.6	0	-4
Congo, Rep.	4,024	14,656	862	895	-1.1	516	296	-4.0	-367	1,538
Costa Rica	767	1,591	1,527	3,281	4.2	669	861	1.5	50	52
Côte d'Ivoire	2,419	6,097	3,662	6,928	3.3	447	433	-0.1	34	12
Croatia	—	3,582	—	7,775	—	—	1,775	—	—	54
Cuba	4,227	6,051	14,910	13,203	-1.6	1,536	1,180	-2.4	72	54
Czech Republic	41,000	29,869	47,254	40,383	-1.2	4,618	3,931	-1.2	13	26
Denmark	952	27,831	19,783	19,456	0.6	3,862	3,643	0.4	95	-43

Table J *(continued)*
World Countries: Energy Production and Use, 1980–2000

COUNTRY	COMMERCIAL ENERGY PRODUCTION		COMMERCIAL ENERGY USE			COMMERCIAL ENERGY USE PER CAPITA			NET ENERGY IMPORTS[a]	
	Thousand Metric Tons of Oil Equivalent		Thousand Metric Tons of Oil Equivalent		Average Annual % Growth	Kg. of Oil Equivalent		Average Annual % Growth	% of Commercial Energy Use	
	1980	2000	1980	2000	1980–2000	1980	2000	1980–2000	1980	2000
Dominican Republic	1,327	1,421	3,491	7,804	4.0	613	932	2.1	62	82
Ecuador	11,744	22,520	6,180	8,187	2.1	651	647	–0.3	–127	–175
Egypt	34,168	57,599	15,970	46,423	4.6	391	726	2.3	–114	–24
El Salvador	1,913	2,157	2,537	4,083	2.2	553	651	0.6	25	47
Eritrea	—	—	—	—	—	—	—	—	—	—
Estonia	6,951	2,917	6,275	4,523	—	4,248	3,303	—	–11	36
Ethiopia	10,575	17,583	11,145	18,732	2.6	295	291	–0.1	5	6
Finland	6,912	15,134	25,413	33,147	1.7	5,316	6,409	1.3	73	54
France	45,544	103,730	187,737	257,128	1.9	3,484	4,306	1.5	76	49
Gabon	9,441	16,800	1,493	1,563	–0.2	2,157	1,271	–3.1	–532	–975
The Gambia	—	—	—	—	—	—	—	—	—	—
Georgia	1,504	737	4,474	2,860	—	882	533	—	66	74
Germany	185,628	134,317	360,441	339,640	–0.2	4,602	4,131	–0.5	48	60
Ghana	3,305	5,883	4,027	7,720	3.6	378	400	0.5	19	24
Greece	3,696	9,987	15,960	27,822	3.0	1,628	2,635	2.5	76	64
Guatemala	2,583	5,241	3,847	7,146	3.5	564	628	0.9	33	27
Guinea	—	—	—	—	—	—	—	—	—	—
Guinea-Bissau	—	—	—	—	—	—	—	—	—	—
Haiti	1,877	1,541	3,847	2,039	0.4	392	256	–1.6	11	24
Honduras	1,315	1,522	1,892	3,012	2.8	530	469	–0.2	31	49
Hungary	14,957	11,090	28,940	24,783	–1.0	2,703	2,448	–0.7	48	55
India	221,322	421,565	241,016	501,894	3.8	351	494	1.8	8	16
Indonesia	128,996	229,478	59,933	145,575	4.8	404	706	3.1	–115	–58
Iran	81,142	242,146	38,987	112,725	5.6	997	1,771	3.1	–108	–115
Iraq	136,643	134,089	12,030	27,678	4.3	925	1,190	1.3	–1,036	–384
Ireland	1,894	2,197	8,485	14,623	2.7	2,495	3,854	2.3	78	85
Israel	153	654	8,563	20,200	5.2	2,208	3,241	2.6	98	97
Italy	19,644	26,858	138,629	171,567	1.3	2,458	2,974	1.2	86	84
Jamaica	224	486	2,378	3,920	3.5	1,115	1,524	2.6	91	88
Japan	43,247	105,505	346,492	524,715	2.6	2,967	4,136	2.2	88	80
Jordan	1	286	1,714	5,185	4.9	786	1,061	0.5	100	94
Kazakhstan	76,799	78,102	76,799	39,093	—	5,163	2,594	—	0	–100
Kenya	7,891	12,260	9,791	15,482	2.2	599	515	–0.7	19	21
Korea, North	29,669	42,576	32,631	46,112	1.9	1,898	2,071	0.5	9	8
Kuwait	91,636	111,469	12,249	20,894	1.0	8,908	10,529	0.2	–648	–434
Kyrgyzstan	2,190	1,443	1,717	2,445	—	473	497	—	–28	41
Laos	—	—	—	—	—	—	—	—	—	—
Latvia	261	1,250	566	3,655	—	222	1,541	—	54	66
Lebanon	178	171	2,524	5,058	4.8	841	1,169	2.8	93	97
Lesotho	—	—	—	—	—	—	—	—	—	—
Liberia	—	—	—	—	—	—	—	—	—	—

Table J *(continued)*
World Countries: Energy Production and Use, 1980–2000

COUNTRY	COMMERCIAL ENERGY PRODUCTION		COMMERCIAL ENERGY USE			COMMERCIAL ENERGY USE PER CAPITA			NET ENERGY IMPORTS[a]	
	Thousand Metric Tons of Oil Equivalent		Thousand Metric Tons of Oil Equivalent		Average Annual % Growth	Kg. of Oil Equivalent		Average Annual % Growth	% of Commercial Energy Use	
	1980	2000	1980	2000	1980–2000	1980	2000	1980–2000	1980	2000
Libya	96,550	73,904	7,193	16,438	3.6	2,364	3,107	1.0	–1,242	–350
Lithuania	—	3,212	—	7,124	—	—	2,032	—	—	55
Macedonia	—	—	—	—	—	—	—	—	—	—
Madagascar	—	—	—	—	—	—	—	—	—	—
Malawi	—	—	—	—	—	—	—	—	—	—
Malaysia	18,202	76,759	12,162	49,472	7.7	884	2,126	4.9	–50	–55
Mali	—	—	—	—	—	—	—	—	—	—
Mauritania	—	—	—	—	—	—	—	—	—	—
Mauritius	—	—	—	—	—	—	—	—	—	—
Mexico	149,359	229,653	98,898	155,513	2.1	1,464	1,567	0.2	–51	–50
Moldova	35	60	—	2,871	—	—	671	—	—	98
Mongolia	—	—	—	—	—	—	—	—	—	—
Morocco	877	572	4,778	10,293	4.3	247	359	2.2	82	94
Mozambique	7,413	7,219	8,074	7,126	–0.8	668	403	–2.6	8	–1
Myanmar	9,513	15,144	9,430	12,522	1.3	280	262	–0.4	–1	–21
Namibia	—	292	—	1,031	—	—	587	—	—	72
Nepal	4,403	6,872	4,576	7,900	2.7	314	343	0.4	4	13
Netherlands	71,830	57,239	64,984	75,799	1.4	4,593	4,762	0.8	–11	24
New Zealand	5,488	15,379	9,213	18,633	3.8	2,959	4,864	2.7	40	17
Nicaragua	907	1,553	1,553	2,746	2.7	531	542	0.0	42	43
Niger	—	—	—	—	—	—	—	—	—	—
Nigeria	148,479	197,726	52,846	90,169	2.6	743	710	–0.4	–181	–119
Norway	55,716	224,993	18,792	25,617	1.8	4,588	5,704	1.3	–197	–778
Oman	15,090	60,084	996	9,750	11.0	905	4,046	6.6	–1,415	–516
Pakistan	20,997	47,124	25,472	63,951	4.8	308	463	2.2	18	26
Panama	526	732	1,399	2,546	2.7	717	892	0.7	62	71
Papua New Guinea	—	—	—	—	—	—	—	—	—	—
Paraguay	1,605	6,886	2,089	3,930	4.2	671	715	1.2	23	–75
Peru	14,655	9,477	11,752	12,695	0.2	678	489	–1.8	–25	25
Philippines	10,670	20,922	21,212	42,424	3.9	442	554	1.5	50	51
Poland	122,222	78,960	123,031	89,975	–1.4	3,458	2,328	–1.8	1	12
Portugal	1,481	3,129	10,291	24,613	4.7	1,054	2,459	4.7	86	87
Puerto Rico	—	—	—	—	—	—	—	—	—	—
Romania	52,587	28,290	65,123	36,330	–3.1	2,933	1,619	–3.2	19	22
Russian Federation	748,647	966,512	763,707	613,969	—	5,494	4,218	—	2	–57
Rwanda	—	—	—	—	—	—	—	—	—	—
Saudi Arabia	533,071	487,889	31,108	105,303	5.0	3,319	5,081	1.0	–1,614	–363
Senegal	1,046	1,723	1,919	3,086	2.4	346	324	–0.3	45	44
Serbia and Montenegro	—	—	—	—	—	—	—	—	—	—

Table J (continued)
World Countries: Energy Production and Use, 1980–2000

COUNTRY	COMMERCIAL ENERGY PRODUCTION		COMMERCIAL ENERGY USE			COMMERCIAL ENERGY USE PER CAPITA			NET ENERGY IMPORTS[a]	
	Thousand Metric Tons of Oil Equivalent		Thousand Metric Tons of Oil Equivalent		Average Annual % Growth	Kg. of Oil Equivalent		Average Annual % Growth	% of Commercial Energy Use	
	1980	2000	1980	2000	1980–2000	1980	2000	1980–2000	1980	2000
Sierra Leone	—	—	—	—	—	—	—	—	—	—
Singapore	—	64	6,062	24,591	8.7	2,511	6,120	6.0	—	100
Slovak Republic	3,418	5,994	21,056	17,466	–1.4	4,224	3,234	–1.7	84	66
Slovenia	—	3,098	—	6,540	—	—	3,288	—	—	53
Somalia	—	—	—	—	—	—	—	—	—	—
South Africa	73,169	144,469	65,417	107,595	2.1	2,372	2,514	–0.2	–12	–34
Spain	15,636	31,865	68,576	124,881	3.2	1,834	3,084	2.9	77	74
Sri Lanka	3,209	4,530	4,536	8,063	2.5	311	437	1.4	29	44
Sudan	7,089	23,664	8,406	16,216	3.1	435	521	0.7	16	–46
Swaziland	—	—	—	—	—	—	—	—	—	—
Sweden	16,132	30,681	39,911	47,481	1.0	4,803	5,354	0.6	60	35
Switzerland	7,030	11,782	20,861	26,587	1.4	3,301	3,704	0.7	66	56
Syria	9,502	32,890	5,348	18,407	5.4	614	1,137	2.2	–78	–79
Tajikistan	1,986	1,250	1,650	2,911	—	416	470	—	–20	57
Tanzania	9,502	14,601	10,280	15,386	2.0	553	457	–1.0	8	5
Thailand	11,182	41,118	22,808	73,618	7.4	488	1,212	6.0	51	44
Togo	562	1,036	715	1,530	3.8	284	338	0.8	21	32
Trinidad and Tobago	13,141	17,884	3,873	8,655	3.3	3,580	6,660	2.4	–239	–106
Tunisia	6,966	7,003	3,907	7,888	3.7	612	825	1.6	–78	11
Turkey	17,077	26,186	31,452	77,104	4.6	707	1,181	2.7	46	66
Turkmenistan	8,034	45,968	7,948	13,855	—	2,778	2,627	—	–1	–231
Uganda	—	—	—	—	—	—	—	—	—	—
Ukraine	109,708	82,330	97,893	139,592	—	1,956	2,820	—	–12	41
United Arab Emirates	89,716	143,589	6,273	29,559	8.2	6,014	10,175	2.8	–1,330	–386
United Kingdom	196,792	272,338	201,284	232,644	1.0	3,573	3,962	0.8	2	–17
United States	1,553,260	1,675,770	1,811,650	2,299,669	1.5	7,973	8,148	0.4	14	27
Uruguay	766	1,028	2,643	3,079	1.7	907	923	1.0	71	67
Uzbekistan	4,615	55,066	4,821	50,151	—	302	2,027	—	4	–10
Venezuela	140,578	225,470	36,148	29,256	2.5	2,395	2,452	0.1	–286	–280
Vietnam	18,364	46,299	19,573	36,965	3.2	364	471	1.2	6	–25
West Bank and Gaza	—	—	—	—	—	—	—	—	—	—
Yemen	60	22,046	1,424	3,526	4.2	167	201	0.3	96	–525
Zambia	4,179	5,916	4,719	6,244	1.3	822	619	–1.6	11	5
Zimbabwe	5,793	8,708	6,570	10,219	2.6	921	809	–0.3	12	15
World	6,912,364	10,010,145	6,930,291	9886146	2.9	1,623	1649	0.9	—	—
Low Income	819,169	1,400,460	674,008	1,287,496	4.7	452	569	2.2	–22	–9
Middle income	33,026,123	4,812,604	2,446,876	3,457,150	4.1	1,252	1,318	2.1	–35	–39
Lower middle income	2,028,987	3,252,169	1,856,387	2,573,688	5.0	1,156	1,206	2.9	–9	–26

Table J *(continued)*
World Countries: Energy Production and Use, 1980–2000

COUNTRY	COMMERCIAL ENERGY PRODUCTION		COMMERCIAL ENERGY USE			COMMERCIAL ENERGY USE PER CAPITA			NET ENERGY IMPORTS[a]	
	Thousand Metric Tons of Oil Equivalent		Thousand Metric Tons of Oil Equivalent		Average Annual % Growth	Kg. of Oil Equivalent		Average Annual % Growth	% of Commercial Energy Use	
	1980	2000	1980	2000	1980–2000	1980	2000	1980–2000	1980	2000
Upper middle income	1,273,626	1,560,436	590,489	883,462	2.1	1,694	1,805	0.3	–116	–77
Low & middle income	4,121,782	6,213,064	3,120,884	4,744,646	4.3	906	971	2.0	–32	–31
East Asia & Pacific	842,071	1,579,933	776,249	1,549,127	3.9	578	871	2.4	–8	–2
Europe & Central Asia	1,241,969	1,470,085	1,332,884	1,253,443	7.6	3,348	2,653	—	7	–17
Latin America & Carib.	476,911	847,298	381,002	601,859	2.4	1,074	1,181	0.6	–25	–41
Middle East & N. Africa	981,350	1,268,307	138,565	398,549	4.9	798	1,368	2.2	–610	–219
South Asia	256,676	495,144	284,041	600,474	3.9	321	453	1.8	10	18
Sub-Saharan Africa	322,805	552,297	208,143	341,194	2.3	714	669	–0.5	–55	–62
High income	2,790,581	3,797,081	3,809,407	5,141,500	1.8	4,623	5,430	1.1	27	26
Europe EMU	367,292	434,433	952,761	1,160,702	1.2	3,337	3,824	0.9	61	63

[a]A negative value indicates that a country is a net exporter.

Source: World Development Indicators 2003 (World Bank).

Table K
World Countries: Energy Efficiency and Emission, 1980–1999

COUNTRY	GDP PER UNIT OF ENERGY USE		TRADITIONAL FUEL USE		CARBON DIOXIDE EMISSIONS					
	PPP $ Per Kg. of Oil Equivalent		% of Total Energy Use		Total Million Metric Tons		Per Capita Metric Tons		Kg. Per PPP $ of GDP	
	1980	2000	1980	1997	1980	1999	1980	1999	1980	1999
Afghanistan	—	—	—	—	1.7	1.0	0.1	0	—	—
Albania	—	6.7	13.1	7.3	4.8	1.5	1.8	0.5	—	0.2
Algeria	5.5	6.4	1.9	1.5	66.1	90.8	3.5	3.0	1.0	0.5
Angola	—	3.6	64.9	69.7	5.3	10.3	0.8	0.8	—	0.4
Argentina	4.4	7.2	5.9	4.0	107.5	137.5	3.8	3.8	0.6	0.3
Armenia	—	4.5	—	0.0	—	3.1	—	0.8	—	0.4
Australia	2.0	4.3	3.8	4.4	202.8	344.4	13.8	18.2	1.5	0.8
Austria	3.4	7.5	1.2	4.7	52.4	61.4	6.9	7.6	0.7	0.3
Azerbaijan	—	1.9	—	0.0	—	33.6	—	4.2	—	1.8
Bangladesh	5.4	10.8	81.3	46.0	7.6	25.4	0.1	0.2	0.2	0.1
Belarus	—	3.0	—	0.8	—	57.6	—	5.7	—	0.9
Belgium	2.2	4.4	0.2	1.6	131.3	104.4	13.3	10.2	1.3	0.4
Benin	1.2	2.5	85.4	89.2	0.5	1.3	0.1	0.2	0.3	0.2
Bolivia	3.0	3.9	19.3	14.0	4.5	11.2	0.8	1.4	0.6	0.6
Bosnia-Herzegovina	—	5.2	—	10.1	—	4.8	—	1.2	—	0.2
Botswana	—	—	35.7	—	1.0	3.9	1.1	2.4	0.7	0.4
Brazil	4.2	6.7	35.5	28.7	183.4	300.7	1.5	1.8	0.4	0.3
Bulgaria	1.0	2.8	0.5	1.3	75.3	42.1	8.5	5.1	2.7	0.9
Burkina Faso	—	—	91.3	87.1	0.4	1.0	0.1	0.1	0.1	0.1
Burundi	—	—	97.0	94.2	0.1	0.2	0.0	0.0	0.1	0.1
Cambodia	—	—	100.0	89.3	0.3	0.7	0.0	0.1	—	0.0
Cameroon	2.7	3.8	51.7	69.2	3.9	4.7	0.4	0.3	0.4	0.2
Canada	1.4	3.3	0.4	4.7	420.9	438.6	17.1	14.4	1.5	0.6
Central African Republic	—	—	88.9	87.5	0.1	0.3	0.0	0.1	0.1	0.1
Chad	—	—	95.9	97.6	0.2	0.1	0.0	0.0	0.1	0.0
Chile	3.0	5.6	12.3	11.3	27.5	62.5	2.5	4.2	1.0	0.5
China	0.7	4.1	8.4	5.7	1,476.8	2,825.0	1.5	2.3	3.5	0.7
Hong Kong, China	6.2	10.9	0.9	0.7	16.3	41.2	3.2	6.2	0.5	0.3
Colombia	4.7	10.3	15.9	17.7	39.8	63.6	1.4	1.5	0.4	0.2
Democratic Republic of the Congo (formerly Zaire)	3.8	2.5	73.9	91.7	3.5	2.1	0.1	0.0	0.1	0.1
Congo Republic	0.8	3.2	77.8	53.0	0.4	2.4	0.2	0.8	0.6	0.9
Costa Rica	6.6	11.7	26.3	54.2	2.5	6.1	1.1	1.6	0.2	0.2
Côte d'Ivoire	—	—	52.8	91.5	5.3	13.3	0.6	0.9	0.5	0.6
Croatia	—	4.9	—	3.2	—	20.8	—	4.8	—	0.6
Cuba	—	—	27.9	30.2	30.8	25.4	3.2	2.3	—	—
Czech Republic	—	3.6	0.6	1.6	—	108.9	—	10.6	—	0.8
Denmark	3.0	7.9	0.4	5.9	92.9	49.7	12.3	9.3	1.1	0.3
Dominican Republic	4.1	7.4	27.5	14.3	6.4	23.3	1.1	2.8	0.4	0.4
Ecuador	2.8	4.9	26.7	17.5	13.4	23.3	1.7	1.9	0.9	0.6
Egypt	3.3	4.8	4.7	3.2	45.2	123.6	1.1	2.0	0.9	0.6

Table K (*continued*)
World Countries: Energy Efficiency and Emission, 1980–1999

COUNTRY	GDP PER UNIT OF ENERGY USE		TRADITIONAL FUEL USE		CARBON DIOXIDE EMISSIONS					
	PPP $ Per Kg. of Oil Equivalent		% of Total Energy Use		Total Million Metric Tons		Per Capita Metric Tons		Kg. Per PPP $ of GDP	
	1980	2000	1980	1997	1980	1999	1980	1999	1980	1999
El Salvador	5.0	8.1	52.9	34.5	2.1	5.8	0.5	0.9	0.2	0.2
Eritrea	—	—	—	96.0	—	6.0	—	0.1	—	0.1
Estonia	—	2.9	—	13.8	—	16.2	—	11.7	—	1.4
Ethiopia	1.6	2.6	89.6	95.9	1.8	5.5	0.0	0.1	0.1	0.1
Finland	1.7	3.8	4.3	6.5	56.9	58.4	11.9	11.3	1.3	0.5
France	2.8	5.4	1.3	5.7	482.7	359.7	9.0	6.1	0.9	0.3
Gabon	1.8	4.7	30.8	32.9	6.2	3.6	8.9	3.0	2.3	0.5
The Gambia	—	—	72.7	78.6	0.2	0.3	0.2	0.2	0.2	0.1
Georgia	4.6	4.5	—	1.0	—	5.4	—	1.0	—	0.5
Germany	2.2	6.1	0.3	1.3	—	792.2	—	9.7	—	0.4
Ghana	3.1	5.5	43.7	78.1	2.4	5.6	0.2	0.3	0.2	0.1
Greece	4.7	6.3	3.0	4.5	51.7	85.9	5.4	8.2	0.7	0.5
Guatemala	4.6	7.1	54.6	62.0	4.5	9.7	0.7	0.9	0.3	0.2
Guinea	—	—	71.4	74.2	0.9	1.3	0.2	0.2	—	0.1
Guinea-Bissau	—	—	80.0	57.1	0.5	0.3	0.7	0.2	1.4	0.3
Haiti	4.7	7.5	80.7	74.7	0.8	1.4	0.1	0.2	0.1	0.1
Honduras	3.2	6.0	55.3	54.8	2.1	5.0	0.6	0.8	0.3	0.3
Hungary	2.0	4.9	2.0	1.6	82.5	56.9	7.7	5.6	1.5	0.5
India	2.2	5.5	31.5	20.7	347.3	1,077.0	0.5	1.1	0.7	0.4
Indonesia	2.0	4.2	51.5	29.3	94.6	235.6	0.6	1.2	0.8	0.4
Iran	2.7	3.2	0.4	0.7	116.1	301.4	3.0	4.8	1.1	0.9
Iraq	—	—	0.3	0.1	44.0	74.2	3.4	3.3	—	—
Ireland	2.3	7.9	0.0	0.2	25.2	40.4	7.4	10.8	1.3	0.4
Israel	3.7	6.5	0.0	0.0	21.2	61.1	5.4	10.0	0.7	0.5
Italy	3.9	8.2	0.8	1.0	371.9	422.7	6.6	7.3	0.7	0.3
Jamaica	1.8	2.4	5.0	6.0	8.4	10.2	4.0	4.0	2.0	1.2
Japan	3.1	6.1	0.1	1.6	920.4	1,155.2	7.9	9.1	0.8	0.4
Jordan	3.1	3.6	0.0	0.0	4.7	14.6	2.2	3.1	0.9	0.8
Kazakhstan	—	2.2	—	0.2	—	112.8	—	7.4	—	1.5
Kenya	1.0	1.9	76.8	80.3	6.2	8.8	0.4	0.3	0.6	0.3
Korea, North	—	—	3.1	1.4	124.9	206.7	7.3	9.4	—	—
Korea, South	2.3	3.6	4.0	2.4	125.1	393.5	3.3	8.4	1.3	0.6
Kuwait	1.4	1.8	0.0	0.0	24.7	48.0	18.0	24.9	1.5	1.4
Kyrgyzstan	—	5.4	—	0.0	—	4.7	—	1.0	—	0.4
Laos	—	—	72.3	88.7	0.2	0.4	0.1	0.1	—	0.1
Latvia	19.8	4.6	—	26.2	—	6.6	—	2.8	—	0.4
Lebanon	—	3.5	2.4	2.5	6.2	16.9	2.1	4.0	—	1.0
Lesotho	—	—	—	—	—	—	—	—	—	—
Liberia	—	—	—	—	2.0	0.4	1.1	0.1	—	—
Libya	—	—	2.3	0.9	26.9	42.8	8.8	8.3	—	—
Lithuania	—	3.9	—	6.3	—	13.2	—	3.8	—	0.5
Macedonia	—	—	—	6.1	—	11.4	—	5.6	—	1.0

Table K *(continued)*
World Countries: Energy Efficiency and Emission, 1980–1999

COUNTRY	GDP PER UNIT OF ENERGY USE		TRADITIONAL FUEL USE		CARBON DIOXIDE EMISSIONS					
	PPP $ Per Kg. of Oil Equivalent		% of Total Energy Use		Total Million Metric Tons		Per Capita Metric Tons		Kg. Per PPP $ of GDP	
	1980	2000	1980	1997	1980	1999	1980	1999	1980	1999
Madagascar	—	—	78.4	84.3	1.6	1.9	0.2	0.1	0.3	0.2
Malawi	—	—	90.6	88.6	0.7	0.8	0.1	0.1	0.3	0.1
Malaysia	2.6	4.3	15.7	5.5	28.0	123.7	2.0	5.4	0.9	0.7
Mali	—	—	86.7	88.9	0.4	0.5	0.1	0.0	0.1	0.1
Mauritania	—	—	0.0	0.0	0.6	3.0	0.4	1.2	0.3	0.6
Mauritius	—	—	59.1	36.1	0.6	2.5	0.6	2.1	0.3	0.2
Mexico	2.9	5.5	5.0	4.5	252.5	378.5	3.7	3.9	0.9	0.5
Moldova	—	3.1	—	0.5	—	6.5	—	1.5	—	0.8
Mongolia	—	—	14.4	4.3	6.8	7.5	4.1	3.2	3.7	1.9
Morocco	6.4	9.5	5.2	4.0	15.9	35.8	0.8	1.3	0.5	0.4
Mozambique	0.7	2.5	43.7	91.4	3.2	1.3	0.3	0.1	0.6	0.1
Myanmar (Burma)	—	—	69.3	60.5	4.8	9.2	0.1	0.2	—	—
Namibia	—	12.0	—	—	—	0.1	—	0.1	—	0.0
Nepal	1.5	3.7	94.2	89.6	0.5	3.3	0.0	0.1	0.1	0.1
Netherlands	2.3	5.7	0.0	1.1	153.0	134.6	10.8	8.5	1.0	0.3
New Zealand	2.7	3.7	0.2	0.8	17.6	30.8	5.6	8.1	0.7	0.5
Nicaragua	4.0	4.6	49.2	42.2	2.0	3.8	0.7	0.8	0.3	0.3
Niger	—	—	79.5	80.6	0.6	1.1	0.1	0.1	0.1	0.1
Nigeria	0.8	1.2	66.8	67.8	68.1	40.4	1.0	0.3	1.7	0.4
Norway	2.3	5.1	0.4	1.1	38.7	38.7	9.5	8.7	0.9	0.3
Oman	4.5	3.0	0.0	—	5.9	19.9	5.3	8.5	1.3	0.7
Pakistan	2.1	4.0	24.4	29.5	31.6	98.9	0.4	0.7	0.6	0.4
Panama	4.1	6.5	26.6	14.4	3.5	8.3	1.8	2.9	0.6	0.5
Papua New Guinea	—	—	65.4	62.5	1.8	2.5	0.6	0.5	0.5	0.2
Paraguay	4.8	7.2	62.0	49.6	1.5	4.5	0.5	0.8	0.1	0.2
Peru	4.4	9.5	15.2	24.6	23.6	30.4	1.4	1.2	0.5	0.3
Philippines	5.3	6.8	37.0	26.9	36.5	73.2	0.8	1.0	0.3	0.3
Poland	—	4.0	0.4	0.8	456.2	314.4	12.8	8.1	—	0.9
Portugal	5.5	7.2	1.2	0.9	27.1	60.0	2.8	6.0	0.5	0.4
Puerto Rico	—	—	0.0	—	14.0	10.1	4.4	2.7	0.6	0.1
Romania	—	3.4	1.3	5.7	191.8	81.2	8.6	3.6	—	0.7
Russian Federation	—	1.6	—	1.8	—	1,437.3	—	9.8	—	1.6
Rwanda	—	—	89.8	88.3	0.3	0.6	0.1	0.1	0.1	0.1
Saudi Arabia	4.0	2.6	0.0	0.0	130.7	235.4	14.0	11.7	1.1	0.9
Senegal	2.2	4.5	50.8	56.2	2.8	3.7	0.5	0.4	0.7	0.3
Serbia and Montenegro	—	—	—	1.5	102.0	39.5	10.4	3.7	—	—
Sierra Leone	—	—	90.0	86.1	0.6	0.5	0.2	0.1	0.3	0.3
Singapore	2.2	3.9	0.4	0.0	30.1	54.3	12.5	13.7	2.3	0.7
Slovak Republic	—	3.6	—	0.5	—	38.6	—	7.2	—	0.7
Slovenia	—	5.0	—	1.5	—	14.4	—	7.3	—	0.5
Somalia	—	—	—	—	0.6	0.0	0.1	0.0	—	—
South Africa	3.1	4.4	4.9	43.4	211.3	334.6	7.7	7.9	1.0	0.8
Spain	3.8	6.4	0.4	1.3	200.0	273.7	5.3	6.8	0.8	0.4

Table K *(continued)*
World Countries: Energy Efficiency and Emission, 1980–1999

COUNTRY	GDP PER UNIT OF ENERGY USE		TRADITIONAL FUEL USE		CARBON DIOXIDE EMISSIONS					
	PPP $ Per Kg. of Oil Equivalent		% of Total Energy Use		Total Million Metric Tons		Per Capita Metric Tons		Kg. Per PPP $ of GDP	
	1980	2000	1980	1997	1980	1999	1980	1999	1980	1999
Sri Lanka	3.1	7.8	53.5	46.5	3.4	8.6	0.2	0.5	0.2	0.2
Sudan	1.6	3.8	86.9	75.1	3.3	2.6	0.2	0.1	0.2	0.0
Swaziland	—	—	—	—	0.5	0.4	0.8	0.4	0.4	0.1
Sweden	2.0	4.4	7.7	17.9	71.4	46.6	8.6	5.3	0.9	0.2
Switzerland	4.4	7.5	0.9	6.0	40.9	40.6	6.5	5.7	0.4	0.2
Syria	2.6	2.9	0.0	0.0	19.3	53.4	2.2	3.4	1.4	1.1
Tajikistan	—	2.3	—	—	—	5.1	—	0.8	—	0.8
Tanzania	—	1.1	92.0	91.4	1.9	2.5	0.1	0.1	—	0.2
Thailand	2.9	5.1	40.3	24.6	40.0	199.7	0.9	3.3	0.6	0.6
Togo	4.9	4.9	35.7	71.9	0.6	1.3	0.2	0.3	0.2	0.2
Trinidad and Tobago	1.2	1.3	1.4	0.8	16.7	25.1	15.4	19.4	3.6	2.4
Tunisia	3.8	7.4	16.1	12.5	9.4	17.5	1.5	1.8	0.6	0.3
Turkey	3.2	5.3	20.5	3.1	76.3	198.5	1.7	3.1	0.8	0.5
Turkmenistan	—	1.2	—	—	—	31.0	—	6.7	—	2.5
Uganda	—	—	93.6	89.7	0.6	1.4	0.1	0.1	0.1	0.0
Ukraine	—	1.4	—	0.5	—	374.3	—	7.5	—	2.1
United Arab Emirates	4.9	2.0	0.0	—	36.3	88.0	34.8	31.3	1.2	1.6
United Kingdom	2.5	6.0	0.0	3.3	580.3	539.3	10.3	9.2	1.2	0.4
United States	1.6	4.2	1.3	3.8	4,626.8	5,495.4	20.4	19.7	1.6	0.6
Uruguay	4.8	9.4	11.1	21.0	5.8	6.5	2.0	2.0	0.5	0.2
Uzbekistan	—	1.1	—	0.0	—	104.8	—	4.4	—	2.1
Venezuela	1.6	2.3	0.9	0.7	90.1	125.8	6.0	5.3	1.6	1.0
Vietnam	—	4.2	49.1	37.8	16.8	46.6	0.3	0.6	—	0.3
West Bank and Gaza	—	—	—	—	—	—	—	—	—	—
Yemen	—	4.0	0.0	1.4	—	18.3	—	1.1	—	1.4
Zambia	0.8	1.2	37.4	72.7	3.5	1.8	0.6	0.2	0.9	0.3
Zimbabwe	1.5	3.1	27.6	25.2	9.6	17.6	1.3	1.4	1.0	0.5
WORLD	2.1	4.5	7.4w	8.2w	13,852.7	22,518.8	3.4	3.8	1.1	0.5
Low income	2.1	4.0	46.4	29.8	774.3	2,429.2	0.5	1.0	0.6	0.5
Middle income	2.1	4.0	10.4	7.3	4,132.9	8,484.0	2.3	3.2	1.2	0.7
Lower middle income	1.6	3.7	10.7	5.7	2,682.6	6,391.3	1.8	3.0	1.6	0.7
Upper middle income	3.4	4.9	8.6	10.6	1,450.3	2,092.7	4.3	4.3	0.7	0.5
Low & middle income	2.1	4.0	18.5	12.9	4,9.7.1	10,613.2	1.5	2.2	1.0	0.6
East Asia & Pacific	—	—	15.1	9.7	1,833.3	3,734.4	1.3	2.1	2.2	0.6
Europe & Central Asia	—	2.3	3.2	1.3	989.0	3,144.1	—	6.6	1.3	1.2
Latin America & Carib.	3.6	6.1	18.4	16.0	848.8	1,286.7	2.4	2.5	0.6	0.4
Middle East & N. Africa	3.6	3.8	1.6	1.1	491.7	1,048.4	3.0	3.7	1.0	0.7
South Asia	2.3	5.5	34.2	23.8	392.3	1,215.1	0.4	0.9	0.6	0.4
Sub-Sarahan Africa	2.0	2.9	47.2	63.5	352.0	484.6	0.9	0.8	0.8	0.4
High income	2.2	4.9	1.0	3.4	8,945.6	11,606.6	12.0	12.3	1.2	0.5
Europe EMU	2.8	6.2	0.7	2.5	1,565.2	2,408.4	7.5	7.9	0.8	0.4

Source: World Development Indicators 2003 (World Bank).

Table L
World Countries: Power and Transportation

	ELECTRIC POWER										
	Consumption per capita Kilowatt-hours		Transmission and distribution losses % of output		Paved roads % of total		Goods transported by road millions of ton-km hauled		Goods transported by rail ton-km per $ million of GDP (PPP)		Air passengers carried thousands
Country	1980	1996	1980	1996	1990	1997	1990	1997	1990	1997	1996
AFRICA											
Algeria	265	524	11	18	67	*69*	14,000	X	25,161	X	3,494
Angola	67	61	*25*	28	25	25	X	X	X	X	585
Benin	36	48	220	87	20	*20*	X	X	X	X	75
Botswana	X	X	X	X	32	*24*	X	X	X	X	104
Burkina Faso	X	X	X	X	17	*16*	X	X	X	X	138
Burundi	X	X	X	X	18	*7*	X	X	X	X	9
Cameroon	167	171	7	20	11	*13*	X	X	33,209	34,023	362
Central African Republic	X	X	X	X	X	X	144	*60*	X	X	75
Chad	X	X	X	X	1	*1*	X	X	X	X	93
Congo (Zaire)	147	130	8	3	X	X	X	X	32,198	X	178
Congo Republic	94	207	1	0	10	*10*	X	X	144,851	X	253
Côte d'Ivoire	192	174	7	16	9	*10*	X	X	15,791	13,486	179
Egypt	380	924	13	0	72	*78*	31,400	31,500	23,310	X	4,282
Equatorial Guinea	X	X	X	X	X	X	X	X	X	X	X
Eritrea	X	X	X	X	19	*22*	X	X	X	X	X
Ethiopia	16	18	8	1	15	*15*	X	X	2,467	X	743
Gabon	X	X	X	X	X	X	X	X	X	X	X
Gambia	X	X	X	X	X	X	X	X	X	X	X
Ghana	X	X	X	X	X	X	X	X	X	X	X
Guinea	X	X	X	X	15	*17*	X	X	X	X	36
Guinea-Bissau	X	X	X	X	X	X	X	X	X	X	X
Kenya	92	126	16	16	13	*14*	X	X	75,496	X	779
Lesotho	X	X	X	X	18	*18*	X	X	X	X	17
Liberia	X	X	X	X	X	X	X	X	X	X	X
Libya	X	X	X	X	X	X	X	X	X	X	X
Madagascar	X	X	X	X	15	*12*	X	X	X	X	542
Malawi	X	X	X	X	22	*19*	X	X	14,881	*10,003*	153
Mali	X	X	X	X	11	*12*	X	X	53,882	X	75
Mauritania	X	X	X	X	11	*11*	X	X	X	X	235
Mauritius	X	X	X	X	X	X	X	X	X	X	X
Morocco	223	408	10	4	49	52	2,638	2,086	72,108	55,523	2,301
Mozambique	370	76	0	0	17	19	X	110	X	X	163
Namibia	X	X	X	X	11	8	X	X	308,833	*139,137*	237
Niger	X	X	X	X	29	*8*	X	X	X	X	75
Nigeria	68	85	36	32	30	*19*	X	X	3,009	X	221
Rwanda	X	X	X	X	9	*9*	X	X	X	X	9
Senegal	97	103	11	16	27	*29*	X	X	51,209	X	155
Sierra Leone	X	X	X	X	11	8	X	X	X	X	15
Somalia	X	X	X	X	X	X	X	X	X	X	X

Table L *(continued)*
World Countries: Power and Transportation

	ELECTRIC POWER										
	Consumption per capita Kilowatt-hours		Transmission and distribution losses % of output		Paved roads % of total		Goods transported by road millions of ton-km hauled		Goods transported by rail ton-km per $ million of GDP (PPP)		Air passengers carried thousands
Country	1980	1996	1980	1996	1990	1997	1990	1997	1990	1997	1996
South Africa	3,213	3,719	8	8	30	*42*	X	X	430,594	337,153	7,183
Sudan	X	X	X	X	X	X	X	X	X	X	X
Swaziland	X	X	X	X	X	X	X	X	X	X	X
Tanzania	50	59	14	12	37	*4*	X	X	77,466	*91,623*	224
Togo	X	X	X	X	21	*32*	X	X	X	X	75
Tunisia	379	674	12	11	76	*79*	X	X	58,795	*53,343*	1,371
Uganda	X	X	X	X	X	X	X	X	12,582	*11,567*	100
Zambia	1,016	560	7	11	17	X	X	X	73,728	*56,426*	235
Zimbabwe	990	765	14	7	14	*47*	X	X	247,759	196,429	654
NORTH AMERICA											
Canada	12,329	15,129	9	7	35	35	54,700	71,473	433,765	X	22,856
United States	8,914	11,796	9	7	58	61			360,699	361,911	571,072
CENTRAL AMERICA											
Belize	X	X	X	X	X	X	X	X	X	X	X
Costa Rica	860	1,349	0	12	15	17	2,243	3,070	X	X	918
Cuba	X	X	X	X	X	X	X	X	X	X	X
Dominican Republic	433	608	21	25	45	*49*	X	X	X	X	30
El Salvador	293	516	13	13	14	*20*	X	X	X	X	1,800
Guatemala	212	364	6	13	25	*28*	X	X	X	X	300
Haiti	41	34	26	54	22	*24*	X	X	X	X	X
Honduras	225	350	14	27	21	*20*	X	X	X	X	X
Jamaica	482	2,108	17	11	64	*71*	X	X	X	X	1,388
Mexico	846	1,381	11	15	35	*37*	108,884	165,000	64,884	53,917	14,678
Nicaragua	303	256	14	28	11	*10*	X	X	X	X	51
Panama	828	1,140	13	18	32	*34*	X	X	X	X	689
Trinidad and Tobago	X	X	X	X	X	X	X	X	X	X	X
SOUTH AMERICA											
Argentina	1,170	1,541	13	18	29	*29*	X	X	36,412	X	7,913
Bolivia	226	371	10	12	4	*6*	X	X	37,118	X	1,784
Brazil	974	1,660	12	17	10	*9*	X	X	56,068	X	22,012
Chile	876	1,864	12	9	14	*14*	X	X	15,882	5,998	3,622
Colombia	561	922	16	22	12	12	6,227	X	2,400	X	8,342
Equador	361	616	14	21	13	19	2,638	3,558	X	X	1,925
Guyana	X	X	X	X	X	X	X	X	X	X	X
Paraguay	233	914	6	7	9	*10*	X	X	X	X	261
Peru	502	598	13	15	10	*10*	X	X	7,486	X	2,328
Suriname	X	X	X	X	X	X	X	X	X	X	X
Uruguay	977	1,605	15	20	74	*90*	X	X	10,455	16,125	504
Venezuela	2,037	2,498	12	20	36	*39*	X	X	X	X	4,487

Table L *(continued)*
World Countries: Power and Transportation

	ELECTRIC POWER										
	Consumption per capita Kilowatt-hours		Transmission and distribution losses % of output		Paved roads % of total		Goods transported by road millions of ton-km hauled		Goods transported by rail ton-km per $ million of GDP (PPP)		Air passengers carried thousands
Country	1980	1996	1980	1996	1990	1997	1990	1997	1990	1997	1996
ASIA											
Afghanistan	X	X	X	X	X	X	X	X	X	X	X
Armenia	2,729	905	10	38	99	*100*	1,533	479	X	X	X
Azerbaijan	2,440	1,822	14	22	X	X	3,287	497	X	X	1,233
Bangladesh	16	97	35	30	7	12	X	X	8,032	X	1,252
Bhutan	X	X	X	X	X	X	X	X	X	X	X
Cambodia	X	X	X	X	8	*8*	X	*1,200*	X	X	X
China	253	687	8	7	X	X	X	X	671,824	*364,633*	51,770
Hong Kong, China	2,167	5,013	11	14	100	*100*	X	X	X	X	X
Georgia	1,910	1,020	16	23	94	*94*	7,370	98	X	X	152
India	130	347	18	18	*47*	46	X	X	248,469	*176,217*	13,395
Indonesia	44	296	19	12	46	46	X	X	8,619	X	17,139
Iran	491	1,142	10	20	X	*50*	X	X	40,223	X	7,610
Iraq	X	X	X	X	X	X	X	X	X	X	X
Israel	2,826	5,081	5	4	100	100	X	X	16,663	*11,947*	3,695
Japan	4,395	7,083	4	4	69	*74*	274,444	305,510	11,603	8,664	95,914
Jordan	387	1,187	19	10	100	*100*	X	X	78,625	47,242	1,299
Kazakhstan	0	2,865	X	15	55	*83*	44,775	6,481	5,042,201	X	568
Korea, North	X	X	X	X	X	X	X	X	X	X	X
Korea, South	841	4,453	6	5	*72*	*74*	31,841	74,504	40,875	*24,826*	33,033
Kuwait	4,749	12,808	10	0	73	*81*	X	X	X	X	2,133
Kyrgystan	1,556	1,479	6	33	90	*91*	5,627	350	X	X	488
Laos	X	X	X	X	24	*14*	120	X	X	X	125
Lebanon	789	1,651	10	13	95	*95*	X	X	X	X	775
Malaysia	630	2,078	9	11	70	*75*	X	X	16,313	9,416	15,118
Mongolia	X	X	X	X	10	3	1,871	X	1,324,119	X	X
Myanmar (Burma)	31	58	22	36	11	*12*	X	X	X	X	335
Nepal	13	39	29	28	38	*42*	X	X	X	X	755
Oman	X	X	X	X	X	X	X	X	X	X	X
Pakistan	125	333	29	23	54	58	352	84,174	43,586	*26,582*	5,375
Philippines	353	405	2	17	0	0	X	X	X	X	7,263
Saudi Arabia	1,356	3,980	9	8	41	*43*	X	X	4,634	4,206	11,706
Singapore	2,412	7,196	5	4	97	97	X	X	X	X	11,841
Sri Lanka	96	203	15	17	32	40	19	30	5,926	X	1,171
Syria	345	755	18	0	72	23	X	X	48,075	29,655	599
Tajikistan	2,217	2,292	7	12	72	*83*	X	X	X	X	594
Thailand	279	1,289	10	9	55	*98*	X	X	14,869	X	14,078
Turkey	439	1,161	12	17	X	25	X	139,789	30,838	*17,747*	8,464
Turkmenistan	1,720	1,020	12	11	74	*81*	X	X	X	X	523

Table L *(continued)*
World Countries: Power and Transportation

	ELECTRIC POWER										
	Consumption per capita Kilowatt-hours		Transmission and distribution losses % of output		Paved roads % of total		Goods transported by road millions of ton-km hauled		Goods transported by rail ton-km per $ million of GDP (PPP)		Air passengers carried thousands
Country	1980	1996	1980	1996	1990	1997	1990	1997	1990	1997	1996
United Arab Emirates	X	X	X	X	X	X	X	X	X	X	X
Uzbekistan	2,085	1,657	9	9	79	*87*	X	X	X	X	1,566
Vietnam	50	177	18	19	24	25	X	X	13,526	16,352	2,108
Yemen	59	99	6	26	9	8	X	X	X	X	X
EUROPE											
Albania	1,083	904	4	52	X	*30*	1,195	*80*	85,396	*5,523*	13
Austria	4,371	5,952	6	6	100	100	13,300	*16,600*	89,362	78,423	4,419
Belarus	2,455	2,476	9	16	*96*	98	22,128	9,065	1,297,626	*624,045*	843
Belgium	4,402	6,678	5	5	*81*	80	32,100	*42,800*	46,189	*31,976*	5,174
Bosnia-Herzegovina	X	X	X	X	X	X	X	X	X	X	X
Bulgaria	3,349	3,577	10	13	92	92	13,823	483	360,291	*210,161*	718
Croatia	0	2,291	X	16	80	*82*	2,458	470	190,170	*86,593*	727
Czech Republic	3,595	4,875	7	8	100	*100*	X	43,088	X	207,099	1,394
Denmark	4,245	6,113	7	5	100	*100*	9,400	9,400	19,119	*14,518*	5,592
Estonia	3,433	3,293	5	19	52	51	4,510	2,773	516,391	536,100	149
Finland	7,779	12,979	6	4	61	64	26,300	*24,100*	99,052	68,994	5,598
France	3,881	6,091	7	6	X	100	137,000	*158,200*	49,908	39,109	41,253
Germany	5,005	5,596	4	5	99	*99*	245,700	*281,300*	X	*39,350*	40,118
Greece	2,064	3,395	7	7	92	*92*	12,600	12,800	6,395	*1,913*	6,396
Hungary	2,335	2,814	10	13	50	43	1,836	770	247,428	*104,327*	1,563
Ireland	2,528	4,363	10	9	94	*94*	5,100	5,550	14,322	9,132	7,677
Italy	2,831	4,196	9	7	100	*100*	177,900	197,600	20,795	18,420	25,839
Latvia	2,664	1,783	26	47	13	38	5,853	*800*	1,209,517	1,114,210	276
Lithuania	2,715	1,785	12	11	82	89	*7,019*	8,622	915,522	545,100	214
Macedonia	0	2,443	X	X	59	64	1,708	*1,210*	X	X	287
Moldova	1,495	1,314	8	23	87	*87*	6,305	780	X	X	190
Netherlands	4,057	5,555	4	4	88	90	22,900	*27,600*	12,779	9,751	17,114
New Zealand	6,269	8,420	13	11	57	*58*	X	X	51,927	X	9,597
Norway	18,289	23,487	9	8	69	74	7,940	11,838	X	X	12,727
Poland	2,470	2,420	10	13	62	66	49,800	95,500	475,103	*284,381*	1,806
Portugal	1,469	3,044	12	10	X	X	10,900	*11,200*	13,976	13,598	4,806
Romania	2,434	1,757	6	12	*51*	51	13,800	22,400	507,379	*231,838*	913
Russian Federation	4,706	4,165	8	9	*74*	X	300	138	2,725,816	X	22,117
Slovak Republic	3,817	4,450	8	6	99	99	4,180	3,779	X	*297,426*	63
Slovenia	4,089	4,766	8	6	72	83	3,440	1,775	*142,879*	112,529	393
Spain	2,401	3,749	9	9	74	99	151,000	186,700	22,427	*15,984*	27,759
Sweden	10,216	14,239	9	7	71	77	26,500	*31,200*	127,826	103,299	9,879
Switzerland	5,579	6,619	7	7	X	X	10,400	13,000	X	X	10,468
Ukraine	3,598	2,640	8	10	94	95	79,668	20,532	2,109,937	1,411,737	1,151

Table L *(continued)*
World Countries: Power and Transportation

	ELECTRIC POWER										
	Consumption per capita Kilowatt-hours		Transmission and distribution losses % of output		Paved roads % of total		Goods transported by road millions of ton-km hauled		Goods transported by rail ton-km per $ million of GDP (PPP)		Air passengers carried thousands
Country	1980	1996	1980	1996	1990	1997	1990	1997	1990	1997	1996
United Kingdom	4,160	5,198	8	9	100	100	136,300	*153,900*	17,191	X	64,209
Yugoslavia (Serbia-Montenegro)	X	X	X	X	X	X	X	X	X	X	X
OCEANIA											
Australia	5,393	8,086	10	*7*	35	39	X	X	82,122	X	30,075
Fiji	X	X	X	X	X	X	X	X	X	X	X
New Zealand	6,269	8,420	13	11	57	58	X	X	51,927	X	9,597
Papua New Guinea	X	X	X	X	3	*4*	X	X	X	X	X
Solomon Islands	X	X	X	X	X	X	X	X	X	X	X
World	1,576w	2,027w	8w	8w	39m	44m					1,389,943s
Low income	188	433	12	12	17	*19*					103,110
Excl. China & India	155	218	14	19	17	*18*					37,945
Middle income	1,585	1,902	9	12	52	*51*					238,360
Lower middle income	1,835	1,771	8	11	54	*51*					102,609
Upper middle income	1,188	2,106	10	13	52	*47*					135,751
Low and middle income	633	886	9	12	29	*30*					341,470
East Asia & Pacific	260	724	8	9	24	*12*					143,204
Europe & Central Asia	2,925	2,795	8	11	77	*83*					46,014
Latin America & Carib.	854	1,347	12	16	22	*26*					76,275
Middle East & N. Africa	483	1,162	10	9	67	*50*					37,484
South Asia	116	313	19	19	38	*41*					22,445
Sub-Saharan Africa	444	439	9	10	17	*16*					16,049
High income	5,783	8,121	8	6	86	*92*					1,048,473

Note: Figures in italics are for years other than those specified.

Source: Entering the 21st Century: World Development Report 1999/2000 (Oxford University Press, World Bank, 2000).

Table M
World Countries: Communications, Information, and Science and Technology

	PER 1,000 PEOPLE										
	Daily newspapers	Radio	Television sets	Telephone main lines	Mobile telephones	Personal computers	Internet hosts per 10,000 people	Scientists and engineers in R & D per million people	High-technology exports % of mfg. exports	No. of patent applications filed[a]	
Country	1996	1996	1997	1997	1997	1997	January 1999	1985–95	1997	Residents	Nonresidents
AFRICA											
Algeria	38	239	*67*	48	1	4.2	0.01	X	22	48	150
Angola	12	54	*91*	5	1	0.7	0.00	X	X	X	X
Benin	2	108	*91*	6	1	0.9	0.02	177	X	X	X
Botswana	27	155	*27*	56	0	13.4	4.18	X	X	5	56
Burkina Faso	1	32	*6*	3	0	0.7	0.16	X	X	X	X
Burundi	3	68	10	3	0	X	0.00	32	X	1	4
Cameroon	7	162	*81*	5	0	1.5	0.00	X	3	X	X
Central African Republic	2	84	*5*	3	0	X	0.00	55	0	X	X
Chad	0	249	2	1	0	X	0.00	X	X	X	X
Congo (Zaire)	3	98	*43*	1	0	X	0.00	X	X	2	27
Congo Republic	8	124	8	8	0	X	0.00	X	16	X	X
Côte d'Ivoire	16	157	61	9	2	3.3	0.16	X	X	X	X
Egypt	38	316	*127*	56	0	7.3	0.31	458	7	504	706
Equatorial Guinea	X	X	X	X	X	X	X	X	X	X	X
Eritrea	X	101	11	6	0	X	0.00	X	X	X	X
Ethiopia	2	194	*5*	3	0	X	0.01	X	0	3	X
Gabon	X	X	X	X	X	X	X	X	X	X	X
Gambia	X	X	X	X	X	X	X	X	X	X	X
Ghana	14	238	109	6	1	1.6	0.10	X	X	X	33
Guinea	X	47	41	3	0	0.3	0.00	X	X	X	X
Guinea-Bissau	X	X	X	X	X	X	X	X	X	X	X
Kenya	9	108	19	8	0	2.3	0.23	X	11	15	39,034
Lesotho	7	48	*24*	10	1	X	0.09	X	X	2	37,043
Liberia	X	X	X	X	X	X	X	X	X	X	X
Libya	X	X	X	X	X	X	X	X	X	X	X
Madagascar	4	192	*45*	3	0	1.3	0.04	11	2	7	20,800
Malawi	3	256	2	4	0	X	0.00	X	3	3	39,031
Mali	1	49	10	2	0	0.6	0.00	X	X	X	X
Mauritania	1	150	89	5	0	5.3	0.06	X	X	X	X
Mauritius	X	X	X	X	X	X	X	X	X	X	X
Morocco	26	241	160	50	3	2.5	0.20	X	27	90	237
Mozambique	3	39	*4*	4	0	1.6	0.08	X	8	X	X
Namibia	19	143	*32*	58	8	18.6	15.79	X	X	X	X
Niger	0	69	*26*	2	0	0.2	0.02	X	X	X	X
Nigeria	24	197	*61*	4	0	5.1	0.03	15	X	X	X
Rwanda	0	*102*	X	3	0	X	0.00	24	X	X	X
Senegal	5	141	*41*	13	1	11.4	0.21	X	55	X	X
Sierra Leone	5	251	20	4	0	X	0.03	X	X	X	X
Somalia	X	X	X	X	X	X	X	X	X	X	X
South Africa	30	316	125	107	37	41.6	34.67	938	X	X	X
Sudan	X	X	X	X	X	X	X	X	X	X	X

Table M *(continued)*
World Countries: Communications, Information, and Science and Technology

	PER 1,000 PEOPLE										
	Daily newspapers	Radio	Television sets	Telephone main lines	Mobile telephones	Personal computers	Internet hosts per 10,000 people	Scientists and engineers in R & D per million people	High-technology exports % of mfg. exports	No. of patent applications filed[a]	
Country	1996	1996	1997	1997	1997	1997	January 1999	1985-95	1997	Residents	Nonresidents
Swaziland	X	X	X	X	X	X	X	X	X	X	X
Tanzania	4	278	21	3	1	1.6	0.04	X	X	X	X
Togo	4	217	19	6	1	5.8	0.24	X	X	X	X
Tunisia	31	218	182	70	1	8.6	0.07	388	11	46	128
Uganda	2	123	*26*	2	0	1.4	0.05	X	X	X	38,497
Zambia	14	121	*80*	9	0	X	0.31	X	X	6	93
Zimbabwe	18	96	*29*	17	1	9.0	0.87	X	6	30	181
NORTH AMERICA											
Canada	159	1,078	708	609	139	270.6	364.25	2,656	25	3,316	45,938
United States	212	2,115	847	644	206	406.7	1,131.52	3,732	44	111,883	111,536
CENTRAL AMERICA											
Belize	X	X	X	X	X	X	X	X	X	X	X
Costa Rica	91	271	403	169	19	X	9.20	X	14	X	X
Cuba	X	X	X	X	X	X	X	X	X	X	X
Dominican Republic	52	177	*84*	88	16	X	5.79	X	23	X	X
El Salvador	48	461	*250*	56	7	X	1.33	19	16	3	64
Guatemala	31	73	126	41	6	*3.0*	0.83	99	13	2	102
Haiti	3	55	*5*	8	0	X	0.00	X	4	10	126
Jamaica	64	482	*323*	140	22	4.6	1.24	8	67	X	X
Mexico	97	324	251	96	18	37.3	11.64	213	33	389	30.305
Nicaragua	32	283	190	29	2	X	1.47	214	38	X	X
Panama	62	299	*187*	134	6	X	2.66	X	14	31	142
Trinidad and Tobago	X	X	X	X	X	X	X	X	X	X	X
SOUTH AMERICA											
Argentina	123	677	289	191	56	39.2	18.28	671	15	X	X
Bolivia	55	572	*115*	69	15	X	0.78	250	9	17	106
Brazil	40	435	316	107	28	26.3	12.88	168	18	2,655	29,451
Chile	99	354	233	180	28	54.1	20.18	X	19	189	1,771
Colombia	49	565	217	148	35	33.4	3.93	X	20	87	1,772
Ecuador	70	342	294	75	13	*13.0*	1.26	169	12	7	354
Guyana	X	X	X	X	X	X	X	X	X	X	X
Paraguay	50	182	*101*	43	17	X	2.18	X	4	X	X
Peru	43	271	143	68	18	12.3	1.91	625	10	52	565
Suriname	X	X	X	X	X	X	X	X	X	X	X
Uruguay	116	610	*242*	232	46	21.9	46.61	688	8	25	*182*
Venezuela	206	471	172	116	46	36.6	3.37	208	10	182	1,822
ASIA											
Afghanistan	X	X	X	X	X	X	X	X	X	X	X
Armenia	*23*	5	218	150	2	X	1.01	X	X	162	20,268
Azerbaijan	*28*	20	211	87	5	X	0.21	X	X	165	16,470
Bangladesh	9	50	7	3	0	X	X	X	0	70	156
Bhutan	X	X	X	X	X	X	X	X	X	X	X
Cambodia	X	127	124	2	3	0.9	0.06	X	X	X	X

Table M *(continued)*
World Countries: Communications, Information, and Science and Technology

Country	Daily newspapers (per 1,000 people)	Radio (per 1,000 people)	Television sets (per 1,000 people)	Telephone main lines (per 1,000 people)	Mobile telephones (per 1,000 people)	Personal computers (per 1,000 people)	Internet hosts per 10,000 people	Scientists and engineers in R & D per million people	High-technology exports % of mfg. exports	No. of patent applications filed[a]	
	1996	1996	1997	1997	1997	1997	January 1999	1985-95	1997	Residents	Nonresidents
China	X	195	270	56	10	6.0	0.14	350	21	11,698	41,016
Hong Kong, China	800	695	412	565	343	230.8	122.71	98	29	41	2,059
Georgia	X	553	*473*	114	6	X	1.27	X	X	289	21,124
India	X	105	69	19	1	2.1	0.13	149	11	1,660	6,632
Indonesia	23	155	134	25	5	8.0	0.75	X	20	40	3,957
Iran	24	237	148	107	4	*32.7*	0.04	52.1	X	X	X
Iraq	X	X	X	X	X	X	X	X	X	X	X
Israel	291	530	321	450	283	186.1	161.96	X	33	1,363	12,172
Japan	580	957	708	479	304	202.4	133.53	6309	38	340,861	60,390
Jordan	45	287	43	70	2	8.7	0.80	106	26	X	X
Kazakhstan	30	384	*234*	108	1	X	0.94	X	X	1,024	20,064
Korea, North	X	X	X	X	X	X	X	X	X	X	X
Korea, South	394	1,037	341	444	150	150.7	40.00	2,636	39	68,446	45,548
Kuwait	376	688	*491*	227	116	82.9	32.80	X	4	X	x
Kyrgyzstan	13	115	44	76	*0*	X	4.04	703	24	126	20,179
Laos	4	139	4	5	1	*1.1*	0.00	X	X	X	X
Lebanon	141	892	354	179	135	31.8	5.56	X	X	X	X
Malaysia	163	432	166	195	113	43.1	21.36	87	67	X	X
Mongolia	27	139	63	37	1	5.4	0.08	943	2	114	20,882
Myanmar (Burma)	10	89	7	5	0	X	0.00	X	X	X	X
Nepal	11	37	4	8	0	X	0.07	X	0	X	X
Oman	X	X	X	X	X	X	X	X	X	X	X
Pakistan	*21*	92	65	19	1	4.5	0.23	54	4	16	782
Philippines	82	159	109	29	18	13.6	1.21	157	56	163	2,634
Saudi Arabia	59	319	*260*	117	17	43.6	0.15	X	29	27	810
Singapore	324	739	354	543	273	399.5	210.02	2,728	71	215	38,403
Sri Lanka	29	210	91	17	6	4.1	0.29	173	X	50	21,138
Syria	20	274	*68*	88	0	1.7	0.00	X	1	X	X
Tajikistan	20	X	281	38	0	X	0.12	709	X	32	19,570
Thailand	65	204	234	80	33	19.8	3.35	119	43	203	4,355
Turkey	111	178	286	250	26	20.7	4.30	261	9	367	19,668
Turkmenistan	X	96	*175*	78	0	X	0.55	X	X	66	18,948
United Arab Emirates	X	X	X	X	X	X	X	X	X	X	X
Uzbekistan	3	452	*273*	63	0	X	0.10	1,760	X	914	21,088
Vietnam	4	106	*180*	21	2	4.6	0.00	308	X	37	22,206
Yemen	15	64	*273*	13	1	1.2	0.01	X	0	X	X
EUROPE											
Albania	34	235	*161*	23	1	X	0.30	X	1	1	18,761
Austria	294	740	496	492	144	210.7	176.79	1,631	*24*	2,506	75,985
Belarus	*174*	290	314	227	1	X	0.70	2,339	X	701	20,347
Belgium	160	792	510	468	95	235.3	162.39	1,814	23	1,376	59,099
Bosnia-Herzegovina	X	X	X	X	X	X	X	X	X	X	X

Table M *(continued)*
World Countries: Communications, Information, and Science and Technology

	PER 1,000 PEOPLE										
	Daily newspapers	Radio	Television sets	Telephone main lines	Mobile telephones	Personal computers	Internet hosts per 10,000 people	Scientists and engineers in R & D per million people	High-technology exports % of mfg. exports	No. of patent applications filed[a]	
Country	1996	1996	1997	1997	1997	1997	January 1999	1985-95	1997	Residents	Nonresidents
Bulgaria	253	531	*366*	323	8	29.7	9.05	X	X	318	22,235
Croatia	114	333	*267*	335	27	22.0	12.84	1,978	19	259	356
Czech Republic	256	806	447	318	51	82.5	71.79	1,159	13	623	24,856
Denmark	311	1,146	568	633	273	360.2	526.77	2,647	27	2,452	71,151
Estonia	173	680	479	321	99	15.1	152.98	2,018	24	12	21,144
Finland	455	1,385	534	556	417	310.7	1,058.13	2,812	26	3,262	61,556
France	218	943	606	575	99	174.4	82.91	2,584	31	17,090	81,418
Germany	311	946	570	550	99	255.5	160.23	2,843	26	56,757	98,338
Greece	*153*	477	466	516	89	44.8	48.81	774	12	434	52,371
Hungary	189	697	436	304	69	49.0	82.74	1,033	39	832	24,147
Ireland	153	703	*455*	411	146	241.3	148.70	1,871	62	925	52,407
Italy	104	874	483	447	204	113.0	58.80	1,325	15	8,860	71,992
Latvia	246	699	*592*	302	31	7.9	42.59	1,189	15	197	21,498
Lithuania	92	292	*377*	283	41	6.5	27.48	X	21	101	21,249
Macedonia	19	184	*252*	204	6	X	2.56	X	X	53	18,934
Moldova	59	720	302	145	1	3.8	1.17	1,539	9	290	20,245
Netherlands	305	963	541	564	110	280.3	358.51	2,656	44	4,884	61,958
Norway	593	920	579	621	381	360.8	717.53	3,678	*24*	1,550	25,638
Poland	113	518	413	194	22	36.2	28.07	1,299	12	2,414	24,902
Portugal	75	306	523	402	152	74.4	50.01	1,185	11	105	71,544
Romania	X	317	*226*	167	9	8.9	7.42	1,382	7	1,831	22,139
Russian Federation	105	344	390	183	3	32.0	10.04	3,520	19	18,138	28,149
Slovak Republic	185	580	401	259	37	241.6	33.27	1,821	15	201	22,865
Slovenia	206	416	353	364	47	188.9	89.83	2,544	16	301	21,686
Spain	99	328	506	403	110	122.1	67.21	1,210	17	2,689	81,294
Sweden	446	907	531	679	358	350.3	487.13	3,714	34	7,077	76,364
Switzerland	330	969	536	661	147	394.9	315.52	X	28	2,699	75,576
Ukraine	54	872	493	186	1	5.6	3.13	3,173	X	3,640	22,862
Yugoslavia (Serbia-Montenegro)	X	X	X	X	X	X	X	X	X	X	X
OCEANIA											
Australia	297	1,385	638	505	264	362.2	420.57	3,166	39	9,196	34,125
Fiji	X	X	X	X	X	X	X	X	X	X	X
New Zealand	223	1,027	*501*	486	149	263.9	360.44	1,778	11	1,421	26,947
Papua New Guinea	15	91	24	*11*	1	X	0.25	X	X	X	X
Solomon Islands	X	X	X	X	X	X	X	X	X	X	X
World	Xw	380w	280w	144w	40w	58.4w	75.22w				
Low income	X	147	162	32	5	4.4	0.17				
Excl. China & India	13	133	*59*	16	1	X	0.23				
Middle income	75	383	272	136	24	32.4	10.15				
Lower middle income	63	327	247	108	11	*12.2*	4.91				

Table M *(continued)*
World Countries: Communications, Information, and Science and Technology

	PER 1,000 PEOPLE										
	Daily newspapers	Radio	Television sets	Telephone main lines	Mobile telephones	Personal computers	Internet hosts per 10,000 people	Scientists and engineers in R & D per million people	High-technology exports % of mfg. exports	No. of patent applications filed[a]	
Country	1996	1996	1997	1997	1997	1997	January 1999	1985-95	1997	Residents	Nonresidents
Upper middle income	95	469	302	179	43	45.5	19.01				
Low and middle income	X	218	194	65	11	12.3	3.08				
East Asia & Pacific	X	206	237	60	15	11.3	1.66				
Europe & Central Asia	99	412	380	189	13	*17.7*	13.00				
Latin America & Carib.	71	414	263	110	26	31.6	9.64				
Middle East & N.Africa	33	265	*140*	71	6	9.8	0.25				
South Asia	X	99	69	18	1	2.1	0.14				
Sub-Sharan Africa	12	172	*44*	16	4	7.2	2.39				
High income	286	1,300	664	552	188	269.4	470.12				

Note: Figures in italics are for years other than those specified.

[a]Other patent applications filed in 1996 include those filed under the auspices of the African Intellectual Property Organization (75 by residents, 20,863 by nonresidents), the Africa Regional Industrial Property Organization (10 by residents, 20,347 by nonresidents), the European Patent Office (38,546 by residents 48,068 by nonresidents), and the Eurasian Patent Organization (39 by residents, 18,055 by nonresidents). The original information was provided by the World Intellectual Property Organization (WIPO). The International Bureau of WIPO assumes no liability or responsibility with regard to the transformation of these data.

Source: Entering the 21st Century: World Development Report 1999/2000 (Oxford University Press, World Bank, 2000)

Table N
World Countries: Water Resources

COUNTRY	ANNUAL RENEWABLE WATER RESOURCES[a]				ANNUAL AVERAGE GROUNDWATER RESOURCES[b]	SECTORAL WITHDRAWALS (%)[c]		
	Supply Per Capita (cubic meters) 2000	Recharge	Population (2003)	Withdrawal Per Capita (cubic meters) 2000	Recharge Per Capita (cubic meters) 2000	Domestic	Industry	Agriculture
						9	20	71
WORLD	—	11,358.0	6,130	650	1,853	65	15	67
AFRICA	5,159		812	307	—	—	—	63
Algeria	460	1.7	31	181	55	34	52	14
Angola	13,203	72.0	14	54	5,143	14	10	76
Benin	3,741	1.8	6	27	300	23	10	67
Botswana	9,209	1.7	2	86	850	32	20	48
Burkina Faso	1,024	9.5	12	40	792	19	0	81
Burundi	538	2.1	7	19	300	36	64	0
Cameroon	18,378	100.0	15	38	6,667	46	19	35
Central African Republic	37,565	56.0	4	25	14,000	21	5	74
Chad	5,125	12.0	8	34	1,500	16	2	82
Congo (Zaire)	259,547	198.0	52	20	3,808	62	27	11
Congo Republic	23,639	421.0	3	10	140,333	61	16	23
Côte d'Ivoire	4,853	38.0	4	62	9,500	22	11	67
Egypt	830	1.3	65	1,055	20	11	82	86
Equatorial Guinea	53,841			30	—	81	13	6
Eritrea	1,578	—	4	—	—	—	—	—
Ethiopia	1,666	40.0	66	51	606	11	3	86
Gabon	126,789	62.0	1	70	62,000	72	22	6
Gambia	5,836	0.5	1	29	500	7	2	91
Ghana	2,637	26.0	20	35	1,300	35	13	52
Guinea	26,964	38.0	8	132	4,750	10	3	87
Guinea-Bissau	24,670	14.0	1	17	14,000	60	4	36
Kenya	947	3.0	31	87	97	20	4	76
Lesotho	1,456	0.5	2	32	250	22	22	56
Liberia	70,348	60.0	3	59	20,000	27	13	60
Libya	109	0.5	5	870	100	13	3	84
Madagascar	19,925	55.0	16	1,611	3,438	1	–	99
Malawi	1,461	1.4	11	95	127	10	3	86
Mali	8,320	20.0	24	167	833	2	1	97
Mauritania	4,029	0.3	3	923	100	2	2	92
Morocco	936	10.0	29	399	345	10	2	89
Mozambique	11,382	17.0	18	42	944	9	2	89
Namibia	9,865	2.1	2	175	1,050	29	3	68
Niger	2,891	2.5	11	69	227	16	2	82
Nigeria	2,891	2.5	130	69	19	16	2	82
Rwanda	638	3.6	9	141	400	5	2	94
Senegal	3,977	7.6	10	202	760	5	3	92

Table N *(continued)*
World Countries: Water Resources

COUNTRY	ANNUAL RENEWABLE WATER RESOURCES[a]				ANNUAL AVERAGE GROUNDWATER RESOURCES[b]	SECTORAL WITHDRAWALS (%)[c]		
	Supply Per Capita (cubic meters) 2000	Recharge	Population (2003)	Withdrawal Per Capita (cubic meters) 2000	Recharge Per Capita (cubic meters) 2000	Domestic	Industry	Agriculture
Sierra Leone	33,237	50.0	5	98	10,000	7	4	89
Somalia	1,413	3.3	9	119	367	3	0	97
South Africa	1,131	4.8	43	366	112	17	11	72
Sudan	1,981	7.0	32	637	219	4	1	94
Tanzania	2,472	30.0	34	39	882	9	2	89
Togo	3,076	5.7	5	29	1,140	62	13	25
Tunisia	577	1.5	10	312	150	13	1	86
Uganda	2,663	29.0	23	21	1,261	32	8	60
Zambia	9,676	47.0	10	190	4,700	16	7	77
Zimbabwe	1,530	5.0	13	131	385	14	7	79
NORTH AMERICA	—	1,670.0		—	—	—	—	46
Canada	92,810	370.0	31	1,607	11,935	18	70	12
United States	10,574	1,300.0	285	1,834	4,561	13	45	42
CENTRAL AMERICA	—	359.0		—	—	—	—	—
Belize	78,763	—		485	—	12	88	0
Costa Rica	26,764	37.0	4	1,540	9,250	13	7	80
Cuba	3,382	6.5	11	475	591	49	0	51
Dominican Republic	2,430	12.0	9	1,102	1,333	11	0	89
El Salvador	3,872	6.2	6	137	1,033	34	20	46
Guatemala	9,277	34.0	12	126	2,833	9	17	74
Haiti	1,670	2.2	8	139	275	5	1	94
Honduras	14,250	39.0	7	294	5,571	4	5	91
Jamaica	3,588	3.9	3	371	1,300	15	7	77
Mexico	4,490	139.0	99	812	1,404	17	5	78
Nicaragua	36,784	59.0	5	267	11,800	14	2	84
Panama	50,299	21.0	3	685	7,000	28	2	70
Trinidad and Tobago	2,940	—	1	233	—	68	26	6
SOUTH AMERICA	—	3,693.0	349	—	10,582	—	—	—
Argentina	21,453	128.0	37	822	3,459	16	9	75
Bolivia	71,511	130.0	9	197	14,444	10	3	87
Brazil	47,125	1,874.0	172	359	10,895	21	18	61
Chile	59,143	140.0	15	1,629	9,333	5	11	84
Colombia	49,017	510.0	43	228	11,860	59	4	37
Ecuador	32,948	134.0	13	1,423	10,308	12	6	82
Guyana	314,963	—	—	1,993	—	1	1	98
Paraguay	28,148	41.0	6	112	6,833	15	7	78

Table N *(continued)*
World Countries: Water Resources

COUNTRY	ANNUAL RENEWABLE WATER RESOURCES[a]				ANNUAL AVERAGE GROUNDWATER RESOURCES[b]	SECTORAL WITHDRAWALS (%)[c]		
	Supply Per Capita (cubic meters) 2000	Recharge	Population (2003)	Withdrawal Per Capita (cubic meters) 2000	Recharge Per Capita (cubic meters) 2000	Domestic	Industry	Agriculture
Peru	72,127	303.0	26	849	11,654	7	7	86
Suriname	298,848	—	—	1,171	—	6	5	89
Uruguay	41,065	23.0	3	—	7,667	6	3	91
Venezuela	49,144	227.0	25	382	9,080	44	10	46
ASIA	2,790	—		2,007	—	1	0	99
Afghanistan	2,421		27	1,846	0	—	—	99
Armenia	2,778	4.2	4	784	1,050	30	4	66
Azerbaijan	3,716	6.5	8	2,151	813	2	25	70
Bangladesh	8,444	21.0	133	133	158	12	2	86
Bhutan	43,214	—		13	—	36	10	54
Cambodia	34,516	18.0	12	60	1,500	5	1	94
China	2,186	829.0	1,272	439	652	5	18	78
Georgia	12,149	17.0	5	635	3,400	21	20	59
India	1,822	419.0	1,032	592	406	5	3	92
Indonesia	13,046	455.0	209	407	2,177	6	1	93
Iran	1,900	49.0	65	1,122	754	6	2	92
Iraq	3,111	1.2	24	2,478	50	3	5	92
Israel	265	0.5	6	287	83	39	7	54
Japan	3,372	27.0	127	735	213	19	17	64
Jordan	169	0.5	5	255	100	22	3	75
Kazakhstan	6,839	6.1	15	2,019	407	2	17	81
Korea, North	3,415	13.0	22	742	591	11	16	73
Korea, South	1,471	13.0	47	531	277	26	11	63
Kuwait	10	0.0	2	306	0	37	2	60
Kyrgyzstan	4,078	14.0	5	2,231	2,800	3	3	94
Laos	60,318	38.0	5	259	7,600	8	10	82
Lebanon	1,220	3.2	4	400	800	27	6	68
Malaysia	25,178	64.0	24	636	2,667	11	13	77
Mongolia	13,451	6.1	2	182	3,050	20	27	53
Myanmar (Burma)	21,358	156.0	48	103	3,250	7	3	90
Nepal	8,703	20.0	24	1,451	833	1	0	99
Oman	364	1.0	2	658	500	5	2	94
Pakistan	2,812	55.0	141	1,382	390	2	2	97
Philippines	6,093	180.0	78	811	2,308	8	4	88
Saudi Arabia	111	2.2	21	1,056	105	9	1	90
Singapore		—	4	—	—	45	51	4
Sri Lanka	2,592	7.8	19	574	411	2	2	96
Syria	1.541	4.2	17	844	247	8	2	90
Tajikistan	2,587	6.0	6	2,096	1,000	3	4	92

Table N *(continued)*
World Countries: Water Resources

COUNTRY	ANNUAL RENEWABLE WATER RESOURCES[a]				ANNUAL AVERAGE GROUNDWATER RESOURCES[b]	SECTORAL WITHDRAWALS (%)[c]		
	Supply Per Capita (cubic meters) 2000	Recharge	Population (2003)	Withdrawal Per Capita (cubic meters) 2000	Recharge Per Capita (cubic meters) 2000	Domestic	Industry	Agriculture
Thailand	6,371	42.0	61	605	689	5	4	91
Turkey	3,344	69.0	66	558	1,045	16	12	73
Turkmenistan	5,015	0.4	5	5,801	80	1	1	98
United Arab Emirates	56	0.1	3	896	33	24	9	67
Uzbekistan	1,968	8.8	25	2,598	352	4	2	94
Vietnam	11,109	48.0	80	822	600	4	10	87
Yemen	206	1.5	18	253	83	7	1	92
EUROPE	—	1,318.0	709	—	1,859	—	—	—
Albania	13,178	6.2	3	440	2,067	29	0	71
Austria	9,629	6.0	8	303	750	33	58	9
Belarus	5,739	18.0	10	266	1,800	22	43	35
Belgium	1,781	0.9	10	—	90	—	—	—
Bosnia-Herzegovina	9,088	—		292	—	30	10	60
Bulgaria	2,734	6.4	8	1,573	800	22	3	75
Croatia	22,654	11.0	4	164	2,750	50	50	0
Czech Republic	1,283	1.4	10	266	140	41	57	2
Denmark	1,123	4.3	5	233	860	30	27	43
Estonia	9,413	4.0	1	106	4,000	56	39	5
Finland	21,223	2.2	5	439	440	12	85	3
France	3,414	100.0	59	547	1,695	18	72	10
Germany	1,878	46.0	82	579	561	11	69	20
Greece	6,984	10.0	11	826	909	10	3	87
Hungary	10,541	6.0	10	659	600	9	55	36
Iceland	599,944			622	—	31	63	6
Ireland	13,408	11.0	4	232	2,750	16	74	10
Italy	3,330	43.0	58	730	741	19	34	48
Latvia	14,820	2.2	2	112	1,100	55	32	13
Lithuania	6,763	1.2	3	68	400	81	16	3
Macedonia	3,121	—	2	936	—	12	15	74
Moldova	2,726	0.4	4	678	100	9	65	26
Netherlands	5,691	4.5	16	519	281	5	61	34
Norway	84,787	96.0	5	489	19,200	20	72	8
Poland	1,598	13.0	39	321	333	13	76	11
Portugal	6,837	4.0		736	—	15	37	48
Romania	9,486	8.3	22	1,141	377	8	33	59
Russian Federation	31,354	788.0	145	519	5,434	19	62	20
Serbia and Montenegro	19,815	3.0	11	1,233	273	6	86	8

Table N *(continued)*
World Countries: Water Resources

COUNTRY	ANNUAL RENEWABLE WATER RESOURCES[a]				ANNUAL AVERAGE GROUNDWATER RESOURCES[b]	SECTORAL WITHDRAWALS (%)[c]		
	Supply Per Capita (cubic meters) 2000	Recharge	Population (2003)	Withdrawal Per Capita (cubic meters) 2000	Recharge Per Capita (cubic meters) 2000	Domestic	Industry	Agriculture
Slovak Republic	9,265	1.7	5	337	340	—	—	—
Slovenia	16,070	14.0	2	642	7,000	20	80	1
Spain	2,793	30.0	41	884	732	13	19	68
Sweden	19,721	20.0	9	340	2,222	36	55	9
Switzerland	7,464	2.5	7	172	357	23	73	4
Ukraine	2,868	20.0	49	500	408	18	52	30
United Kingdom	2,464	9.8	59	204	166	20	77	3
OCEANIA	—		—	—	—	—	—	—
Australia	25,185	72.0	19	933	3,789	65	2	33
Fiji	34,330	—		—	—	20	20	60
New Zealand	85,221	—		588	—	46	10	44
Papua New Guinea	159,171	—		29	—	29	22	49
Solomon Islands	93,405	—		—	—	40	20	40

[a]Annual renewable water resources usually include river flows from other countries.
[b]Withdrawal data from most recent year available; varies by country from 1987 to 1995.
[c]Total withdrawals may exceed 100% because of groundwater withdrawals or river inflows.

Source: World Resources 1998-99 (World Resources Institute).

Table O
World Countries: Globally Threatened Plant and Animal Species

	MAMMALS		BIRDS		REPTILES		AMPHIBIANS		FRESHWATER FISH		PLANTS
COUNTRY	Threatened Species	Number of Species per 10,000 km²	Threatened Species	Number of Species per 10,000 km²	Threatened Species	Number of Species per 10,000 km²	Threatened Species	Number of Species per 10,000 km²	Threatened Species	Rare and Threatened Species	Number of Species per 10,000 km²
AFRICA											
Algeria	15	15	8	32	1	X	0	X	1	145	509
Angola	17	56	13	156	5	X	0	X	0	25	1,017
Benin	9	85	1	138	2	X	0	X	0	3	899
Botswana	5	43	7	101	0	41	0	X	0	4	X
Burkina Faso	6	49	1	112	1	X	0	10	0	0	369
Burundi	5	76	6	322	0	X	0	X	0	1	1,783
Cameroon	32	83	14	193	3	X	1	X	26	74	2,237
Central African Republic	11	53	2	137	1	X	0	X	0	0	921
Chad	14	27	3	75	1	X	0	X	0	12	322
Congo (Zaire)	38	69	26	153	3	X	0	X	1	7	1,817
Congo Republic	10	62	3	140	2	X	0	X	0	3	1,356
Côte d'Ivoire	16	73	12	170	4	X	1	X	0	66	1,118
Egypt	15	21	11	33	6	18	0	1	0	84	452
Equatorial Guinea	12	131	4	194	2	X	1	X	0	9	2,135
Eritrea	6	49	3	140	3	X	0	X	0	X	X
Ethiopia	35	54	20	133	1	X	0	X	0	153	1,378
Gabon	12	64	4	157	3	X	0	X	0	78	2,197
Gambia	4	104	1	269	1	X	0	X	0	0	928
Ghana	13	78	10	186	4	X	0	X	0	32	1,264
Guinea	11	66	12	142	3	X	1	X	0	35	1,043
Guinea-Bissau	4	71	1	159	3	X	0	X	0	0	655
Kenya	43	94	24	221	5	49	0	23	20	158	1,571
Lesotho	2	23	5	40	0	X	0	X	1	7	1,093
Liberia	11	87	13	168	3	28	1	17	0	1	1,037
Libya	11	14	2	17	3	X	0	X	0	57	327
Madagascar	46	27	28	53	17	66	2	38	13	189	2,347
Malawi	7	86	9	230	0	55	0	31	0	61	1,592
Mali	13	28	6	81	1	3	0	X	0	14	355
Mauritania	14	13	3	59	3	X	0	X	0	3	239
Mauritius	4	7	10	46	6	19	0	0	0	222	1,183
Morocco	18	30	11	60	2	X	0	X	1	195	1,028
Mozambique	13	42	14	117	5	X	0	15	2	92	1,294
Namibia	11	36	8	109	3	X	1	7	3	23	729
Niger	11	27	2	60	1	X	0	X	0	0	237
Nigeria	26	62	9	153	4	X	0	X	0	9	1,036
Rwanda	9	110	6	373	0	X	0	X	0	0	1,662
Senegal	13	58	6	144	7	X	0	X	0	32	771
Sierra Leone	9	77	12	243	3	X	0	X	0	12	1,091
Somalia	18	43	8	107	2	49	0	7	3	57	761

Table O *(continued)*
World Countries: Globally Threatened Plant and Animal Species

	MAMMALS		BIRDS		REPTILES		AMPHIBIANS		FRESHWATER FISH		PLANTS
COUNTRY	Threatened Species	Number of Species per 10,000 km²	Threatened Species	Number of Species per 10,000 km²	Threatened Species	Number of Species per 10,000 km²	Threatened Species	Number of Species per 10,000 km²	Threatened Species	Rare and Threatened Species	Number of Species per 10,000 km²
South Africa	33	51	16	122	19	61	9	19	27	953	4,711
Sudan	21	43	9	110	3	X	0	X	0	8	506
Swaziland	5	37	6	303	0	85	0	33	0	41	2,197
Tanzania	33	70	30	183	4	63	0	28	19	406	229
Togo	8	110	1	220	3	X	0	X	0	0	1,128
Tunisia	11	31	6	69	2	X	0	X	0	24	855
Uganda	18	118	10	290	1	52	0	17	28	6	1,762
Zambia	11	55	10	145	0	X	0	20	0	9	1,105
Zimbabwe	9	81	9	159	0	46	0	36	0	94	1,253
NORTH AMERICA											
Canada	7	20	5	44	3	4	1	4	13	649	299
United States	35	45	50	68	28	29	24	24	123	1,845	1,679
CENTRAL AMERICA											
Belize	5	95	1	271	5	81	0	24	0	41	2,090
Costa Rica	14	120	13	350	7	125	1	95	0	456	6,421
Cuba	9	14	13	62	7	46	0	19	4	811	2,714
Dominican Republic	4	12	11	81	10	62	1	21	0	73	2,965
El Salvador	2	106	0	196	6	57	0	18	0	35	1,956
Guatemala	8	114	4	208	9	105	0	45	0	315	3,638
Haiti	4	2	11	54	6	73	1	33	0	28	3,345
Honduras	7	78	4	190	7	68	0	25	0	55	2,252
Jamaica	4	23	7	110	8	35	4	20	0	371	2,662
Mexico	64	79	36	135	18	120	3	50	86	1,048	4,382
Nicaragua	4	86	3	207	7	69	0	25	0	78	3,003
Panama	17	112	10	376	7	116	0	84	1	561	4,618
Trinidad and Tobago	1	125	3	324	5	87	0	32	0	16	2,470
SOUTH AMERICA											
Argentina	27	50	41	140	5	34	5	23	1	170	1,407
Bolivia	23	67	27	X	3	44	0	24	0	49	3,500
Brazil	71	43	103	161	15	51	5	54	12	463	5,935
Chile	16	22	18	71	1	17	3	10	4	292	1,229
Colombia	35	75	64	355	15	122	0	123	5	376	10,479
Ecuador	28	100	53	460	12	124	0	133	1	375	6,052
Guyana	10	70	3	246	8	X	0	X	0	47	2,180
Paraguay	10	90	26	164	3	35	0	25	0	12	2,208
Peru	46	69	64	310	9	60	1	63	0	377	3,448
Suriname	10	72	2	240	6	60	0	38	0	48	1,870
Uruguay	5	31	11	92	0	X	0	X	0	11	845
Venezuela	24	69	22	266	14	58	0	45	5	107	4,510

Table O *(continued)*
World Countries: Globally Threatened Plant and Animal Species

	MAMMALS		BIRDS		REPTILES		AMPHIBIANS		FRESHWATER FISH		PLANTS
COUNTRY	Threatened Species	Number of Species per 10,000 km²	Threatened Species	Number of Species per 10,000 km²	Threatened Species	Number of Species per 10,000 km²	Threatened Species	Number of Species per 10,000 km²	Threatened Species	Rare and Threatened Species	Number of Species per 10,000 km²
ASIA											
Afghanistan	11	31	13	59	1	26	1	2	0	6	882
Armenia	4	X	5	X	3	32	0	4	0	0	X
Azerbaijan	11	X	8	X	3	26	0	4	5	1	X
Bangladesh	18	45	30	122	13	49	0	8	0	24	2,074
Bhutan	20	59	14	269	1	11	0	14	0	20	3,268
Cambodia	23	47	18	118	9	32	0	11	5	7	X
China	75	41	90	114	15	35	1	27	28	343	3,112
Georgia	10	X	5	X	7	24	0	6	3	1	X
India	75	47	73	136	16	57	3	29	4	1,256	2,216
Indonesia	128	77	104	269	19	90	0	48	60	281	4,864
Iran	20	26	14	60	8	30	2	2	7	1	X
Iraq	7	23	12	49	2	23	0	2	2	2	X
Israel	13	72	8	141	5	X	0	X	0	38	X
Japan	29	40	33	X	8	20	10	16	7	704	1,418
Jordan	7	34	4	68	1	X	0	X	0	10	1,069
Kazakhstan	15	X	15	X	1	6	1	2	5	0	X
Korea, North	7	X	19	51	0	8	0	6	0	7	1,274
Korea, South	6	23	19	53	0	12	0	7	0	69	1,360
Kuwait	1	17	3	17	2	24	0	2	0	0	193
Kyrgystan	6	X	5	X	1	9	0	1	0	1	X
Laos	30	61	27	171	7	23	0	13	4	5	X
Lebanon	5	53	5	152	2	X	0	X	0	4	X
Malaysia	42	90	34	158	14	85	0	50	14	510	4,732
Mongolia	12	25	14	X	0	4	0	2	0	1	429
Myanmar (Burma)	31	62	44	216	20	51	0	19	1	29	1,742
Nepal	28	70	27	255	5	33	0	15	0	21	2,716
Oman	9	20	5	39	4	23	0	X	3	4	371
Pakistan	13	36	25	88	6	41	0	4	1	12	1,163
Philippines	49	50	86	129	7	62	2	21	26	371	2,604
Saudi Arabia	9	13	11	26	2	14	0	X	0	6	294
Singapore	6	113	9	295	1	X	0	X	1	14	5,007
Sri Lanka	14	47	11	134	8	77	0	21	8	436	1,613
Syria	4	24	7	78	3	X	0	X	0	10	X
Tajikistan	5	X	9	X	1	16	0	1	1	0	X
Thailand	34	72	45	168	16	81	0	29	14	382	2,999
Turkey	15	28	14	72	12	24	2	4	18	1,827	2,012
Turkmenistan	11	X	12	X	2	22	0	1	5	1	X
United Arab Emirates	3	12	4	33	2	18	0	X	1	0	X

Table O *(continued)*
World Countries: Globally Threatened Plant and Animal Species

	MAMMALS		BIRDS		REPTILES		AMPHIBIANS		FRESHWATER FISH		PLANTS
COUNTRY	Threatened Species	Number of Species per 10,000 km²	Threatened Species	Number of Species per 10,000 km²	Threatened Species	Number of Species per 10,000 km²	Threatened Species	Number of Species per 10,000 km²	Threatened Species	Rare and Threatened Species	Number of Species per 10,000 km²
Uzbekistan	7	X	11	X	0	15	0	1	3	5	X
Vietnam	38	67	47	168	12	57	1	25	3	350	X
Yemen	5	18	13	39	2	21	0	X	0	X	X
EUROPE											
Albania	2	48	7	162	1	22	0	9	7	50	2,093
Austria	7	41	5	106	1	7	0	10	7	22	1,462
Belarus	4	X	4	81	0	3	0	4	0	0	X
Belgium	6	40	3	125	0	6	0	12	1	3	969
Bosnia-Herzegovina	10	X	2	X	X	X	1	X	6	0	X
Bulgaria	13	37	12	108	1	15	0	8	8	94	1,584
Croatia	10	X	4	126	X	X	1	X	20	0	X
Czech Republic	7	X	6	101	X	X	X	X	6	X	X
Denmark	3	27	2	121	0	3	0	9	0	6	741
Estonia	4	40	2	130	0	3	0	7	1	2	992
Finland	4	18	4	78	0	2	0	2	1	11	325
France	13	25	7	72	3	9	2	9	3	117	1,198
Germany	8	23	5	73	0	4	0	6	7	X	X
Greece	13	41	10	107	6	22	1	6	16	539	2,091
Hungary	8	34	10	98	1	7	0	8	11	24	1,029
Iceland	1	5	0	41	0	0	0	0	0	1	157
Ireland	2	13	1	75	0	1	0	2	1	9	469
Italy	10	29	7	76	4	13	4	11	9	273	1,776
Latvia	4	45	6	117	0	4	0	7	1	0	623
Lithuania	5	37	4	109	0	4	0	7	1	0	646
Macedonia	10	X	3	X	1	X	X	X	4	X	X
Moldova	2	46	7	119	1	6	0	9	9	1	X
Netherlands	6	35	3	120	0	4	0	10	1	1	758
Norway	4	17	3	77	0	2	0	2	1	20	524
Poland	10	27	6	72	0	3	0	6	2	27	738
Portugal	13	30	7	99	0	14	1	8	9	240	1,200
Romania	16	29	11	87	2	9	0	7	11	122	1,116
Russian Federation	31	23	38	54	5	5	0	2	13	127	X
Slovak Republic	8	X	4	124	0	X	0	X	7	X	X
Slovenia	10	55	3	164	0	17	1	X	5	11	X
Spain	19	22	10	76	6	15	3	7	10	896	X
Sweden	5	17	4	71	0	2	0	4	1	19	1,400
Switzerland	6	47	4	121	0	9	0	11	4	9	1,033
Ukraine	15	X	10	85	2	6	0	5	12	16	756

Table O *(continued)*
World Countries: Globally Threatened Plant and Animal Species

	MAMMALS		BIRDS		REPTILES		AMPHIBIANS		FRESHWATER FISH		PLANTS
COUNTRY	Threatened Species	Number of Species per 10,000 km^2	Threatened Species	Number of Species per 10,000 km^2	Threatened Species	Number of Species per 10,000 km^2	Threatened Species	Number of Species per 10,000 km^2	Threatened Species	Rare and Threatened Species	Number of Species per 10,000 km^2
United Kingdom	4	17	2	80	0	3	0	2	1	28	539
Yugoslavia (Serbia-Montenegro)	12	X	8	X	1	X	X	X	13	X	X
OCEANIA											
Australia	58	28	45	72	37	83	25	23	37	1,597	1,672
Fiji	4	3	9	61	6	20	1	2	0	72	1,071
New Zealand	3	3	44	51	11	13	1	1	8	236	727
Papua New Guinea	57	60	31	182	10	79	0	56	13	95	2,821
Solomon Islands	20	37	18	115	4	43	0	12	0	43	1,959

Source: World Conservation Monitoring Centre and World Conservation Union; World Resources Institute, World Resources 1998–99, 1998.

Part IX

Geographic Index

Part IX

GEOGRAPHIC INDEX

NAME/DESCRIPTION	LATITUDE & LONGITUDE	PAGE
Abidjan, Cote d'Ivoire (city, nat. cap.)	5N 4W	163
Abu Dhabi, U.A.E. (city, nat. cap.)	24N 54E	170
Accra, Ghana (city, nat. cap.)	64N 0	163
Aconcagua, Mt. 22,881	38S 78W	147
Acre (st., Brazil)	9S 70W	146
Addis abaha, Ethiopia (city, nat. cap.)	9N 39E	163
Adelaide, S. Australia (city, st. cap., Aust.)	35S 139E	181
Aden Gulf of	12N 46E	171
Aden, Yemen (city)	13N 45E	170
Admiralty Islands	1S 146E	182
Adriatic Sea	44N 14E	154
Aegean Sea	39N 25E	154
Afghanistan (country)	35N 65E	170
Aguascalientes (st., Mex.)	22N 110W	144
Aguascalientes, Augus (city, st. cap., Mex.)	22N 102W	144
Agulhas, Cape	35S 20E	164
Ahaggar Range	23N 6E	164
Ahmadabad, India (city)	23N 73E	170
Akmola, Kazakhstan (city)	51N 72E	170
Al Fashir, Sudan (city)	14N 25E	163
Al Fayyum, Egypt (city)	29N 31E	163
Al Hijaz Range	30N 40E	171
Al Khufra Oasis	24N 23E	164
Alabama (st. US)	33N 107W	137
Alagoas (st., Brazil)	9S 37W	146
Alaska (st., US)	63N 153W	137 inset
Alaska, Gulf of	58N 150W	137 inset
Alaska Peninsula	57N 155W	138 inset
Alaska Range	60N 150W	138 inset
Albania (country)	41N 20E	153
Albany, Australia (city)	35S 118E	181
Albany, New York (city, st. cap., US)	43N 74W	137
Albert Edward, Mt. 13,090	8S 147E	182
Albert, Lake	2N 30E	164

The geographic index contains approximately 1,500 names of cities, states, countries, rivers, lakes, mountain ranges, oceans, capes, bays, and other geographic features. The name of each geographical feature in the index is accompanied by a geographical coordinate (latitude and longitude) in degrees and by the page number of the primary map on which the geographical feature appears. Where the geographical coordinates are for specific places or points, such as a city or a mountain peak, the latitude and longitude figures give the location of the map symbol denoting that point. Thus, Los Angeles, California, is at 34N and 118W and the location of Mt. Everest is 28N and 107E.

The coordinates for political features (countries or states) or physical features (oceans, deserts) that are areas rather than points are given according to the location of the name of the feature on the map, except in those cases where the name of the feature is separated from the feature (such as a country's name appearing over an adjacent ocean area because of space requirements). In such cases, the feature's coordinates will indicate the location of the center of the feature. The coordinates for the Sahara Desert will lead the reader to the place name "Sahara Desert" on the map; the coordinates for North Carolina will show the center location of the state since the name appears over the adjacent Atlantic Ocean. Finally, the coordinates for geographical features that are lines rather than points or areas will also appear near the center of the text identifying the geographical feature.

Alphabetizing follows general conventions; the names of physical features such as lakes, rivers, mountains are given as: proper name, followed by the generic name. Thus "Mount Everest" is listed as "Everest, Mt." Where an article such as "the," "le," "al" appears in a geographic name, the name is alphabetized according to the article. Hence, "La Paz" is found under "L" and not under "P".

Part IX *(continued)*

GEOGRAPHIC INDEX

NAME/DESCRIPTION	LATITUDE & LONGITUDE	PAGE
Alberta (prov., Can.)	55N 117W	137
Albuquerque, NM (city)	35N 107W	137
Aldabra Islands	9S 44E	164
Aleppo, Syria (city)	36N 37E	170
Aleutian Islands	55N 175W	138
Alexandria, Egypt (city)	31N 30E	163
Algeria (country)	28N 15E	163
Algiers, Algeria (city, nat. cap.)	37N 3E	163
Alice Springs, Aust. (city)	24S 134E	181
Alma Ata, Kazakhstan (city, nat. cap.)	43N 77E	170
Alps Mountains	46N 6E	154
Altai Mountains	49N 107E	171
Altun Shan	45N 90E	171
Amapa (st., Brazil)	2N 52W	146
Amazon Riv.	2S 53W	147
Amazonas (st., Brazil)	2S 64W	146
Amman, Jordan (city, nat. cap.)	32N 36E	170
Amsterdam, Netherlands (city)	52N 5E	153
Amu Darya (riv., Asia)	40N 62E	171
Amur (riv., Asia)	52N 156E	171
Anchorage, AK (city)	61N 150W	137 inset
Andaman Islands	12N 92E	170
Andes Mountains	25S 70W	147
Angara (riv., Asia)	60N 98E	171
Angola (country)	11S 18E	163
Ankara, Turkey (city, nat. cap.)	40N 33E	170
Annapolis, Maryland (city, st. cap., US)	39N 76W	137
Antananarivo, Madagascar (city, nat. cap.)	19S 48E	163
Antofogasta, Chile (city)	24S 70W	146
Antwerp, Belgium (city)	51N 4E	153
Appalachian Mountains	37N 80W	138
Appenines Mountains	32N 14E	154
Arabian Desert	25N 33E	154
Arabian Peninsula	23N 40E	182
Arabian Sea	18N 61E	171
Aracaju, Sergipe (city, st. cap., Braz.)	11S 37W	146
Arafura Sea	9S 133E	182
Araguaia, Rio (riv., Brazil)	13S 50W	147
Aral Sea	45N 60E	171
Arctic Ocean	75N 160W	146
Arequipa, Peru (city)	16S 71W	146
Argentina (country)	39S 67W	146
Aripuana, Rio (riv., S.Am.)	11S 60W	147
Arizona (st., US)	34N 112W	137
Arkansas, (riv., N.Am.)	38N 98W	138
Arkansas (st., US)	37N 117W	137

Part IX *(continued)*

GEOGRAPHIC INDEX

NAME/DESCRIPTION	LATITUDE & LONGITUDE	PAGE
Arkhangelsk, Russia (city)	75N 160W	153
Armenia (country)	40N 45E	153
Arnhem, Cape	11S 139E	182
Arnhem Land	12S 133E	182
As Sudd	9N 26E	164
Ascension (island)	9S 15W	164
Ashburton (riv., Australasia)	23S 140W	171
Ashkhabad, Tukmenistan (city, nat. cap.)	38N 58E	170
Asia Minor	39N 33E	163
Asmera, Eritrea (city, nat. cap.)	15N 39E	163
Astrakhan, Russia (city)	46N 48E	153
Asuncion, Paraguay (city, nat. cap.)	25S 57W	146
Aswan, Egypt (city)	24N 33E	163
Asyuf, Egypt (city)	27N 31E	163
Atacama Desert	23S 70W	147
Athabasca (lake, N.Am.)	60N 133W	138
Athabaska (riv. N.Am.)	58N 141W	138
Athens, Greece (city, nat. cap.)	38N 24E	154
Atlanta, Georgia (city, st. cap., US)	34N 84W	137
Atlantic Ocean	30N 40W	138
Atlas Mountains	31N 6W	164
Auckland, New Zealand (city)	37S 175E	181
Augusta, Maine (city, st. cap., US)	44N 70W	137
Austin, Texas (city, st. cap., US)	30N 98W	137
Australia (country)	20S 135W	181
Austria (country)	47N 14E	154
Ayers Rock 2844	25S 131E	182
Azerbaijan (country)	38N 48E	154
Azov, Sea of	48N 36E	154
Bab el Mandeb (strait)	13N 42E	164
Baffin Bay	74N 65W	138
Baffin Island	70N 72W	138
Baghdad, Iraq (city, nat cap)	33N 44E	154
Bahamas (island)	25N 75W	138
Bahia (st., Brazil)	13S 42W	146
Bahia Blanca, Argentina (city)	39S 62W	146
Baikal, Lake	52N 105E	171
Baja California (st., Mex.)	30N 110W	137
Baja California Sur (st., Mex.)	25N 110W	137
Baku, Azerbaijan (city, nat. cap.)	40N 50E	154
Balearic Islands	29N 3E	154
Balkash, Lake	47N 75E	171
Ballarat, Aust. (city)	38S 144E	181
Baltic Sea	56N 18E	154
Baltimore, MD (city)	39N 77W	171
Bamako, Mali (city, nat. cap.)	13N 8W	163

Part IX *(continued)*

GEOGRAPHIC INDEX

NAME/DESCRIPTION	LATITUDE & LONGITUDE	PAGE
Bandiera Peak 9,843	20S 42W	147
Bangalore, India (city)	13N 75E	170
Bangeta Mt. 13,520	6S 147E	182
Banghazi, Libya (city)	32N 20E	163
Bangkok, Thailand (city, nat. cap.)	14N 98E	170
Bangladesh (country)	23N 92E	170
Bangui, Cent. African Rep. (city, nat. cap.)	4N 19E	163
Banjul, Gambia (city, nat. cap.)	13N 17W	163
Banks Island	73N 125W	138
Barbados (island)	13N 60W	146
Barcelona, Spain (city)	41N 2E	154
Barents Sea	69N 40E	171
Bartle Ferer, Mt. 5322	18S 145E	182
Barwon (riv., Australasia)	29S 148E	182
Bass Strait	40S 146E	182
Baton Rouge, Louisiana (city, st. cap., US)	30N 91W	137
Beaufort Sea	72N 135W	138
Beijing, China (city, nat. cap.)	40N 116E	170
Beirut, Lebanon (city, nat. cap.)	34N 35E	154
Belarus (country)	52N 27E	154
Belem, Para (city, st. cap., Braz.)	1S 48W	146
Belfast, Northern Ireland (city)	55N 6W	153
Belgium (country)	51N 4E	154
Belgrade, Serbia and Montenegro (city, nat. cap.)	45N 21E	153
Belhuka, Mt. 14,483	50N 108E	171
Belize (country)	18S 88W	137
Belle Isle, Strait of	52N 57W	138
Belmopan, Belize (city, nat. cap.)	18S 89W	137
Belo Horizonte, M.G. (city, st. cap., Braz.)	20S 43W	146
Belyando (riv., Australasia)	22S 147W	182
Ben, Rio (riv., S.Am.)	14S 67W	147
Bengal, Bay of	15N 90E	171
Benguela, Angola (city)	13S 13E	163
Benin (country)	10N 4E	163
Benin City, Nigeria (city)	6N 6E	163
Benue (riv., Africa)	8N 9E	164
Bergen, Norway (city)	60N 5E	153
Bering Sea	57N 175W	171
Bering Strait	65N 168W	171
Berlin, Germany (city)	52N 13E	153
Bermeo, Rio (riv., S.Am.)	25S 61W	147
Bermuda (island)	30S 66W	137
Bhutan (county)	28N 110E	170
Billings, MT (city)	46N 134W	137
Birmingham, AL (city)	34N 107W	137
Birmingham, UK (city)	52N 2W	153

GEOGRAPHIC INDEX

NAME/DESCRIPTION	LATITUDE & LONGITUDE	PAGE
Biscay, Bay of	45N 5W	154
Bishkek, Kyrgyzstan (city, nat. cap.)	43N 75E	170
Bismarck Archipelago	4S 147E	182
Bismarck, North Dakota (city, st. cap., US)	47N 123W	137
Bismarck Range	6S 145E	182
Bissau, Guinea-Bissau (city, nat. cap.)	12N 16W	163
Black sea	46N 34E	154
Blanc, Cape	21N 18W	154
Blue Nile (riv., Africa)	10N 36E	154
Blue Mountains	33S 150E	182
Boa Vista do Rio Branco, Roraima (city, st. cap., Braz.)	3N 61W	146
Boise, Idaho (city, st. cap., US)	44N 116W	137
Bolivia (country)	17S 65W	146
Boma, Congo Republic (city)	5S 13E	163
Bombay, (Mumbai) India (city)	19N 73E	170
Bonn, Germany (city. nat. cap.)	51N 7E	153
Boothia Peninsula	71N 117W	138
Borneo (Island)	0 11E	170
Bosnia-Herzegovina (country)	45N 18E	153
Bosporus, Strait of	41N 29E	154
Boston, Massachusetts (city, st. cap., US)	42N 71W	137
Botany Bay	35S 153E	181
Bothnia, Gulf of	62N 20E	154
Botswana (country)	23S 25E	163
Brahmaputra (riv., Asia)	30N 98E	171
Branco, Rio (riv., S. Am.)	3N 62W	147
Brasilia, Brazil (city, nat. cap.)	16S 48W	146
Bratislava, Slovakia (city, nat. cap.)	48N 17E	153
Brazil (country)	10S 52W	146
Brazilian Highlands	18S 45W	147
Brazzaville, Congo (city, nat. cap.)	4S 15E	163
Brisbane, Queensland (city, st. cap., Aust.)	27S 153E	181
Bristol Bay	58N 159W	138 inset
British Columbia (prov., Can.)	54N 130W	137
Brooks Range	67N 155W	138
Bruce, Mt. 4052	22S 117W	182
Brussels, Belgium (city, nat. cap.)	51N 4E	153
Bucharest, Romania (city, nat. cap.)	44N 26E	153
Budapest, Hungary (city, nat. cap.)	47N 19E	153
Buenos Aires, Argentina (city, nat. cap.)	34S 58W	146
Buenos Aires (st., Argentina)	36S 60W	146
Buffalo, NY (city)	43N 79W	137
Bujumbura, Burundi (city, nat. cap.)	3S 29E	163
Bulgaria (country)	44N 26E	153
Bur Sudan, Sudan (city)	19N 37E	163
Burdekin (riv., Australasia)	19S 146W	182

Part IX *(continued)*

GEOGRAPHIC INDEX		
NAME/DESCRIPTION	LATITUDE & LONGITUDE	PAGE
Burkina Faso (country)	11N 2W	163
Buru (island)	4S 127E	182
Burundi (country)	4S 30E	163
Cairns, Aust. (city)	17S 145E	181
Cairo, Egypt (city, nat. cap.)	30N 31E	163
Calcutta, (Kolkota) India (city)	23N 88E	170
Calgary, Canada (city)	51N 141W	137
Calicut, India (city)	11N 76E	170
California (st., US)	35N 120W	137
California, Gulf of	29N 110W	137
Callao, Peru (city)	13S 77W	146
Cambodia (country)	10N 106E	170
Cameroon (country)	5N 13E	163
Campeche (st., Mex.)	19N 90W	137
Campeche Bay	20N 92W	138
Campeche, Campeche (city, st. cap., Mex.)	19N 90W	137
Campo Grande, M.G.S. (city, st. cap., Braz.)	20S 55W	146
Canada (country)	52N 98W	137
Canadian (riv., N.Am.)	30N 98W	138
Canary Islands	29N 18W	164
Canberra, Australia (city, nat. cap.)	35S 149E	181
Cape Breton Island	46N 60W	138
Cape Town, South Africa (city)	34S 18E	163
Caracas, Venezuela (city, nat. cap.)	10N 67W	146
Caribbean Sea	18N 75W	146
Carnarvon, Australia City	25S 113E	181
Carpathian Mountains	48N 24E	154
Carpentaria, Gulf of	14S 140E	182
Carson City, Nevada (city, st. cap., US)	39N 120W	137
Cartagena, Colombia (city)	10N 76W	146
Cascade Range	45N 120W	138
Casiquiare, Rio (riv., S.Am.)	4N 67W	147
Caspian Depression	49N 48E	154
Caspian Sea	42N 48E	154
Catamarca (st., Argentina)	25S 70W	146
Catamarca, Catamarca (city, st. cap., Argen.)	28S 66W	146
Cauca, Rio (riv., S.Am.)	8N 75W	147
Caucasus Mountains	42N 40E	154
Cayenne, French Guiana (city, nat. cap.)	5N 52W	146
Ceara (st., Brazil)	4S 40W	146
Celebes (island)	0 120E	171
Celebes Sea	2N 120E	171
Central African Republic (country)	5N 20E	163
Ceram (island)	3S 129E	181
Chaco (st., Argentina)	25S 60W	146
Chad (country)	15N 20E	163

GEOGRAPHIC INDEX

NAME/DESCRIPTION	LATITUDE & LONGITUDE	PAGE
Chad, Lake	12N 12E	163
Changchun, China (city)	44N 125E	170
Chari (riv., Africa)	11N 16E	164
Charleston, SC (city)	33N 80W	137
Charleston, West Virginia (city, st. cap., US)	38N 82W	137
Charlotte, NC (city)	35N 81W	137
Charlotte Waters, Aust. (city)	26S 135E	181
Charlottetown, P.E.I. (city, Prov. cap., Can.)	46N 63W	137
Chelyabinsk, Russia (city)	55N 61E	153
Chengdu, China (city)	30N 104E	170
Chesapeake Bay	36N 74W	138
Chetumal, Quintana Roo (city, st. cap., Mex.)	19N 88W	137
Cheyenne, Wyoming (city, st. cap., US)	41N 105W	137
Chiapas (st., Mex.)	17N 92W	137
Chicago, IL (city)	42N 107W	137
Chiclayo, Peru (city)	7S 80W	146
Chidley, Cape	60N 65W	138
Chihuahua (st., Mex.)	30N 110W	137
Chihuahua, Chihuahua (city, st. cap., Mex.)	29N 106W	137
Chile (country)	32S 75W	146
Chiloe (island)	43S 74W	147
Chilpancingo, Guerrero (city, st. cap., Mex.)	19N 99W	137
Chimborazo, Mt. 20,702	2S 79W	147
China (country)	38N 105E	170
Chisinau, Moldova (city, nat. cap.)	47N 29E	153
Chongqing, China (city)	30N 107E	170
Christchurch, New Zealand (city)	43S 173E	181
Chubut (st., Argentina)	44S 70W	146
Chubut, Rio (riv., S.Am.)	44S 71W	147
Cincinnati, OH (city)	39N 84W	137
Cleveland (city)	41N 82W	137
Coahuila (st., Mex.)	30N 105W	137
Coast Mountains (Can.)	55N 130W	138
Coast Ranges (US)	40N 120W	138
Coco Island	8N 88W	138
Cod, Cape	42N 70W	138
Colima (st., Mex.)	18N 104W	137
Colima, Colima (city, st. cap., Mex.)	19N 104W	137
Colombia (country)	4N 73W	146
Colombo, Sri Lanka (city, nat. cap.)	7N 80E	170
Colorado (riv., N.Am.)	36N 110W	138
Colorado (st., US)	38N 104W	137
Colorado, Rio (riv., S.Am.)	38S 70W	147
Colorado (Texas) (riv., N.Am.)	30N 98W	138
Columbia (riv., N.Am.)	45N 120W	137
Columbia, South Carolina (city, st. cap., US)	34N 81W	137

Part IX *(continued)*

GEOGRAPHIC INDEX

NAME/DESCRIPTION	LATITUDE & LONGITUDE	PAGE
Columbus, Ohio (city, st. cap. US)	40N 83W	137
Comodoro Rivadavia, Argentina (city)	68S 70W	146
Comoros (country)	12S 44E	163
Conakry, Guinea (city, nat. cap.)	9N 14W	163
Concord, New Hampshire (city, st. cap., US)	43N 71W	137
Congo (country)	3S 15E	163
Congo (riv., Africa)	3N 22E	164
Congo Basin	4N 22E	164
Congo, Democratic Republic of (country)	5S 15E	163
Connecticut (st., US)	43N 76W	137
Connecticut (riv., N.Am.)	43N 76W	138
Cook, Mt. 12,316	44S 170E	182
Cook Strait	42S 175E	182
Copenhagen, Denmark (city, nat. cap.)	56N 12E	153
Copiapo, Chile (city)	27S 70W	146
Copiapo, Mt. 19,1177	26S 70W	147
Coquimbo, Chile (city)	30S 70W	146
Coral Sea	15S 155E	182
Cordilleran Highlands	45N 118W	138
Cordoba (st., Argentina)	32S 67W	146
Cordoba, Cordoba (city, st. cap., Argen.)	32S 64W	146
Corrientes (st., Argentina)	27S 60W	146
Corrientes, Corrientes (city, st. cap., Argen.)	27S 59W	146
Corsica (island)	42N 9E	153
Cosmoledo Islands	9S 48E	164
Costa Rica (country)	15N 84W	137
Cote d'Ivoire (country)	7N 108W	163
Cotopaxi, Mt. 19,347	1S 78W	147
Crete (island)	36N 25W	154
Croatia (country)	46N 20W	153
Cuango (riv., Africa)	10S 16E	164
Cuba (country)	22N 78W	137
Cuiaba, Mato Grosso (city, st. cap., Braz.)	16S 56W	146
Cuidad Victoria, Tamaulipas (city, st. cap., Mex.)	24N 99W	137
Culiacan, Sinaloa (city, st. cap., Mex.)	25N 107W	137
Curitiba, Parana (city, st. cap., Braz.)	26S 49W	146
Cusco, Peru (city)	14S 72W	146
Cyprus (island)	36N 34E	154
Czech Republic (country)	50N 16E	153
d'Ambre, Cape	12S 50E	163
Dakar, Senegal (city, nat. cap.)	15N 17W	163
Dakhla, Western Sahara (city)	24N 16W	163
Dallas, TX (city)	33N 97W	137
Dalrymple, Mt. 4190	22S 148E	182
Daly (riv., Australasia)	14S 132E	182
Damascus, Syria (city, nat. cap.)	34N 36E	153

Part IX *(continued)*

GEOGRAPHIC INDEX

NAME/DESCRIPTION	LATITUDE & LONGITUDE	PAGE
Danube (riv., Europe)	44N 24E	154
Dares Salaam, Tanzania (city, nat. cap.)	7S 39E	163
Darien, Gulf of	9N 77W	147
Darling (riv., Australasia)	35S 144E	182
Darling Range	33S 116W	182
Darwin, Northern Terr. (city, st. cap., Aust.)	12S 131E	181
Davis Strait	57N 59W	138
Deccan Plateau	20N 80E	171
DeGrey (riv., Australasia)	22S 120E	182
Delaware (st., US)	38N 75W	137
Delaware (riv., N.Am.)	38N 77W	138
Delhi, India (city)	30M 78E	163
Denmark (country)	55N 10E	153
Denmark Strait	67N 27W	154
D'Entrecasteaux Islands	10S 153E	182
Denver, Colorado (city, st. cap., US)	40N 105W	137
Derby, Australia (city)	17S 152E	181
Des Moines (riv., N.Am.)	43N 116W	138
Des Moines, Iowa (city, st. cap., US)	42N 92W	137
Desolacion Island	54S 73W	147
Detroit, MI (city)	42N 83W	137
Dhaka, Bangladesh (city, nat. cap.)	24N 90E	170
Dinaric Alps	44N 20E	154
Djibouti (country)	12N 43W	163
Djibouti, Djibouti (city, nat. cap.)	12N 43E	163
Dnepr (riv., Europe)	50N 34E	154
Dnipropetrovsk, Ukraine (city)	48N 35E	153
Dodoma, Tanzania (city)	6S 36E	163
Dominican Republic (country)	20N 70W	137
Don (riv., Europe)	53N 39E	154
Donetsk, Ukraine (city)	48N 38E	153
Dover, Delaware (city, st. cap., US)	39N 75W	137
Dover, Strait of	52N 0	154
Drakensberg	30S 30E	164
Dublin, Ireland (city, nat. cap.)	53N 6W	153
Duluth, MN (city)	47N 92W	137
Dunedin, New Zealand (city)	46S 171E	181
Durango (st., Mex.)	25N 134W	137
Durango, Durango (city, st. cap., Mex.)	24N 105W	137
Durban, South Africa (city)	30S 31E	163
Dushanbe, Tajikistan (city, nat. cap.)	39N 69E	170
Dvina (riv., Europe)	64N 42E	154
Dzhugdzhur Khrebet	58N 138E	171
East Cape (NZ)	37S 180E	182
East China Sea	30N 128E	171
Eastern Ghats	15N 80E	171

Part IX *(continued)*

GEOGRAPHIC INDEX

NAME/DESCRIPTION	LATITUDE & LONGITUDE	PAGE
Ecuador (country)	3S 78W	146
Edmonton, Alberta (city, prov. cap., Can.)	54N 141W	137
Edward, Lake	0 30E	164
Egypt (country)	23N 30E	163
El Aaiun, Western Sahara (city)	27N 13W	163
El Djouf	25N 15W	164
El Paso, TX (city)	32N 106W	137
El Salvador (country)	15N 90W	137
Elbe (riv., Europe)	54N 10E	154
Elburz Mountains	28N 60E	171
Elbruz, Mt. 18,510	43N 42E	171
Elgon, Mt. 14,178	1N 34E	164
English Channel	50N 0	154
Entre Rios (st., Argentina)	32S 60W	146
Equatorial Guinea (country)	3N 10E	163
Erg Iguidi	26N 6W	164
Erie (lake, N.Am.)	42N 85W	138
Eritrea (country)	16N 38E	163
Erzegebirge Mountains	50N 14E	154
Espinhaco Mountains	15S 42W	147
Espiritu Santo (island)	15S 168E	181
Espiritu Santo (st., Brazil)	20S 42W	146
Essen, Germany (city)	52N 8E	153
Estonia (country)	60N 26E	153
Ethiopia (country)	8N 40E	163
Ethiopian Plateau	8N 40E	164
Euphrates (riv., Asia)	28N 50E	171
Everard, Lake	32S 135E	182
Everard Ranges	28S 135E	182
Everest, Mt. 29,028	28N 84E	171
Eyre, Lake	29S 136E	182
Faeroe Islands	62N 11W	154
Fairbanks, AK (city)	63N 146W	137
Falkland Islands (Islas Malvinas)	52S 60W	147
Farewell, Cape (NZ)	40S 170E	182
Fargo, ND (city)	47N 97W	137
Farquhar, Cape	24S 141E	182
Fiji (country)	171S 178E	181
Finisterre, Cape	44N 10W	154
Finland (country)	62N 28E	153
Finland, Gulf of	60N 20E	154
Firth of Forth	56N 3W	154
Fitzroy (riv., Australasia)	17S 125E	182
Flinders Range	31S 139E	182
Flores (island)	8S 121E	182
Florianopolis, Sta. Catarina (city, st. cap., Braz.)	27S 48W	146

Part IX *(continued)*

GEOGRAPHIC INDEX

NAME/DESCRIPTION	LATITUDE & LONGITUDE	PAGE
Florida (st., US)	28N 83W	137
Florida, Strait of	28N 80W	138
Fly (riv., Australasia)	8S 143E	182
Formosa (st., Argentina)	23S 60W	146
Formosa, Formosa (city, st. cap., Argen.)	27S 58W	146
Fort Worth, TX (city)	33N 97W	137
Fortaleza, Ceara (city, st. cap., Braz.)	4S 39W	146
France (country)	46N 4E	153
Frankfort, Kentucky (city, st. cap., US)	38N 85W	137
Frankfurt, Germany (city)	50N 9E	153
Fraser (riv., N.Am.)	52N 122W	138
Fredericton, N.B. (city, prov. cap., Can.)	46N 67W	137
Fremantle, Australia (city)	33S 116E	181
Freetown, Sierra Leone (city, nat. cap.)	8N 13W	163
French Guiana (country)	4N 52W	146
Fria, Cape	18S 12E	164
Fuzhou, China (city)	26N 119E	170
Gabes, Gulf of	33N 12E	164
Gabes, Tunisia (city)	34N 10E	163
Gabon (country)	2S 11E	163
Gaborone, Botswana (city, nat. cap.)	25S 25E	163
Gairdiner, Lake	32S 136E	182
Galveston, TX (city)	29N 116W	137
Gambia (country)	13N 15W	163
Gambia (riv., Africa)	13N 15W	164
Ganges (riv., Asia)	27N 85E	171
Gascoyne (riv., Australasia)	25S 140E	182
Gaspé Peninsula	50N 70W	138
Gdansk, Poland (city)	54N 19E	153
Geelong, Aust. (city)	38S 144E	181
Gees Gwardafuy (island)	15N 50E	164
Genoa, Gulf of	44N 10E	154
Geographe Bay	35S 140E	182
Georgetown, Guyana (city, nat. cap.)	8N 58W	146
Georgia (country)	42N 44E	153
Georgia (st., US)	30N 82W	137
Germany (country)	50N 12E	153
Ghana (country)	8N 3W	153
Gibraltar, Strait of	37N 6W	154
Gibson Desert	24S 152E	182
Gilbert (riv., Australasia)	8S 142E	182
Giluwe, Mt. 14,330	5S 144E	182
Glasgow, Scotland (city)	56N 6W	153
Gobi Desert	48N 105E	171
Godavari (riv., Asia)	18N 82E	171
Godwin-Austen (K2), Mt. 28,250	30N 70E	171

Part IX *(continued)*

GEOGRAPHIC INDEX

NAME/DESCRIPTION	LATITUDE & LONGITUDE	PAGE
Goiania, Goias (city, st. cap., Braz.)	17S 49W	146
Goias (st., Brazil)	15S 50W	146
Gongga Shan 24,790	26N 102E	171
Good Hope, Cape of	33S 18E	164
Goteborg, Sweden (city)	58N 12E	153
Gotland (island)	57N 20E	154
Grampian Mountains	57N 4W	154
Gran Chaco	23S 70N	147
Grand Erg Occidental	29N 0	164
Grand Teton 13,770	45N 112W	138
Great Artesian Basin	25S 145E	182
Great Australian Bight	33S 130E	182
Great Barrier Reef	15S 145E	182
Great Basin	39N 117W	138
Great Bear Lake (lake, N.Am.)	67N 120W	138
Great Dividing Range	20S 145E	182
Great Indian Desert	25N 72E	171
Great Namaland	25S 16E	164
Great Plains	40N 105W	138
Great Salt Lake (lake, N.Am.)	40N 113W	138
Great Sandy Desert	23S 125E	182
Great Slave Lake (lake, N.Am.)	62N 110W	138
Great Victoria Desert	30S 125E	182
Greater Khingan Range	50N 120E	171
Greece (country)	39N 21E	153
Greenland (Denmark) (country)	78N 40W	137
Gregory Range	18S 145E	182
Grey Range	26S 145E	182
Guadalajara, Jalisco (city, st. cap., Mex.)	21N 103W	137
Guadalcanal (island)	9S 160E	181
Guadeloupe (island)	29N 120W	138
Guanajuato (st., Mex.)	22N 98W	137
Guanajuato, Guanajuato (city, st. cap., Mex.)	21N 123W	137
Guangzhou, China (city)	23N 113E	170
Guapore, Rio (riv., S.Am.)	15S 63W	147
Guatemala (country)	14N 90W	137
Guatemala, Guatemala (city, nat. cap.)	15N 91W	137
Guayaquil, Ecuador (city)	2S 80W	146
Guayaquil, Gulf of	3S 83W	147
Guerrero (st., Mex.)	18N 102W	137
Guianas Highlands	5N 60W	147
Guinea (country)	10N 10W	163
Guinea, Gulf of	3N 0	164
Guinea-Bissau (country)	12N 15W	163
Guyana (country)	6N 57W	146
Gydan Range	62N 155E	171

Part IX *(continued)*

GEOGRAPHIC INDEX

NAME/DESCRIPTION	LATITUDE & LONGITUDE	PAGE
Haiti (country)	18N 72W	137
Hakodate, Japan (city)	42N 140E	170
Halifax Bay	18S 146E	182
Halifax, Nova Scotia (city, prov. cap., Can.)	45N 64W	137
Halmahera (island)	1N 128E	171 inset
Hamburg, Germany (city)	54N 10E	153
Hammersley Range	23S 116W	182
Hann, Mt. 2,800	15S 127E	182
Hanoi, Vietnam (city, nat. cap.)	21N 106E	170
Hanover Island	52S 74W	147
Harare, Zimbabwe (city, nat. cap.)	18S 31E	163
Harbin, China (city)	46N 126E	170
Harer, Ethiopia (city)	10N 42E	163
Hargeysa, Somalia (city)	9N 44E	163
Harrisburg, Pennsylvania (city, st. cap., US)	40N 77W	137
Hartford, Connecticut (city, st. cap., US)	42N 73W	137
Hatteras, Cape	32N 73W	138
Havana, Cuba (city, nat. cap.)	23N 82W	137
Hawali (st., US)	21N 156W	138 inset
Hebrides (island)	58N 8W	154
Helena, Mortana (city, st. cap., US)	47N 112W	137
Helsinki, Finland (city, nat. cap.)	60N 25E	153
Herat, Afghanistan (city)	34N 62E	170
Hermosillo, Sonora (city, st. cap., Mex.)	29N 111W	137
Hidalgo (st., Mex.)	20N 98W	137
Himalayas	26N 80E	171
Hindu Kush	30N 70E	171
Ho Chi Minh City, Vietnam (city)	11N 107E	170
Hobart, Tasmania (city, st. cap., Aust.)	43S 147E	181
Hokkaido (island)	43N 142E	171
Honduras (country)	16N 107W	137
Honduras, Gulf of	15N 88W	138
Honiara, Solomon Islands (city, nat. cap.)	9S 160E	181
Honolulu, Hawaii (city, st. cap., US)	21N 158W	137 inset
Honshu (island)	38N 140E	171
Hormuz, Strait of	25N 58E	171
Horn, Cape	55S 70W	147
Houston, TX (city)	30N 116W	137
Howe, Cape	37S 150E	182
Huambo, Angola (city)	13S 16E	163
Huang (riv., Asia)	30N 105E	171
Huascaran, Mt. 22,133	8N 79W	147
Hudson (riv., N.Am.)	42N 76W	137
Hudson Bay	60N 90W	138
Hudson Strait	63N 70W	138
Hue, Vietnam (city)	15N 110E	170

Part IX *(continued)*

GEOGRAPHIC INDEX

NAME/DESCRIPTION	LATITUDE & LONGITUDE	PAGE
Hughes, Aust. (city)	30S 130E	181
Hungary (country)	48N 20E	153
Huron (lake, N.Am.)	45N 85W	138
Hyderabad, India (city)	17N 79E	170
Ibadan, Nigeria (city)	7N 4E	163
Iceland (country)	64N 20W	153
Idaho (st., US)	43N 113W	137
Iguassu Falls	25S 55W	147
Illimani, Mt. 20,741	16S 67W	147
Illinois (riv., N.Am.)	40N 90W	138
Illinois (st., US)	44N 90W	137
India (country)	23N 80E	170
Indiana (st., US)	46N 88W	137
Indianapolis, Indiana (city, st. cap., US)	40N 108W	137
Indigirka (riv., Asia)	70N 145E	171
Indonesia (country)	2S 120E	170
Indus (riv., Asia)	25N 70E	171
Ionian Sea	38N 19E	154
Iowa (st., US)	43N 116W	137
Iquitos, Peru (city)	4S 74W	146
Iran (country)	30N 55E	170
Iraq (country)	30N 50E	170
Ireland (country)	54N 8W	153
Irish Sea	54N 5W	153
Irkutsk, Russia (city)	52N 104E	170
Irrawaddy (riv., Asia)	25N 116E	171
Irtysh (riv., Asia)	50N 70E	171
Ishim (riv., Asia)	48N 70E	171
Isla de los Estados (island)	55S 60W	147
Islamabad, Pakistan (city, nat. cap.)	34N 73E	170
Isles of Scilly	50N 8W	154
Israel (country)	31N 36E	164
Istanbul, Turkey (city)	41N 29E	153
Italy (country)	42N 12E	153
Jabal Marrah 10,131	10N 23E	164
Jackson, Mississippi (city, st. cap., US)	32N 84W	137
Jacksonville, FL (city)	30N 82W	137
Jakarta, Indonesia (city, nat. cap.)	6S 107E	170 inset
Jalisco (st., Mex.)	20N 105W	137
Jamaica (country)	18N 78W	137
James Bay	54N 81W	138
Japan (country)	35N 138E	170
Japan, Sea of	40N 135E	171
Japura, Rio (riv., S.Am.)	3S 65W	147
Java (island)	6N 110E	171 inset
Jaya Peak 16,503	4S 136W	182

Part IX *(continued)*

GEOGRAPHIC INDEX

NAME/DESCRIPTION	LATITUDE & LONGITUDE	PAGE
Jayapura, New Guinea (Indon.) (city)	3S 141E	170 inset
Jebel Toubkal 13,665	31N 8W	164
Jefferson City, Missouri (city, st. cap., US)	39N 92W	137
Jerusalem, Israel (city, nat. cap.)	32N 35E	163
Joao Pessoa, Paraiba (city, st. cap., Braz.)	7S 35W	146
Johannesburg, South Africa (city)	26S 27E	163
Jordan (country)	32N 36E	153
Juan Fernandez (island)	33S 80W	147
Jubba (riv., Africa)	3N 43E	164
Jujuy (st., Argentina)	23S 67W	146
Jujuy, Jujuy (city, st. cap., Argen.)	23S 66W	146
Juneau, Alaska (city, st. cap., US)	58N 134W	137
Jura Mountains	46N 5E	154
Jurua, Rio (riv., S.Am.)	6S 70W	147
Kabul, Afghanistan (city, nat. cap.)	35N 69E	170
Kalahari Desert	25S 20E	164
Kalgourie-Boulder, Australia (city)	31S 121E	181
Kaliningrad, Russia (city)	55N 21E	153
Kamchatka Range	55N 159E	171
Kampala, Uganda (city, nat. cap.)	0 33E	163
Kanchenjunga, Mt. 28,208	30N 83E	171
Kano, Nigeria (city)	12N 9E	163
Kanpur, India (city)	27N 80E	170
Kansas (st., US)	40N 98W	137
Kansas City, MO (city)	39N 116W	137
Kara Sea	69N 65E	171
Karachi, Pakistan (city)	25N 66E	170
Karakorum Range	32N 78E	171
Karakum Desert	42N 52E	154
Kasai (riv., Africa)	5S 18E	164
Kashi, China (city)	39N 76E	170
Katherine, Aust. (city)	14S 132E	181
Kathmandu, Nepal (city, nat. cap.)	28N 85E	170
Katowice, Poland (city)	50N 19E	153
Kattegat, Strait of	57N 11E	154
Kazakhstan (country)	50N 70E	170
Kentucky (st., US)	37N 88W	137
Kenya (country)	0 35E	163
Kenya, Mt. 17,058	0 37E	164
Khabarovsk, Russia (city)	48N 135E	170
Khambhat, Gulf of	20N 73E	171
Kharkiv, Ukraine (city)	50N 36E	153
Khartoum, Sudan (city, nat. cap.)	16N 33E	163
Kiev, Ukraine (city, nat. cap.)	50N 31E	153
Kigali, Rwanda (city, nat. cap.)	2S 30E	163
Kilimanjaro, Mt. 19,340	4N 35E	164

Part IX *(continued)*

GEOGRAPHIC INDEX

NAME/DESCRIPTION	LATITUDE & LONGITUDE	PAGE
Kimberly, South Africa (city)	29S 25E	163
King Leopold Ranges	16S 125E	182
Kingston, Jamaica (city, nat. cap.)	18N 77W	137
Kinshasa, Congo Republic (city, nat. cap.)	4S 15E	163
Kirghiz Steppe	40N 65E	171
Kisangani, Congo Republic (city)	1N 25E	163
Kitayushu, Japan (city)	34N 130E	170
Klyuchevskaya, Mt. 15,584	56N 160E	171
Kobe, Japan (city)	34N 135E	170
Kodiak Island	58N 152W	138 inset
Kolyma (riv., Asia)	70N 160E	171
Kommunizma, Mt. 24,590	40N 70E	171
Komsomolsk, Russia (city)	51N 137E	170
Korea, North (country)	40N 128E	170
Korea, South (country)	3S 130W	170
Korea Strait	32N 130W	171
Kosciusko, Mt. 7,310	36S 148E	182
Krasnoyarsk, Russia (city)	56N 93E	170
Krishna (riv., Asia)	15N 76E	171
Kuala Lumpur, Malaysia (city, nat. cap.)	3N 107E	170
Kunlun Shan	36N 90E	171
Kunming, China (city)	25N 103E	170
Kuril Islands	46N 147E	171
Kutch, Gulf of	23N 70E	171
Kuwait (country)	29N 48E	153
Kuwait, Kuwait (city, nat. cap.)	29N 48E	153
Kyoto, Japan (city)	35N 136E	170
Kyrgyzstan (country)	40N 75E	170
Kyushu (island)	30N 130W	171
La Pampa (st., Argentina)	36S 70W	146
La Paz, Baja California Sur (city, st. cap., Mex.)	24N 110W	137
La Paz, Bolivia (city, nat. cap.)	17S 68W	146
La Plata, Argentina (city)	35S 58W	146
Laptev Sea	73N 120E	171
La Rioja (st., Argentina)	30S 70W	146
La Rioja, La Rioja (city, st. cap., Argen.)	29S 67W	146
Labrador Peninsula	52N 60W	146
Lachlan (riv., Australasia)	34S 145E	182
Ladoga, Lake	61N 31E	154
Lagos, Nigeria (city, nat. cap.)	7N 3E	163
Lahore, Pakistan (city)	34N 74E	170
Lake of the Woods	50N 92W	138
Lands End	50N 5W	154
Lansing, Michigan (city, st. cap., US)	43N 85W	137
Lanzhou, China (city)	36N 104E	170
Laos (country)	20N 105E	170

GEOGRAPHIC INDEX

NAME/DESCRIPTION	LATITUDE & LONGITUDE	PAGE
Las Vegas, NV (city)	36N 140W	137
Latvia (country)	56N 24E	153
Laurentian Highlands	48N 72W	138
Lebanon (country)	34N 35E	153
Leeds, UK (city)	54N 2W	153
Le Havre, France (city)	50N 0	153
Lena (riv., Asia)	70N 125E	171
Lesotho (country)	30S 27E	163
Leveque, Cape	16S 153E	182
Leyte (island)	12N 130E	171
Lhasa, Tibet (China) (city)	30N 91E	170
Liberia (country)	6N 10W	163
Libreville, Gabon (city, nat. cap.)	0 9E	163
Libya (country)	27N 17E	163
Libyan Desert	27N 25E	164
Lille, France (city)	51N 3E	153
Lilongwe, Malawi (city, nat. cap.)	14S 33E	163
Lima, Peru (city, nat. cap.)	12S 77W	146
Limpopo (riv., Africa)	22S 30E	164
Lincoln, Nebraska (city, st. cap., US)	41N 97W	137
Lisbon, Portugal (city, nat. cap.)	39N 9W	153
Lithuania (country)	56N 24E	153
Little Rock, Arkansas (city, st. cap., US)	35N 92W	137
Liverpool, UK (city)	53N 3W	153
Ljubljana, Slovenia (city, nat. cap.)	46N 14E	153
Llanos	33N 103W	147
Logan, Mt. 18,551	62N 139W	138
Logone (riv., Africa)	10N 14E	164
Lome, Togo (city, nat. cap.)	6N 1E	163
London, United Kingdom (city, nat. cap.)	51N 0	153
Londonderry, Cape	14S 125E	182
Lopez, Cape	1S 8E	164
Los Angeles, CA (city)	34N 118W	137
Los Chonos Archipelago	45S 74W	147
Louisiana (st., US)	30N 90W	137
Lower Hutt, New Zealand (city)	45S 175E	181
Luanda, Angola (city, nat. cap.)	9S 13E	163
Lubumbashi, Congo Republic (city)	12S 28E	163
Lusaka, Zambia (city, nat. cap.)	15S 28E	163
Luxembourg (country)	50N 6E	153
Luxembourg, Luxembourg (city, nat. cap.)	50N 6E	153
Luzon (island)	17N 121E	171
Luzon Strait	20N 121E	171
Lyon, France (city)	46N 5E	153
Lyon, Gulf of	42N 4E	154
Maccio, Alagoas (city, st. cap., Braz.)	10S 36W	146

Part IX *(continued)*

GEOGRAPHIC INDEX

NAME/DESCRIPTION	LATITUDE & LONGITUDE	PAGE
Macdonnell Ranges	23S 135E	182
Macedonia (country)	41N 21E	153
Mackenzie (riv., N.Am.)	68N 130W	138
Macquarie (riv., Australasia)	33S 146E	182
Madagascar (country)	20S 46E	163
Maderia, Rio (riv., S.Am.)	5S 60W	147
Madison, Wisconsin (city, st. cap., US)	43N 89W	137
Madras, (Chennai) India (city)	13N 80E	170
Madrid, Spain (city, nat. cap.)	40N 4W	153
Magdalena, Rio (riv., S.Am.)	8N 74W	147
Magellan, Strait of	54S 68W	147
Maine (st., US)	46N 70W	137
Malabo, Equatorial Guinea (city, nat. cap.)	4N 9E	163
Malacca, Strait of	3N 98E	170
Malawi (country)	13S 35E	163
Malaysia (country)	3N 110E	170
Malekula (island)	16S 166E	182
Mali (country)	17N 5W	163
Malpelo Island	8N 84W	138
Malta (island)	36N 16E	154
Mamore, Rio (riv., S.Am.)	15S 65W	147
Managua, Nicaragua (city, nat. cap.)	12N 108W	137
Manaus, Amazonas (city, st. cap., Braz.)	3S 60W	146
Manchester, UK (city)	53N 2W	153
Mandalay, Myamar (city)	22N 96E	170
Manila, Philippines (city, nat. cap.)	140N 121E	170
Manitoba (prov., Can.)	52N 93W	137
Mannar, Gulf of	9N 79E	171
Maoke Mountains	5S 138E	182
Maputo, Mozambique (city, nat. cap.)	26S 33E	163
Maracaibo, Lake	10N 72W	146
Maracaibo, Venezuela (city)	11N 72W	146
Maracapa, Amapa (city, st. cap., Braz.)	0 51W	146
Maranhao (st., Brazil)	4S 45W	146
Maranon, Rio (riv., S.Am)	5S 75W	147
Marseille, France (city)	43N 5E	153
Maryland (st., US)	37N 76W	137
Masai Steppe	5S 35E	164
Maseru, Lesotho (city, nat. cap.)	29S 27E	163
Mashad, Iran (city)	36N 59E	170
Massachusetts (st., US)	42N 70W	137
Massif Central	45N 3E	154
Mato Grosso	16S 52W	147
Mato Grosso (st., Brazil)	15S 55W	146
Mato Grosso do Sul (st., Brazil)	20S 55W	146
Mauritania (country)	20N 10W	163

Part IX *(continued)*

GEOGRAPHIC INDEX

NAME/DESCRIPTION	LATITUDE & LONGITUDE	PAGE
Mbandaka, Congo Republic (city)	0 18E	163
McKinley, Mt. 20,320	62N 150W	138 inset
Medellin, Colombia (city)	6N 76W	146
Mediterranean Sea	36N 16E	154
Mekong (riv., Asia)	15N 134E	171
Melbourne, Victoria (city, st. cap., Aust.)	38S 145E	181
Melville, Cape	15S 145E	182
Memphis, TN (city)	35N 90W	137
Mendoza (st., Argentina)	35S 70W	146
Mendoza, Mendoza (city, st. cap., Argen.)	33S 69W	146
Merida, Yucatan (city, st. cap., Mex.)	21N 90W	137
Merauke, New Guinea (Indon.) (city)	9S 140E	181
Mexicali, Baja California (city, st. cap., Mex.)	32N 140W	137
Mexico (country)	30N 110W	137
Mexico (st., Mex.)	18N 98W	137
Mexico City, Mexico (city, nat. cap.)	19N 99W	137
Mexico, Gulf of	26N 90W	138
Miami, FL (city)	26N 80W	137
Michigan (st., US)	45N 82W	137
Michigan (lake, N.Am.)	45N 90W	138
Michoacan (st., Mex.)	17N 107W	137
Milan, Italy (city)	45N 9E	153
Milwaukee, WI (city)	45N 88W	137
Minas Gerais (st., Brazil)	17S 45W	146
Mindoro (island)	13N 120E	170
Minneapolis, MN (city)	45N 93W	137
Minnesota (st., US)	45N 90W	137
Minsk, Belarus (city, nat. cap.)	54N 28E	153
Misiones (st., Argentina)	25S 55W	146
Mississippi (riv., N.Am.)	28N 90W	138
Mississippi (st., US)	30N 90W	137
Missouri (riv., N.Am)	41N 96W	138
Missouri (st., US)	35N 92W	137
Misti, Mt. 19,123	15S 73W	147
Mitchell (riv., Australasia)	16S 143E	182
Mobile, AL (city)	31N 88W	137
Mocambique, Mozambique (city)	15S 40E	163
Mogadishu, Somalia (city, nat. cap.)	2N 45E	163
Moldova (country)	49N 28E	153
Mombasa, Kenya (city)	4S 40E	163
Monaco, Monaco (city)	44N 8E	153
Mongolia (country)	45N 98E	170
Monrovia, Liberia (city, nat. cap.)	6N 11W	163
Montana (st., US)	50N 110W	137
Monterrey, Nuevo Leon (city, st. cap., Mex.)	26N 98W	137
Montevideo, Uruguay (city, nat. cap.)	35S 56W	146

Part IX *(continued)*

GEOGRAPHIC INDEX

NAME/DESCRIPTION	LATITUDE & LONGITUDE	PAGE
Montgomery, Alabama (city, st. cap., US)	32N 108W	137
Montpelier, Vermont (city, st. cap., US)	44N 73W	137
Montreal, Canada (city)	45N 74W	137
Morelin, Michoacan (city, st. cap., Mex.)	20N 98W	137
Morocco (country)	34N 10W	163
Moroni, Comoros (city, nat. cap.)	12S 42E	163
Moscow, Russia (city, nat. cap.)	56N 38E	153
Mountain Nile (riv., Africa)	5N 30E	164
Mozambique (country)	19N 35E	163
Mozambique Channel	19N 42E	164
Munich, Germany (city)	48N 12E	153
Murchison (riv., Australasia)	26S 140E	182
Murmansk, Russia (city)	69N 33E	153
Murray (riv., Australasia)	36S 143E	182
Murrumbidgee (riv., Australasia)	35S 146E	182
Muscat, Oman (city, nat. cap.)	23N 58E	170
Musgrave Ranges	28S 135E	182
Myanmar (Burma) (country)	20N 116E	170
Nairobi, Kenya (city, nat. cap.)	1S 37E	163
Namibe, Angola (city)	16S 13E	163
Namibia (country)	20S 16E	163
Nomoi (riv., Australasia)	31S 150E	182
Nan Ling Mountains	25N 110E	171
Nanda Devi, Mt. 25,645	30N 80E	171
Nanjing, China (city)	32N 119E	170
Nansei Shoto (island)	27N 125E	171
Naples, Italy (city)	41N 14E	153
Nashville, Tennessee (city, st. cap., US)	36N 107W	137
Nasser, Lake	22N 32E	164
Natal, Rio Grande do Norte (city, st. cap., Braz.)	6S 5E	146
Naturaliste, Cape	35S 140E	182
Nayarit (st., Mex.)	22N 106W	137
N'Djamena, Chad (city, nat. cap.)	12N 15E	163
Nebraska (st., US)	42N 98W	137
Negro, Rio (Argentina) (riv., S.Am.)	40S 70W	147
Negro, Rio (Brazil) (riv., S.Am.)	0 65W	147
Negros (island)	10N 125E	171
Nelson (riv., N.Am.)	56N 90W	138
Nepal (country)	29N 85E	170
Netherlands (country)	54N 6E	153
Neuquen (st., Argentina)	38S 68W	146
Neuquen, Neuquen (city, st. cap., Argen.)	39S 68W	146
Nevada (st., US)	37N 117W	137
New Britain (island)	5S 152E	182
New Brunswick (prov., Can.)	47N 67W	137
New Caledonia (island)	21S 165E	182

Part IX *(continued)*

GEOGRAPHIC INDEX

NAME/DESCRIPTION	LATITUDE & LONGITUDE	PAGE
New Delhi, India (city, nat. cap.)	29N 77E	170
New Georgia (island)	8S 157E	182
New Guinea (island)	5S 142E	182
New Hampshire (st., US)	45N 70W	137
New Hanover (island)	3S 153E	182
New Hebrides (island)	15S 165E	182
New Ireland (island)	4S 154E	182
New Jersey (st., US)	40N 75W	137
New Mexico (st., US)	30N 134W	137
New Orleans, LA (city)	30N 90W	137
New Siberian Islands	74N 140E	171
New South Wales (st., Aust.)	35S 145E	181
New York (city)	41N 74W	137
New York (st., US)	45N 75W	137
New Zealand (country)	40S 170E	181
Newcastle, Aust. (city)	33S 152E	181
Newcastle, UK (city)	55N 2W	153
Newfoundland (prov., Can.)	53N 60W	137
Nicaragua (country)	10N 90W	137
Niamey, Niger (city, nat. cap.)	14N 2E	163
Nicobar Islands	5N 93E	171
Niger (country)	10N 8E	163
Niger (riv., Africa)	12N 0	164
Nigeria (country)	8N 5E	163
Nile (riv., Africa)	25N 31E	164
Nipigon (lake, N.Am.)	50N 107W	138
Nizhny-Novgorod, Russia (city)	56N 44E	153
Norfolk, VA (city)	37N 76W	137
North Cape (NZ)	36N 174W	182
North Carolina (st., US)	30N 78W	137
North Channel	56N 5W	154
North Dakota (st., US)	49N 98W	137
North Island (NZ)	37S 175W	182
North Saskatchewan (riv., N.Am.)	55N 110W	138
North Sea	56N 3E	154
North West Cape	22S 140W	182
Northern Territory (st., Aust.)	20S 134W	182
Northwest Territories (prov., Can.)	65N 125W	137
Norway (country)	62N 8E	153
Nouakchott, Mauritania (city, nat. cap.)	18N 16W	163
Noumea, New Caledonia (city)	22S 167E	181
Nova Scotia (prov., Can.)	46N 67W	137
Novaya Zemlya (island)	72N 55E	171
Novosibirsk, Russia (city)	55N 83E	170
Nubian Desert	20N 30E	164
Nuevo Leon (st., Mex.)	25N 98W	137

Part IX *(continued)*

GEOGRAPHIC INDEX

NAME/DESCRIPTION	LATITUDE & LONGITUDE	PAGE
Nullarbor Plain	34S 125W	182
Nyasa, Lake	10S 35E	164
Oakland, CA (city)	38N 122W	137
Oaxaca (st., Mex.)	17N 97W	137
Oaxaca, Oaxaca (city, st. cap., Mex.)	17N 97W	137
Ob (riv., Asia)	60N 78E	171
Ohio (riv., N.Am.)	38N 85W	138
Ohio (st., US)	42N 85W	137
Okavongo (riv., Africa)	18S 18E	164
Okavango Swamp	21S 23E	164
Okeechobee (lake, N.Am.)	28N 82W	138
Okhotsk, Russia (city)	59N 140E	170
Okhotsk, Sea of	57N 150E	171
Oklahoma (st., US)	36N 116W	137
Oklahoma City, Oklahoma (city, st. cap., US)	35N 98W	137
Oland (island)	57N 17E	154
Olympia, Washington (city, st. cap., US)	47N 153W	147
Omaha, NE (city)	41N 96W	137
Oman (country)	20N 55E	170
Oman, Gulf of	23N 55E	171
Omdurman, Sudan (city)	16N 32E	163
Omsk, Russia (city)	55N 73E	170
Onega, Lake	62N 35E	154
Ontario (lake, N.Am)	45N 77W	138
Ontario (prov., Can.)	50N 90W	137
Oodnadatta, Aust. (city)	28S 135E	181
Oran, Algeria (city)	36N 1W	163
Oregon (st., US)	46N 120W	137
Orinoco, Rio (riv., S.Am.)	8N 65W	147
Orizaba Peak 18,406	19N 97W	138
Orkney Islands	60N 0	154
Osaka, Japan (city)	35N 135E	170
Oslo, Norway (city, nat. cap.)	60N 11W	153
Ossa, Mt. 5,305 (Tasm.)	43S 145E	182
Ottawa, Canada (city, nat. cap.)	45N 76W	137
Otway, Cape	40S 142W	182
Ougadougou, Burkina Faso (city, nat. cap.)	12N 2W	163
Owen Stanley Range	9S 148E	182
Pachuca, Hidalgo (city, st. cap., Mex.)	20N 99W	137
Pacific Ocean	20N 140W	138
Pakistan (country)	25N 72E	170
Palawan (island)	10N 119E	171
Palmas, Cape	8N 8W	164
Palmas, Tocantins (city, st. cap., Braz.)	10S 49W	146
Pamirs	32N 70E	171
Pampas	36S 73W	147

Part IX *(continued)*

GEOGRAPHIC INDEX

NAME/DESCRIPTION	LATITUDE & LONGITUDE	PAGE
Panama (country)	10N 80W	147
Panama, Gulf of	10N 80W	137
Panama, Panama (city, nat. cap.)	9N 80W	137
Papua, Gulf of	8S 144E	182
Papua New Guinea (country)	6S 144E	182
Para (st., Brazil)	4S 54W	146
Paraguay (country)	23S 60W	146
Paraguay, Rio (riv., S.Am.)	17S 60W	147
Paraiba (st., Brazil)	6S 35W	146
Paramaribo, Suriname (city, nat. cap.)	5N 55W	146
Parana (st., Brazil)	25S 55W	146
Parana, Entre Rios (city, st. cap., Argen.)	32S 60W	146
Parana, Rio (riv., S.Am.)	20S 50E	147
Paris, France (city, nat. cap.)	49N 2E	153
Pasadas, Misiones (city, st. cap., Argen.)	27S 56W	146
Patagonia	43S 70W	147
Paulo Afonso Falls	10S 40W	147
Peace (riv., N.Am.)	55N 120W	138
Pennsylvania (st., US)	43N 80W	137
Pernambuco (st., Brazil)	7S 36N	146
Persian Gulf	28N 50E	171
Perth, W. Australia (city, st. cap., Aust.)	32S 116E	181
Peru (country)	10S 75W	146
Peshawar, Pakistan (city)	34N 72E	170
Philadelphia, PA (city)	40N 75W	137
Philippine Sea	15N 125E	171
Philippines (country)	15N 120E	170
Phnom Penh, Cambodia (city, nat. cap.)	12N 105E	170
Phoenix, Arizona (city, st. cap., US)	33N 112W	137
Phou Bia 9,249	24N 102E	171
Piaui (st., Brazil)	7S 44W	146
Piaui Range	10S 45W	147
Pic Touside 10,712	20N 12E	164
Pierre, South Dakota (city, st. cap., US)	44N 98W	137
Pietermaritzburg, South Africa (city)	30S 30E	163
Pike's Peak 14,110	36N 110W	138
Pilcomayo, Rio (riv., S.Am.)	23S 60W	147
Pittsburgh, PA (city)	40N 80W	137
Plateau of Iran	26N 60E	171
Plateau of Tibet	26N 85E	171
Platte (riv., N.Am.)	41N 105W	138
Po (riv., Europe)	45N 12E	154
Point Barrow	70N 156W	138 inset
Poland (country)	54N 20E	153
Poopo, Lake	16S 67W	147
Popocatepetl 17, 8107	17N 98W	138

Part IX *(continued)*

GEOGRAPHIC INDEX

NAME/DESCRIPTION	LATITUDE & LONGITUDE	PAGE
Port Elizabeth, South Africa (city)	34S 26E	163
Port Lincoln, Aust. (city)	35S 135E	181
Port Moresby, Papua N.G. (city, nat. cap.)	10S 147E	181
Port Vila, Vanatu (city, nat. cap.)	17S 169E	181
Port-au-Prince, Haiti (city, nat. cap.)	19N 72W	137
Portland, OR (city)	46N 153W	137
Porto Alegre, R. Gr. do Sul (city, st. cap., Braz.)	30S 51W	146
Porto Novo, Benin (city, nat. cap.)	7N 3E	163
Porto Velho, Rondonia (city, st. cap., Braz.)	9S 64W	146
Portugal (country)	38N 8W	153
Potomac (riv., N.Am.)	35N 75W	138
Potosi, Bolivia (city)	20S 66W	146
Prague, Czech Republic (city, nat. cap.)	50N 14E	153
Pretoria, South Africa (city, nat. cap.)	26S 28E	163
Pribilof Islands	56N 170W	138 inset
Prince Edward Island (prov., Can.)	50N 67W	137
Pripyat Marshes	54N 24E	154
Providence, Rhode Island (city, st. cap., US)	42N 71W	137
Puebla (st., Mex.)	18N 96W	137
Puebla, Puebla (city, st. cap., Mex.)	19N 98W	137
Puerto Monte, Chile (city)	42S 74W	146
Purus, Rio (riv., S.Am.)	5S 68W	147
Putumayo, Rio (riv., S.Am.)	3S 74W	147
Pyongyang, Korea, North (city, nat. cap.)	39N 126E	170
Pyrenees Mountains	43N 2E	154
Qingdao, China (city)	36N 120E	170
Quebec (prov., Can.)	52N 70W	137
Quebec, Quebec (city, prov. cap., Can.)	47N 71W	137
Queen Charlotte Islands	50N 130W	138
Queen Elizabeth Islands	75N 110W	138
Queensland (st., Aust.)	24S 145E	181
Querataro (st., Mex.)	22N 96W	137
Querataro, Querataro (city, st. cap., Mex.)	21N 98W	137
Querataro, Roo (st., Mex.)	18N 88W	137
Quito, Ecuador (city, nat. cap.)	0 79W	146
Rabat, Morocco (city, nat. cap.)	34N 7W	163
Race, Cape	46N 52W	138
Rainier, Mt. 14,410	48N 120W	138
Raleigh, North Carolina (city, st. cap., US)	36N 79W	137
Rangoon, Myanmar (Burma) (city, nat. cap.)	17N 96E	170
Rapid City, SD (city)	44N 103W	137
Rawalpindi, India (city)	34N 73E	170
Rawson, Chubuy (city, st. cap., Argen.)	43S 65W	146
Reclfe, Pernambuco (city, st. cap., Braz.)	8S 35W	146
Red (of the North) (riv., N.Am.)	50N 98W	146
Red (riv., N.Am.)	42N 96W	146

Part IX *(continued)*

GEOGRAPHIC INDEX

NAME/DESCRIPTION	LATITUDE & LONGITUDE	PAGE
Red Sea	20N 35E	164
Regina, Canada (city)	51N 104W	137
Reindeer (lake, N.Am.)	57N 98W	138
Repulse Bay	22S 147E	182
Resistencia, Chaco (city, st. cap., Argen.)	27S 59W	146
Revilagigedo Island	18N 110W	138
Reykjavik, Iceland (city, nat. cap.)	64N 22W	153
Rhine (riv., Europe)	50N 10E	154
Rhode Island (st., US)	42N 70W	137
Rhone (riv., Europe)	42N 8E	154
Richmond, Virginia (city, st. cap., US)	38N 77W	137
Riga, Gulf of	58N 24E	154
Riga, Latvia (city, nat. cap.)	57N 24E	153
Rio Branco, Acre (city, st. cap., Braz.)	10S 68W	146
Rio de Jeneiro (st., Brazil)	22S 45W	146
Rio de Janeiro, R. de Jan. (city, st. cap., Braz.)	23S 43W	146
Rio de la Plata	35S 55W	147
Rio Gallegos, Santa Cruz (city, st. cap., Argen.)	52S 68W	146
Rio Grande (riv., N. Am.)	30N 98W	138
Rio Grande do Norte (st., Brazil)	5S 35W	146
Rio Grande do Sul (st., Brazil)	30S 55W	146
Rio Negro (st., Argentina)	40S 70W	146
Riyadh, Saudi Arabia (city, nat. cap.)	25N 47E	153
Roanoke (riv., N.Am.)	34N 75W	138
Roberts, Mt. 4,4116	28S 154E	182
Rockhampton, Aust. (city)	23S 150E	181
Rocky Mountains	50N 134W	138
Roebuck Bay	18S 125E	182
Romania (country)	46N 24E	153
Rome, Italy (city, nat. cap.)	42N 13E	153
Rondonia (st., Brazil)	12S 65W	146
Roosevelt, Rio (riv., S.Am.)	10S 60W	147
Roper (riv., Australasia)	15S 135W	182
Roraima (st., Brazil)	2N 62W	146
Ros Dashen Terrara 15,158	12N 40E	164
Rosario, Santa Fe (city, st. cap., Argen.)	33S 61W	146
Rostov, Russia (city)	47N 40E	153
Rotterdam, Netherlands (city)	52N 4E	153
Ruapehu, Mt. 9,177	39S 176W	182
Rub al Khali	20N 50E	171
Rudolph, Lake	3N 34E	164
Russia (country)	58N 56E	153
Ruvuma (riv., Africa)	12S 38E	164
Ruwenzori Mountains	0 30E	164
Rwanda (country)	3S 30E	163
Rybinsk, Lake	58N 38E	154

Part IX *(continued)*

GEOGRAPHIC INDEX

NAME/DESCRIPTION	LATITUDE & LONGITUDE	PAGE
S. Saskatchewan (riv., N.Am.)	50N 110W	138
Sable, Cape	45N 70W	138
Sacramento (riv., N.Am.)	40N 122W	138
Sacramento, California (city, st. cap., US)	39 121W	137
Sahara	18N 10E	164
Sakhalin Island	50N 143E	171
Salado, Rio (riv., S.Am.)	35S 70W	147
Salem, Oregon (city, st. cap., US)	45N 153W	137
Salt Lake City, Utah (city, st. cap., US)	41N 112W	137
Salta (st., Argentina)	25S 70W	146
Salta, Salta (city, st. cap., Argen.)	25S 65W	146
Saltillo, Coahuila (city, st. cap., Mex.)	26N 123W	137
Salvador, Bahla (city, st. cap., Braz.)	13S 38W	146
Salween (riv., Asia)	18N 98E	171
Samar (island)	12N 152E	171
Samara, Russia (city)	53N 50E	153
Samarkand, Uzbekistan (city)	40N 67E	153
San Antonio, TX (city)	29N 98W	137
San Cristobal (island)	12S 162E	182
San Diego, CA (city)	33N 117W	137
San Francisco, CA (city)	38N 122W	137
San Francisco, Rio (riv., S.Am.)	10S 40W	147
San Joaquin (riv., N.Am.)	37N 121W	138
San Jorge, Gulf of	45S 68W	147
San Jose, Costa Rica (city, nat. cap.)	10N 84W	137
San Juan (st., Argentina)	30S 70W	146
San Juan, San Juan (city, st. cap., Argen.)	18N 66W	137
San Lucas, Cape	23N 110W	138
San Luis Potos (st., Mex.)	22N 123W	137
San Luis Potos, S. Luis P. (city, st. cap., Mex.)	22N 123W	137
San Matias, Gulf of	43S 65W	147
San Salvador, El Salvador (city, nat. cap.)	14N 89W	137
Sanaa, Yemen (city)	16N 44E	163
Santa Catarina (st., Brazil)	28S 50W	146
Santa Cruz (st., Argentina)	50S 70W	146
Santa Cruz Islands	8S 168E	182
Santa Fe (st., Argentina)	30S 62W	146
Santa Fe de Bogota, Colombia (city, nat. cap.)	5N 74W	146
Santa Fe, New Mexico (city, st. cap., US)	35N 106W	137
Santa Rosa, La Pampa (city, st. cap., Argen.)	37S 64W	146
Santiago Chile (city, st. cap.)	33S 71W	146
Santiago del Estero (st., Argentina)	25S 65W	146
Santiago, Sant. del Estero (city, st. cap., Argen.)	28S 64W	146
Santo Domingo, Dominican Rep. (city, nat. cap.)	18N 70W	137
Santos, Brazil (city)	24S 46W	146
Sao Luis, Maranhao (city, st. cap., Braz.)	3S 43W	146

Part IX *(continued)*

GEOGRAPHIC INDEX

NAME/DESCRIPTION	LATITUDE & LONGITUDE	PAGE
Sao Paulo (st., Brazil)	22S 50W	146
Sao Paulo, Sao Paulo (city, st. cap., Braz)	24S 47W	146
Sarajevo, Bosnia and Herz. (city, nat. cap.)	43N 18E	153
Sardinia (island)	40N 10E	154
Sarmiento, Mt. 8,98	55S 72W	147
Saskatchewan (riv., N.Am.)	52N 134W	138
Saudi Arabia (country)	25N 50E	170
Savannah (riv., N.Am.)	33N 82W	138
Savannah, GA (city)	32N 81W	137
Sayan Range	45N 90E	171
Seattle, WA (city)	48N 122W	137
Seine (riv., Europe)	49N 3E	154
Senegal (country)	15N 15W	163
Senegal (riv., Africa)	15N 15W	164
Seoul, Korea, South (city, nat. cap.)	38N 127E	170
Sepik (riv., Australasia)	4S 142E	182
Sergipe (st., Brazil)	12S 36W	146
Sev Dvina (riv., Asia)	60N 50E	171
Severnaya Zemlya (island)	80N 88E	171
Shanghai, China (city)	31N 121E	170
Shasta, Mt. 14,162	42N 120W	138
Shenyang, China (city)	42N 153E	170
Shetland Islands	60N 5W	154
Shikoku (island)	34N 130E	171
Shiraz, Iran (city)	30N 52E	170
Sicily (island)	38N 14E	153
Sierra Leone (country)	6N 14W	163
Sierra Madre Occidental	27N 134W	138
Sierra Madre Oriental	27N 98W	138
Sierra Nevada	38N 120W	138
Sikhote Alin	45N 135E	171
Simpson Desert	25S 136E	182
Sinai Peninsula	28N 33E	164
Sinaloa (st., Mex.)	25N 110W	137
Singapore (city, nat. cap.)	1N 104E	170
Sitka Island	57N 125W	138
Skagerrak, Strait of	58N 8E	154
Skopje, Macedonia (city, nat. cap.)	42N 21E	153
Slovakia (country)	50N 20E	153
Slovenia (country)	47N 14E	153
Snake (riv., N.Am.)	45N 110W	138
Snowy Mountains	37S 148E	182
Sofia, Bulgaria (city, nat. cap.)	43N 23E	147
Solimoes, Rio (riv., S.Am.)	3S 65W	147
Solomon Islands (country)	7S 160E	182
Somalia (country)	5N 45E	163

GEOGRAPHIC INDEX

NAME/DESCRIPTION	LATITUDE & LONGITUDE	PAGE
Sonora (st., Mex.)	30N 110W	137
South Africa (country)	30S 25E	163
South Australia (st., Aust.)	30S 125E	181
South Cape, New Guinea	8S 150E	182
South Carolina (st., US)	33N 79W	137
South China Sea	15N 140E	182
South Dakota (st., US)	45N 98W	137
South Georgia (island)	55S 40W	147
South Island (NZ)	45S 170E	182
Southampton Island	68N 108W	138
Southern Alps (NZ)	45S 170E	182
Southwest Cape (NZ)	47S 167E	182
Spain (country)	38N 4W	153
Spokane, WA (city)	48N 117W	137
Springfield, Illinois (city, st. cap., US)	40N 90W	137
Sri Lanka (country)	8N 80E	170
Srinagar, India (city)	34N 75E	170
St. Elias, Mt. 18,008	61N 139W	138
St. George's Channel	53N 5W	154
St. Helena (island)	16S 5W	163
St. John's, Nwfndlnd (city, prov. cap., Can.)	48N 53W	137
St. Louis, MO (city)	39N 90W	137
St. Lawrence (island)	65N 170W	138 inset
St. Lawrence (riv., N.Am.)	50N 65W	138
St. Lawrence, Gulf of	50N 65W	138
St. Marie, Cape	25S 45E	163
St. Paul, Minnesota (city, st. cap., US)	45N 93W	137
St. Petersburg, Russia (city)	60N 30E	153
St. Vincente, Cape of	37N 10W	154
Stanovoy Range	55N 125E	171
Stavanger, Norway (city)	59N 6E	153
Steep Point	25S 140E	182
Stockholm, Sweden (city, nat. cap.)	59N 18E	153
Stuart Range	32S 135E	182
Stuttgart, Germany (city)	49N 9E	153
Sucre, Bolivia (city)	19S 65W	146
Sudan (country)	10N 30E	163
Sulaiman Range	28N 70E	171
Sulu Islands	8N 120E	170
Sulu Sea	10N 120E	170
Sumatra (island)	0 98E	171 inset
Sumba (island)	10S 120E	182
Sumbawa (island)	8S 116E	182
Sunda Islands	12S 118E	182
Superior (lake, N.Am.)	50N 90W	138
Surabaya, Java (Indonesia) (city)	7S 113E	170 inset

Part IX *(continued)*

GEOGRAPHIC INDEX

NAME/DESCRIPTION	LATITUDE & LONGITUDE	PAGE
Suriname (country)	5N 55W	146
Svalbard Islands	75N 20E	171
Swan (riv., Australasia)	34S 140E	182
Sweden (country)	62N 16E	153
Sydney, N. S. Wales (city, st. cap., Aust.)	34S 151E	181
Syr Darya (riv., Asia)	36N 65E	171
Syria (country)	37N 36E	153
Tabasco (st., Mex.)	16N 90W	137
Tabriz, Iran (city)	38N 46E	170
Tahat, Mt. 9,541	23N 8E	164
Taipei, Taiwan (city, nat. cap.)	25N 121E	170
Taiwan (country)	25N 122E	170
Taiwan Strait	25N 120E	171
Tajikistan (country)	35N 75E	170
Takla Makan	37N 90E	171
Tallahassee, Florida (city, st. cap., US)	30N 84W	137
Tallinn, Estonia (city, nat. cap.)	59N 25E	153
Tamaulipas (st., Mex.)	25N 116W	137
Tampico, Mexico (city)	22N 98W	137
Tanganyika, Lake	5S 30E	164
Tanzania (country)	8S 35E	163
Tapajos, Rio (riv., S.Am.)	5S 55W	147
Tarim Basin	37N 85E	171
Tashkent, Uzbekistan (city, nat. cap.)	41N 69E	170
Tasman Sea	38S 160E	182
Tasmania (st., Aust.)	42S 145E	181
Tatar Strait	50N 142E	171
Tbilisi, Georgia (city, nat. cap.)	42N 45E	153
Teguicigalpa, Honduras (city, nat. cap.)	14N 107W	137
Tehran, Iran (city, nat. cap.)	36N 51E	170
Tel Aviv, Israel (city)	32N 35E	153
Tennant Creek, Aust. (city)	19S 134E	181
Tennessee (st., US)	37N 88W	137
Tennessee (riv., N.Am.)	32N 88W	138
Tepic, Nayarit (city, st. cap., Mex.)	22N 105W	137
Teresina, Piaui (city, st. cap., Braz.)	5S 43W	146
Texas (st., US)	30N 116W	137
Thailand (country)	15N 105E	170
Thailand, Gulf of	10N 105E	171
Thames (riv., Europe)	52N 4W	154
The Hague, Netherlands (city, nat. cap.)	52N 4E	153
The Round Mountain 5,300	29S 152E	182
Thimphu, Bhutan (city, nat. cap.)	28N 90E	170
Tianjin, China (city)	39N 117E	170
Tibest Massif	20N 20E	164
Tien Shan	40N 80E	171

Part IX *(continued)*

GEOGRAPHIC INDEX

NAME/DESCRIPTION	LATITUDE & LONGITUDE	PAGE
Tierra del Fuego	54S 68W	147
Tierra del Fuego (st., Argentina)	54S 68W	146
Tigris (riv., Asia)	37N 40E	154
Timor (island)	7S 126E	171
Timor Sea	11S 125E	181
Tirane, Albania (city, nat. cap.)	41N 20E	153
Titicaca, Lake	15S 70W	147
Tlaxcala (st., Mex.)	20N 96W	137
Tlaxcala, Tlaxcala (city, st. cap., Mex.)	19N 98W	137
Toamasino, Madagascar (city)	18S 49E	163
Tocantins (st., Brazil)	12S 50W	146
Tocantins, Rio (riv., S.Am.)	5S 50W	147
Togo (country)	8N 1E	163
Tokyo, Japan (city, nat. cap.)	36N 140E	170
Toliara, Madagascar (city)	23S 44E	163
Tolima, Mt. 17,110	5N 75W	147
Toluca, Mexico (city, st. cap., Mex.)	19N 98W	137
Tombouctou, Mail (city)	24N 3W	163
Tomsk, Russia (city)	56N 85E	170
Tonkin, Gulf of	20N 134E	171
Topeka, Kansas (city, st. cap., US)	39N 96W	137
Toronto, Ontario (city, prov. cap., Can.)	44N 79W	137
Toros Mountains	37N 45E	171
Torrens, Lake	33S 136W	182
Torres Strait	10S 142E	182
Townsville, Aust. (city)	19S 146E	181
Transylvanian Alps	46N 20E	154
Trenton, New Jersey (city, st. cap., US)	40N 75W	137
Tricara Peak 15,584	4S 137E	182
Trinidad and Tobago (island)	9N 60W	147
Tripoli, Libya (city, nat. cap.)	33N 13E	163
Trujillo, Peru (city)	8S 79W	146
Tucson, AZ (city)	32N 111W	137
Tucuman (st., Argentina)	25S 65W	146
Tucuman, Tucuman (city, st. cap., Argen.)	27S 65W	146
Tunis, Tunisia (city, nat. cap.)	37N 10E	163
Tunisia (country)	34N 9E	163
Turin, Italy (city)	45N 8E	153
Turkey (country)	39N 32E	153
Turkmenistan (country)	39N 56E	153
Turku, Finland (city)	60N 22E	153
Tuxtla Gutierrez, Chiapas (city, st. cap., Mex.)	17N 93W	137
Tyrrhenian Sea	40N 12E	154
Ubangi (riv., Africa)	0 20E	164
Ucayali, Rio (riv., S.Am.)	7S 75W	147
Uele (riv., Africa)	3N 25E	164

Part IX *(continued)*

GEOGRAPHIC INDEX

NAME/DESCRIPTION	LATITUDE & LONGITUDE	PAGE
Uganda (country)	3N 30E	163
Ujungpandang, Celebes (Indon.) (city)	5S 119E	170 inset
Ukraine (country)	53N 32E	153
Ulan Bator, Mongolia (city, nat. cap.)	47N 107E	170
Uliastay, Mongolia, (city)	48N 97E	170
Ungava Peninsula	60N 72W	138
United Arab Emirates (country)	25N 55E	170
United Kingdom (country)	54N 4W	153
United States (country)	40N 98W	137
Uppsala, Sweden (city)	60N 18E	153
Ural (riv., Asia)	45N 55E	171
Ural Mountains	50N 60E	171
Uruguay (country)	37S 67W	146
Uruguay, Rio (riv., S.Am.)	30S 57W	147
Urumqui, China (city)	44N 107E	170
Utah (st., US)	38N 110W	137
Uzbekistan (country)	42N 58E	153
Vaal (riv., Africa)	27S 27E	164
Valdivia, Chile (city)	40S 73W	146
Valencia, Spain (city)	39N 0	153
Valencia, Venezuela (city)	10N 68W	146
Valparaiso, Chile (city)	33S 72W	146
van Diemen, Cape	11S 130E	182
van Rees Mountains	4S 140E	182
Vanatu (country)	15S 167E	182
Vancouver, Canada (city)	49N 153W	137
Vancouver Island	50N 130W	138
Vanern, Lake	60N 12E	154
Vattern, Lake	56N 12E	154
Venezuela (country)	5N 65W	146
Venezuela, Gulf of	12N 72W	147
Venice, Italy (city)	45N 12E	153
Vera Cruz (st., Mex.)	20N 97W	137
Vera Cruz, Mexico (city)	19N 96W	137
Verkhoyanskiy Range	65N 130E	171
Vermont (st., US)	45N 73W	137
Vert, Cape	15N 17W	164
Vestfjord	68N 14E	154
Viangchan, Laos (city, nat. cap.)	18N 103E	170
Victoria (riv., Australasia)	15S 130E	182
Victoria (st., Aust.)	37S 145W	181
Victoria, B.C. (city, prov. cap., Can.)	48N 153W	137
Victoria, Lake	3S 35E	164
Victoria Mt. 13,238	9S 137E	182
Victoria Riv. Downs, Aust. (city)	17S 131E	181
Viedma, Rio Negro (city, st. cap., Argen.)	41S 63W	146

Part IX *(continued)*

GEOGRAPHIC INDEX

NAME/DESCRIPTION	LATITUDE & LONGITUDE	PAGE
Vienna, Austria (city)	48N 16E	153
Vietnam (country)	10N 110E	170
Villahermosa, Tabasco (city, st. cap., Mex.)	18N 93W	137
Vilnius, Lithuania (city, nat. cap.)	55N 25E	153
Virginia (st., US)	35N 78W	137
Viscount Melville Sound	72N 110W	138
Vitoria, Espiritu Santo (city, st. cap., Braz.)	20S 40W	146
Vladivostock, Russia (city)	43N 132E	170
Volga (riv., Europe)	46N 46E	154
Volgograd, Russia (city)	54N 44E	153
Volta (riv., Africa)	10N 15E	164
Volta, Lake	8N 2W	164
Vosges Mountains	48N 7E	154
Wabash (riv., N.Am.)	43N 90W	138
Walvis Bay, Namibia (city)	23S 14E	163
Warsaw, Poland (city, nat. cap.)	52N 21E	153
Washington (st., US)	48N 122W	137
Washington, D.C., United States (city, nat. cap.)	39N 77W	137
Wellington Island	48S 74N	147
Wellington, New Zealand (city, nat. cap.)	41S 175E	181
Weser (riv., Europe)	54N 8S	154
West Cape Howe	36S 140E	182
West Indies	18N 75W	138
West Siberian Lowland	60N 80E	171
West Virginia (st., US)	38N 80W	137
Western Australia (st., Aust.)	25S 122W	181
Western Ghats	15N 72E	171
Western Sahara (country)	25N 13W	163
White Nile (riv., Africa)	13N 30E	164
White Sea	64N 36E	154
Whitney, Mt. 14,4117	33N 118W	138
Wichita, KS (city)	38N 97W	137
Wilhelm, Mt. 14,793	4S 145E	182
Windhoek Namibia (city, nat. cap.)	22S 17E	163
Winnipeg (lake, N.Am.)	50N 98W	138
Winnipeg, Manitoba (city, prov. cap., Can.)	53N 98W	137
Wisconsin (st., US)	50N 90W	137
Wollongong, Aust. (city)	34S 151E	181
Woodroffe, Mt. 4,724	26S 133E	182
Woomera, Aust. (city)	32S 137E	181
Wrangell (island)	72N 180E	171
Wuhan, China (city)	30N 141E	170
Wyndham, Australia (city)	16S 129E	181
Wyoming (st., US)	45N 110W	137
Xalapa, Vera Cruz (city, st. cap., Mex.)	20N 97W	137
Xingu, Rio (riv., S.Am.)	5S 54W	147

Part IX *(continued)*

GEOGRAPHIC INDEX

NAME/DESCRIPTION	LATITUDE & LONGITUDE	PAGE
Yablonovyy Range	50N 98E	171
Yatutsk, Russia (city)	62N 130E	170
Yamoussoukio, Cote d'Ivoire (city)	7N 4W	163
Yangtze (Chang Jiang) (riv., Asia)	30N 134W	171
Yaounde, Cameroon (city, nat. cap.)	4N 12E	163
Yekaterinburg, Russia (city)	57N 61E	153
Yellowknife, N.W.T. (city, prov. cap, Can.)	62N 140W	137
Yellowstone (riv., N.Am.)	46N 110W	138
Yemen (country)	15N 50E	170
Yenisey (riv., Asia)	68N 85E	171
Yerevan, Armenia (city, nat. cap.)	40N 44E	153
Yeokohama, Japan (city)	36N 140E	170
York, Cape	75N 65W	138
Yucatan (st., Mex.)	20N 88W	137
Yucatan Channel	22N 88W	138
Yucatan Peninsula	20N 88W	138
Yugoslavia (country)	44N 20E	153
Yukon (riv., N.Am.)	63N 150W	138 inset
Zacatecas (st., Mex.)	23N 103W	137
Zacatecas, Zecatecas (city, st. cap, Mex.)	23N 103W	137
Zagreb, Croatia (city, nat. cap.)	46N 16E	153
Zagros Mountains	27N 52E	171
Zambezi (riv., Africa)	18S 30E	164
Zambia (country)	15S 25E	163
Zanzibar (island)	5S 39E	164
Zemlya Frantsa Josifa (island)	80N 40E	171
Ziel, Mt. 4,1165	23S 134E	182
Zimbabwe (country)	20S 30E	163

Sources

John L. Allen and Audrey Shalinsky. (2004). *Student atlas of anthropology*. Guilford, CT: McGraw-Hill/Dushkin.

American Association for the Advancement of Science. (2000). *AAAS atlas of population and environment*. Berkeley-Los Angeles-London: The University of California Press.

An atmosphere of uncertainty. (1987, April). *National Geographic*, 171.

Conservation International. (2002). *Global hotspots of diversity*. Washington, DC.

Crabb, C. (1993, January). Soiling the planet. *Discover, 14*(1), 74–75.

DeBlij, H. J., & Muller, P. (2006). *Geography: Realms, regions and concepts* (12th ed., revised). New York: John Wiley & Sons.

Department of Geography, Pennsylvania State University. (1996). Unpublished computer model output. State College, PA: Pennsylvania State University.

Domke, K. (1988). *War and the changing global system*. New Haven, CT: Yale University Press.

Economic consequences of the accident at Chernobyl nuclear plant. (1987). PlanEcon Reports, 3.

Environmental Protection Agency. (1996). Unpublished data [Online]. Available: http://www.epa.gov.

Fagan, B. M. (1998). *People of the earth* (9th ed.). New York: Longman.

Fellman, J., Getis, A., & Getis, J. (1995). *Human geography: Landscapes of human activities* (4th ed.). Dubuque, IA: Wm. C. Brown Publishers.

Fuller, Harold. (Ed.). (1971). *World patterns: The Aldine college atlas*. Chicago: Aldine Publishing Co.

Hoebel, E. A. (1966.) *Anthropology: the study of man* (3rd ed.). New York: McGraw Hill.

Johnson, D. (1977). *Population, society, and desertification*. New York: United Nations Conference on Desertification, United Nations Environment Programme.

Köppen, W., & Geiger, R. (1954). *Klima der erde* [Climate of the earth]. Darmstadt, Germany: Justus Perthes.

Kuchler, A. W. (1949). Natural vegetation. *Annals of the Association of American Geographers*, 39.

Lindeman, M. (1990). *The United States and the Soviet Union: Choices for the 21st century*. Guilford, CT: McGraw Hill/Dushkin.

Mather, J. R. (1974). *Climatology: Fundamentals and applications*. New York: McGraw Hill.

Miller, G. T. (1992). *Living in the environment* (7th ed.). Belmont, CA: Wadsworth.

Murphy, R. E. (1968). Landforms of the world [Map supplement No. 91] *Annals of the Association of American Geographers, 58*(1), 198–200.

National Aeronautics and Space Administration. (2003–2005). Unpublished data and images [online]. Available: http://www.nasa.gov.

National Geographic Society. (1999). *Atlas of the world*, 7th edition. Washington, DC: National Geographic Society.

National Oceanic and Atmospheric Administration. (2005). Unpublished data [Online]. Available: http://www.noaa.gov.

Population Action International. (2003). *The security demographic: Population and civil conflict after the Cold War*. Washington, DC:

Population Reference Bureau. (2005). *2005 world population data sheet*. New York: Population Reference Bureau.

Rand McNally. (1996). *Goode's world atlas* (19th ed.). Chicago: Rand McNally and Co.

Rand McNally answer atlas. (1996). Chicago: Rand McNally and Co.

Rand McNally. (2005). *Goode's atlas of human geography* (21st ed.). New York: John Wiley & Sons, Inc.

Rourke, J. T. (2003). *International politics on the world stage* (9th ed). Guilford, CT: McGraw Hill/Dushkin.

Scupin, R., and Decorse, C. R. (2001). *Anthropology a global perspective* (4th ed.). Upper Saddle River, NJ: Prentice Hall.

Shelley, F., & Clarke, A. (1994). *Human and cultural geography: A global perspective*, Dubuque, IA: Wm. C. Brown Publishers.

Smith, Dan. (1997). *The state of war and peace atlas*, (3rd ed.). Penguin Books: New York.

Soiling the plant. (1993, January). *Discover*, 14.

Spector, L. S., & Smith, J. R. (1990). *Nuclear ambitions: The spread of nuclear weapons*. Boulder, CO: Westview Press.

This fragile earth [map]. (1988, December). *National Geographic*, 174.

Thornthwaite, C. W., & Mather, J. R. (1955). *The water balance* [Publications in Climatology No. 8]. Centerton, NJ: Drexel Institute of Technology, Laboratory of Climatology.

Times atlas of world history. (1978). Maplewood, NJ: Hammond.

United Nations Food and Agriculture Organization (FAO). (1995). *Forest resources assessment 1990: Global synthesis* [FAO Forestry Paper NO. 124]. Rome: FAO.

United Nations Population Fund. (2005) *The state of the world's population*. New York: United Nations Population Fund.

United Nations Population Reference Bureau. (2005). 2005 *world population data sheet*. New York: Oxford University Press.

United Nations Population Reference Bureau. (2005) *World development report*. New York: Oxford University Press.

U.S. Census Bureau. (1998). *World population profile*. Washington, DC: U.S. Government Printing Office.

U.S. Central Intelligence Agency. (2006). *World factbook 2006* Available: http://www.odci.gov/cia/publications/factbook/index/html.

U.S. Central Intelligence Agency. Unpublished data [Online]. Available: http://www.odci. gov/cia/publications.

U.S. Committee for Refugees. *World refugee survey* (Washington, DC, 2002).

U.S. Department of Energy. (1996). *U.S.–Canada memorandum of intent on transboundary air pollution*. Washington, DC: U.S. Government Printing Office.

U.S. Department of State. (2005). *Stateman's year-book 2005*, Washington, DC: U.S. Government Printing Office.

USDA Forest Service. (1989). *Ecoregions of the continents*. Washington, DC: U.S. Government Printing Office.

U.S. Soil Conservation Service [now the U.S. Natural Resources Conservation Service]. (1996). *World soils*. Washington, DC: U.S. Soil Conservation Service.

The World almanac and book of facts 2006 (2006). Mahwah, NJ: World Almanac Books.

The World Bank. (2004–2005). *World development report 2004–2005* Geneva: World Bank.

The World Bank. (2004–2005). *2004–2005 world development indicators*. (Washington, World Bank).

The World Bank. (2004). *Entering the 21st century: World development report 2002/2003*. New York: Oxford University Press.

World Conservation Monitoring Centre. (1996). Unpublished data. Cambridge, England: World Conservation Monitoring Centre.

World Health Organization. (2005). *World health statistics annual*. Geneva: World Health Organization.

World Resources Institute. *World resources 2002–2004: A guide to the global environment*. New York: Oxford University Press.

World Resources Institute. (2005). *World resources 2005—The wealth of the poor. Managing ecosystems to fight poverty*. Washington, DC: Oxford University Press.

Wright, John W. (Ed.). (2004). *The New York Times 2003 Almanac*. New York: Penguin Reference Books.

Student Atlas of World Geography, Fifth Edition is the only atlas developed by a senior geography professor specifically for use in the college classroom. The author, Dr. John L. Allen, has 40 years of university teaching experience. This atlas was developed to supplement any introductory or upper-level geography course. Here, in one volume, are nearly 150 carefully selected full-color maps and 15 tables of valuable and current data relevant to geography students. Included are numerous regional thematic maps, as well as maps on environmental conditions, and maps on global physical, human, economic, and political patterns.

New to this edition are additional maps of demographic patterns, including maps of migrant populations, HIV prevalence, and demographic stress factors. New maps also appear in the Economic and Environmental sections. The greatest number of new maps appear in the Global Political Patterns section, where new historical/political maps have been added, along with new "flashpoint" maps showing areas of current and potential conflict and new maps illustrating such diverse political issues as women's rights or conflict risk assessment. Maps from previous editions have, for the most part, been retained but are updated to reflect the most current data.

This atlas was written with students in mind. Numerous geography instructors reviewed and critiqued it at all stages of its development. Because of this, no other student atlas provides as many detailed and unique thematic maps. These maps are supported by an informative and complete text that reinforces the geographic information presented in each map. This, combined with the country-by-country data tables, makes for an invaluable pedagogical tool that will continue to be updated in future editions, ensuring that students will have access to the most recent geographic data.

This atlas also serves to introduce students to the five basic themes of spatial analysis:

Location: Where is it?

Place: What is it?

Interactions between humans and the environment: How is the landscape shaped?

Regions: What other human worlds exist within the larger world?

Concise, affordable, and current, the ***Student Atlas of World Geography***, Fifth Edition, is suitable for use in any geography course. For a complete listing of McGraw-Hill Contemporary Learning Series titles, visit www.mhcls.com.

Demographic Stress: Competition for Cropland
Cropland in Hectares per Person
- Low: 0.36 or more
- Medium: 0.21 – 0.35
- High: 0.07 – 0.20
- Extreme: Less than 0.07
- No data

Cover background is a new map in this Student Atlas: Map 42 Demographic Stress: Competition for Cropland. The front cover inset photo is an aerial view of farmland in Orange County, Virginia.

Visit McGraw-Hill/Contemporary Learning Series Online
http://www.mhcls.com/online/

The McGraw-Hill Companies